A HISTORY OF THE BRAIN

A History of the Brain tells the full story of neuroscience, from antiquity to the present day. It describes how we have come to understand the biological nature of the brain, beginning in prehistoric times, and the progress to the twentieth century with the development of modern neuroscience.

This is the first time a history of the brain has been written in a narrative way, emphasising how our understanding of the brain and nervous system has developed over time, with the development of a number of disciplines including anatomy, pharmacology, physiology, psychology and neurosurgery.

The book covers:

- beliefs about the brain in ancient Egypt, Greece and Rome
- the Medieval period, Renaissance and Enlightenment
- the 19th century
- the most important advances in the 21st century and future directions in neuroscience.

The discoveries leading to the development of modern neuroscience have given rise to one of the most exciting and fascinating stories in the whole of science. Written for readers with no prior knowledge of the brain or history, the book will delight students, and will also be of great interest to researchers and lecturers with an interest in understanding how we have arrived at our present knowledge of the brain.

Andrew P. Wickens is Senior Lecturer in Psychology and Neuroscience at the University of Central Lancashire, UK. His main area of expertise is in biological psychology and neuroscience.

A HISTORY OF THE BRAIN

From Stone Age surgery to modern neuroscience

Andrew P. Wickens

Psychology Press
Taylor & Francis Group
LONDON AND NEW YORK

First published 2015
by Psychology Press
27 Church Road, Hove, East Sussex BN3 2FA

and by Psychology Press
711 Third Avenue, New York, NY 10017

Psychology Press is an imprint of the Taylor and Francis Group, an Informa business.

© 2015 Andrew P. Wickens

The right of Andrew P. Wickens to be identified as the author of this work has been asserted by him in accordance with sections 77 and 78 of the Copyright, Designs and Patents Act 1988.

All rights reserved. No part of this book may be reprinted or reproduced or utilized in any form or by any electronic, mechanical, or other means, now known or hereafter invented, including photocopying and recording, or in any information storage or retrieval system, without permission in writing from the publishers.

Trademark notice: Product or corporate names may be trademarks or registered trademarks, and are used only for identification and explanation without intent to infringe.

British Library Cataloguing in Publication Data
A catalogue record for this book is available from the British Library

Library of Congress Cataloging in Publication Data
Wickens, Andrew P.
A history of the brain: from stone age surgery to modern neuroscience/Andrew P. Wickens.
pages cm
Includes bibliographical references and index.
1. Brain – History. 2. Neurosciences – History. I. Title.
QP376.W485 2015
612.8'2 – dc23
2014021500

ISBN: 978-1-84872-364-1 (hbk)
ISBN: 978-1-84872-365-8 (pbk)
ISBN: 978-1-315-79454-9 (ebk)

Typeset in Bembo and Stone Sans
by Florence Production Ltd, Stoodleigh, Devon, UK

Printed and bound in the United States of America by Publishers Graphics, LLC on sustainably sourced paper.

CONTENTS

Preface — xi
Acknowledgements — xiv

1 Head or heart? The ancient search for the soul — 1

The emergence of the human mind — 2
Stone Age surgery: trepanation — 4
The earliest reference to a brain disorder: epilepsy — 8
The brain in ancient Egypt — 9
The first word for brain — 10
The causes of behaviour in Homeric times — 11
The brain as an organ of sensation: Alcmaeon of Crotona — 12
Hippocrates and the brain — 13
Plato appears to confirm the importance of the brain — 16
Aristotle's alternative view of the psyche — 18
Aristotle and the brain — 20

2 The discovery of the nervous system — 27

A centre of academic excellence: Alexandria — 28
Herophilus of Chalcedon — 29
The discovery of the nervous system — 31
Erasistratus of Chios — 32
The founder of experimental physiology: Galen of Pergamon — 33
Galen's account of the brain — 35
Galen's account of the cranial and spinal nerves — 37
Galen and the squealing pig — 39
Galen's account of the spinal cord — 41
The role of psychic pneuma in nervous function — 41

The first experiments on the brain	43
Galen's great legacy	44

3 From late antiquity to the Renaissance: the Cell Doctrine 48

The origins of the Cell Doctrine	49
The Cell Doctrine in the West	51
The rise of scholasticism and later accounts of the Cell Doctrine	52
The earliest illustration of the brain	53
Other early illustrations of the brain	58
Mondino de Luzzi: the restorer of anatomy	61
Leonardo da Vinci: Renaissance Man	64
Leonardo's search for the soul	65
Visualising the ventricles	69
The first printed anatomical text with illustrations	70
The reawakening: Andreas Vesalius	72
De humani corporis Fabrica	73
The Fabrica's *depiction of the brain*	76
The impact of the Fabrica	78

4 Searching for the ghost in the machine 83

Descartes: a new foundation for science	84
Cartesian dualism	86
The first account of the reflex	88
The site of mind–body interaction: the pineal gland	90
The early years of the scientific revolution	91
Thomas Willis: the father of modern neurology	93
Cerebri anatome	94
Localising functions in different brain areas	97
The rational soul, animal spirits and nervous activity	99
The microscopic world revealed	99
A test of the animal spirits theory	102
Albrecht von Haller: irritability and sensibility	103
Robert Whytt and the sentient principle	106

5 A new life force: animal electricity 111

The early history of electricity	112
Luigi Galvani and the discovery of animal electricity	114
A rejection of animal electricity: Alessandro Volta	117
Volta invents the battery	120
Animating the dead: Giovanni Aldini	120
The inspiration for Frankenstein	123
The invention and use of the galvanometer	123

Differing views: Johannes Müller and Emil Du Bois-Reymond	125
The discovery of the action potential	127
Hermann von Helmholtz	128
Measuring the speed of the nerve impulse	129

6 The rise and fall of phrenology — 134

Franz Joseph Gall: the founder of phrenology	135
Gall in Paris	138
The neuroanatomy of Gall and Spurzheim	138
Exorcising the soul from the brain	140
Feel the bumps, know the man	141
Craniology and social reform	143
Spurzheim and the popularisation of phrenology	143
George Combe	145
What happened to phrenology?	147
Gall's legacy	149
Lesioning the cerebral cortex: Pierre Flourens	150
Lesioning the cerebellum and medulla	152
Phrenology re-examined	153

7 The nerve cell laid bare — 157

The microscope at the turn of the nineteenth century	158
The cradle of histology: Jan Purkinje	159
The first depiction of a nerve cell	160
A landmark in modern biology: the cell theory	163
The nerve cell begins to take shape: Robert Remak	165
Nerve cells are incorporated into cellular theory: Albert von Kölliker	167
The isolated nerve cell: Otto Deiters	170
The silver impregnation stain: Camillo Golgi	172
Objections to the nerve net theory	175
The founder of modern neuroscience: Santiago Ramón y Cajal	176
The nerve cell starts to give up its secrets	178
The neuron doctrine	181
The synapse is named	182
Golgi and Cajal win the Nobel Prize	184

8 The return of the reflex — 189

Early accounts of involuntary action	190
The reflex as the mediator of sympathy: Robert Whytt	191
The Bell–Magendie law	193
The excito-motory reflex: Marshall Hall	196
Psychic reflexes: Ivan Mikhailovich Sechenov	198
The integrated reflex: Charles Scott Sherrington	200

The discovery of proprioception	202
An examination of more complex reflexes	202
Sherrington's legacy	203
Conditioned reflexes: Ivan Pavlov	205
Pavlov's impact on psychology	208
Searching for the engram: Karl Lashley	209
Reflexes as cell assemblies: Donald Hebb	211

9 Mapping the cerebral cortex — 216

Language and the frontal lobes: Jean-Baptiste Bouillaud	217
Paul Broca and his speech centre	218
Right-handedness and the dominant hemisphere	220
The minor hemisphere: John Hughlings Jackson	221
A second language centre: Carl Wernicke	223
The discovery of a motor area in the frontal cortex	226
Evidence for localisation builds: David Ferrier	229
Stimulating the human brain	231
Cortical areas for hearing and sight	232
Ferrier's mistake	233
A test of cortical localisation	234
The story of Phineas Gage	235

10 The rise of psychiatry and neurology — 242

A new concept of disease: Giovanni Battista Morgagni	243
The rise of nervous disorders in the eighteenth century	245
The discovery of Parkinson's disease	247
The Napoleon of the Salpêtrière: Jean Martin Charcot	248
The anatomo-clinical method	249
The discovery of multiple sclerosis	251
Other neurological conditions discovered at the Salpêtrière	252
Charcot's investigations into hysteria	254
Charcot's legacy	257
A new way of classifying mental illness: Emil Kraepelin	258
The discovery of schizophrenia	260
A new type of dementia: Alois Alzheimer	261

11 Solving the mystery of the nerve impulse — 267

Action potentials and ions: Julius Bernstein	268
The all-or-nothing law	270
Thermionic valves and amplifiers	272
Recording from single nerve fibres: Edgar Adrian	273

The giant squid comes to the rescue	276
Testing Bernstein's membrane hypothesis	277
The role of sodium in the action potential	279
The action potential explained	280
The role of calcium	282
The quantal release of neurotransmitter	283
The discovery of ion channels	285

12 The discovery of chemical neurotransmission — 289

The curious effects of curare	290
The first statement regarding chemical neurotransmission	292
Mapping the sympathetic ('involuntary') nervous system	292
Defining the sympathetic and parasympathetic systems: John Newport Langley	294
A claim for neurotransmitters and receptors	297
Henry Dale investigates the properties of ergot	299
The discovery of acetylcholine	301
Demonstrating chemical neurotransmission: Otto Loewi	303
Confirming acetylcholine as a neurotransmitter	305
The soup versus sparks debate	306
Neurotransmitters in the brain	309
The first drug treatments for mental illness	310
The first antidepressants	312

13 Neurosurgery and clinical tales — 317

Recovery from brain injury: Francois Quesnay	318
The beginnings of modern neurosurgery	319
The first official neurosurgeon: Victor Horsley	322
Learning from epilepsy: Wilder Penfield	324
Stimulating the human brain	326
Exploring the stream of consciousness	328
Penfield in later life	330
The Wada technique	331
Mental illness under the surgeon's knife	332
Ice pick surgery	334
The split-brain technique	335
Presenting information to individual hemispheres: Roger Sperry	336
The two different personalities of the brain	338
The man who instantly forgot: the case of HM	339

14 Surveying the last 50 years and looking ahead — 345

Neuroscience comes of age: Parkinson's disease	346
L-dopa therapy	347

The beginning of the genetic revolution	348
The rise of molecular neuroscience	349
Localising the mutation in Huntington's disease	350
The focus turns to Alzheimer's disease	352
Drug treatment for Alzheimer's disease	353
Genetic engineering in neuroscience	355
Historical landmarks in stem cell biology	356
Stem cells in neuroscience	356
Stem cells for research	358
Computers and the development of computerised tomography	359
Visualising the brain in real-time	361
Magnetic resonance imaging	361
Where are brain scanning techniques taking us?	363
The promise of artificial intelligence	364
Brain–computer interfaces	367
A new era begins as big investment targets the brain	368
The final frontier	370
Author Index	378
Subject Index	383

PREFACE

> *'The longer you can look back, the further you can look forward'*
> Winston Churchill addressing the Royal College
> of Physicians in March, 1944.

The idea behind this book first began to germinate around 2001 after I finished writing a textbook on biological psychology. In its first chapter I had tried to give some historical overview of the subject, highlighting the major landmarks in our understanding of the brain and its intricate workings. Not only was this meant to provide an introduction to biopsychology, but encourage new students by showing them how they could be part of a great mystery story whose origins went far back into our ancient past and whose future held no bounds. Yet, I knew my understanding of historical matters was at best patchy. The history of neuroscience is not regularly taught in school or university, and like most others I had picked up my knowledge piecemeal from incidental study. I was also aware of many gaps in my knowledge, with certain pieces of the jigsaw missing. For someone who took their role as an educator seriously, it made me feel uneasy and I began to look around for texts that would help me broaden my knowledge. To my surprise there were not many. The first I came across was Stanley Finger's *Origins of Neuroscience*[1] – a lavishly illustrated tome meant for the more specialist reader. Although not a book to be read from cover to cover, it was great to dip into and I devoured its pages. But then I discovered Finger's other book *Minds Behind the Brain*[2] written for a more general audience, focusing on the work of individuals across the great sweep of history from ancient Egypt to the twentieth century. The book was inspirational for me since it showed that history was not only educational, but could be entertaining, interesting and enjoyable.

It was at this juncture the idea for this book probably began to materialise. I also began to ponder whether it was possible to write a narrative that would continuously trace the flow of historical development from the ancient past to the present. I knew instinctively it was an ambitious idea. Yet, it excited me since I knew if all the threads were joined up in a meaningful way, the emerging picture would be one giving a new insight into the development of neuroscience. Armed with little more than enthusiasm I set sail into new waters with my compass. I must confess the voyage has taken me far longer than I imagined. Admittedly, I have written two further editions of my textbook during this time, led a far from settled existence, and become extremely disenchanted working at a modern university unconcerned with the merits of writing popular educational books. Even so, had I known this book was to take some ten years to write, I may have been more cautious in choosing my destination. Nonetheless, the endeavour has not been without its rewards and at this time of writing I am pleased with my journey.

As far as we can tell, spinning around on this small planet of ours, the human brain is the only thing in the cosmos capable of consciousness, free will and self-reflection. It has also been described as the most complex object in the universe – a biological machine that transcends itself to become greater than the sum of its parts. Getting across some of this complexity to the student is a challenge. The brain's anatomical composition is certainly impressive, containing around one hundred billion or 10^{11} tiny nerve cells – a figure so great that if you counted one per second, it would take over 30,000 years. Each is not only a tiny generator of electrical impulses that move at speeds approaching 100 miles per hour, but do so in a system that has in the region of 150,000 km of nerve fibres. Yet, it is the chemical events taking place at the endings of each fibre, or what is called the *synapse*, which is the site where the real information processing of the brain takes place. It is believed the human brain has around five hundred trillion of these tiny junctions (5×10^{14}), a figure that may come close to equalling the number of sand grains on our planet. How many nerve cells and their synapses are at work in your brain now as you read these words is anybody's guess. But one thing is certain: at every given second of your existence, your brain abounds with billions of rapidly moving electrical signals in an incredible weave of changing patterns, causing it to be awash with a large variety of chemicals. This book, at least in part, will try to explain how we have arrived at this biological fact.

Yet, all this great complexity does not get close to explaining the really special property of the mass resembling congealed porridge inside our skulls. In 1942, the English neurophysiologist Charles Sherrington referred to the human brain as an enchanted loom – a machine with physical bits-and-bobs, which was supplemented with the very special added mysterious ingredient of consciousness. In fact, these two types of 'thing' are scientifically irreconcilable. He could not conceive how the mechanical loom, no matter how complex, became enchanted, and over 60 years later we still do not know the answer. It is arguably the greatest scientific mystery we face today, although one we should theoretically be able to solve, since the answer lies inside our heads. The solution to this perplexing puzzle is crucially important for it will not only ultimately explain who we are, but also likely give profound insights into other conundrums such as the nature of the soul, the relationship between mind and body, and perhaps even a greater understanding of the physical universe.

I have tried to write this book in such a way that it describes the routes by which human beings have come to understand their own brain. In doing this, I want to show how the adventure has a long history that goes back to the dawn of human civilisation, and one that has taxed the minds of many great philosophers and scientists ever since. The story, I am convinced, is as exciting and fascinating as any other in the cannon of science and needs to be told. I will also be impertinent enough to claim this is the first time a historical narrative of the brain has been attempted. For this reason I feel justified in calling this a story, although one whose intention is educational.

Writing a cohesive history of the brain has been a challenging task. Indeed, the fact that this type of book has not been attempted before is testament to its difficulties. For example, one has to be a specialist in many different areas including anatomy, biochemistry, physiology, pharmacology, philosophy and psychology (to name a few). And if one is totally honest, no suitably qualified person probably exists – or if they do, they are few in number. Further, to weave all these disciplines together in a coherent and integrated account is a challenge of Rubik Cube proportions. This was all the more worrying for when I began writing this book, it increasingly dawned on me that history is less about hard facts, and more to do with reconstruction and interpretation. Such a manufacture, however, is necessary for the complexity of the past to be understood in a more simple, meaningful and comprehensible way. Thus, the telling of history must inevitably reflect the prejudices, biases and ignorances of its author.

So what is the alternative? There isn't one. Despite its inevitable mixture of fact and subjectivity, history is important. Indeed, it would be difficult, if not impossible, to fully understand our world, if it had no place in history. This is also true for any academic subject you care to mention. In writing this book, I have tried to keep things as simple as possible by expressing ideas in a clear and jargon-free way. Hence, the book is intended for anybody with an interest in the human brain. However, I have also sought to make it rigorous enough to provide a useful learned introduction to the history of neuroscience for the more serious student or researcher. This makes the book highly ambitious. Although I am convinced the story can be told and needs to be told, whether I am up to the task is another matter. If I am not, then I am hopeful others will try to improve upon my attempt. Whatever the verdict, I am at least comforted that I have planted the seeds of something new and worthwhile.

Andrew Wickens
April 2014

Notes

1 Finger, S. (1994) *Origins of Neuroscience*. Oxford University Press: Oxford.
2 Finger, S. (2000) *Minds Behind the Brain*. Oxford University Press: Oxford.

ACKNOWLEDGEMENTS

Writing a book is not unlike writing a symphony. Although the creative act of putting ink to paper, or typing a computer keyboard, inevitably requires a lonely and obsessive existence, the score could not be played without the help of many other skilled people. I am no exception and would like to acknowledge the help of many others who have made this book possible. First of all, I want to thank Lucy Kennedy, who as part time Editor for Psychology Press, saw the merits of my work and set into motion the process that would ultimately lead to its publication. I would also like to thank her successor Ceri Griffiths, and especially Michael Fenton who were always an email away to help me with my inquiries and difficulties. This book has also required me to track down a number of pictures to illustrate the text, and in this regard I would particularly like to thank the Wellcome Library in London for their marvellous archive of material that they allowed me to use for free. A cursory glance through my book will show that I have depended heavily on their generosity. Other organisations that have helped me illustrate this text include the Royal Society, the National Library of Medicine, the Warren Museum, the Max Planck Institute and the Wylie Agency. I would also like to thank Professor Antonio DiIeva at the University of Milan, and Professor Manuel Graeber at the University of Sydney who were very kind in responding to my request for material. Others have assisted me by reading my chapters and offering advice. Special mention must go to Andrew Hedgecock who not only carefully scrutinised the first four chapters and helped iron out some of its grammatical idiosyncrasies, but is developing a website for myself as I write these words. I would also like to thank Frances Kirby for reading some of the early chapters, Charlotte Caswell for her work drawing several of the diagrams, Carolyn Holleyman for her proofreading and Kelly Derrick for her project management.

1

HEAD OR HEART? THE ANCIENT SEARCH FOR THE SOUL

It ought to be generally known that the source of our pleasure, merriment, laughter and amusement, as of our grief, pain, anxiety, and tears, is none other than the brain.

Hippocrates

And of course, the brain is not responsible for any of the sensations at all. The correct view is that the seat and source of sensation is the region of the heart.

Aristotle

Summary

Much of this chapter covers a period from about 3000 BC when the first seeds of Western civilisation were sown in Mesopotamia and Egypt, to the great flowering of ancient Greece with the rise of Athens as its cultural centre during the fifth century BC. It is by far the longest period of history covered in this book, stretching over two millennia. Curiously for much of it, people attributed different causes to their behaviour than we do today. They saw their thoughts, desires and actions arising not from the brain, but the heart. This belief is clearly seen in ancient Egypt where bodies were preserved for the after-life. The process of mummification tells us a great deal about how the Egyptians regarded the soul for they scrupulously preserved the heart as the repository of its earthly deeds, and saw its 'goodness' as the key to eternity. In contrast, the brain was extracted through the nose with an iron hook and thrown away! Nor were they alone in their disregard for the brain. The primacy of the heart was maintained throughout Biblical times. Indeed, it is a revealing fact that the 'brain' is never mentioned in the Bible as a cause of behaviour, whereas it is replete with references to the heart. Even when we come to the ancient Greeks who replaced myth and superstition with logic and reason, the heart is still attributed with mental powers. This is shown, for example, in their earliest stories, namely those of Homer, whose characters are frequently depicted as puppets of fate, propelled into action by divine will and emotional impulses originating from the heart.

> Although Homer was aware of the brain's existence, he never attributed it as a cause of action. In fact, we have to wait until the fifth century BC, before the earliest reference to the brain as the site of reasoning is found in the fragmented writings of the Ionian philosopher Alcmaeon. However, this view is more strongly endorsed in the great collection of works by Hippocrates (also in the fifth century) who leaves the reader with no doubt that the brain provides the individual with all his mental faculties and intelligence. Yet, this belief was not universally accepted. Although supported by the great philosopher Plato, the importance of the brain was dismissed by his pupil Aristotle who continued to attribute the heart with sensory and mental powers. This may appear perplexing to a modern reader, although it must be remembered that the ancient Greeks up to Aristotle had not yet discovered the nervous system. Consequently, they were forced to attribute the body's principal site of communication to the heart, and its great multitude of blood vessels. This system was also seen as containing the *psyche*, a spiritual entity, which provided the animating and life-giving forces of the person.

The emergence of the human mind

It is often said that nothing in biology makes sense unless understood in evolutionary terms, and this must also surely be true of the human brain. Arguably, the human brain is evolution's greatest achievement – a journey that can be said to have begun on Earth some three to four billion years ago when the first simple forms of life such as single cells first appeared. In this great time span, however, the human brain is a relatively recent event. In fact, modern humans are basically the latest in a long line of two-legged apes, known as hominins, which diverged from a chimpanzee-like ancestor some seven million years ago. Many types of this creature would exist, although the first to have begun the hominoid lineage to human beings is believed to be *Australopithecus afarensis* (a famous example is "Lucy"), which appeared over three million years ago. These animals were about 1.3 metres tall (four feet) with relatively small brains of about 450 cubic centimetres (cc) not dissimilar to a chimpanzee. However, their upright posture and bipedalism allowed them to free their hands and walk long distances, a characteristic leading them to have a very different lifestyle. Although *Australopithecus afarensis* remained unchanged for one million years, they evolved into the larger brained human-like *Homo habilis* (literally 'handy man') some 2.3 million years ago in eastern Africa.[1] In turn, this ape was the ancestor of several other hominin species including one appearing in the fossil record some 500,000 years later called *Homo erectus*, with a more robust and taller skeleton accompanied by prominent brow ridges above its eyes. However, perhaps the most striking characteristic of this ape was the size of its brain, which almost doubles *Homo habilis* at about 1100 cc. Thus, in just over two million years, beginning with *Australopithecus*, the hominin brain had almost trebled in size.

Around 600,000 years ago the fossil record begins to show the emergence of a new ape in Africa lacking the prominent brow of *habilis*. Instead they had a higher vaulted cranium containing a brain similar in size to our own at around 1400 cc. Although little evidence survives about the rest of their skeleton, these creatures are generally recognised as archaic *Homo sapiens* (meaning 'wise man'). And, as far as we can tell, they developed into modern day human beings (or *Homo sapiens sapiens*) about 200,000 years ago. Originating in places such as Ethiopia, Sudan, Tanzania and South Africa, these creatures had a slight and nimble skeletal frame, with opposable thumbs that could manipulate objects dexterously. While they are the only surviving branch of the hominin family, for much of their early evolutionary history they were not alone. Around 150,000 years ago and surviving to about 30,000 years ago, we also see evidence of the Neanderthals.[2] Robust, heavily muscled, and short in stature, these creatures were common in Europe during the Last Ice Age. While Neanderthals are a side branch of the hominoid evolutionary tree with extinction as their fate, they remain of great interest, not least because their brains were bigger than our own with a volume of around 1550 cc. Why they became extinct is not clear. It is most likely they were not able to compete for resources as effectively as *Homo sapiens*, although recent genetic evidence suggests some degree of inter-breeding with humans may also have occurred.

The earliest *Homo sapiens* were skilled in their use of stone technology and produced a diverse range of tools including sharpened blades, small portable axes, and spears. This helped them be skilful nomadic hunter gatherers who roamed and explored their environment in small groups to search for game and food – a pursuit also encouraging them to leave their African homeland in various migrations, including one 60,000 years ago when they crossed the Arabian Peninsula to reach Asia and Australia. While more intelligent than their ancestors some anthropologists have pointed out that these first humans, nonetheless, do not show any appreciably different behaviour from their ancestors. Nor do they put their technology to any radically new use. But some 40,000 years ago in the Late Stone Age this situation changed abruptly in a cultural shift, which some have dubbed 'the mind's big bang'. The earliest evidence of this change is found in Europe with the behaviour of so-called Cro-Magnon man who inhabited the caves of France and northern Spain.[3] Suddenly, these early humans are seen to be constructing shelters and simple dwellings, which they lit and warmed by fire, allowing them to cook and create permanent settlements. We also see Cro-Magnon man making tools from bone, ivory and antler, and wearing clothes from furs adorned with beads and buttons. He adds to his fashion spectacle by wearing carved amulets, pendants and bracelets. Cro-Magnon was also the first human to create art as shown by magnificent cave paintings that include depictions of horses, mammoths, reindeer and other mammals. His artistic talents even extended to music as shown by flute-like instruments made from bone. The reasons for this change in human behaviour are highly contentious in anthropological circles with some arguing it represents a sudden major advance in human thinking,[4] perhaps driven on by social pressures, or the development of language encouraging new thought processes, while others see it as something that only appeared gradually. Whatever the cause, it is nonetheless clear at this point in our ancestry, we are witnessing the start of something very special: a being who is much like us today.

However, there is another remarkable change associated with Late Stone Age man: they buried their dead, and often left a wide range of personal belongings in their graves including necklaces, bracelets, hunting weapons and animal bones. Early humans also often went to considerable trouble to position the body and arrange stones in a careful manner around the grave. Although Neanderthals are known to have buried their dead in shallow pits some 100,000 years ago, and may have occasionally left bodies with simple remnants, their mortuary or ritualistic behaviours were not as sophisticated as the early humans whose burials often involved red ochre (a form of iron oxide that yields a pigment when heated), which was used to paint the animal bones placed in graves. This practice shows that our own ancestors associated this pigment with death, making the important mental leap of using it as part of a symbolic burial ritual. The burial of the dead is a highly revealing change in behaviour for it shows one of our earliest beliefs, perhaps beginning some 27,000 to 23,000 years ago, was in a spiritual existence beyond our earthly senses where souls depart following the death of the individual. In other words, man is beginning to question the world and become self-aware of his own existence. It is arguably the first evidence that consciousness is emerging from the human brain.

The contemplation of life after death is a crucial turning point in our human journey as it surely entails some type of spiritual belief – encouraging an awareness of mortal existence and a puzzling of one's place in the greater scheme of things leading to religious practice. Early man must also have noticed that the dead stopped breathing and realised breath (later to become the *psyche* or *pneuma* in the writings of the ancient Greeks) was vital to life. This would have inevitably encouraged a belief in two different worlds: one associated with physical reality where the living exist, and another that was non-material and spiritual. Extending this belief to its natural conclusion, it is a small step to regard living things as fundamentally different from non-living things since they contain some non-physical element, or spiritual force, providing them with life and animation.[5] The adoption of such beliefs was to have several far reaching consequences. One was a belief in animism in which natural phenomena such as animals, plants, rocks, thunder etc were imbued with a wilful or life-like presence. Another was the idea that body and spirit exist as separate entities. This was to become a central tenet of all the world's great religions, and provided the idea that there is a difference between the spiritual mind and physical body. Despite its great antiquity, this is still a concept strongly influencing the way we think of ourselves today. Indeed, there are some, including several eminent scientists, who remain tied to the belief that the human brain is greater than the sum of its biological parts and governed by a spiritual force.

Stone Age surgery: trepanation

The Late Stone Age was an epoch that began some 50,000 years ago in Europe and progressed into the New Stone Age (or Neolithic) period, roughly corresponding to an era spanning from 3000 to 1800 BC.[6] During this time we see the development of intelligent farming practices with people planting crops and domesticating animals. The invention of polished stone tools, creation of pottery and the formation of small settlements were also signs of new creative behaviour. However, something else also emerged during this time that shows the ingenuity of the early human mind, and which is quite unexpected: the first perforations of the human skull performed by surgical procedures with specialised stone tools. Thus, Neolithic man was using scrapers and sharp instruments to perform an operation known as trepanation (from the Greek *trypanon* meaning 'to bore'), which is the earliest known surgical procedure to have been undertaken by mankind.

The first skull recognised as being trepanned was discovered in an Inca cemetery in Peru during the 1830s. It was given to the American diplomat Ephraim George Squier who presented the skull to the New York Academy of Medicine in 1865. Afterwards, it was sent to Paris in 1867 to be inspected by Paul Broca, founder of the world's first Anthropological Society. To Broca's great surprise the skull's perforation had clearly been produced by a specialised cutting tool and involved advanced surgery – something he had previously considered not feasible for whom he regarded as 'primitive people'. The skull also encouraged others to look for similar examples in France, and within a few years a colleague of Broca called Dr Prunières

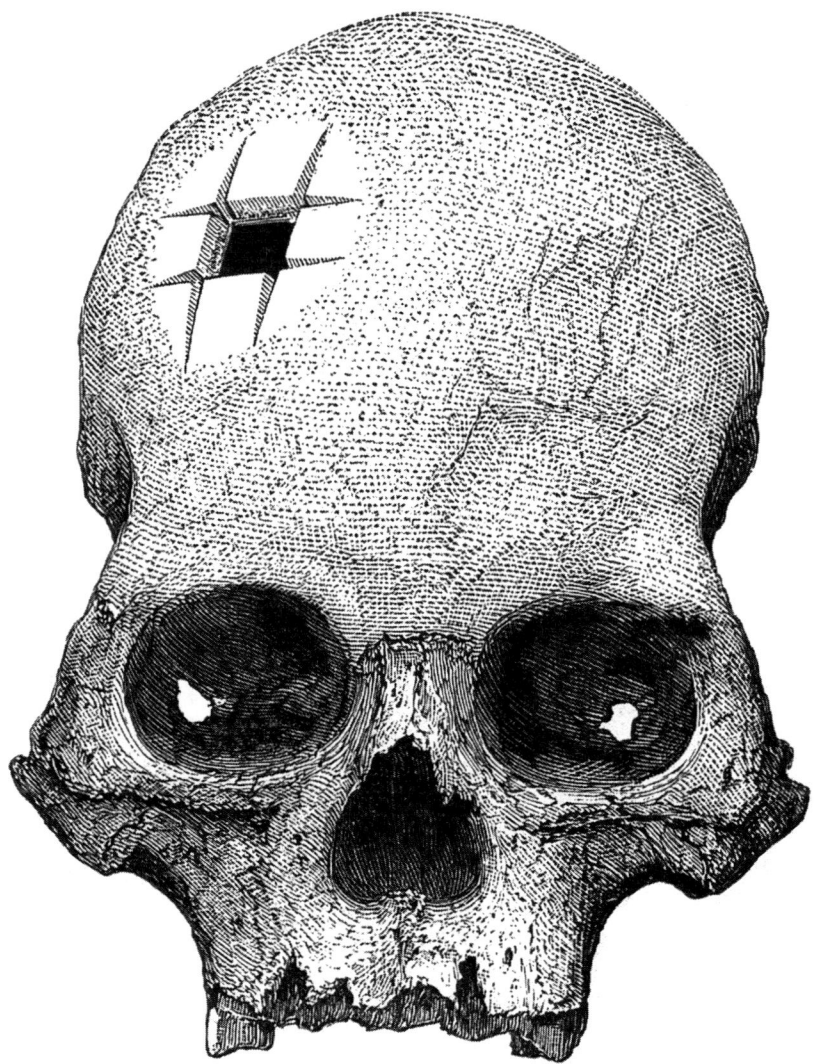

FIGURE 1.1 The Peruvian skull showing evidence of trepanation presented to the New York Academy of Medicine by Ephraim George Squier in 1865 and later sent to Paul Broca. Actually, this skull is relatively recent and dates from between 1400 and 1530 AD. It now resides in the American Museum of Natural History.

Source: Squier 1877, p. 457.

had unearthed a much older perforated cranium from a Neolithic burial site. It was the first of many trepanned crania to be discovered, most of which revealed evidence of bony regrowth around the wounds indicating the patients had survived the procedure by several months or even years. Accompanying these skulls in many of the graves were smaller pieces of sculpted oval-shaped cranial bone, called *rondelles* with bored holes. These apparently served as amulets and were worn around the neck like a modern pendant.

At first these skulls generated considerable scepticism not least because trepanation was a surgical technique performed in the hospitals of the time associated with a high rate of mortality. But as more skulls were found, it became clear trepanation was a viable operation for Neolithic surgeons. Moreover, palaeontologists began to realise that holes could be drilled through bone quite quickly. Indeed, Broca proved to himself it was possible to bore a hole in a deceased two-year old child's skull in just four minutes, although taking 50 minutes in a thicker adult one. Broca also performed trepanation on living dogs, showing it was relatively easy to avoid damaging the brain's protective dura mater and surface. In fact, we now know trepanation was practised throughout Europe during Neolithic times and was surprisingly common in certain locations. Skulls have been found in France (where the highest numbers occur) along with those in Spain, Denmark, Germany, England, Italy, Russia and the Balkans. Although most have been dated by estimating the age of their grave sites, a few have been assessed with radio-carbon dating. This has revealed that some Ukrainian skulls are from 7300–6200 BC, and others from France date from 5100–4900 BC. Most trepanations, however, appear to have been performed between 3000 and 2000 BC. Interestingly, the earliest may well go back to around 10,000 BC, although these derive from northern Africa and the Middle East, indicating that trepanation did not originate in Europe.

The tools used to perform trepanation vary according to period and geographical location. In the oldest sites, the skull bone was scraped away using a sharpened piece of stone with fragments of bone removed by hand. But as time progressed, more advanced techniques were adopted, including grooving, boring and cutting with sharp edged tools. Although the instruments were crude, the surgeons were often able to cut holes with sloping edges into the skull, which helped them safely remove debris from the operation. In most cases of Neolithic trepanation the holes are round or oval in shape and several centimetres in diameter. However, one perforation has been found measuring 13 × 10 cm that almost obliterated the left side of the cranium. These holes are also predominantly found on the left side of the skull situated between the frontal and posterior axis. While skulls with single openings are most common, there are cases where multiple holes from several operations have been found.

The Neolithic surgeons who performed these operations must have been highly skilled medical practitioners. Undertaking the painful procedure of cutting through the scalp to expose the skull's surface would have been a great challenge, especially without effective anaesthetic agents. During the operation it was also imperative the surgeon avoided damaging the brain's surface with its rich supply of blood vessels – the rupture of which would have led to uncontrollable bleeding and death. This would have required very careful scraping as the brain sits just under the cranial vault by a few millimetres. In addition the patient must have been nursed with some form of post-operative care as open perforated wounds are highly susceptible to infection. Although we can not be sure how Neolithic surgeons viewed the brain, its vital importance to life must have been recognised. Despite the obvious dangers of the procedure, estimates suggest between 50 per cent and 90 per cent of Neolithic patients survived trepanation, with some making a long-term recovery.

We can only speculate why Neolithic man went to such lengths to undertake this painful and dangerous surgery. For Broca, trepanation was most likely performed to remove evil spirits from the head, a view supported by the fact that primitive operations for broadly similar reasons still take place in certain parts of the world today (e.g. Africa and Polynesia). Broca also hypothesised that a high percentage of trepanned patients were children suffering from infantile

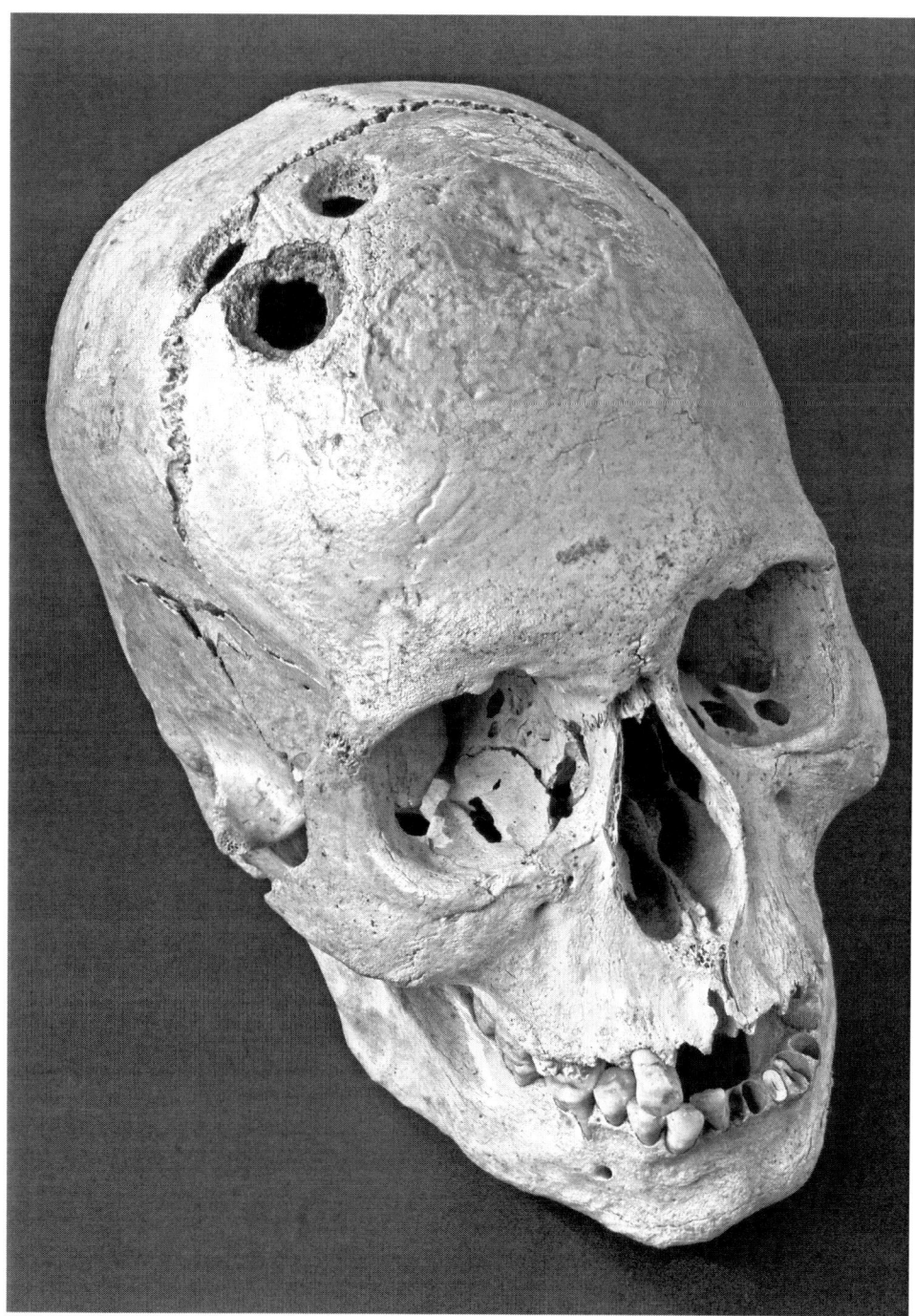

FIGURE 1.2 An example of a Bronze Age skull from Jericho, Palestine, dated from between 2200–2000 BC.

Source: Wellcome Library, London.

convulsions – a condition most likely to have been interpreted by Neolithic doctors as resulting from demonic possession. The Victorian English neurosurgeon Victor Horsley broadly agreed with Broca, although he thought it more likely trepanation was used to treat seizures resulting from depressed fractures of the skull. It is unlikely we will ever know for sure. Nonetheless, trepanation continued to be used as a surgical procedure throughout history and performed for a great variety of reasons. These include madness, idiocy, moral degeneration, headache, the removal of foreign bodies and the release of pressures, airs, vapours, and humours.

The earliest reference to a brain disorder: epilepsy

If there is a point that distinguishes prehistory from antiquity, it must be the invention of writing. Although this practice arose in several different cultures throughout the ancient world, the earliest dates back to around 3300 BC in Mesopotamia and Egypt. Prior to this time the only form of meaningful communication, with the possible exception of art, was the spoken word – a highly subjective medium for conveying knowledge. Everything changed with the use of writing, which enabled information to be recorded and scrutinised. Most importantly, it encouraged scepticism and set into motion new types of intellectual activity. This advance can be seen, for example, with the early Babylonians whose invention of writing enabled them to compile detailed accounts of their celestial observations – information, which could never have been expressed precisely in oral form. Consequently, the motions of the stars and planets were recorded and their future position predicted. However, writing as we know it today was much slower to develop, taking about 3000 years to fully emerge. The earliest forms used simple pictographs, such as an ear of barley for grain, inscribed onto wet clay tablets. This type of cuneiform script was replaced around 3000 BC by the Egyptians who etched marks into papyrus reed, leading to the production of scrolled manuscripts. Over time the use of pictographs, however, would become more refined, with some signs representing the actual sound of the word. Not only did this lead to grapheme writing, but it also allowed the elements of human speech to be represented. Finally, around 800 BC, a fully alphabetic form of writing appeared in Greece using a different sign for consonants and vowels. This forms the basis of Latin, English, and most other modern written languages spoken today. Such a development was crucial in laying the later foundations of philosophy and science in the West.

A further benefit of writing was it allowed a legal code to be constructed and justice administrated over large areas of land. One of the earliest sets of law is attributed to the Babylonian King Hammurabi (1792–1750 BC) who created the city state of Babylon. To publicise his laws, King Hammurabi had them inscribed into a giant stone pillar, which he erected in the city of Sippar. This stone, which is now housed in the Louvre museum in Paris, lists a total of 282 laws – the most famous being the proclamation that the penalty for causing loss of an eye should be the loss of one's own (from where we get our modern expression 'an eye for an eye, a tooth for a tooth', and which was also reiterated in the Old Testament of the Bible. Another interesting law states that slaves can be taken back to their sellers and money returned if they develop *bennu* within a month of being sold. While subject to different interpretations, the Babylonian historian Marten Stol has made a strong case that *bennu* is a condition that occurred when a God took sudden possession of a person. Put into modern terms, it refers to a type of seizure – the likeliest type being one caused by an infection produced from the larvae of the tapeworm (cysticercosis), which was known to be prevalent at the time. If so, this makes epilepsy the oldest disorder of the brain to be recorded in written form.

The brain in ancient Egypt

For many, Egypt ranks as the most interesting place in the early civilised world, not only because of its rich archaeological treasures, but because its people recorded things in hieroglyphs, which have enabled us to understand their world-view and beliefs. It also has a long fascinating history, covering almost 3000 years, beginning around 3100 BC when Menes became the first Pharaoh to rule a land dominated by the tides of the Nile. This was the start of a succession of thirty dynasties, governed by several hundred kings, who built a civilisation of great complexity until conquered by Alexander the Great in 332 BC. The Egyptians also saw themselves as a divine people, where religion was central to society and their Pharaohs were worshipped as deities. They further lived in a world in which the supernatural and natural were closely entwined. However, the most conspicuous belief of the ancient Egyptians, as shown by the great pyramids and their remarkable variety of funerary remains, was that of an after-life.

To provide a successful passage into the next world, the Egyptians believed it was necessary to preserve the body so that its soul had a safe haven in which to dwell before judgement day. In the early dynastic periods this was done by preserving the corpse in layers of linen impregnated with resin. By the time of the Middle Kingdom (2040–1675 BC) this practice had evolved into a more elaborate procedure where various organs were taken from the body, leaving its cavities to be dried out and packed with cloth. The body was then wrapped, masked and placed in a wooden coffin – a task undertaken by a guild of highly trained craftsmen. How the Egyptians treated the various parts of the body during this process tells us a great deal about their beliefs. By far the most important part of the body to be preserved was the heart, which was always carefully placed back into the mummy. The Egyptians did this partly because it represented the person's self, but also because the heart was seen as holding the key to eternity, for it was to be weighed in a special ceremony by God Anubis to determine its goodness. If heavy with guilt, it was consumed by a monster called Ammit: if lighter than the feather of Ma'at, representing the goddess of truth and justice, the person passed into the spiritual world. The intestines, lungs, liver and stomach were also deemed important – being embalmed and stored in canopic jars next to the mummy. However, all of this reverence was in marked contrast to the brain, which was extracted through the nostrils by an iron hook and thrown away![7] As far as we can tell, the ancient Egyptians attributed little importance to the brain, other than seeing it as a vesicle for passing wet mucus or snot to the nose.

It is clear from Egyptian writings, especially those adorning the walls of tombs, that the heart was not only seen as the repository of the soul's earthly deeds, but attributed with powers of thought, reasoning and emotion. This was a view shared by other ancient people of the time, including the Mesopotamians who imbued the liver with soul-like properties. This was probably influenced by the fact that the liver is the largest organ in the body and heavily laden with blood. The primacy of the heart was further maintained into Biblical times. Indeed, when the Old Testament was translated into Greek (a task completed during the second century BC) it was accepted that man's intellect and emotions resided in the heart and not the brain. In fact, the brain is not mentioned in the Bible, although it contains hundreds of references to the heart.[8] It is easy to dismiss these ancient views as being rooted in a long history of superstitious and magical beliefs. However, it should not be forgotten that the ancient Egyptians, in particular, were renowned as skilled physicians – even to the later Greeks.[9] Yet, in spite of their medical knowledge, the Egyptians disregarded the brain's importance. They can be forgiven, however, for it was strongly influenced by several factors. This included the heart's location in the centre of the body; its hot nature (a sure sign of life); and its associated blood vessels, which provided the only known means of communication to various parts of the body.[10]

The first word for brain

The ancient Egyptians also provide the earliest written reference to the brain. This appears in a medical work known as the Edwin Smith papyrus, purchased by an American of that name in Luxor in 1862, and translated by James Henry Brested during the 1920s (Brested 1930). Although believed to have been written around 1700 BC, it is probably a copy of a much older text, possibly a manual for military surgeons, dating from around 3000 BC. Forming a scroll measuring over 15 feet in length, the papyrus describes 48 types of head and neck injury, along with advice on treatment and surgical intervention. The brain is first mentioned near the beginning (case six) where it is represented by a four-part hieroglyph, composed of a 'vulture', 'reed', 'folded cloth' and a final suffix which means 'little'. Despite their pictorial nature, the first three of these hieroglyphs are actually phonemes making a sound, which is believed to have resembled ' ah-i-s'. This is also translatable as 'skull-offal' or 'skull-marrow'. Interestingly, the origin of our modern word 'brain' has a similar derivation for it is believed to come from 'braegen', an Old English word (c. 1000) that appears to be related to a yet older French word 'bran' meaning refuse.[11]

The Edwin Smith papyrus also provides another description of the brain. Here it is reported as being similar to the corrugations of molten copper, which 'throbs' and 'flutters' when exposed. This is clearly a likely reference to an open skull wound, through which the pulse-like movements of surface blood vessels can be observed. There is also mention of a sack covering the brain (i.e. the meninges) and fluid underneath which may be a reference to the cerebrospinal fluid. The authors of the papyrus also appear to have some understanding of the brain's function, since they describe the effects of head injury, including abnormal eye

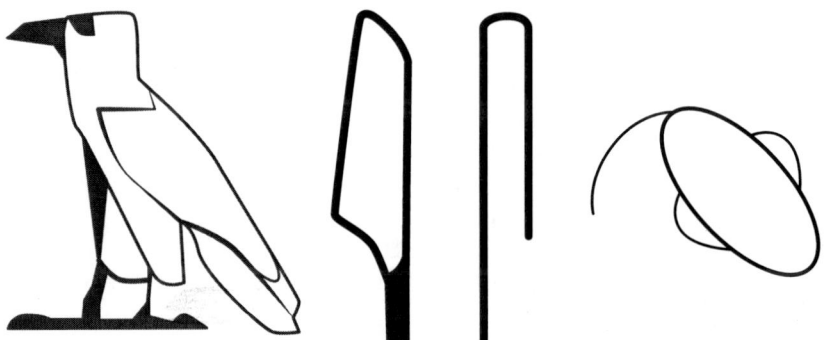

FIGURE 1.3 The ancient Egyptian hieroglyph for brain as shown in the Edwin Smith papyrus (Brested 1930, volume 1, Plate IIA).

Source: Wikimedia Commons http://Wikepedia.org/wiki/file:Hieroglyphic-brain.jpg.

movements, paralysis and speech loss. The Edwin Smith papyrus also points out that if a finger is placed into a wound of the brain, the person will shudder violently. Nonetheless, it is an interesting fact that the word 'brain' is only known to occur eight times in Egyptian history – with six of these instances occurring in the Edwin Smith papyrus. This provides further proof, if needed, that the Egyptians attached little importance to the brain as a cause of behaviour.

The causes of behaviour in Homeric times

The decline of Egypt overlaps with the rise of ancient Greece, which began on the islands of Crete between 1900 and 1450 BC. Although towns with grand buildings were built there, this civilisation mysteriously disappeared around 800 BC and replaced by settlements on the mainland. This new Greek world came to consist of about 300 city states including colonies in southern Italy and on the coasts of the Aegean and Black Sea. It was also a time when an advanced Greek culture began to form encouraging literacy and learning. Two famous stories invented during this period are the epic poems the *Iliad* and *Odyssey*, which were probably conceived in the ninth and eighth centuries BC by Homer (Homer 2003a, 2003b), and kept alive by oral tradition until written down around 400 BC. The *Iliad* and *Odyssey* are based on the events of the Trojan Wars and tell the adventures of Greek heroes such as Odysseus, Achilles, and Hector who have to confront many challenges in a world full of myth and magic inhabited by gods with human-like weaknesses. The stories, realistic and fantastic by turns, contain gruesome and unsettling passages, which are still enjoyed by students today who continue an unbroken tradition of two and a half millennia of Greek education. Perhaps often overlooked are the insights Homer provides about how the ancient Greeks understood themselves during the first throes of Western civilisation.

It is clear, even from a cursory reading of Homer, that the self is composed of many different spiritual agents and forces. One of these is the *psyche*, a term etymologically linked with the word *psychēin* meaning to breathe, which specifically refers to the vital life force that keeps the person alive. It also leaves the body after death. For example, Homer describes it flying away from the limbs, or leaving the body through the mouth or wound in the chest. This force also appears to be immortal, as shown when Hector speaks his last words to Achilles, stating that 'His *psyche* fled from his limbs and went to Hades.' Yet, surprisingly, the *psyche* does not have any intellectual function. Instead Homer attributes mental ability to several other soul-forms that reside in the chest. Perhaps the most important of these is the *thymos* (mentioned over 400 times in the *Illiad* alone), which lies in the diaphragm where it exists as a mixture of air and vapour exhaled from the hot blood. The *thymos* is perhaps best described as the source of emotion that drives the person into action. For example, friendship, joy, sadness, grief and revenge all spring from the *thymos* as seen when Achilles is angered: 'His brave *thymos* roused him.' Nonetheless, the *thymos* is also capable of rational action as shown when Odysseus enquires: 'but why does my *thymos* consider that?'

The type of soul-like entity that most closely resembles intelligence is the *noos* (from where we get our word *nous*). For example, Odysseus (the main hero of the *Odyssey*) who is renowned for his guile and acumen is described as having a *noos* that surpasses all other mortals. The *noos* is also referred to as having a plan of action or the capacity to see the true measure of things. Yet, despite its intellectual nature, Homer again locates the *noos* in the chest. In fact, he attaches no obvious importance to the brain as a cause of behaviour anywhere in his writings.

In contrast, he always regards the chest as the site where all responses to life's joys, sorrows and challenges occur. Trying to understand Homer's conception of human action presents a great challenge, for he seems to attribute little importance, if any, to self-determining mental processes such as introspection or self-awareness. Instead, humans are depicted as puppets of fate. At best, Homer may explain behaviour in terms of divine will and human impulse at the same time. This can be seen, for example, when Diomedes reports that Achilles will return to fighting, 'when his heart tells him to, and when the god moves him.' However, in many other instances, mental states are fostered upon the individual by external forces over which they have no control.

This lack of self-determination has puzzled some experts and even led them to make the claim that the early Greeks were without mental awareness or any form of organised conscious life. The scholar Bruno Snell, for example, has argued it was only after Homer that people begin to consistently write as if they were the 'owners' of their mental states. Before this, the Greeks attributed their actions to external forces. The American psychologist Julian Jaynes (1976) has gone even further and claimed that before 1200 BC all humans were *preconscious*, according to our own modern standards. Instead of having self-determination Jaynes postulates the ancient human mind was like that of a schizophrenic who heard the God-like voices telling it what to do. Thus, according to Jaynes, ancient people had no 'awareness of awareness', or even a sense of self-volition, believing themselves to be controlled instead by divine forces. This, he claims, was due to the dominance of the brain's right hemisphere in governing behaviour, rather than the language-mediated left hemisphere that has control today.[12] It is, however, a highly controversial theory. In *The God Delusion*, Richard Dawkins has gone so far as to say that it is either complete rubbish or the work of consummate genius. Nonetheless, one should perhaps be at least open to the possibility that ancient people may well have thought differently from us today, resulting in their attributing different causes to their behaviour. If so, this may have been another factor in the heart and not the brain being seen as the driving force of human action.

It is also relevant for this book to note that the *Iliad* and *Odyssey* introduce a number of words that have since become part of our modern vocabulary including those describing various parts of the body. In fact, more than 150 anatomical terms are found in Homer, although only three relate to the brain. These are (1) *enkephalos*, which literally refers to the material inside our head (i.e. the brain) and is the source of our words 'encephalon' and 'cerebrum'; (2) *muelos* referring to the marrow inside the spine, from where we get the word 'medulla'; and (3) *sinew* describing a ligament or tendon. Although 'sinew' may not appear to have any obvious relationship with the brain, it would later become connected with the nervous system by forming the origin of our word *neuron* (i.e. nerve cell).

The brain as an organ of sensation: Alcmaeon of Crotona

The point at which the ancient Greeks abandoned gods and myths as the way of making sense of their world, and turned instead to analysing human experience through reason, was one of the most profound transitions in the course of human history. This new orientation can be first found in the teachings of a small band of thinkers who lived in the city of Miletus in Ionia (now located in west Turkey) during the sixth century BC. At the time this city was a great port located at the hub of several important sea-trading routes enabling the Ionians to become familiar with a wide variety of people and new ideas. They were also highly literate, and this heady brew of learning, exploration and commerce encouraged them to pose new questions and seek rational answers. We can see this occurring in the writings of Thales (*c.* 652–548 BC) who discovered a number of new mathematical laws, explained how the constellations could be used for navigation, and predicted a solar eclipse. He also extended Homer's notion of the *psyche* by recognising it as the animating force that moved in the universe including the stars and planets – a position he seems to have arrived at when contemplating the attraction of magnets! Other Ionian philosophers included Anaximander who drew the first map of the world, and Anaximenes who realised that air (*pneuma*) was essential for life.

The Ionian interest in philosophy also extended to other parts of the Greek world over the next two centuries, making this period a remarkably fertile one for intellectual advancement. It also produced a number of important philosophers including the famous mathematician Pythagoras (*c.* 570–500 BC) who associated the *psyche* with rationality and thinking; Empedocles (*c.* 500–430 BC) who listed the four cardinal elements as earth, air, fire and water; and Democritus (460–370 BC) who held that the world was composed of tiny particles called atoms. Remarkably, all of these great thinkers still continued to embrace the tradition of viewing the heart as the most important reasoning organ of the body. In fact, Empedocles who is credited with being the first to postulate that the blood vessels carry *pneuma* around the body, seemingly believed the blood surrounding the heart was the seat of a man's soul, and the origin of his thought and reasoning. He even went as far to say that intelligence depended on the blood's composition.

However, it was only a matter of time before someone had the daring audacity to recognise the brain as having mental powers, and the first philosopher to make this giant leap appears to have been Alcmaeon (*c.* 510–440 BC). Born in Crotona (now Calabria in southern Italy) Alcmaeon is an obscure figure who is known to have written several important works, although none have survived. Consequently we are dependent on others to learn of his work. The most important of these is Theophrastus[13] who informs us that Alcmaeon was the first to have studied anatomy by undertaking animal dissection. In doing this, Alcmaeon discovered two channels leading from the back of the eye to the brain, which we can recognise today as the optic nerves. From this simple observation, Alcmaeon also hypothesised similar passages called '*poroi*' (or porous tubes) for the ears, nose and tongue. It would lead him to state that, 'All senses are connected with the brain.' This was a bold statement and Alcmaeon supported it by noting that 'movements' of the brain (a possible reference to head injury or concussion) can cause sensations to be lost. Another commentator on Alcmaeon is Aetius who lived around the same time. He goes even further than Theophrastus by attributing Alcmaeon with the statement, 'The governing facility of intelligence is in the brain.' If Theophrastus and Aetius are correct, then Alcmaeon's insight marks a revolutionary transformation in mankind's understanding of himself. Some modern commentators have even gone so far to say that Alcmaeon's achievements are comparable to those of Copernicus and Darwin.[14]

Hippocrates and the brain

While we only have second hand accounts of Alcmaeon's exact writings on the role of the brain in mental processes, there is no uncertainty on the matter when we come to the father of modern medicine Hippocrates (*c.* 460 to 370 BC). A historical figure who also became a myth, Hippocrates more than any other individual in antiquity freed medical practice from mysticism and superstition by replacing it with the idea that health is a physical process amenable

14 Head or heart? The ancient search for the soul

FIGURE 1.4 A bust of Hippocrates (*c.* 460–370 BC) on show in the British Museum, believed to be a Roman copy from the second or third century BC.

Source: Wellcome Library, London.

to understanding through observation and reasoning. Born and raised on the Greek island of Kos, Hippocrates learned medicine from his father and founded a school that taught all students whether rich or poor. Greatly revered during his own lifetime, his writings were collected by scholars at the Library of Alexandria after his death, where they would form a body of work known as the *Corpus Hippocraticum*. The compilation of this work has generated much speculation since it is now known much of it was not written by Hippocrates himself, but by many of his students and followers. Nonetheless, the corpus, which contains around seventy treatises on a wide range of medical matters comprising books, lectures, essays and clinical notebooks, emphasises a number of Hippocratic beliefs and practices. These include, for example, the humoral theory of disease that explains illness as an imbalance in the four humours (blood, phlegm, yellow bile and black bile); the importance of diagnosis based on sound observation; and an ethical doctor–patient relationship encapsulated in the famous oath that every practising physician takes today.

References to the brain abound throughout the *Corpus Hippocraticum*, with the most famous occurring in a book entitled '*On the Sacred Disease*' which deals with epilepsy. This disorder was widely seen at the time as a form of possession, where the person was seized by a demon as divine punishment for some sin (hence the term 'sacred disease'). Hippocrates goes to great lengths to refute this idea by pointing out it is a notion associated with witch-doctors, faith-healers, quacks and charlatans. Instead, he builds a convincing case that epilepsy is an illness of the brain caused by an excess of phlegm acting to block air flowing through its blood vessels – a situation only rectified through a sudden and violent seizure. More importantly, the first part of *On the Sacred Disease* informs the reader in the most explicit way that the brain is responsible for all our mental activity including intelligence and madness. This can be seen in the following quote, the earliest and most forthright assertion of a simple fact, that all of us now take for granted:

> It ought to be generally known that the source of our pleasure, merriment, laughter and amusement, as of our grief, pain, anxiety, and tears, is none other than the brain. It is specially the organ which enables us to think, see, and hear, and to distinguish the ugly and the beautiful, the bad and the good, pleasant and unpleasant . . . it is the brain too which is the seat of madness and delirium, of the fears and frights which assail us . . . it is the thing where lies the cause of insomnia and sleep-walking, of thoughts that will not come, forgotten duties and eccentricities . . . so long as the brain is still, a man is in his right mind.[15]

One can not fail to be impressed by this remarkably coherent and modern sounding account of the brain's involvement in sense perception, judgement and emotion, as well as its association with mental disturbance. Written some four hundred years before the birth of Christ, it can be said to be the historical point at which the brain is finally recognised as the organ of consciousness and cause of our unique human behaviour.

On the Sacred Disease also provides an excellent anatomical description of the human brain. For example, it is said to resemble that of all other animals in being 'double' with a thin vertical membrane dividing it through the middle – a reference to the brain's two cerebral hemispheres. The brain's blood supply is also mentioned with many of its vessels being described as slender, although two are stout – one coming from the liver and the other from the spleen. Although this anatomy is confused, it is still an extraordinary observation, for Hippocrates is surely describing the carotid and vertebral arteries that supply the brain with blood. Despite this, the Hippocratic writers had a very different conception of the brain from the modern one we have today. Since they had no idea of the nervous system's existence, this inevitably led to an understanding of physiological function very different from our own. In fact, Hippocrates believed that the blood was filled with a type of 'vital air', which was alive with a spiritual force and self-perpetuating. Indeed, Hippocrates believed this substance provided intelligence as well as causing the movement of the body's organs and limbs. This is most clearly stated in *Breaths* when he writes, 'No constituent of the body . . . contributes more to intelligence than does the blood.'

Despite this, it is certain that the Hippocratic writers observed nerve pathways during their dissections – although they did not recognise what they saw. For example, in the *Nature of*

Bones, two 'stout cords' are described leaving the brain to travel down on either side of the windpipe to terminate between the backbone and diaphragm. This would appear to be the vagus nerve, or possibly the sympathetic trunks running from the base of the skull to the coccyx. Similar cord-like pathways are also described passing to the ribs, liver and spleen. Although referred to as *tonoi* (strings) it is clear the Hippocratic writers regarded their function as similar to veins. Similarly, in *Fleshes*, two fine 'veins' running from the brain to the eyes are described which must have been the optic nerves (as observed by Alcmaeon). Again, their true significance was not realised. In fact, it would be another hundred years or so, before anatomists in Alexandria who were allowed to practise human dissection for the first time, finally gained an understanding of the nervous system (see next chapter).

Plato appears to confirm the importance of the brain

The most glorious days of ancient Greece were to fall in the fifth century BC. By this time Athens had become the cultural centre of the Greek world, the birth place of democracy and a haven for writers, artists and philosophers. Although its heyday was to be brief, lasting for only about 50 years or so, Athens saw the emergence of the two greatest philosophers of antiquity: Plato (424–347 BC) and Aristotle (384–322 BC). Their works were to have a major impact on the course of Western thinking despite being 'lost' or unavailable in Europe until the twelfth and thirteenth centuries. Both philosophers wrote extensively on a broad range of subjects with much of it surviving intact. The writings of Plato, for example, are almost complete, making up around five modern volumes; whereas Aristotle's consist of 31 treatises, although these only represent a small proportion of what he actually wrote. Together these works cover a staggeringly huge amount of different subjects, and introduce a rich variety of theoretical issues that remain central in philosophical debate today. It is sometimes said that Plato attempted to understand the world through pure reasoning, whereas Aristotle was more empirical and willing to observe. While this is far from the truth, it is perhaps fair to say both excelled in different areas of thought. Plato had a greater impact on politics, theology, ethics, and aesthetics, while Aristotle's influence was stronger in science and biology.

Born into an aristocratic Athenian family, Plato was expected to follow a political career. However, as a young man he came under the influence of Socrates, a profound thinker who left no written work, but was famed for his oratories and philosophical interrogation. Following the trial and forced suicide of Socrates in 399 BC for allegedly corrupting the youth of Athens, Plato resolved to become a philosopher. Determined to continue in the style of his mentor, most of Plato's writings take the form of dialogues: one of the speakers normally being Socrates and the other a student after which the work is named (e.g. the *Phaedro*, the *Timaeus* etc). Although it is difficult to characterise, one of the things that distinguishes Plato is his desire to understand what lies behind the immediate reality of our everyday experience. That is, he seeks to reveal the true underlying nature of things rather than their superficial appearance. It would lead Plato to believe in an unchanging, non-physical and perfect world existing beyond our immediate senses, comprised of ideas (or forms) that can only be revealed through deep philosophical study. To support his approach, Plato points out that humans have an intuitive knowledge of things they could not have possibly learned from experience, including ideas about beauty, goodness, courage, justice and love. For Plato they could only derive from a divine entity belonging to a non-material world that is also responsible for creating our spiritual self.

FIGURE 1.5 A marble bust of Plato (424– 347 BC) believed to be a Roman copy taken from an ancient portrait.

Source: Wellcome Library, London.

Plato's theory of forms leads him to a position where he sees human beings as having an earthly perishable body occupied by a spiritual or soul-like self that he calls the *psyche*. This idea, otherwise known as dualism (i.e. the body and soul are regarded as different entities) is made most strongly in a dialogue called the *Phaedo*, which takes place between Socrates and several of his students just prior to his death. In this work, Socrates argues that the *psyche* is immortal and destined to leave his body at the moment of death. Its destination is the realm of true knowledge where the *psyche* had resided before birth. Socrates also describes the *psyche* as a force that imparts life to the body, which also provides it with self-volition, enabling bodily movement and wilful action. However, perhaps most important of all, Socrates holds the true purpose of the *psyche* in the earthly physical world is to reason and establish (or more accurately 'recollect') what is true. Thus, the *psyche* is much more than a life force in the

Homeric tradition. Rather, it is the organ of thought that is the true self of the individual. This is an important change, for the *psyche* now becomes recognisable as the entity many people and religions accept today as the soul.[16]

Plato's concept of the soul is, however, very different from our own, for he would attribute it with three distinct parts: the *epithymetikon*, the *thymos*, and the *logistikon*. The *epithymetikon* was associated with the liver and gut, where it fulfilled the basic vegetative needs of the individual, while the *thymos* located in the heart instigated emotions such as anger, fear, pride and courage. Both of these were also believed to exist in other animals and perish at the point of death. However, the *logistikon* was different: this was a spiritual force unique to humans, providing the individual with thought and intelligence. It was also immortal, capable of reincarnation, and resided in the brain. Plato's conception of the rational soul was to have a lasting impact on western culture, because many aspects of it would be embraced by Christian thinkers. In particular, Saint Augustine (354–430 AD) took many of Plato's ideas and wove them into the fabric of Christianity. By doing this, Plato's authority would be strengthened and his idea that the brain is the sole organ of reasoning was given official sanction.

Plato can never be considered a biologist, or a founding father of neuroscience; nonetheless he writes about the intellectual powers of the head and brain in a number of his dialogues. For example, in the *Phaedo* he states that: 'the brain is the originating power of the perceptions of hearing and sight and smell, and memory and opinion may come from them' (Jowett 1892). A similar endorsement also apparently occurs in the *Timaeus* where Plato writes: 'the head . . . is the most divine part and dominates over the rest of the body in us' (Plato 1937). However, perhaps Plato's most famous description of the brain's purpose is given in his chariot allegory, which is introduced in the *Phaedrus*. Here, he characterises the soul as two horses: one black and ugly, which represents the baser appetitive components of the soul; and the other white and noble characterising a more honourable animal. Both are controlled by the charioteer, representing intellect and reason who must do his best to stop the two horses pulling in separate directions. The brain (presumably) is therefore seen as being responsible for exercising executive control over our baser desires.

Aristotle's alternative view of the psyche

One would have imagined with the weight of authority provided by Hippocrates and Plato that the brain would be widely accepted as the organ of reasoning and intelligence by the fourth century BC. However, even in Athens, this was not the case. Many were still not prepared to discard traditional beliefs that had existed from the beginning of early civilisation, and one thinker who continued regarding the heart as an organ of the mind was Aristotle. Arguably the greatest of all philosophers, Aristotle was born into an aristocratic family with royal connections (his father was court physician to King Amyntas III of Macedonia) who was sent to study at Plato's Academy at the age of 17. Here he was to stay for nearly 20 years to become its most eminent pupil and teacher. Said to be a dandy who bought attention on himself by wearing rings on his fingers and cutting his hair unfashionably short, Aristotle was nonetheless the most obvious person to succeed Plato after his death in 347 BC. However, the position went to Plato's nephew Speusippus. Embittered and humiliated, Aristotle left Athens to journey through Asia Minor, spending some of his time employed as personal tutor to Alexander – later to become 'great' as a result of his conquests.[17] Returning to Athens at the age of 49, Aristotle founded his own school called the Lyceum. It was during these years that Aristotle completed many of his great works – a corpus that contained a far greater amount of factual knowledge than Plato. In fact, it has been said that Aristotle was the last man to know everything there was to be known in his time. However, following the death of Alexander in 323 BC, strong anti-Macedonian feelings arose in Athens, forcing Aristotle to flee after a charge of impiety. He died a year later, at the age of 62, in Chalcis (now Khalkis).

Although Aristotle's conception of the soul shares some features with that of Plato, it is fundamentally different in several important respects. For example, like Plato, Aristotle believed the *psyche* consisted of three parts that were hierarchically arranged. The lowest was the *nutritive psyche*, found also in plants, which governed nourishment, growth and decay. The intermediate part was the *sensitive psyche*, common to all animals and capable of perception, desire and locomotion. And, finally, humans alone were considered to have an *intellectual psyche* enabling them to think, reason and understand. However, crucially, unlike Plato, Aristotle believed these souls did not exist as individual entities – rather they combined to work in

FIGURE 1.6 A stone sculpture of Aristotle (348–322 BC) carved in the walls of Chartres Cathedral, France.

Source: Wellcome Library, London.

unison, making the *psyche* one thing rather than three. In other words, they could not be separated. Aristotle also had the same idea when it came to the form of the body. While accepting that the body is composed of two types of substance: a material element called *hyle*, and a non-physical one that produces its 'form' called *morphê*, he did not think these two substances could be detached. They were, in other words, facets of the same thing, which combine to create a soul-like unity. It is a very different idea to the one proposed by Plato who believed in a separate spiritual soul imparting action on a physical body.

Aristotle attempts to explain his theory more clearly in *De anima* ('*On the Soul*') when he compares the relationship of the body and the *psyche* to a lump of wax that has a pattern stamped into it. He points out the absurdity of considering the wax and its imprint as two separate things. Extending the same idea to the soul, Aristotle argues the *psyche* can not exist as a purely spiritual entity because it is nothing without a physical body. In other words, the *psyche* is a type of animating intelligent life force wholly dependent on the configuration of the body to provide its dynamism. This can be a difficult concept for students of Aristotle to understand, and perhaps a better way of explaining it, is to give a modern day example from vision (not from Aristotle). So, for example, we can argue the physical body needs the *psyche* to see, but the *psyche* cannot see without an eye (a physical organ).

Not surprisingly, Aristotle's theory of the *psyche* has led to different interpretations. Some have argued since Aristotle did not propose a mental substance then he must have believed in a form of materialism that denies a spiritual component to the mind. This leads to the doctrine of monism (the opposite of dualism) as it proposes the physical structure of the brain must give rise to the mind. Others, however, have pointed out that Aristotle's concept of the *psyche* retains some 'spiritual-like' properties since it can not be reduced to any material substance. Thus, it can be best regarded as a form of vitalism – the view that living organisms contain a special non-physical force that distinguishes them from inanimate objects. The later theory, in various manifestations, was to be a popular one up until the nineteenth century. Either way, Aristotle's concept of the soul is different from the dualist one proposed by Plato.[18]

Aristotle and the brain

Aristotle is generally regarded as the greatest biologist of antiquity. Indeed, his most extensive researches, comprising about one-fifth of his extant work, concern the nature of plant and animal life. Of course, Aristotle was not the first to ask questions of a biological nature. However, he was the first to make the investigation of all living things an academic pursuit. In doing this, Aristotle made many important advances including an hierarchical classification of animal types, which he divided into 'red blooded' and 'non-red blooded' (a taxonomy that roughly corresponds to what we know today as mammals and invertebrates). He also distinguished different species (*eidos*) and groupings (*genus*). In addition, Aristotle described the appearance and habits of over 500 types of animal, which makes him the legitimate founder of zoology. However, Aristotle's greatest biological achievement was his study of comparative anatomy. Fascinated by how the body worked, Aristotle dissected an astonishing variety of creatures from simple cephalopods, molluscs and fishes, to reptiles, birds and mammals including an elephant.[19] This enabled Aristotle to make several important advances including a description of the gall-bladder, the chambered stomach of a cow, and the heart's great artery. He also gave an accurate account of the female womb with nomenclature we still use today. Despite this, Aristotle was still prone to making a number of mistakes. One of his most famous being his belief in the possibility of spontaneous generation – an idea partly supported from his observation that flies could emerge from dead animals. Aristotle also discerned no difference between the veins and the arteries, a mistake that would later cause considerable confusion for investigators who relied on his authority.

Aristotle also extended his dissection to the brain (*encephalos*), which is mentioned in several works including *History of Animals* (*Historia animalium*) and *Parts of Animals* (*De partibus animalium*). With regard to the vertebrate brain, Aristotle notes it is covered by two membranes that are richly innervated with fine blood vessels, a strong one near the surface of the skull and the other being much finer. It is likely that Aristotle is referring to the meninges here – or more specifically the dura mater and pia mater. Aristotle also describes the brain as 'paired' providing a clear reference to the two cerebral hemispheres. He also recognises a separate structure that he calls the *parencephalis* located 'at the back' that is markedly different in appearance and texture (i.e. the cerebellum). Not content to observe, Aristotle also cut into the brain for he refers to 'a small hollow' in its centre (one of the ventricles), and 'liquidity surrounding it' (most likely the cerebrospinal fluid). However, other observations are more confusing, especially when Aristotle asserts that the brain does not fill the cranial cavity, which leads to a large cavity in its posterior region. This is a statement that certainly does not apply to the human brain, or indeed the vertebrate one. Furthermore, because the brain itself was cold to touch, Aristotle mistakenly believed it was bloodless and devoid of all blood vessels.

However, Aristotle's gravest error would concern the brain's function. Seemingly unimpressed by its cold and uniform structure, he emphasised the importance of the heart for sensation, movement and mental thought. In Aristotle's defence, it can be said he used a number of arguments to back up his claims. One was his belief that warmth is essential for life. Because the heart was seen as the furnace of the body, it made sense to Aristotle to regard this as the place where the *psyche* existed, for it was the same force that gave life and motion to the individual. This view was supported when Aristotle examined the development of the embryonic chick inside the egg, and observed the tiny beating heart as the first organ to emerge after just four days. The heart also stood out in contrast to the brain, which Aristotle regarded as the coldest and wettest part of the body. This led Aristotle to view the brain as little more than a cooling organ. Because heat rises, Aristotle reasoned that the network of blood vessels covering the brain's surface must act to cool down 'the heat and seething' of the heart's blood. Aristotle even adapted this logic to explain the large size of the human brain. Instead of attributing its greater mass to increased intelligence, he argued that humans were warmer than other animals and therefore needed a larger brain to cool down the blood.

By rejecting the brain, Aristotle came to regard the heart as the site where all the senses came together, or what he called the *sensus communae*. Again, he had evidence to support his theory. For example, Aristotle reasonably pointed out that if one touches the brain of a living animal it does not give rise to any sensation. Thus, the brain is apparently senseless. In addition, Aristotle noted that very simple creatures such as worms and sea urchins do not have a brain, although they are obviously capable of sensation. Although Aristotle was aware that others such as Alcmaeon and Hippocrates had described hollow ducts leading from the eyes to the brain, he dismissed such observations by arguing that these pathways actually passed to the rich vascular membranes on the brain's outer surface. This then allowed them direct communication via the blood with the heart. One might ask if this is the case, then why are the eyes closer to the head than the heart? Again Aristotle had an answer: the eyes are composed

22 Head or heart? The ancient search for the soul

of water and need to be kept cool. Regarding the ears, another organ he had carefully dissected, Aristotle again denied any connection with the brain, but was aware of a pathway to the roof of the mouth, which we now recognise as the Eustachian tubes.

A problem facing Aristotle's *sensus commune* theory was to explain how the heart received sensations through the blood. He did this by resorting to a breath-like substance called the *pneuma*. According to Aristotle there were two types: one breathed in through the lungs (i.e. air) that heated the blood in the heart's furnace; and a second more mysterious one passed into the embryo's blood from the sperm of the father at conception. This second type is crucial to understanding the full extent of Aristotle's physiology, for its purpose was to serve as the intermediary between the *psyche* and the body's mechanical operation. It was, in effect, the instrument by which the soul carried out its commands in controlling the physiological activities of the body. And, among its many functions was the conveying of sensory impressions to the heart. A similar explanation was also used by Aristotle to explain movement. This time Aristotle emphasised the heart as the prime mediator of all voluntary action, with

FIGURE 1.7 Cover of a book that includes several works by Aristotle published in the eighteenth century.

Source: Wellcome Library, London.

its rhythmical beating causing the *pneuma* to contract and expand in the blood, thereby making the limbs move.[20] In addition, Aristotle believed that the heart had sinews, called *neura*, which radiated out to the bone joints of the skeleton. These too were given an important role in movement by Aristotle who likened them to tiny strings that pulled on the bones – much like the action of a marionette. It must be stressed, however, that Aristotle's understanding of the *neura* does not refer to nerves. Rather it refers to several kinds of fibrous tissues holding the body together, which include tendons, ligaments and sinews.

Aristotle's influence on Western thought has been immense. Although his works were lost during the Dark Ages in the West, they were rediscovered in the Medieval period to become greatly revered and idolised. Indeed, up until the sixteenth century his teachings formed the greatest synthesis of physics and biology, and were widely considered to be beyond criticism or reproach. In fact, their authority was to prevent further enquiry for centuries. Yet, one can not help but be disappointed, even shocked, at Aristotle's mistaken understanding of the brain, even though his view was one with a long tradition that had begun in ancient Egypt, and was an established part of Greek thinking. Today, our language is still replete with hundreds of expressions such as 'to break someone's heart', which reveal the vestiges of this ancient thinking.[21] Perhaps we can forgive Aristotle to a certain degree, for he had no conception of a nervous system, or a method allowing him to recognise the true nature of the brain. Yet, Aristotle's cardiocentralist position still feels somewhat bizarre. And, this is not just from our modern perspective, for the Roman physician Galen in the second century AD confessed he 'blushed' to quote Aristotle's teachings on the brain. However, in the years following Aristotle's death, anatomical advances in Alexandria would finally begin to prove the existence of a nervous system and see the true nature of the heart. It is to these great discoveries we now turn in the next chapter.

Notes

1. The British anthropologist Richard Wrangham in his book *Catching Fire* has also suggested that *Homo erectus* may have been the first to use fire to cook food – a development that would have allowed his brain to grow and digestive tract to shrink.
2. So called because they were first discovered in the Neander Valley in Germany.
3. The name derives from the rock shelter of Cro-Magnon in southwest France where the first skeletons were found.
4. The anthropologist Richard Klein has called this change 'the most dramatic behavioural shift that anthologists will ever detect.'
5. In fact, this became a concept known as vitalism and it was to prove a popular explanation of biological processes up until the nineteenth century.
6. The Neolithic period occurred at different times in different parts of the world. For example, in China it ran from about 10,000 BC to 2000 BC; in northern India, from 8000 to 4500 BC; in Egypt, from 7000 to 4500 BC; and in the Eastern Mediterranean, from about 10,000 to 3300 BC.
7. A practice observed and recorded by the Greek historian Herodotus who visited Egypt around 450 BC.
8. The King James Version of the Bible, for example, lists 830 occurrences of the word 'heart' in over 762 verses.
9. In the *Odyssey* Homer writes 'In Egypt, the men are more skilled in medicine than any of human kind.'
10. The Egyptians believed that the blood vessels not only delivered blood to all the body parts, but were channels for the flow of air, saliva, mucus, tears, sperm, nutrients and even body waste.
11. In medieval Germany the term 'bregen' was used to refer to the brains of slaughtered animals and was edible material used in the making of 'Bregenwurst' sausages.

12 This is not the place to consider the functions of the two cerebral hemispheres of the brain. It will be addressed elsewhere in this book (e.g. see Chapters 8 and 13). However, there is considerable evidence to show that in most people, the function of language is localised in the left hemisphere of the brain.
13 Theophrastus (c. 371–287 BC) was a Greek philosopher of the Peripatetic school and the immediate successor of Aristotle in the leadership of his Lyceum.
14 Doty (2007) *Neuroscience*, 147, 561–568.
15 Chadwick, J. and Mann, N.W. (eds) (1983) *Hippocratic Writings*. Penguin: London.
16 Plato did not use the word 'soul' but this is an appropriate moment to introduce it. The word is believed to derive from the Old English *sáwol* or *sáwel*, which apparently first appeared in the eighth century poem Beowulf. It may well have originated as the Greek 'ensouled' meaning alive and introduced to Germanic peoples such as the Goths by early missionaries.
17 Little is known of this relationship except that Alexander always took with him a copy of the *Illiad* annotated by Aristotle.
18 To make matters more complex, there is a part in *De anima*, which has been described as the most perplexing that Aristotle ever wrote, which describes a type of 'active intellect' (or *nous*) that can be separated from the body, and is immortal and eternal. This is, however, at odds with the rest of Aristotle's teachings on the matter.
19 There is no evidence Aristotle ever performed dissection on human cadavers since this was a practice strictly taboo in the Greek world of his times.
20 In Aristotle's time the muscles were merely seen as sense organs for the sensation of touch.
21 It is also interesting to note that The Oxford Dictionary devotes 16 columns to the word heart, and all the expressions that include it, whereas the word brain occupies only three.

Bibliography

Allen, J.S. (2009) *The Lives of the Brain*. Belknap Press: Cambridge, MA.
Alt, K.W., Jeunesse, C., Buitrago-Téllez, C.H., Wächter, R., Boës, E. and Pichler, S.L. (1997) Evidence for stone age cranial surgery. *Nature*, 387, 360.
Aristotle (1986) *De Anima* (H. Lawson-Tancred, Trans). Penguin: London.
Arnott, R., Finger, S. and Smith C.U.M. (eds) (2003) *Trepanation: History, Discovery, Theory*. Swets & Zeitlinger: Lisse.
Barcia Goyanes, J.J. (1995) Notes on the historical vocabulary of neuroanatomy. *History of Psychiatry*, vi, 471–482.
Barnes, J. (1984) *The Complete Works of Aristotle: The Revised Oxford Translation*. Princeton University Press: Princeton, NJ.
Barnes, J. (ed.) (1995) *The Cambridge Companion to Aristotle*. Cambridge University Press: Cambridge.
Barnes, J. (2000) *Aristotle: A Very Short Introduction*. Oxford University Press: Oxford.
Bremmer, J.N. (1983) *The Early Greek Concept of the Soul*. Princeton University Press: Princeton, NJ.
Brested, J.H. (1930) *The Edwin Smith Papyrus*. University of Chicago Press: Chicago, IL.
Bruyn, G.W. (1982) The seat of the soul. In Clifford Rose, F. and Brnum, W.F. (eds) *Historical Aspects of the Neurosciences*. Raven Press: New York.
Chadwick, J. and Mann, N.W. (1950) *The Medical Works of Hippocrates*. Blackwell: London.
Clarke, E. (1963) Aristotelian concepts of the form and function of the brain. *Bulletin of the History of Medicine*, 37, 1–14.
Clarke, E. and Stannard, J. (1963) Aristotle on the anatomy of the brain. *Journal of the History of Medicine*, 18, 130–148.
Clifford Rose, F. (1994) The neurology of Ancient Greece – an overview. *Journal of the History of Neuroscience*, 3, 237–260.
Clifford Rose, F. (2009) Cerebral localisation in antiquity. *Journal of the History of the Neurosciences*, 18, 239–247.
Codellas, P.S. (1932) Alcmaeon of Croton: His life, work and fragments. *Proceedings of the Royal Society of Medicine*, 25 (7), 1041–1046.
Critchley, M. (1966) *The Divine Banquet of the Brain: The Harveian Oration 1966*. Harrison: London.

Crivellato, E. and Ribatti, D. (2007) Soul, mind, brain: Greek philosophy and the birth of neuroscience. *Brain Research Bulletin*, 71, 327–336.

Dawson, W.R. (1967) The Egyptian medical papyri. In Brothwell, D. and Sandison A. (eds) *Diseases in Antiquity*. C. Thomas: Springfield, IL.

Doty, R.W. (2007) Alkmaion's discovery that brain creates mind: A revolution in human knowledge comparable to that of Copernicus and Darwin. *Neuroscience*, 147, 561–568.

Elsberg, C.A. (1931) The Edwin Smith surgical papyrus. *Annals of Medical History*, 3, 271–279.

Fales, F.M. (2010) Mesopotamia. In Finger, S., Boller, F. and Tyler, K.L. (eds) *Handbook of Clinical Neurology*, vol. 95. Elsevier: Amsterdam.

Feinsod, M. (2010) Neurology in the Bible and the Talmud. In Finger, S., Boller, F. and Tyler, K. L. (eds) *Handbook of Clinical Neurology*, vol. 95. Elsevier: Amsterdam.

Finger, S. (1994) *Origins of Neuroscience*. Oxford University Press: Oxford.

French, R.K. (1978) The thorax in history 1: From ancient times to Aristotle. *Thorax*, 33, 1–8.

Garrison, F.H. (1969) *History of Neurology*. Revised and enlarged by McHenry, L.C. Thomas: Springfield, IL.

Green, C.D. and Groff, P.R. (2003) *Early Psychological Thought: Ancient Accounts of Mind and Soul*. Praeger: London.

Goetz, S. and Taliaferro, C. (2011) *A Brief History of the Soul*. Wiley-Blackwell: Chichester.

Gross, C.G. (1995) Aristotle on the Brain. *The Neuroscientist*, 1 (4), 245–250.

Gross, C.G. (2009) *A Hole in the Head*. MIT Press: Cambridge, MA.

Hergenhahn, B.R. (2001) *An Introduction to the History of Psychology*. Wadsworth: Belmont, CA.

Homer (2003) *Iliad*. (E.V. Rieu, P. Jones and D.C.H. Rieu, Trans). Penguin: London.

Homer (2003) *Odyssey*. (E.V. Rieu, D.C.H. Rieu, and P. Jones, Trans). Penguin: London.

Jaynes, J. (1976) *The Origins of Consciousness in the Breakdown of the Bicameral Mind*. Houghton-Mifflin: Boston, MA.

Jowett, B. (1892) *The Dialogues of Plato*. Clarendon Press: Oxford.

Karenberg, A. (2010) The Greco-Roman world. In Finger, S., Boller, F. and Tyler, K.L. (eds) *Handbook of Clinical Neurology*, vol. 95. Elsevier: Amsterdam.

Katona, G. (2002) The evolution of the concept of psyche from Homer to Aristotle. *Journal of Theoretical and Philosophical Psychology*, 22, 28–44.

Kenny, A. (1998) *A Brief History of Western Philosophy*. Blackwell: Oxford.

Lewin, R. (2005) *Human Evolution: An Illustrated Introduction*. Blackwell: Oxford.

Lillie, M.C. (1998) Cranial surgery dates back to the Mesolithic. *Nature*, 391, 853–854.

Lindberg, D.C. (2007) *The Beginnings of Western Science*. University of Chicago Press: Chicago, IL.

Lloyd, G. (1975) Alcmaeon and the early history of dissection. *Sudhoffs Archives*, 59, 113–147.

MacDonald, P.S. (2004) *History of the Concept of the Mind*. Ashgate: Aldershot.

Magner, L.N. (1994) *A History of the Life Sciences*. Marcel Dekker: New York.

Malomo, A.O., Idowu, O.E. and Osuagwu, F.C. (2006) Lessons from history: Human anatomy, from the origin to the Renaissance. *International Journal of Morphology*, 24, 99–104.

Margetts, E.L. (1967) Trepanation of the skull by the medicine-men of primitive cultures, with particular reference to present-day native East African practice. In Brothwell, D. and Sandison A. (eds) *Diseases in Antiquity*. C. Thomas Publications: Springfield, IL.

Mazzone, P., Banchero, M.A. and Esposito, S. (1987) Neurological sciences at their origin: Neurological surgery in the medicine on Ancient Egypt. *Pathologica*, 79, 787–800.

Mithen, S. (1996) *The Prehistory of the Mind*. Phoenix Books: Guernsey.

Mumford, D.B. (1996) Somatic symptoms and psychological distress in the Illiad of Homer. *Journal of Psychosomatic Research*, 41, 139–148.

Nutton, V. (1988) The legacy of Hippocrates. *Transactions of the Medical Society of London*, 21–30.

Osmond, R. (2003) *Imaging the Soul: A History*. Sutton Publishing: Stroud.

Philips, E.D. (1957) The beginnings of medical and biological science among the Greeks. *The Irish Journal of Medical Science*, 373, 1–14.

Philips, E.D. (1957) The brain and nervous phenomena in the Hippocratic writings. *The Irish Journal of Medical Science*, 381, 377–390.

Plato (1937) *Plato's cosmology: The Timaeus of Plato* (F.M. Cornford, Trans). Harcourt Brace: New York.

Santoro G., Wood M.D., Merlo L., Anastasi G.P. and Tomasello, F.G.A. (2009) The anatomic location of the soul from the heart, through the brain, to the whole body and beyond: A journey through Western history, science and philosophy. *Neurosurgery*, 65, 633–643.

Saul, F.P. and Saul, J.M. (1997) Trepanation: Old world and new world. In Greenblatt, S.H. (ed.) *A History of Neurosurgery*. American Association of Neurosurgeons: Park Ridge.

Singer, C. (1957) *A Short History of Anatomy and Physiology from the Greeks to Harvey*. Dover: New York.

Spencer, A.J. (1991) *Death in Ancient Egypt*. Penguin Books: London.

Squier, E.G. (1877) *Peru: Incidents of Travel and Exploration in the Land of the Incas*. Harper and Brothers: New York.

Stol, M. (1993) *Epilepsy in Babylonia (Cuneiform Monographs 2)*. Brill Academic Press: Boston, MA.

Striedter, G.F. (2005) *Principles of Brain Evolution*. Sinauer: Sunderland, MA.

Tattersall, I. (1998) *Becoming Human*. Oxford University Press: Oxford.

Verano, J.W. and Finger, S. (2010) Ancient Trepanation. In Finger, S., Boller, F. and Tyler, K. L. (eds) *Handbook of Clinical Neurology*, vol. 95. Elsevier: Amsterdam.

Walshe, T.M. (1997) Neurological concepts in archaic Greece: What did Homer know? *Journal of the History of the Neurosciences*, 6, 72–81.

Weber, J. and Wahl, J. (2006) Neurosurgical aspects of trepanations from Neolithic times. *International Journal of Osteoarchaeology*, 16, 536–545.

Wright, J.P. and Potter, P. (eds) (2000) *Psyche and Soma*. Clarendon Press: Oxford.

York, G.K. and Steinberg, D.A. (2010) Neurology in Ancient Egypt. In Finger, S., Boller, F. and Tyler, K.L. (eds) *Handbook of Clinical Neurology*, vol. 95. Elsevier: Amsterdam.

2

THE DISCOVERY OF THE NERVOUS SYSTEM

The muscles move certain organs, but they themselves require, in order to be moved, certain nerves from the brain, and if you intercept one of these with a ligature, immediately the muscle in which the nerve is inserted and the organ moved are rendered motionless.

Galen of Pergamon

I have done as much for medicine as Trajan did for the Roman Empire when he built bridges and roads through Italy. It is I, and I alone, who have revealed the true path of medicine.

Galen of Pergamon

Summary

The concept of the nervous system with its executive centre in the brain, receiving impressions from the senses and sending information to the musculature, is a relatively recent one in human history. In fact, the earliest observation of nerve pathways occurred in the fifth century BC when the Croton philosopher Alcmaeon performed animal dissection and noticed two tube-like channels running from the eyes to the brain. Although it would enable him to be the first philosopher of antiquity to attribute the brain with the power of sensation and intelligence, Alcmaeon did not realise their true significance. Less than a century later the Hippocratic writers also noticed two 'stout cords' (or *tonoi*), which left the brain and ran alongside the oesophagus on either side of the windpipe to reach the body where they ramified into branches in the chest. However, these tracts did not seem to fit into the humoral physiology of the times and the Hippocratic authors were unable, or reluctant, to allocate them a function. This is hardly surprising for the only ramifying system of channels known at the time, known to reach all parts of the body, were the blood vessels. As a result, these were believed to convey both sensation and motion, usually in the form of *pneuma* contained in the blood, which also provided the body with its animating life force. Consequently, many thinkers such as Aristotle regarded the heart and not the brain as the controlling centre of the body.

> The big breakthrough that changed this long-held cardiocentralist model of the human body would occur in the Greek outpost of Alexandria in Egypt where the practice of human dissection was sanctioned for a short period around 300 BC. Two anatomists to take full advantage of this situation were Herophilus and Erasistratus who recognised the existence of sensory nerves entering the brain, and motor ones leaving it to reach the muscles of the body. They also arrived at a pneumatic, or pump-like, theory of nervous function. However, the greatest neuroscientist of antiquity was undoubtedly the Roman physician Galen. A committed researcher, Galen famously discovered the cranial nerves, and mapped out the peripheral nervous system, which he examined by using vivisection and experimental methods. Galen's work represented the pinnacle of anatomical achievement in the ancient world, establishing beyond doubt the importance of the nervous system and brain in behaviour. His opinions on all things medical would not be seriously challenged for at least another 1500 years.

A centre of academic excellence: Alexandria

In 336 BC, the Macedonian King (Philip II) was assassinated and succeeded by his son Alexander aged just twenty years old. It was an event to change the course of history, for over the next decade Alexander 'the Great' would single-handedly conquer more or less the whole world known to antiquity without suffering a single defeat. Against overwhelming odds, the entire area from Greece in the West, North Africa in the south, the Danube in the north, and India in the east, succumbed to Alexandrian rule. It was a vast empire, covering around two million square miles, which Alexander was keen to link in an international network of trade and commerce. Although the independence of the Greek city states, including Athens, came to an end during this period, the Macedonians had great respect for the cultural achievements of their neighbours. They also encouraged the Greeks to colonise new territories – a policy resulting in many cities of the Mediterranean world becoming multiracial and multilingual with a distinctly Greek ethos. This is known as the Hellenistic period and it lasted some three hundred years, until the rise of the Roman Empire in the first century BC.

In his attempt to build an empire, Alexander founded over 70 cities, including one in his own name on Egypt's northern coast in 322 BC. He chose this location partly because it contained two sea harbours situated at the crossroads of several important trading routes including one linking Macedonia with the rich Nile Valley. A few months later he left Egypt to undertake new military conquests in the East. Sadly for him, Alexander never saw Africa again as he died from natural causes on campaign in Babylon in 323 BC. He was just 32 and without an heir. It was a situation that inevitably caused his empire to fragment with many of Alexander's conquered lands taken over by his former generals. However, some of these proved surprisingly capable, including Ptolemy, a possible half-brother to Alexander, who declared himself king of Egypt. Determined to make his capital a fitting place for the entombment of Alexander's body,[1] Ptolemy went about erecting a grand city of stone buildings, which included the Pharos lighthouse, one of the seven wonders of the ancient world. More importantly, Ptolemy also authorised the construction of a grand museum and library. These buildings, completed in the reign of his son Ptolemy II, were to form a remarkable institution devoted to learning. The museum, for example, is said to have been equipped with lecture halls, covered walks and seating for informal conversations, while also boasting a botanical garden, observatory and zoo. It also attracted a number of full-time scholars who lived in boarding houses nearby, including Euclid, Archimedes and Eratosthenes.[2] The library was no less magnificent, soon eclipsing the one in Athens, and becoming so extensive it is said to have housed over 700,000 scrolls.

For almost 300 years, the museum and library at Alexandria ruled as the most glorious centre of learning in the ancient world, with texts said to be taken off of every ship that came into port and swiftly copied by official scribes. Unfortunately, this situation would not last. The first event leading to its downfall occurred during the Roman conquest of Egypt in 48 BC, when the museum and library suffered serious fire damage after Julius Caesar set enemy ships alight in the port. A good idea of the carnage can be gleamed from the Roman stoic philosopher Seneca who reports that over 40,000 books were lost. Although its sister library, the Temple of Sarapis, located in another part of the town, continued as a place of study for several more centuries, this too would be destroyed, in this case by fanatical Christians after the 'pagan' temples of Alexandria were forced to close down by religious decree in 391 AD. Even so, the museum may have been in use until 642 AD when it was finally razed to the ground by invading Arabs. The destruction of the library at Alexandria is one of the greatest tragedies of all time, in which many of the great accomplishments of antiquity, covering several hundred years of philosophy, science, medicine and literature were lost for ever. This history of the Western world may well have been very different and certainly more enlightened had it not fallen.

Herophilus of Chalcedon

Alexandria also became famous for another reason: it was the first and only place in the ancient world to sanction dissection of the human body. How this extraordinary decision came about has never been fully explained by historians, since there was a powerful taboo against the desecration of human remains in Egypt.[3] It should not be forgotten that the whole point of mummification was to preserve the body, enabling it to enter the after-world resembling its earthly form. This reverence to the dead was also shared by the rest of the Greek world. Consequently, up until the beginning of the third century BC, all knowledge of internal human anatomy had been extrapolated from examining animals. However, it seems the Ptolemic dynasty in Alexandria was powerful or authoritarian enough to overrule these longstanding taboos in their zeal to advance medical knowledge. There is even some suggestion they provided live criminals for vivisection to further their aims. Despite this, human dissection seems to have been only performed by a small group of anatomists of whom the most famous were Herophilus (c. 335–280 BC) and Erasistratus (c. 310–250 BC). Their work continued for perhaps 20 or 30 years, after which human dissection fell back into disrepute and not practised again until the fourteenth century in northern Italy. Even then, it was not widely performed in Europe until the time of Vesalius in the sixteenth century – a hiatus of over 1800 years.

Herophilus appears to have been the first person to learn anatomy from post-mortem examination of human corpses. Although he is known to have written a number of books on a broad range of topics, including heart physiology and midwifery, these have all been lost. Thus we are dependent on others for a fragmented account of his life and work. The most important sources are Galen, Rufus and Celsus. From their accounts we can establish that Herophilus was born in the small and impoverished town of Chalcedon on the Asiatic side of the Bosporus, just across from modern day Istanbul, in around 335 BC. We also know he learned medicine from Praxagoras on the Greek island of Kos where Hippocrates had established his medical school. A follower of Aristotle, Praxagoras's greatest claim to fame was to distinguish between veins and arteries – believing the former contained blood while the later provided channels for the movement of air (*pneuma*). It is said that Praxagoras came to this conclusion after observing animals with their throats cut revealed arteries drained of any blood (unlike veins). This error would not be disproven until the work of Galen in the second century AD. After being taught by Praxagoras, Herophilus left Kos around 300 BC to work in Alexandria where he became personal physician to Ptolemy.

Herophilus was the inventor of the 'clepsydra', or portable water clock, which he used to measure the pulse, correctly deducing this movement was caused by the beating of the heart. However, it would be his pursuit of human dissection that made him legendary. Accused by the early Christian father Tertullian of 'butchering' 600 men in his relentless quest for anatomical knowledge, Herophilus's investigations resulted in an extraordinary number of new discoveries. These included, for example, the first accurate description of the digestive system including the pancreas gland and liver. His understanding of female reproductive biology was equally impressive, which he helped demystify by showing the womb consisted of just one chamber. Nor did it move as Hippocratic writers had claimed. He also realised the womb was connected to the ovaries whose function he insightfully likened to the male's testicles. Herophilus also described the male genital system that led him to understand how spermatozoa were formed and transported to the seminal vesicles. And if this is not impressive enough, Herophilus is also known to have written a highly influential treatise entitled *On Eye Diseases*, which explained the eye in detail, and identified many of its parts including the *cornea*, *iris*, *choroid* and *retina*.

Herophilus was also the first person to elucidate the internal anatomy of the human brain. Perhaps his greatest achievement here was to describe the ventricular system – a series of interconnected cavities inside its mass.[4] Although parts of this system had been noticed by others, Herophilus provided the first complete picture of its overall structure. He described, for example, how the two anterior ventricles of the brain (one in each cerebral hemisphere) were connected to the bottom-most (posterior) ventricle close to the spinal cord by a small passage that he called the *aqueduct*. In addition, he identified a long groove in this lower ventricle that he likened to a reed pen (*kalamos*). Herophilus also noticed that the inside surface of the ventricles were uneven and contained small bumpy protrusions, which he called the *choroid plexus* (meaning delicate knots). Today we now know these structures produce cerebrospinal fluid. Herophilus also outlined the external features of the brain. Although he did not go beyond Aristotle in describing the cerebrum (*enkephalos*) and cerebellum (*parenkephalis*), Herophilus did provide a much better account of the dural sinuses on its surface. In fact, he recognised the confluence by which the four great cranial venous sinuses come together which today is still known as the *torcular Herophili*. Curiously, Herophilus also describes a large network of arteries and veins at the base of the brain, which he called the *rete mirable* (the great net). This vascular complex is not found in humans and shows that Herophilus must have supplemented his anatomical knowledge from examining the brains of animals including pigs and oxen. The *rete mirable* would later become an integral part of Galen's physiology who unquestioningly accepted Herophilus' account by stating it existed in man. It was to be a mistake with far-reaching consequences that was not recognised until the late Renaissance.

The discovery of the nervous system

However, Herophilus made his greatest contribution to anatomy by discovering the true nature of the nervous system – possibly the single most profound insight into the workings of the body ever made by any individual. Sadly, we have scant knowledge of how this momentous discovery came about because we are reliant on second-hand accounts of others, most notably Rufus of Ephesus, and Galen who lived some 500 years after Herophilus.[5] We can infer from these authors, however, that Herophilus recognised the body contained thread-like pathways, which did not originate from the heart or blood vessels as Aristotle maintained, but from the brain and spinal cord. Although he called these pathways *neura* due to their resemblance to sinews, he importantly realised they carried out the functions previously attributed to the blood vessels. In fact, according to Rufus, Herophilus went much further by distinguishing two types of nerve: 'porous' ones carrying sensation to the brain (or rather to its surface or meninges) and 'solid' ones passing out to the muscles involved in producing movement. Thus, Herophilus clearly came to recognise the brain as the main control centre of the nervous system.[6] After realising the existence of the nervous system, Herophilus also undertook a careful anatomical examination of its pathways by tracing them to the bones, ligaments and muscles. This must have been a considerable feat taking many hours of meticulous dissection. It is said by Galen that Herophilus described seven pairs of nerves arising from the brain (now known as cranial nerves) and established six of their destinations: these were apparently the optic, oculomotor, trigeminal, facial, auditory and hypoglossal nerves.

FIGURE 2.1 No original busts or depictions of Herophilus or Erasistratus exist from antiquity. This is a woodcut from 1532, created by Lorenz Fries showing Herophilus on the left and Erasistratus on the right.

Source: Wellcome Library, London.

Herophilus also attempted to explain how the nervous system worked. In doing this, he made another revolutionary breakthrough: he located the *pneuma*, not in the heart as Aristotle had done, but in the ventricles of the brain. This was a significant change for two reasons. The first was that Herophilus now attributed the *neura* with a pneumatic function. That is, the nerves were regarded as hollow tubes, which received an air-like substance (*pneuma*) pumped into them from the ventricles. The second implication was perhaps even more profound. Herophilus localised the soul in the ventricles of the brain. It was much the same as saying that the ventricles of the brain were responsible for all our higher psychological functions including intelligence. Although he does not seem to attribute the four ventricles with different functions, Galen does imply that Herophilus recognised the posterior-most ventricle as the most important – perhaps because of its close proximity to the spinal cord. These theories provided a giant leap forward. Now anatomists had a new explanation of brain function (the ventricular theory) and nervous activity (the *pneuma* theory) and it would prove a very powerful combination. Indeed, in the hands of the early Christian fathers, the ventricular theory would later form the basis of the Cell Doctrine, one of the most influential and enduring theories in the history of the brain (see next chapter).

Erasistratus of Chios

Like Herophilus, Erasistratus was an anatomist who dissected the human body and wrote a number of books that have all been lost. This means we are again dependent on others, most notably Galen for knowledge of his work. Nor can the dates in the life of Erasistratus be stated with any certainty, although we can be reasonably confident he worked in Alexandria around the same time as Herophilus. Nonetheless, it appears that Erasistratus was born around 310 BC on the island of Chios in the Aegean Sea and studied medicine in Athens where he became associated with the Lyceum.[7] About 280 BC Erasistratus travelled to Kos, some 20 years after the departure of Herophilus, to study under Praxagoras. Presumably, Erasistratus travelled to Alexandria in the years after his medical education in Kos, although it is also known he spent some time in Syria employed as royal physician to King Seleucus, who then ruled over a vast empire. Here, Erasistratus would achieve fame for curing the King's son and heir (Antiochus) of a wasting illness that had bought him to the brink of death caused by love sickness[8] – a story later recounted by Plutarch and other writers of the ancient world. In fact, this is sometimes said to be the first recorded instance of a physical illness being attributed to a psychological cause – making Erasistratus the father of psychosomatic medicine. Perhaps because of his fame, Erasistratus would be invited to Alexandria by Ptolemy II, with the promise of being allowed to perform human dissection. Whether he ever worked with Herophilus, or even knew him, is not clear. At Alexandria, Erasistratus become recognised for his attempt to use mechanistic principles to explain physiological processes – an approach which rejected the idea there were hidden forces in the body. This was at odds with the humoral theories of Hippocrates and often attracted criticism, even ridicule, from subsequent Greek authorities including Galen.

We have been told Erasistratus paid close attention to the brain, and improved upon the observations of Herophilus by providing a more thorough account of the ventricular system by describing four main cavities. In fact, Herophilus had only mentioned three ventricles, and was seemingly unaware of the central one (the third) located in the midbrain. Erasistratus also paid greater attention to the convoluted shape of the cerebral hemispheres, likening its surface to the coils of the small intestine. Believing these gyrations to be more extensive in humans than in animals, Erasistratus concluded they were related to intelligence. This theory was criticised by Galen, however, who pointed out that the donkey has a similarly convoluted brain, yet was known for being stupid. Erasistratus, like Herophilus, also recognised different nerves for sensation and movement, although according to Galen his early works state they originate in the brain's outer membranes and not the brain itself (an idea he may have got from Aristotle during his time at the Lyceum). Erasistratus also appears to have been the first to realise the marrow of the brain and spinal cord were continuous with each other.

However, perhaps Erasistratus' greatest legacy was his explanation of the way blood and *pneuma* are generated in the body. His anatomical observations had caused him to conclude that every part of the body received a 'triple-weaving of vessels' composed of veins, arteries and nerves. From this, Erasistratus deduced that every organ received blood from the veins, an air like substance (*vital pneuma*) from the arteries, and *psychic pneuma* from the nerves. These substances were also attributed with different functions. The blood, according to Erasistratus, was created in the liver to absorb the nutrients of the stomach, which was conveyed to the right ventricle of the heart for distribution in the veins, thereby providing the body with nourishment. In parallel with this process, air was drawn into the lungs and passed to the heart where it was transformed into *vital pneuma*. This provided the body's innate heat and was pumped into the arteries to support functions such as digestion and nutrition. Both of these ideas were not new, but the third component of Erasistratus' theory was: he hypothesised that some of the *vital pneuma* flowed to the brain as *psychic pneuma*. This was stored in the ventricles where it could encounter movements from the sensory nerves to produce sensation and awareness. In turn, the *psychic pneuma* flowing through the hollow nerves from the ventricles produce movement by causing the muscles to swell and contract, much like a balloon. This theory of muscular movement remained popular up until the seventeenth century and was even endorsed by Descartes.

The founder of experimental physiology: Galen of Pergamon

Some two centuries after Alexander's death, the Greek world underwent a period of transition in which its increasingly warring and fractious city states were annexed and absorbed into the Roman Republic. Like the Macedonians before them, the Romans assimilated Greek culture into their own way of life, which they then spread through their conquests over the Mediterranean, North Africa, Asia and most of Western Europe, to create the largest and most populous empire the world had yet seen. Indeed, at its height it has been estimated that one in every four people of the world lived and died under Roman law. Whereas the genius of the Greeks had lay in intellectual pursuits such as art, literature and philosophy, the Romans were superior in practical based skills, establishing a highly effective military organisation, and building an infrastructure of roads, bridges and fortified outposts from which they could administer their provinces, impose taxation and encourage trade. Because of the Empire's vast extent (it reached a size of over 2.5 million square miles) and long endurance (it lasted roughly 500 years), Roman institutions and culture were to have an indelible and long-lasting impact on the development of language, religion, architecture, philosophy, law, and government in many foreign lands. This was particularly the case in Europe.

As the Greek empire declined, many of its scholars left to settle in Rome, and one person to make this journey was Claudius Galen (AD 129–200).[9] The son of a wealthy architect,

Galen was born in the thriving city of Pergamon (now Bergama in Turkey) famed for a library rivalling that of Alexandria and an imposing temple dedicated to medicine and healing. Sufficiently rich to be independent and with his father's blessing, Galen chose to become a doctor. At the age of 20, Galen left Pergamon, perhaps in no small part as a means of getting away from his cantankerous mother,[10] to broaden his medical knowledge in the cities of Smyrna, Corinth, and Alexandria. Returning to Pergamon in AD 157, at the age of 28, Galen was appointed chief physician to the gladiators – a position requiring him to treat a variety of severe injuries. Although Galen claimed no gladiator ever cared by him died of his injuries, he acknowledged that their wounds provided him with a 'window into the body' by which he could extend his anatomical and surgical knowledge. After serving in this post for about four years, Galen left for Rome where he enhanced his reputation after curing the gravely ill philosopher Eudemus of malaria. Several years later Galen was appointed as physician to the Emperor. At first, this also required him to serve in a number of military campaigns in the north east of the country. However, Galen's martial service was short-lived as he was called back to Rome in AD 169 due to an outbreak of plague. Now with a settled position in the imperial court, Galen would spend the remainder of his life as a Roman citizen, serving four Emperors and becoming Rome's most famous doctor. His fame was such that the Emperor Marcus Aurelius referred to him as '*Primum sane medicorum esse, philosophorum autem solum*' (first among doctors and unique among philosophers).

Galen was a prolific author who wrote on a surprisingly wide variety of topics, including medicine (of course), but also logic, philosophy and literary criticism. It is said by some commentators that he kept 20 scribes to write down his every word, which if true, would help to explain how he managed to write over 300 major works during his time in Rome. More than 170 of these have survived and they run to over 2.5 million words. Arguably the greatest is *De usu partium* (*On the Usefulness of the Parts of the Body*), which is a treatise that shows how every structure of the body has its own specialised function. A modern translation of this book takes up two volumes and over 800 pages. This outpouring of writing means Galen's works provide the most extensive surviving body of work of any ancient author. In fact, it comprises about 10 per cent of the ancient Greek knowledge we possess today. However, Galen is also an important historical figure for his attempts to summarise and synthesise the work of his predecessors – much of it now lost. For example, Galen was keen to acknowledge the humoral teachings of Hippocrates (his hero), the anatomy and physiology of Herophilus and Erasistratus, and the philosophy of Plato and Aristotle. However, his dedication to work may well have been at the expense of his personal life: Galen never married, is said to have had an ascetic distaste for sexual excess and lived a frugal and simple life.

Galen's true greatness, however, lay in his determination to supplement what was already known with his own facts, discoveries and theories. This pursuit of the truth led Galen to perform what was rare in the antique world: scientific experiments, including the first ever vivisectional studies on live animals. Thus Galen can be considered the founder of experimental physiology, who attempted to obtain hard factual information about the workings of the body. It was an endeavour that enabled him to make many new discoveries. For example, Galen was the first to realise the arteries were filled with blood and not air. And, by tying the ureter of living animals Galen discovered that urine was created by the kidneys and not the bladder as was then believed. He also realised muscle only has the ability to contract, and from his observation of the opening and closing of eyelids correctly deduced the relaxation of a given

The discovery of the nervous system 35

FIGURE 2.2 Galen of Pergamon (AD 129–200). No busts of Galen have survived from antiquity. This portrait is from a picture in the Juliana Anicia Manuscript (now in Vienna) from AD 487. Although the original is badly deteriorated, it was re-drawn by Mr T.L. Poulton at University College, London.

Source: Wellcome Library, London.

muscle was always produced by the contraction of an opposing one. Such was the authority of Galen, and the quality of his empirical work, that his influence reigned supreme over medicine for 15 centuries after his death. However, as we shall see next, it is for his discoveries on the nervous system, and especially the spinal cord, that Galen is most renowned.

Galen's account of the brain

Galen's most thorough description of the brain is given in the ninth chapter of *De anatomicis adminstrationibus* (*On Anatomical Procedures*), which is an account based on verbatim notes delivered during his demonstrations of dissection. This provides a chronicle that far surpasses

FIGURE 2.3 Title page from Galen's *De anatomicis adminstrationibus* (*On Anatomical Procedures*) published in 1531.

Source: Wellcome Library, London.

anything that had preceded it, although Galen does not appear to have observed anything new. The 'talk' begins with Galen providing instructions on how to remove an oxen brain from the casing of its skull by using iron knives and other tools. Following this, Galen labels many of its surface features including the distribution of blood vessels and sinuses. He also describes the *meninges* protecting the brain including their fibrous outermost *dura mater*, and the inner more delicate *pia mater*. Galen also explains how to cut the brain in order to reveal its internal organs. One of the first structures to be revealed using his method is the *corpus callosum* whose appearance Galen likens to hardened or white callused skin. This is the most striking of the brain's internal structures and today we know it is a pathway connecting the two cerebral hemispheres. Galen also provided a detailed account of the brain's ventricles. These, he demonstrated, are composed of two anterior ventricles connected by 'holes' (the *intraventricular foramina*) to the middle ventricle. This ventricle is in turn connected by an *aqueduct* to the posterior ventricle feeding into the spinal cord. Galen also recognised the *choroid plexus* within the ventricles, which he recognised as a collection of veins and arteries woven together.

Delving deeper into the brain, Galen comes across a singular pine-cone shaped body or *conarium* (from which we get our modern term 'pineal gland') that rests close to the aqueduct connecting the third and fourth ventricles. Another conspicuous structure is a white coloured long arch whose two arms are found below the lateral ventricles, which Galen refers to as the *fornix*. He also likens this to a vaulted roof because in structural terms it seems to be holding up the rest of the brain above it, including the cerebral hemispheres.[11] Although Galen admits that anatomists had seen both the pineal gland and fornix before, he does not give their names. Turning the brain over, and looking at it from below, Galen describes a funnel-shaped stalk called the *infundibulum*, which is connected with the *pituitary gland* (its name being derived from *pituita* meaning 'phlegm'). Galen was well aware of this structure for in *De usu partium*, following Hippocrates, he writes that it was responsible for removing phlegm and mucus from the brain and passing it into the nasal passages.

Another part of the brain described by Galen was its 'buttocks'. These are most likely are rounded protrudences located in the upper brainstem, which today are known as the *superior* and *inferior colliculi*. Galen also refers to a narrow worm-like structure located between the two halves of the *cerebellum* which he calls the *vermiform process*. This was shortened to *vermis* in the work of later writers who would recognise it as a channel regulating the flow of *pneuma* into the fourth ventricle. The *vermis* was also to play an important role in what became known as the Cell Doctrine during the medieval period (see next chapter). It is also relevant to note that the terms given in italics in this section are Latinised. Although Galen was a member of the Roman court he shunned Latin, preferring to speak and write in Greek – a tongue that was actually quite popular in Rome at the time.

Galen's account of the cranial and spinal nerves

Few would disagree that Galen's greatest anatomical achievement was his account of the nervous system.[12] Despite being forced to use animals in his investigations as human dissection was strictly forbidden in Rome, he went far beyond Herophilus and Erasistratus by showing that the nervous system was composed of two different systems: (1) a group of nerve pathways which enter and leave the brain through holes (called foramina) in the base of the skull (now known as cranial nerves); and (2) a more diffuse set connected with the spinal cord. And as we shall see below, he described both systems in astonishing detail. On top of this, Galen was also able to distinguish the nerves from ligaments (which he recognised were attached to bones) and also from the tendons attached to muscle. Thus, from now on, nerves were never again confused with the body's sinews. However, as Galen admitted, this work had been 'a toilsome and difficult matter' especially as the hot climate of Rome was hardly conducive to anatomical investigation. Considering the complexity of the nervous system with its many fine and intermingling branches coursing through the body (most would have not been easy to observe within its blood and gore) and the primitive dissecting tools at his disposal, Galen's success in outlining the general distribution of nerve pathways surely has to be one of the greatest anatomical accomplishments of all time.

Galen's description of the cranial nerves leaves the reader in no doubt that the source of all nerves, of all sensation, and of voluntary motion is the encephalon (the brain). The discovery of these nerves, however, did not begin with Galen, who admitted they were first recognised by the Alexandrian anatomist Marinus who also gave them numbers. Galen must have been well aware of the work of Marinus, since he also confesses to copying 20 of his books. However, since none of these writings survive, we are forced to rely on Galen's observations which are in accordance with his Alexandrian predecessor. Today, we know that the human brain has 12 pairs of cranial nerves (I–XII). Galen lists seven of these, although it is clear from his account that he actually recognised ten, since he confused the pathways of certain others (see below). He also disregarded the olfactory nerve, which is now recognised as the first by most authorities, for he believed it was a hollow tube which allowed odours to pass directly into the ventricles. For the record, a simplified account of Galen's cranial nerves and their destinations, using the numbering system adopted by Marinus, is as follows: (I) the optic passing to the back of the eyes; (II) the oculomotor to the muscles of the eyes; (III) the trigeminal extending to the jaw, teeth, lips and tongue; (IV) a pathway passing to the palette;[13] (V) the facial nerve to ear and cheeks;[14] (VI) the vagus nerve passing down into the body including the lungs, heart, larynx, oesophagus and stomach;[15] and (VII) the hypoglossal nerve passing to the tongue.

Although these nerves give the impression of passing from the brain to innervate various organs of the body, Galen also realised that some of them had sensory functions. Indeed, following Herophilus and Erasistratus, Galen believed the sensory nerves were 'soft' (a characteristic that made them pliable and suitable to receive impressions); and the motor nerves were 'hard' (a characteristic that gave them greater strength which was necessary to move the muscles). Galen also believed that the soft sensory nerves proceeded directly to the front parts of the brain, whereas the hard nerve pathways left from its posterior-most areas including the cerebellum.[16] This was not entirely new. Herophilus and Erasistratus had also identified the motor nerves as arising from the posterior part of the brain, leading them to regard the posterior ventricle as most important for behaviour. However, Galen's idea of attributing the anterior part of the brain with sensation was entirely new. It was also to become a highly influential one in the centuries to come. Interestingly, Galen also believed that the nature of the nerves could influence one's personality. For example, he held that brave people had a preponderance of hard nerves, which is where we get our modern expression 'nerves of steel'.

Galen also gave the first accurate account of the spinal cord and its nerve pathways. For example, he described the spinal cord as 'hard, hollow and articulated' and regarded it as an extension of the brain by way of its marrow which passed through the inside of the vertebral column. Importantly, he also saw this system as having a very different function from the cranial nerves. In fact, he referred to it as the 'foundation stone for the instruments necessary to life', thereby apparently recognising the brain, through its action on the spinal cord, as being responsible for controlling the life-governing processes of the body. Galen also observed nerve pathways exiting from each side of the spinal cord 'like the branches of a tree' at regular intervals from between the vertebrae. In fact, he listed 29 pairs of spinal nerves and followed many of them to muscles and ligaments. He was particularly successful at doing this with the upper parts of the spinal cord (i.e. the cervical and first part of the thoracic), tracing these nerves to the shoulder and upper arm, and the lower ones to the forearm and hand. However, this was not all for Galen also outlined a second nervous appendage passing down on either side of the spinal cord. We now know this to be the sympathetic trunks – two long chains of ganglia (i.e. bundles of nerve fibres) lying on the outside of the vertebral column which extend all the way down to the coccyx forming a rich array of connections with the spinal nerves on its way via the *rami communicantes*.[17] Again Galen was remarkably perceptive, for he believed this system of nerves enabled animal spirits to travel freely around the body. More specifically, Galen attributed it with giving the body a functional unity, allowing its various parts to interact together, or what he called physiological 'sympathy'.

Galen and the squealing pig

Although Aristotle is known to have removed a heart from a tortoise, and Erasistratus is said to have inserted a tube into the arteries to determine if their walls caused the pulse, these practices were relatively uncommon in the ancient world. With Galen we see a much more sophisticated form of experimentation being undertaken on living animals, which today would surely qualify as 'scientific'. Galen also knew his procedures were ground breaking, for he was highly scornful of his predecessors and complained: 'Did they ever take the trouble to make a section themselves, or . . . put a ligature around the parts in the living animal in order to learn what function is injured?' But Galen's investigations were not only innovative, they were exhaustive with just about every part of the body's anatomy subject coming under his scrutiny. This included the reproductive system (e.g. he measured the pulse of the umbilical cord and removed the ovaries from pigs), and digestive system (e.g. he examined the contents of the stomach and intestines). He also examined the cardiovascular system by famously placing a ligature around the artery at two places and then cutting open the closed-off section to show that it contained blood – thereby disproving the ancient notion that arteries only contained *pneuma*. However, Galen's most sophisticated and celebrated experiments would examine the function of the nervous system.

To perform his experiments Galen required a constant supply of animals, and this led him to use just about every type of creature he could find in the Roman markets, including on one occasion a war elephant purchased from the Circus Maximus. His preferred choice of animal, however, was the North African Barbury ape (*Macaca sylvana*) which most closely resembled man with his round face, upright posture and lack of a tail. However, it appears when working on the nervous system, Galen chose the pig because its facial expression was less expressive in response to pain. This is easy to understand for Galen had no way of anaesthetising his animals, and was forced to strap them down on an operating table before cutting their bodies open. Perhaps his most famous experiment took place around 160 AD when working in Pergamon as physician to the gladiators. On this occasion Galen was attempting to sever the branch of the vagus nerve going to the lungs in order to examine its effects on respiration. Not surprisingly, the pig struggled violently and squealed in pain as

FIGURE 2.4 Woodcut of the squealing pig demonstration taken from a front page of one of Galen's books published in Venice in 1586.

Source: Wellcome Library, London.

the operation was begun. Then something unexpected happened: the animal suddenly fell silent at the moment the nerve was cut. Galen quickly realised he had made a mistake. Instead of severing the branch going to the lungs, he had cut the (laryngeal) nerve going to the vocal apparatus. In other words, he had discovered the 'nerves of the voice'.

The significance of this discovery was not lost on Galen who realised he had just proved the voice was controlled by the brain. Yet, there were many who were sceptical of his claims, especially the Aristotelians who still believed the heart was the centre of intellect, but Galen had the means of disproving his critics and was confident enough to repeat his experiment in a public demonstration – a feat no doubt increasing his reputation beyond Pergamon. For Galen, the experiment provided irrefutable proof that the brain was the site of reason and intellect. Or, as he put it: 'The voice is the most important of all psychic operations. Since it announces the thoughts of the rational soul, it must be produced by organs which receive nerves from the brain'. Galen even had human clinical evidence to back up this claim as he reported the case of two patients who suffered damage to the laryngeal nerves during surgery of the neck and experienced loss of vocalisation as a result.

Galen went on to perform this experiment in a variety of other animals, including dogs, goats, bears, cows, monkeys and lions. He is also believed to have dramatised the spectacle in public arenas for the people of Rome. Here, Galen would start by cutting open a pig's neck, and then proceed by severing the exposed nerve pathways one by one, until he got to the laryngeal. Then with a single climatic cut of his knife, the animal stopped its distressed squealing. To conclude the exhibition, it is said Galen patched the nerves back together, upon which the animal cried out again. It was the most famous anatomical demonstration of its time, and one inspiring later investigators, for many of Galen's books up until the

Renaissance were illustrated with front pieces depicting the experiment. One person to be impressed was Leonardo da Vinci who drew the human laryngeal nerve. Curiously, none of Galen's own work shows any evidence of anatomical drawing, although simple diagrams of the body are believed to have been made by Aristotle some 500 years earlier.

Galen's account of the spinal cord

Galen experimental investigations also extended to examining the function of the spinal cord. Prior to Galen, the best account of spinal cord injuries had come from Hippocrates who had written: 'Such patients are more apt to lose the power of their legs and arms, to have torpor of the body, and retention of urine.' However, Galen went much further than his predecessor. This work is most fully covered in *Anatomical Procedures* in which Galen examines the spinal cord by cutting it at various levels, and then observing how the various parts of the body became affected and paralysed. One of his main findings was that lesions always produced a loss of movement and sensation below the level of the cut. Thus, a cut to the bottom part of the spinal cord paralysed the legs but not the arms. More importantly, Galen also found that the deficits of spinal cord lesions became less severe as one moved down the cord. For example, a cut made between the first and second vertebrae caused general paralysis and death – whereas a lesion made to the third and fourth vertebrae restored some movement, but stopped the movements of the lungs. However, cutting the sixth and seventh vertebrae only paralysed the chest muscles and had no effect on movements of the diaphragm. If the lesion was made lower in the spinal cord, then paralysis was confined to the lower limbs and bladder.

Another technique used by Galen, this time to investigate the function of the nerve pathways leaving the spinal cord, was to tie them off with fine threads made of wool – a clever procedure allowing their functions to return after the ligature was loosened. Galen not only used this procedure to (again) show the importance of the laryngeal nerve in vocalisation, and the glossopharyngeal in respiration, but also to investigate specific nerve pathways controlling the musculature. Indeed, by using this technique, Galen demonstrated the interruption of a nerve caused the muscle to become paralysed and lose all sensation. Interestingly, an exception to this rule was the vagus nerve controlling the heart. Severing this pathway failed to stop its beating, a finding which led to Galen stating that it must have its own intrinsic 'pulsatile force'. Galen even managed to get some idea of where the motor pathways existed in the spinal cord, for he discovered that an up-and-down incision made along its central axis did not produce paralysis. Thus, he showed the spinal nerves innervating the body arise from the cord's lateral regions. Today, we know that the ascending (sensory) and descending (motor) tracts connecting the spinal cord with the brain do indeed pass through its white matter found in its outer regions. It is also interesting to note that Galen also realised that the spinal nerves did not innervate all parts of the body, for those passing to the face and head were always derived from the brain (i.e. cranial nerves).

The role of psychic pneuma in nervous function

As great as Galen's anatomical achievements were, they were only a means to attempting a grander explanation of how the nervous system worked. In other words, anatomy was used by Galen as a vehicle to better understand nervous physiology including the brain's operation.

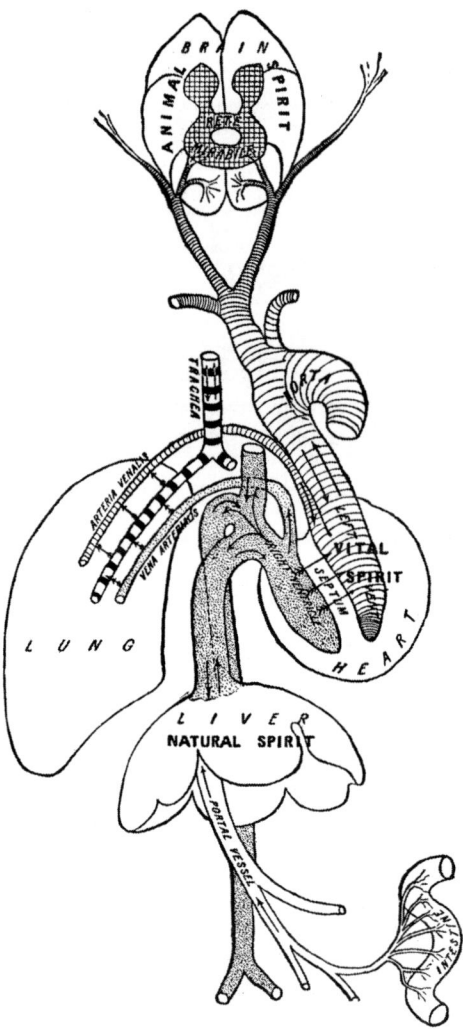

FIGURE 2.5 The physiological system of Galen showing how the three types of spirit (*natural, vital and animal*) are formed in the body.

Source: Wellcome Library, London.

In this regard, Galen was tied by the prevailing beliefs of his time, which recognised an air-like substance called the *pneuma* as the vital breath-like principle of life. The concept of *pneuma* has a long history in Greek thought going back at least to Homer. It also formed an important component of Aristotle's philosophy for he regarded it as the instrument of the *psyche* (or soul) that carried out all the body's biological processes. However, the physiological importance of the *pneuma* was greatly elaborated upon by Erasistratus, who distinguished between *vital pneuma* essential for life, and *psychic pneuma* which played a more important role in neurological function. This later substance was thought to be stored in the brain's ventricles and to flow through the nerves to move the muscles. Galen continued this long Greek tradition by recognising the existence of *pneuma* as a prerequisite of animal life. However, he went much further in elaborating the importance of the *psychic pneuma* in his physiological system. In fact, Galen based his physiology on the triadic plan first conceived by Erasistratus. This system was based on three main organs (liver, heart and brain); three vessels (veins, arteries and nerves) and three spirits (natural, vital and animal). However, it was different in significant respects to Erasistratus – especially with regards to the brain.

As we have seen, Erasistratus developed a physiological system where blood formed in the liver was passed to the heart where it was mixed with *pneuma* obtained from the lungs to form a *vital spirit*. This generated the body's life-giving heat, although some of it also flowed to the brain where it was transformed into *psychic pneuma* and stored in the ventricles. Here the *psychic pneuma* essentially acted as the soul and the nervous force at the same time. Although Galen broadly agreed with this theory, he differed in one important respect: for him, the *psychic pneuma* was not created in the brain, but in a large network of fine arteries at the base of the brain called the *rete mirable* (or great net). Once formed, the *psychic pneuma* then passed to the blood vessels of the *choroid plexus* protruding into the fourth ventricle where further refinement of the *pneuma* took place. Only then was the *psychic pneuma* stored in the ventricles. Although this change may not appear to be a radical departure from Erasistratus, it was to have serious repercussions later for the human brain does not contain a *rete mirable* and Galen probably inferred its existence from the ox brain. While this error would not be recognised until the work of Flemish anatomist Vesalius in the sixteenth century, it would seriously undermine Galen's authority which up to that point was untarnished.

Galen also disagreed with Erasistratus in several other important respects. For example, Erasistratus had confined the *psychic pneuma* to the ventricles where it served as the *psyche*, but Galen insisted the *psychic pneuma* was able to permeate into the actual substance of the brain itself. One of the reasons for this change was Galen's belief that the 'hard' motor nerves of the brain arose from the cerebellum, and not the ventricles as Erasistratus maintained. Consequently, in Galen's opinion, the *psychic pneuma* had to find its way through the brain's substance to gain access to the motor nerve pathways. Galen also contradicted Erasistratus on the nature of the *psyche* (i.e. soul) too. Whereas Erasistratus believed the *psychic pneuma* to be synonymous with the soul, Galen relegated it to a subservient role, calling it the soul's 'primary instrument' for acting on the body.[18] In other words, the *psychic pneuma* was not spirit, but some type of intermediate substance, acting between the soul and the physical world. In fact, Galen went as far to say he had no knowledge of the nature, or essence of the soul. This was quite an omission for such a self-proclaimed expert considering the soul's importance, although a remarkably frank one.

The first experiments on the brain

As always, Galen was not content to simply theorise about the ventricles, or the nature of the substance believed to be inside them. Instead, he attempted to answer such questions by performing the first ever experimental studies on the brain. Galen reasoned that if the ventricles housed the soul as Erasistratus had insisted, a penetrating knife cut into them would allow the *pneuma* to escape. Since the *pneuma* was also a life force, then this should invariably prove fatal to the animal. However, Galen's investigations, performed mainly on goats, showed that this did not always occur. For example, when Galen made cuts through the overlying cortex to pierce the anterior (lateral) ventricles, he found animals quickly recovered from their ordeal. He also examined the effects of ventricular compression which produced little effect other than to stop the eyes blinking. Galen even had a human case to support his claims for he had once treated a youth with a ruptured anterior ventricle, who suffered no long-term consequences from the injury. However, lesions to the other two remaining ventricles (middle and posterior) were not so benign. An incision into the middle ventricle caused animals to go into a stupor from which they rarely recovered. The same type of cut into the posterior ventricle was even more devastating causing coma and quick death. A similar effect was also produced when a cut severed the brain from the spinal cord.

Galen also attempted to understand the nature of the substance that flowed through the nerves to activate the muscles. Erasistratus had said that this nervous substance was *psychic pneuma* derived from the ventricles of the brain. However, Galen could find little evidence to support this idea. Indeed, such a theory assumed that the nerves should be hollow, but Galen's observations indicated that they were hard and non-porous. This led Galen to speculate that the nerves might use some other type of substance although he could do little better than refer to it as a 'power'. For example, he writes, 'There exists then in the nerves a considerable force that runs down from above, from the great principle, for this power is not innate in the muscles and does not originate in them'. On another occasion he even likened this force to 'sunlight'. However, in other accounts, he resorts back to the idea of a pneumatic substance, or liquid, being forced out of the brain by hydraulic pressure. He would also support this idea by arguing the brain has a pumping mechanism since it can be seen to 'contract and expand' when an injury to the head allows its surface to be exposed.[19] The pneumatic theory of nerve function would not be overthrown until the eighteenth century when it was replaced by ideas concerned with animal electricity.

Galen's great legacy

As Galen's life came to a close around AD 200, a key chapter in the history of medicine ended. He had synthesised his own work with the best taken from the Greek medical tradition, and reported everything of significance that could be said on anatomy. More importantly, he had supported his claims through rigorous and careful experimentation. There was no one to match him and consequently Galen exercised an authority in medical matters for more than 1500 years equalled only by Hippocrates. During this time there would be virtually no further study of the brain and nervous system. Galen also wrote voluminously on a variety of other subjects and while a fire in AD 192 destroyed much of his personal library, some works remained unscathed. These were later translated from Greek (Galen did not write in Latin) into Syriac and Arabic, forming the basis of medicine throughout the Islamic world in the centuries after Rome's fall. Indeed, by the ninth century, through the endeavours of a Christian physician in Baghdad called Hunain ibn Ishaq (d. 873), we know that 129 versions of Galen's works had been translated in this way. When the Dark Ages finally began to lift in the West during the twelfth and thirteenth centuries, Galen was among the first of the classical authors to be translated into Latin from texts bought back from the East. Although he was not a Christian, Galen nevertheless stressed the intricacy and beauty of the body, claiming its design could not be due to chance. This made the teachings of Galen amenable to the Christian church whose endorsement helped raise him to the greatest medical authority in the fourteenth and fifteenth centuries. The combination of Galen's robust methodology, his comprehensive and prolific investigations, and the theological acceptance of his work, also conveyed the impression that there was little else to learn about anatomy and physiology.

Galen, who was sometimes accused of being over-bearing and proud, also contributed to his own immortality by claiming on many occasions he had bought the practice of medicine to perfection. Or as he put it in his own words:

> I have done as much for medicine as Trajan did for the Roman Empire when he built bridges and roads through Italy. It is I, and I alone, who have revealed the true path of medicine. It must be admitted that Hippocrates already staked out the path. He prepared the way, but I have made it possible.

Galen may have been the greatest physician of antiquity, but he was not perfect or infallible. Although it would take well over a millennium for his flaws to be exposed, by the sixteenth century the first clear errors in his anatomical and physiological writings began to be discovered. In particular his use of animals including dogs, pigs and macaques, to understand and describe the human body, would bring much misunderstanding which was slow to be resolved. Nonetheless, Galen's demise (covered in the next chapter) was to be an important development in mankind's quest to understand his nature better, for it finally showed there was much more to learn about the body after all. Even so, his authority continued to be respected and revered long after aspects of his work were called into question. It is no exaggeration to label him a genius whose determination to understand the workings of the body, brain and nervous system through experimental methodology was way ahead of its time. Such work would only begin again in earnest with the scientific revolution. Galen can be acknowledged as the earliest neuroscientist.

Notes

1 It is said that Ptolemy wanted the body to be buried in Alexandria in order to fulfil a prophecy made by Alexander's favourite soothsayer Aristander. She had predicted that the country in which Alexander the Great would lay would become the most prosperous in the world.
2 Eratosthenes is famous for his attempt at measuring the circumference of the Earth, which he estimated at 24,662 miles – a remarkably accurate attempt as the true figure is 24,817 miles.
3 Some historians even doubt that dissection took place in Alexandria, although these are clearly in the minority – see Von Staden (1989) page 142.
4 Today, it is recognised that the brain has four connected ventricles. The first two (known as the anterior or lateral ventricles) feed into the middle (or third) ventricle lying medially in the brain. The posterior (or fourth) ventricle is found in the brainstem.
5 Actually, Galen calls Herophilus and his contemporary, the anatomist, Eudemus, the first persons after Hippocrates to record carefully their dissections of the nerves. However, little is known about Eudemus or his work.
6 To confuse matters somewhat, according to Galen, Herophilus also assigned the power of moving to the veins and arteries.
7 According to Pliny, Erasistratus was actually a grandson of Aristotle by his daughter Pythias.
8 In fact, Antiochus had fallen in love with his father's young beautiful wife – a problem diagnosed by Erasistratus after he noted that his pulse quickened when Antiochus entered the same room as her. When the King found out about the truth of the situation, he graciously gave his wife to his son. He also awarded Erasistratus a generous fee, said to be 100 talents, which would have made him wealthy for the rest of his life.
9 The name Galen is derived from the Greek word *galenos* meaning calm and serene. This does not seem to fit Galen very well who was often regarded as arrogant, exhibitionist, and somewhat pompous in his opinion.
10 Galen's mother was famous for her temper. It is said her screaming and nagging could be heard in the street outside her house, and she sometimes bit her housemaids when she got angry.

11 It is believed that our modern word 'fornication' (i.e. to have intercourse) derives from fornix (i.e. 'an archway' or 'vaulted roof') because prostitutes in Rome solicited for sex in the vaults underneath the city.
12 This work is largely covered in *De anatomicis adminstrationibus*, although some of it is also discussed in *De usu partium*, and a shorter treatise entitled *De nervorum dissectione* (The dissection of the nerves).
13 This pathway is recognised today as a branch of the trigeminal nerve.
14 We now know that Galen was describing two cranial nerves here – the facial and auditory.
15 Here, Galen confuses three cranial nerves together – the vagus, accessory and glossopharyngeal.
16 Galen also appears to have had a notion of a mixed type of nerve (i.e. one with both a sensory and motor function), such as the one that innervates the tongue, as it had an intermediate 'hardness'.
17 The rami communicantes are communicating branches that exist between the spinal nerves and the sympathetic trunk.
18 Galen wrote: 'Whatever the substances of the *psyche* may be it uses the *psychic pneuma* as its first instrument and various bodily parts as its second instrument.'
19 This observation had been first made by the ancient Egyptians in the earliest ever recorded reference to the brain from around 1700 BC.

Bibliography

Acar, F., Naderi, S., Guvencer, M., Ture, U. and Arda, M.N. (2005) Herophilus of Chalcedon: A pioneer in Neuroscience, *Neurosurgery*, 4, 861–867.
Christie, R.V. (1987) Galen on Erasistratus. *Perspectives in Biology and Medicine*, 30, 440–449.
Clifford Rose, F. (1993) European neurology from the beginnings until the 15th century: an overview. *Journal of the History of Neuroscience*, 2, 21–44.
Cosans, C.E. (1998) The experimental foundations of Galen's teleology. *Studies in History and Philosophy of Science*, 29, 63–80.
Dobson, J.F. (1925) Herophilus of Alexandria. *Proceedings of the Royal Society of Medicine*, 18, 19–32.
Dobson, J.F. (1927) Erasistratus. *Proceedings of the Royal Society of Medicine*, 20, 21–28.
Freeman, F.R. (1994) Galen's ideas on neurological function. *Journal of the History of Neuroscience*, 3, 263–271.
Goss, C.M. (1966) On anatomy of nerves by Galen of Pergamon. *American Journal of Anatomy*, 118, 327–336.
Gross, C.G. (1988) Galen and the squealing pig. *The Neuroscientist*, 4, 216–221.
Hall, T.S. (1975) *History of General Physiology: 600 BC to AD 1900*. University of Chicago Press: Chicago, IL.
Jones, D.E.H. (1971) The great museum at Alexandria: its ascent to glory. *Smithsonian* (Dec), 53–60.
Karenberg, A. (2010) The Greco-Roman world. In Finger, S., Boller, F. and Tyler, K.L. (eds) *Handbook of Clinical Neurology*, vol. 95. Elsevier: Amsterdam.
Keele, K.D. (1961) Three early masters of experimental medicine – Erasistratus, Galen and Leonardo da Vinci. *Proceedings of the Royal Society of Medicine*, 54, 577–588.
Kilgour, F.G. (1957) Galen. *Scientific American*, 196, 105–114.
Lloyd, G. (1975) Alcmaeon and the early history of dissection. *Sudhoffs Archives*, 59, 113–147.
Longrigg, J. (1988) Anatomy in Alexandria in the third century B.C. *British Journal of History and Science*, 21, 455–488.
Major, R.H. (1961) Galen as neurologist. *World Neurology*, 2, 372–380.
May, M. (1968) *Galen: On the Usefulness of the Parts of the Body*. Cornell University Press: Ithaca, NY.
Nutton, V. (2004) *Ancient Medicine*. Routledge: London.
Prendergast, J.S. (1930) The background of Galen's life and activities, and its influence on his achievements. *Proceedings of the Royal Society of Medicine*, 23, 1131–1149.
Riese, W. (1959) *A History of Neurology*. MD Publications: New York.
Rocca, J. (1997) Galen and the ventricular system. *Journal of the History of Neuroscience*, 6, 227–239.
Rocca, J. (1998) Greek and Greek neuroscience. *Early Science and Medicine*, 3, 216–240.
Rothschuh, K.E. (1973) *History of Physiology*. Krieger: Huntington, NY.

Sarton, G. (1954) *Galen of Pergamon*. University of Kansas Press: Kansas, KS.
Siegel, R.E. (1968) *Galen's System of Physiology and Medicine*. Karger: Basal.
Sigerist, H.E. (1935) *Great Doctors: A Biographical History of Medicine*. George Allen & Unwin: London.
Singer, C. (1956) *Galen on Anatomical Procedures*. Oxford University Press: Oxford.
Singer, C. (1957) *A Short History of Anatomy and Physiology from the Greeks to Harvey*. Dover: New York.
Smith, E.S. (1971) Galen's account of the cranial nerves and the autonomic nervous system: Part 1. *Clio Medicine*, 6, 77–98.
Smith, E.S. (1971) Galen's account of the cranial nerves and the autonomic nervous system: Part 2. *Clio Medicine*, 6, 173–194.
Spillane, J.D. (1981) *The Doctrine of the Nerves: Chapters in the History of Neurology*. Oxford University Press: Oxford.
Stahnisch, F.W. (2010) On the use of animal experimentation in the history of neurology. In Finger, S., Boller, F. and Tyler, K. L. (eds) *Handbook of Clinical Neurology*, Vol. 95. Elsevier: Amsterdam.
Temkin, O. (1951) On Galen's pneumatology. *Gesnerus*, 8, 180–189.
Temkin, O. (1973) *Galenism: Rise and Decline of a Medical Philosophy*. Cornell University Press: Ithaca, NY.
Von Staden, H. (1992) The discovery of the body: human dissection and its cultural contexts in ancient Greece. *The Yale Journal of Biology and Medicine*, 65, 223–241.
Wills, A. (1999) Herophilus, Erasistratus, and the birth of neuroscience. *The Lancet*, 354, 1719–1720.
Wilson, L. (1959) Erasistratus, Galen and the Pneuma. *Bulletin of the History of Medicine*, 33, 293–314.
Wilson, N. (2009) *Encyclopaedia of Ancient Greece*. Routledge: London.
Wiltse, L.L. and Pait, T.G. (1998) Herophilus of Alexandria (325–255 B.C.): The father of anatomy. *Spine*, 23, 1904–1914.
Woollam, D.M.H. (1958) Concepts of the brain and its functions in classical antiquity. In F.M.L. Poynter (ed.) *The Brain and its Functions*. Blackwell: Oxford.
Woollam, D.H.M. (1962) The historical significance of the cerebrospinal fluid. *Medical History*, 6, 91–114.

3
FROM LATE ANTIQUITY TO THE RENAISSANCE

The Cell Doctrine

The knowledge of all things is possible.

Leonardo da Vinci

I venture to ascribe no more to the ventricles than that they are cavities and spaces in which the inhaled air, added to the vital spirit from the heart, is, by power of the peculiar substance of the brain, transformed into animal spirit.

Vesalius

Summary

The death of Galen at the end of the second century AD, and Rome's fall some 200 years later, marks a point where new anatomical and physiological research was put on hold for more than 1200 years. Moreover, during much of this time, practically all ancient Greek knowledge, including Aristotle and Galen, was lost to the West, leading to a time of scientific apathy, which some have referred to as the Dark Ages. In the East, the situation was different with many great works of antiquity being copied down, translated into new languages, and dispersed through Byzantine and Islamic lands. These transcriptions were also helpfully commented upon by Eastern scholars. Although this did not necessarily lead to any fundamentally new advances in knowledge, in some cases it led to new ideas. One innovation was the Cell Doctrine where the ventricles of the brain were attributed with different intellectual functions such as sensation, memory and movement. The idea of the ventricles containing psychic pneuma had originated with the Alexandrian anatomists Herophilus and Erasistratus, and was supported to some degree by Galen. None of these investigators, however, had identified any of the ventricles with a specific role. This new development appears to have begun with pagan thinkers in the fourth and fifth centuries BC, and it proved amenable to both Christian and Islam scholars in the centuries that followed. While many versions of the Cell Doctrine were to be formulated, its basic concept was to provide the dominant theory of brain function well into the Renaissance. In fact, no other psychological theory of the brain can be said to have had such an enduring impact. The relative intellectual stagnation of the West would only begin to show the seeds of change in the thirteenth and fourteenth centuries. An important development was the return of travellers back from places such as Byzantine with copies of ancient texts for translation into Latin, resulting in a renewed respect for Greek values and thinking. Another was the practice of human dissection, which became more acceptable especially in places such as northern Italy – a change in attitude perhaps influenced by the Crusades where parts of the body such as the heart was sometimes returned home for entombment. However, the most important advance would be the emergence of intaglio technology, most notably woodcut illustrations and printing in the fifteenth century. With this development we begin to see the publication of the first dissection manuals, books on anatomy and depictions of the brain. The culmination of this new change would be the publication of the *Fabrica* by Andreas Vesalius in 1543, a pivotal point in man's new understanding of himself, which helped overthrow the long held dogmatic medieval beliefs and mistakes of the past, and point to a new way forward.

The origins of the Cell Doctrine

Following the fall of Rome in AD 476, it is often said the West entered a long period of intellectual stagnation, commonly known as the Dark Ages, which only began to change during the Renaissance over a millennium later. In some ways, however, this paints a misleading picture. Although innovation may not have been one of the key features of the medieval period, it was a time when ideas about the spiritual nature of human beings and their relationship with God underwent important change. Further, it was a more civilised time than many suppose, where people lived in feudal societies governed in large part by the authority of the church. Nonetheless, with Rome's decline, the centre of the Roman Empire shifted to Byzantium (Constantinople), a city providing a vital link with the past, for it was one of the cradles of Greek philosophy with its heritage kept alive by scholars. In fact, the Eastern Roman Empire would thrive for more than a thousand years until the fall of Constantinople in 1453. The Muslims who occupied many of the surrounding lands during this period also had a strong desire for knowledge,[1] and they too helped preserve Greek knowledge by translating it into Arabic. The teachings of Galen were particularly revered and synthesised into their medicine. However, while the study of Greek works flourished in the East, the same did not occur in the West. Such a situation was hardly surprising since most great works of antiquity had been lost to the West after Rome's demise.[2] Thus, learning was centred in the monasteries and focussed more on seeking divine revelation – with theology dominating scholastic activity. This plight only showed the first signs of change in the thirteenth century when European scholars returned from the East with copies of ancient texts for translation into Latin. It would be the first step in the 'rebirth of knowledge' in the West leading to the Italian Renaissance.

The Eastern scholars had not only preserved the writings of the great Greek philosophers, but often extended them with additional commentary. And in other cases they even incorporated their own beliefs into the work. One instance where this occurred is when certain authors began to attribute the ventricles of the brain with different mental functions

– a theory that became known as the Cell Doctrine (so called, because the ventricles were imagined to resemble hollow or spherical cells). The ventricles had, of course, played an essential part in ancient physiology. They were the repositories in the brain for the storage of psychic pneuma which Galen had referred to as the souls 'instrument' for controlling the actions of the body, but importantly, he had not localised psychological activities to any of them. Although Galen linked the frontal part of the brain with sensation (believed to receive sensory information from the 'soft' nerves), and movement to its posterior part (the region giving rise to the 'hard' motor nerves), he never ascribed these functions to the ventricles. However, 200 years after Galen's death, some Byzantine scholars would begin to do just that. It was to prove a highly influential theory that dominated medieval thinking about the brain, although no standard version of the Cell Doctrine ever became established. In fact, at least 25 variations of the Cell Doctrine theory have been found in medieval manuscripts showing that its exact formulation was subject to various interpretations.

The earliest account linking the ventricles with mental functions is provided by an obscure pagan Christian doctor and native of Apamea (now Syria) called Posidonius around AD 370.[3] Little is known about Posidonius, although it is reported by a sixth century writer called Aëtius who was a physician at the Imperial Court in Constantinople, that he placed the faculty of thought and reasoning in the middle ventricle of the brain. In a less clear passage of text, Aëtius also seems to suggest that Posidonius localised imagination in the front of the brain and memory in its posterior regions. However, according to some modern commentators, it is unclear whether he actually placed these functions in the ventricles, or the substance of the brain. Nonetheless, even if the theory was limited to the middle ventricle, Posidonius had gone beyond Galen and opened up a new chapter in brain localisation. The ventricular theory would also have undoubtedly been an attractive theory to early Christian thinkers such as Posidonius, for the clear fluid-filled hollows of the brain were a much more suitable repository for the spiritual or incorporeal soul than its 'earthly' grey matter.

It was not long before the remaining ventricles were linked with mental processes, and by a more influential authority: Nemesius (c. AD 390) the Christian Bishop of Emesa (now Homs in Syria). The author of a treatise written in Greek entitled *De natura hominis* (*On the Nature of Man*), Nemesius attempted to integrate Platonic concepts of the soul into Christian philosophy, while accepting the basic premises of Galenic physiology. In doing this, Nemesius speculated on how the brain could give rise to mental thought and arrived at a position similar to Posidonius by placing sensation in the anterior ventricles, intellect in the middle cavity, and memory in the posterior cell.[4] Interestingly, Nemesius also supported his theory by referring to the effects of brain damage – noting that lesions to the front ventricles caused impairments of sensation, while those to the middle ventricle 'deranged the mind'. However, according to a much later twelfth century text from Salerno called the *Anatomia nicolai physici*, this version of the Cell Doctrine was strongly influenced by the way in which the law courts went about making their decisions. These places typically contained three main chambers – the vestibulum where declarations were heard, the consistorium where evidence was discussed, and the apotheca where the verdict was heard. It was also a small step to extend this notion to the brain. As the *Anatomia nicolai physici* states ' First we gather ideas into the cellular phantisca, in the second cell we think them over, in the third we lay down our thought, that is, we commit to memory'.

The Cell Doctrine in the West

The Cell Doctrine received an important endorsement from Saint Augustine (AD 354–430). One of the most influential of the early Christian fathers and a towering figure in medieval philosophy, his writings were largely preserved in the West to form the backbone of early Christian theology. In fact, some Christians today might be surprised to learn that many of their beliefs, including the idea of original sin, derives not from the Bible, but Augustine. Born in the town of Hippo (now in Algeria), Augustine travelled widely as a young man and studied in Carthage, Rome and Milan. Strongly influenced by the works of Plato, Augustine converted to Christianity in his early thirties, and became Bishop of Hippo in 395. Here he developed his own approach to philosophy and theology based on his insistence that his true desire was to have knowledge of God and the soul. Although prepared to use philosophical reasoning to fulfil his aims, Augustine nonetheless believed divine wisdom could only be obtained through faith. Or in his words, 'Unless thou believe thou shalt not understand.' Thus, philosophy was considered as subservient to religious revelation. His views would be expressed in a large number of treatises. But interestingly, in *The Literal Meaning of Genesis* written in AD 401, Augustine makes a surprisingly clear statement about the localisation of mental functions in the brain's ventricles:

> The medical writers point out that there are three ventricles in the brain. One of these, which is in the front near the face, is the one from which all sensations come; the second, which is in the back of the brain near the neck, is the one from which all motions comes; the third, which is between the first two, is where the medical writers place the seat of memory. Since movement follows sensation, a man without this seat of memory would be unable to know what he ought to do if he should forget what he has done.

This short description of the Cell Doctrine is considerably different from the one proposed by Posidonius and Nemesius. Although sensation remains localised to the anterior ventricles, Augustine (or rather the medical writers he refers to) has removed the function of intellect from the middle ventricle and replaced it with memory – which Posidonius and Nemesius had placed posteriorly. Moreover, Augustine now places a new faculty, namely movement, in the remaining last ventricle. In fact, he justifies this by pointing out that it is more sensible to place memory between the faculties of sensation and movement. Interestingly, Augustine's positioning of movement in the posterior ventricle is also closely akin to the teachings of Galen who had localised the motor nerves and power of motion to the rear-part of the brain.

Although Augustine equates the ventricles with cognitive functioning, he rejected the idea that they provided the repository of the soul as some ancient writers such as Erasistratus had done. Rather, following in the footsteps of Plato, Augustine believed the soul to be a substance of divine origin with the power of intellect and present in every part of the body. Moreover, the purpose of the soul on its earthly journey was to become more like God, so that after death it can enter into the heavenly realm, to live for eternity. In other words, it was the soul which provided the intellect – not the middle ventricle. Augustine also provides a clever argument to support his case for he writes: 'When there is any pain in the foot, the eye looks, the tongue speaks, the hands move, and this would not occur unless what of the soul is in those parts felt also in the foot.' Nonetheless, Augustine's tacit acceptance was an important endorsement of the Cell Doctrine – not least because he would be the most authoritative voice of the church for the next eight centuries.

The rise of scholasticism and later accounts of the Cell Doctrine

The Cell Doctrine not only proved acceptable to Christians, but Muslims and Hebrews as well. Indeed, there are numerous accounts of the theory in the eighth and ninth centuries by various Eastern writers, although they would not begin to reach the West until the eleventh and twelfth centuries. The story of how the West came to obtain these works, along with those of the great philosophers of antiquity is a complex one. But one factor was the founding of the first medical schools and universities – the later being self-autonomous organisations offering degrees.[5] An important institution in this respect was the School of Salerno in southern Italy, which became renowned as a centre of medical excellence in much the same way as Alexandria had been regarded in the antique world. The emergence of Salerno is generally attributed to a scholar from Carthage called Constantinus Africanus, who arrived at the Abbey of Monte Cassino some hundred miles north of Salerno in 1060, where he began the task of translating Arabic medical texts into Latin. These then served as educational texts for the students of Salerno, and later at other universities, which inevitably led to a demand for new books and experts who could translate them. Following the death of Constantinus, much of this work would take place in Spain where there was contact between Moorish and European scholars. Perhaps the most famous translator of this period was Gerald of Cremona (1114–1187) whose greatest achievement would be the translation of the monumental 14 volume Canon of Medicine first written in 1025 by the most renowned of all Arabian medical scholars, Ali al-Husain ibn 'Abdullah Ibn Sīnā (otherwise known as Avicenna). This was an enormous undertaking since the book ran to more than a million words on over a thousand pages, which provided a thorough synthesis of all medical knowledge from antiquity. It relied heavily on the teachings of Hippocrates, Aristotle and Galen, and was enriched with important commentaries and new details. It would serve as the standard text in university medical education in Europe for the next 600 years.

The beginnings of the educational revival taking place in the eleventh century accelerated in the twelfth. A trickle of translations now became a steady stream, and Western scholars found themselves struggling to assimilate and organise this new corpus of knowledge. One philosopher who stood out from the rest due to the great breadth of his work was Aristotle. His writings were translated into Latin and provided scholars with a powerful new means of understanding their world. He not only offered a cosmology where the Earth was at the centre of the heavens (which was acceptable to Christian theology), but introduced revolutionary new concepts such as form, matter, motion and substance. Aristotle also had important things to say about the soul and its psychological facilities including sensory perception, memory and imagination. And, he provided a new way of classifying and understanding the naturalistic world. The work of Aristotle was unsurpassed in scope and it led to the rise of scholasticism – essentially a type of scholarly pursuit that attempted to reconcile Christian theology with Aristotelian knowledge. This activity also became associated with the monks of the Dominican and Franciscan orders who lived lives of poverty, and placed a far greater emphasis on reasoning and argument than other religious scholars.

Around this time there was a further proliferation of ideas concerning the Cell Doctrine with scholars attributing the ventricles of the brain with functions derived from Aristotelian psychology. One notable person to do this was Avicenna. While most famed for his Canon, Avicenna was a polymath who wrote extensively on a wide range of subjects which included commentating on Aristotle. No writer enjoyed greater authority during the thirteenth and fourteenth centuries, and his account of the Cell Doctrine would be the best known. In fact, Avicenna advocated a five-cell doctrine. The front ventricle of the brain was described as housing two of the cells including the *sensus communis*[6] along with an image store called *imaginatio* where sensory impressions were retained. These two components combined to create the faculty of *fantasia* which enables us to imagine things. From here, information is passed to the third and fourth cells (the middle ventricle) providing us with the faculty of *cogitativa* and *seu estimativa*. The first of these (*cogitativa*) can be equated with intellect, a trait which Avicenna believed was distinctly human. However, *seu estimativa* (also known as the apprehensive faculty) is perhaps best defined as instinct, or something we share with animals. The final cell was responsible for *memorativa* and deemed to be the storehouse for our thoughts and recollections.

Avicenna also added a further element to his ventricular model by placing a valve-like organ called the *vermis* between the anterior and middle cells whose function was to regulate the 'flow' of sensations reaching the reasoning part of the brain. The *vermis* can be found in the work of Galen who likened it to a 'worm-like appendage' which was located between the two cerebella. Following Galen, it also appears to have been regarded as a structure regulating the passage of pneuma between the middle and posterior ventricles. However, some 800 years later in the writings of Avicenna, it had moved forward, seemingly to act as a filter for the intellectual hub of the middle ventricle. Avicenna was not the first to attribute the *vermis* with this operation and he may have got the idea from a translator called Costa ben Luca (AD 820–912) who lived in Baalbek, Syria. The *vermis* was also described by Constantinus Africanus who believed the head's position could control the flow of spirit through the valves. Thus, throwing the head back when trying to remember, opened the vermis allowing the passage of spirits from the anterior ventricles to the posterior where memory was located. In contrast, bending the head forward closed the *vermis*, enabling greater concentration to take place.

The earliest illustration of the brain

As new translations appeared in the West, the reproduction of illustrations also increased. Of course, humans have long attempted to depict representations of their world, as evidenced by the art of Cro-Magnon man some 40,000 years ago, who etched the pictures of various animals on the walls of his caves. It would also be hard to imagine the emergence of civilisation without mankind's compulsion to draw, which ultimately led to him inventing pictograms and alphabets, along with symbols and numbers to allow mathematical calculation. Yet, medical illustration presents a great puzzle. Although portrayals of the human form are found on various media such as clay, brick, papyri silk and bamboo from ancient civilisations such as Babylonia, Egypt, China and India, there is nothing to show they attempted to draw anatomy or provide medical illustration. This fact is perhaps most surprising in relation to the ancient Egyptians when considering their interest in mummification and medical papyri.

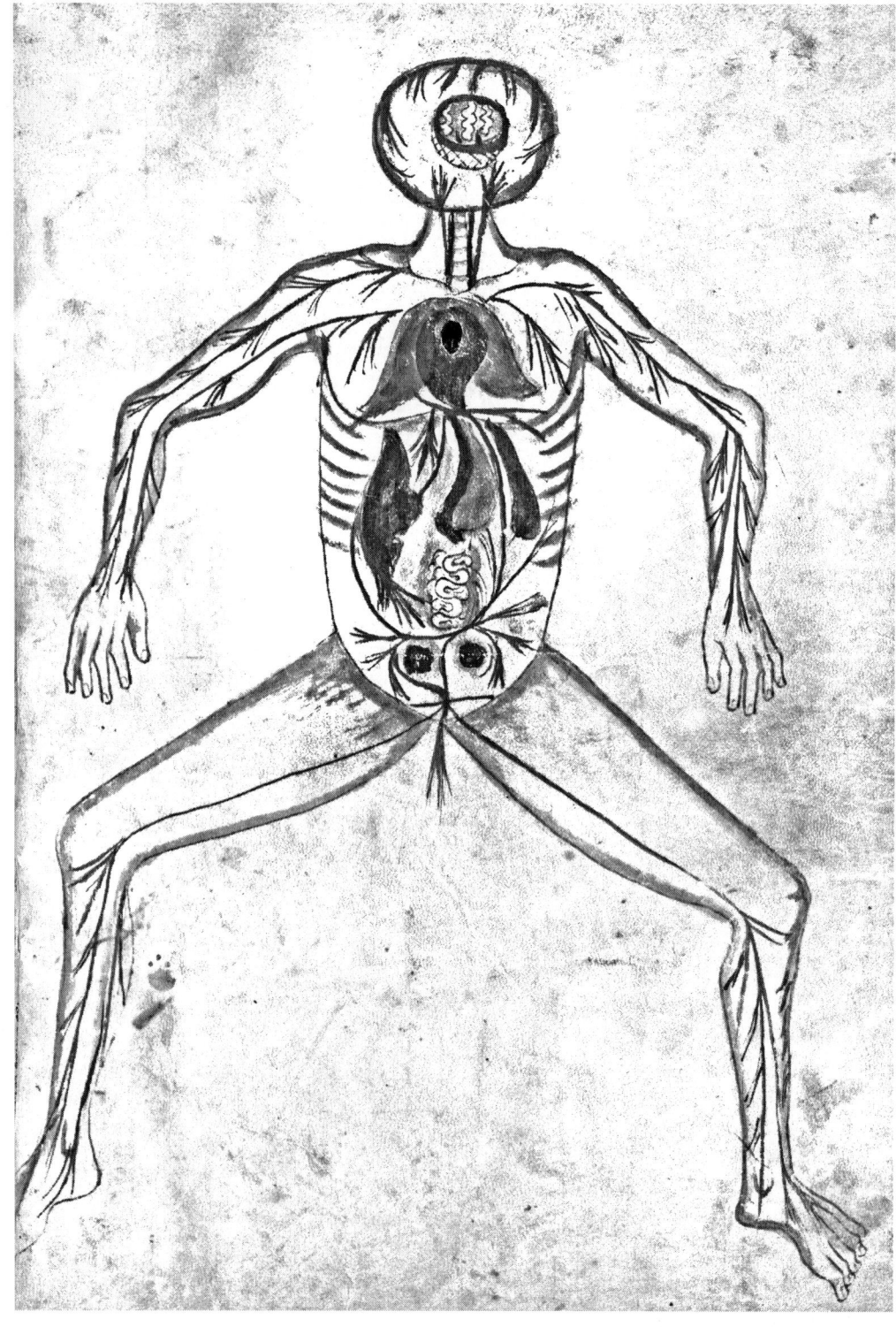

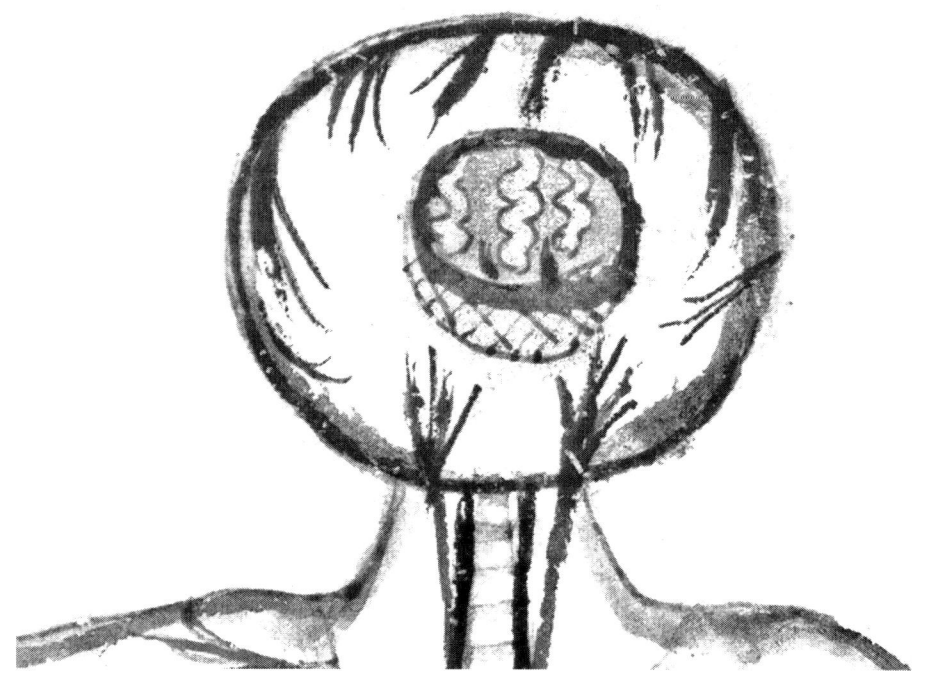

FIGURE 3.1 *(A) Facing page and (B) above.* One of a series of five drawings found in a number of Eastern and Western medical manuscripts which are believed to be copied from a much earlier source – possibly in Alexandria circa 300 BC. The insert shows possibly the earliest depiction of the brain.

Source: From Sudhoff 1908, plate 2. With permission from University of Basel (Basel University Library, Sign D II, fol 170).

Even when we come to the ancient Greeks, anatomical illustration does not appear to have been deemed especially important. For example, the *Corpus Hippocraticum* compiled in the fourth century BC, provides the single largest collection of medical manuscripts, yet contains nothing to suggest it was illustrated. In fact, it is only when we come to Aristotle the first clues that drawings might have been utilised are found when he writes in *De generatione animalium* of the importance of teaching anatomy through 'paradigms, schemata and diagrams.' Unfortunately, none of his illustrations have survived and one suspects they were not naturalistic or life-like. Nor did Galen encourage his students to rely on illustrations since he taught direct visualisation and handling of anatomical structures was the only way to appreciate their form and relationship. This reluctance may also in part be due to the difficulties of drawing for the non-specialist, and the further challenges facing the scribes who copied their work. Indeed, according to Pliny the Elder (AD 23–79) the Greeks always sought to substitute words for images because of the difficulties in reproducing visual content.

The first true anatomical illustrations were probably produced in Alexandria around 300 BC. This was the only place in the ancient world where human dissection took place and it is possible anatomists such as Herophilus and Erasistratus attempted to draw what they observed. However, no original drawings exist from this time, and the proof they ever existed is far

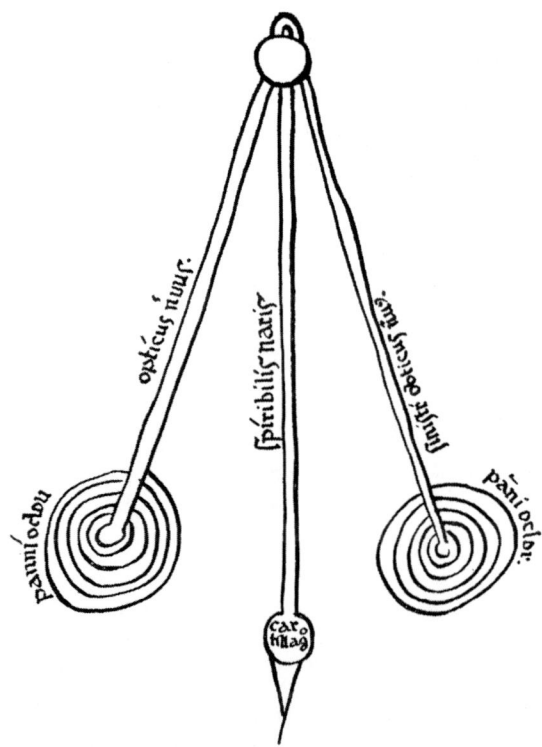

FIGURE 3.2 (A) Schematic diagram of eyes and nose connected with their respective nerve pathways, which all enter an unspecified centre (presumably in the brain). This drawing probably dates from the late twelfth or early thirteenth century.

Source: From Sudhoff 1914, p. 370.

from certain. Some evidence has come from a group of drawings known as the five figure series (or fünfbilderserie) found in a number of Eastern and Western medical manuscripts, which are possibly copied from a much earlier common source. According to the German medical historian Karl Sudhoff the originals were composed in Alexandria (Sudhoff 1908).[7] These drawings depict the nervous, skeletal, muscular, arterial and venous systems respectively, and all show a two dimensional figure in a curious frog-like squatting posture with splayed limbs (see Figure 3.1). The reason for their flat symbolic form has been much debated. Of course, it may be the originals were simply drawn in this curious diagrammatic way. Supporting this possibility is the fact that realistic anatomical illustration and portraiture did not appear in Western art until the fourteenth and fifteenth centuries. Consequently, medieval texts before this date contain highly simplified and abstract schematic representations (examples are given in Figure 3.2). In addition, Arabic illustrators were prohibited by Islamic law to portray life like depictions of the body. And, to confound matters, it is certain that artists copying the Alexandrian pictures would never have seen inside the human body as dissection was banned elsewhere – a fact also encouraging their simplistic interpretation.

Whatever the reason for their odd appearance, the alleged Alexandria drawings, or rather their reproductions, provide the earliest known illustrations of the brain. Figure 3.1 is

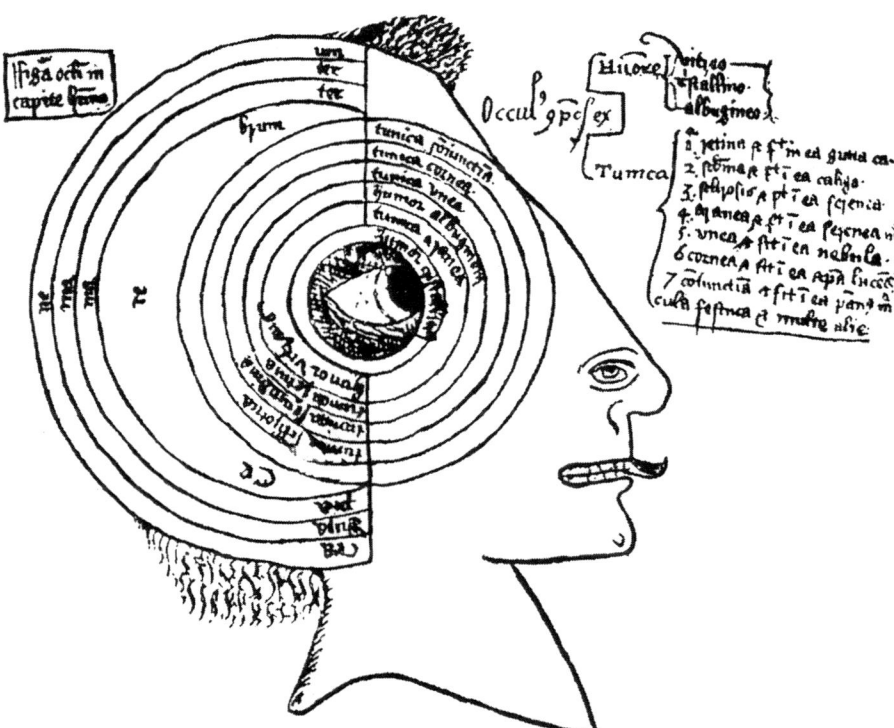

FIGURE 3.2 (B) The anatomy of the eye from a late fourteenth century manuscript held in the British Museum which depicts seven layers, including the retina (1) and cornea (6). On the left, the cranium, dura mater, pia mater and cerebellum are also named.

Source: From Choulant 1920, p. 76.

believed to have been copied from an ancient text in Salerno around 1250. It clearly reveals the practice of human dissection. Although the main purpose of this drawing was to show the body's venous system, it shows two vessels entering the brain, located centrally inside the head which is identifiable by the wrinkled outer surface of its cortex. The text accompanying the drawing also attributes the brain with *imaginativa*, *logistica* and *memoria*, thereby showing knowledge of the Cell Doctrine – no doubt added at a later date from the original.

Other early illustrations of the brain

The oldest extant (original) illustration of the brain in existence today is from an eleventh century manuscript belonging to Caius College in Cambridge. This drawing shows a strong Anglo-Saxon influence (Figure 3.3) and depicts the four most important organs of the body (liver, heart, testes and brain) which are illustrated as one goes clockwise. Closer examination of the brain, or what is referred to as the *cerebrum*, shows it has three lines going across it that provide the boundaries for '*fantasia*' (imagination), '*intellectus*' (reasoning) and '*memoria*' (memory). The artist also seems to be well acquainted with Aristotle, since the accompanying text describes the brain as being cold and moist while the heart is hot and dry.

A change from schematic anatomical illustration to more naturalistic representations was to occur in the late Middle Ages when artists began to adopt more sophisticated approaches to their subject matter. Why this sudden change came about is far from clear, but it appears to have been initiated, or at least strongly influenced, by Italian painters of the late thirteenth and early fourteenth centuries who began to understand the intricacies of perspective and three-dimensional depth – a skill also enabling them to manipulate light and shading to greater effect. While providing one of the great driving forces behind the development of the early Renaissance, it also influenced illustration. We can also begin to see this greater realism being applied to the Cell Doctrine. One of the earliest examples (author unknown) is shown in Figure 3.4. This is a drawing dating from around 1310, depicting five cells in the brain showing the influence of Avicenna. A number of similar diagrams are found in other medieval manuscripts including the great scholastical works of Albertus Magnus (1193–1280), Thomas Aquinas (1225–1274) and Roger Bacon (1214–1292).

Another important artistic development took place at the beginning of the fifteenth century when woodcuts were used to produce printed illustrations. Here a carved image was engraved into a block of wood and stained with ink. A similar concept would be used by Johannes Gutenberg in his invention of the printing press in 1450. Despite being a simple invention, it enabled illustrations to be replicated accurately, quickly and most important cheaply. Again this technique was used to demonstrate the Cell Doctrine with the most famous example produced by the German Carthusian prior Gregor Reisch. This appears in his *Margarita philosophica* of 1503 which is generally recognised to be the first modern encyclopaedia, and widely adopted as a university textbook during the sixteenth century. It contained a number of beautiful illustrations produced by woodcuts – one of which shows a human head with three freely communicating cells. Because of the book's popularity the illustration was widely plagiarised with reproductions of it being found into the nineteenth century.

FIGURE 3.3 The oldest known original illustration of the brain (shown in the top left-hand corner) taken from an eleventh century manuscript owned by Cambridge University (Ms. 428/428, folio 50).

Source: Reproduced by kind permission of the Master and Fellows of Gonville and Caius College, Cambridge.

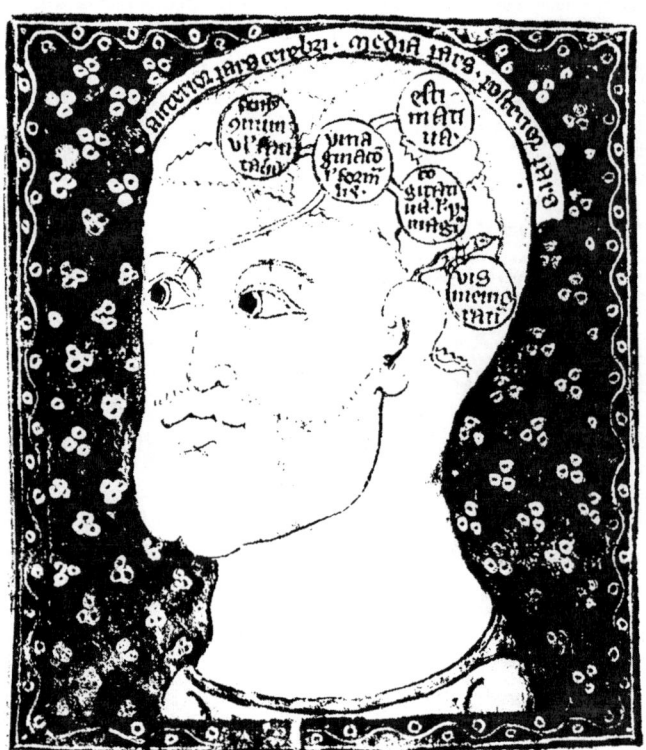

FIGURE 3.4
A fourteenth century drawing of the Cell Doctrine (*c*.1310) showing the influence of Avicenna. The cells are labelled (right-to-left) *sensus communis*, *yimaginatio*, *estimativa*, *cogitavia vel yimaginatio*, and *vis memorativa*. Interestingly, the diagram places the *vermis* (depicted here as containing an eye, presumably representing a form of inner awareness) between the fourth and fifth cells.

Source: From De Lint 1926, p. 27.

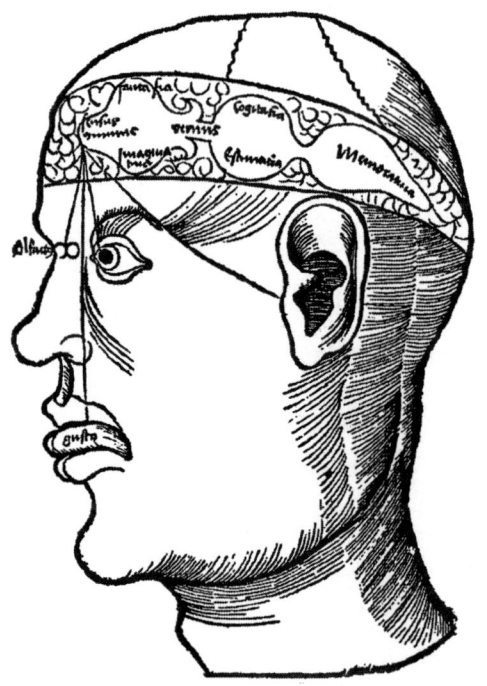

FIGURE 3.5
Perhaps the best known Cell Doctrine illustration (produced from a woodcut), which first appeared in Gregor Reisch's *Margarita philosophica* of 1503. The first cell is inscribed with *sensus communis*, *fantasia* and *imaginativa*, the second with *cogitativa estimatava* and the third with *memorativa*. The *vermis* in this instance is shown connecting the first and second cells.

Source: Wellcome Library, London.

Mondino de Luzzi: the restorer of anatomy

The dissection of human corpses had only taken place in Alexandria, and even then its practice was brief, perhaps lasting no more than a hundred years. After this there was no more human anatomisation until the thirteenth century – a hiatus of 1500 years. One of the most important figures to reinstate this procedure was Mondino de Luzzi (1275–1326). Born and educated in the northern Italian city of Bologna, famous for its university, Mondino would rise to become its professor of surgery. He also lived at a time when the taboos surrounding human dissection were beginning to weaken. One reason for this lay with the Crusades, who were attempting to restore Christian control of the Holy Land. The inevitable deaths of those fallen in battle, especially those belonging to nobility, often led their families to seek ways of bringing the remains of the deceased back home for burial. The intent was often to preserve the heart for internment in a shrine, or in some cases boiling the body to obtain skeletal remains. Although this had led to a Papal proclamation (*De sepolturis*) in 1300 prohibiting the desecration of human bodies, partly to stop the lucrative trading of relics and bones by Neapolitan sailors, this proved ineffective. A second force encouraging human dissection was an increasing desire by many civic authorities of the time to investigate homicide and causes of death. In fact, this had led to autopsies being performed, perhaps as early as 1250, by certain doctors in a small number of places in parts of Italy and northern Europe. Clearly, attitudes to the dead were changing and it would lead to the first documented public human dissection by Mondino de Luzzi in 1315. The cadaver is believed to have been of an executed criminal and the operation sanctioned by the Vatican. A second public dissection may have been undertaken the same year.

Following this breakthrough, the study of anatomy through human dissection began to enter the medical curriculum – a change which also gave Mondino the opportunity to perform many more operations as part of his university duties. A complete anatomical demonstration at Bologna is believed to have taken Mondino four days and involved him removing the digestive, respiratory, circulatory and muscular systems. Because Mondino had no way of preserving the body, the procedure would have taken place away from sunlight in a cool university building. It is also said that he dissected from the inside out as the internal organs decompose more quickly. The main purpose of the lesson was to teach anatomy, and to do this Mondino directed the operation from a high chair overlooking the scene. He instructed a 'barber' (apparently the only type of person who was permitted by law to dissect) to cut open and dismember the corpse with various knives and cleavers. A second assistant pointed out the organs with a stick to the students assembled around the corpse. This was a new way of teaching anatomy and it attracted much interest, putting Bologna and other Italian universities at the forefront of the medical world.

In 1316, Mondino wrote a dissection manual for his students entitled *Anathomia corporis humani*. It would later be published in 1478, over a hundred years after his death, as the first anatomical text and dissection manual in the West. Containing just 22 leaves (i.e. 44 pages) and printed in gothic characters, the influence of the *Anathomia* was profound, being used to teach anatomy for the next 250 years, with over 40 editions in various languages published during this period. While the book offered nothing new over Galen or Avicenna (apparently it was largely copied from their works and made the same mistakes) it was still a landmark in medical history, helping to establish anatomy as a major academic discipline. Although the original version of *Anathomia* lacked illustrations, this situation was rectified in 1345 by one

FIGURE 3.6 (A) Dissection scene from the Italian *Fasciculo di Medicina* printed in Venice in 1493. The picture forms the first page of an Italian translation of the Anathomia of Mondino.

Source: From Singer 1957.

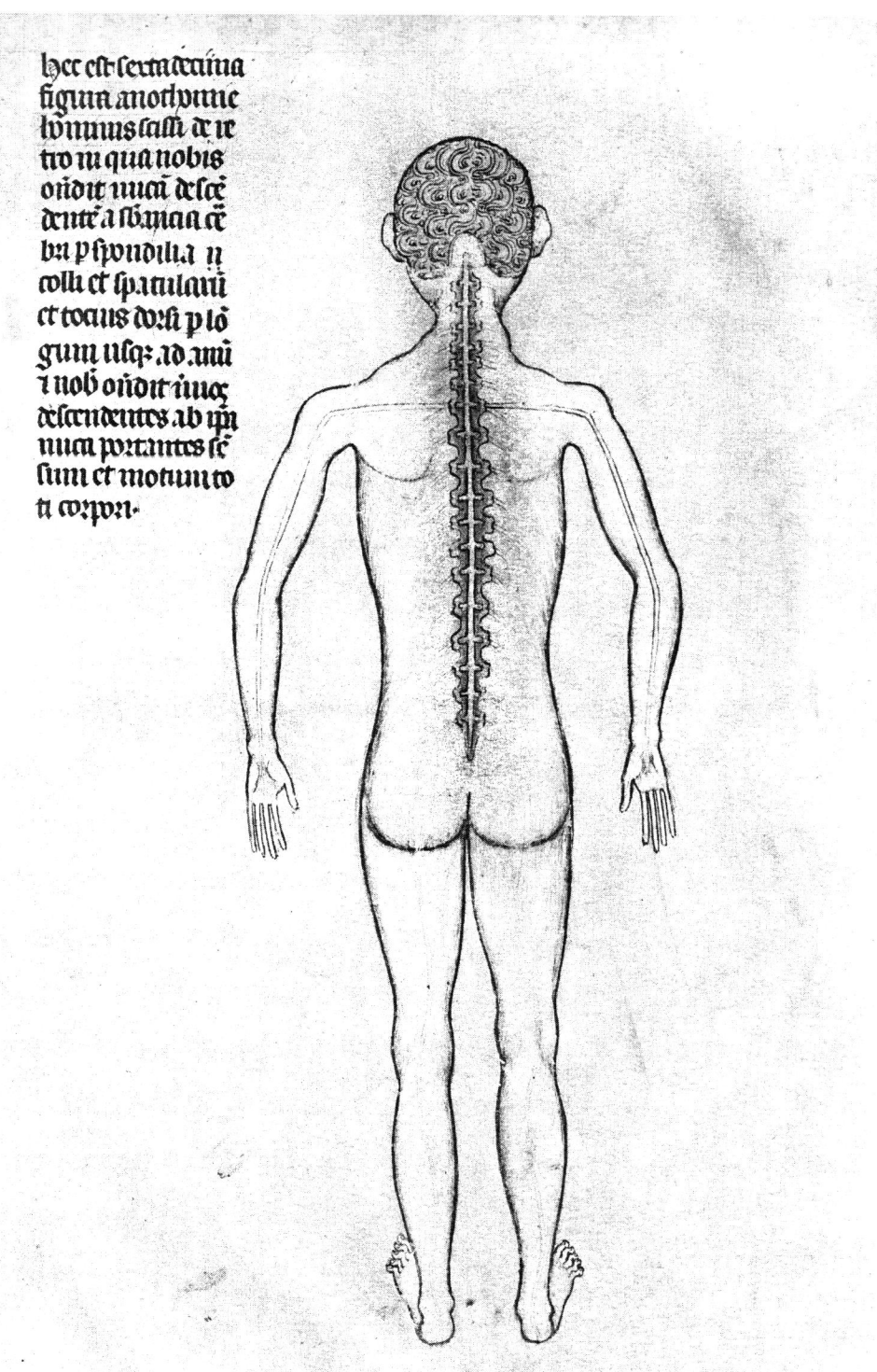

FIGURE 3.6 (B) Plate XVI from *Anathomia designata per figures* by Guido da Vigevano (1345) showing the spinal cord and origin of the spinal nerves.

Source: Kindly provided by Professor Antonio Di Ieva at the Medical University of Vienna.

of Mondino's pupils, Guido da Vigevano, who wrote *Anathomia designata per figures* – a text known to have contained 24 colour plates. Sadly, 18 of these have been lost. Nonetheless, the ones that remain include six showing the head, brain and spine. These pictures form the first ever modern anatomical drawings of the nervous system.

Leonardo da Vinci: Renaissance Man

For many historians, the Renaissance (the term derives from the Italian *rinascita* meaning 'rebirth') originated in Italy and covers a period spanning the fourteenth to seventeenth centuries. Strongly influenced by the rediscovery of humanist values based on the Greek ideal, the Renaissance also corresponded to a period of economic growth, especially after the ravages of the Black Death, leading to new developments in art, architecture, politics, world exploration and academic pursuits. While many influences contributed to this change, special mention must be made of two. The first was the fall of Constantinople to the Ottoman Turks in 1453 – a massive blow for Christendom causing many of its scholars with their books and ideas to seek refuge in the West. The second was the invention of the printing press in 1450 by Johann Guttenberg. Prior to this, most books were authorised by the church and copied laboriously by hand, or printed using blocks of engraved wood. Consequently, they were expensive and most likely studied by monks sufficiently educated in Latin to read them. The printing press initiated a true revolution, shifting intellectual life and literacy to a much wider and non-secular population. One result was an emergence of a new intelligentsia able to publish their own work. This change also took place relatively quickly. It is estimated that by 1500 over ten million books had been published in northern Europe comprising some 40,000 different titles.

The Renaissance was also a time when a number of artists took an active interest in drawing the human form. The greatest of these was Leonardo da Vinci (1472–1519), the illegitimate son of a lawyer and peasant woman, and widely recognised as a genius even in his own lifetime. Famous for paintings such as the Mona Lisa and The Last Supper, Leonardo (as he is generally known) also excelled in many other disciplines including engineering, mathematics and mechanics. Indeed, his sketch books reveal a fascination bordering on obsession with a wide range of subjects far ahead of their time, which include the invention of armoured vehicles, parachutes and flying machines. These drawings are extraordinary for they provide the first real signs of radical new thinking in the West since the fall of Rome. Leonardo's innovation also extended to drawing the human body, including its dissected forms, in a modern naturalistic and realistic way.[8] However, this was more than art for Leonardo who was intent on finding out for himself more about the deeper secrets of the body. For him, to draw was to understand. Or rather, when people asked Leonardo the secret of his creative genius, he replied *saper vedere* meaning to see what others don't. This 'new vision' also extended to the brain, which he examined on several occasions throughout his anatomical researches.

Growing up in the small Tuscan hamlet of Vinci, a place that was already steeped in painting tradition, Leonardo's interest in human anatomy first arose as a teenager when he moved to Florence to become a pupil under the famous artist and sculptor Andrea del Verrocchio. Here he was introduced to many different types of art, including the drawing of the human body, which required him to carefully examine its outer muscular form and features. His training also involved drawing flayed limbs where the skin had been removed to expose the muscles and sinews – a situation probably leading him to observe the practice of human dissection for the first time. At the end of his apprenticeship, sometime around 1482, Leonardo moved to Milan, the site of one of the largest medical teaching schools in Italy. For reasons that have never been fully explained, he was granted permission to perform dissection of human corpses. It would be the start of a life-long quest to illustrate the complete structure of the human body. Although drawing primarily with chalk and crayons, Leonardo also introduced a number of new techniques including the use of transparencies, cross-sections and three-dimensional shading. But Leonardo's main aim was to understand how the body worked, especially in terms of physical mechanics (i.e. as a perfect machine). Consequently, his folios contain a large amount of written commentary attempting to explain what he is depicting.[9]

The task Leonardo set himself was formidable. It was also highly unpleasant, for the hot climate of Italy meant that the corpses soon decomposed. To make matters worse, Leonardo was forced to conduct his investigations in a small room at night under a flickering candle light away from prying eyes. It was a ghastly situation with Leonardo surrounded at close quarters by 'quartered and flayed corpses' which were too 'horrible to behold'.[10] He also had to confront this unpleasant scene many times for Leonardo's quest took him many years, involving over a hundred corpses, and filling more than 120 notebooks. Sadly, this large corpus of work, known to include 779 anatomical drawings, was never published in his lifetime. Although Leonardo had intended to present his findings in collaboration with a noted professor of anatomy called Marc Antonio del Torre – his colleague died prematurely at the age of 30. The tragic consequence was that the work became 'lost' for several more centuries. Despite this, Leonardo's drawings may have had some impact on the development of anatomy for after his death in 1519, the notebooks passed to pupil Francesco Melzi who kept them at his villa. It appears he allowed other anatomists to view them and even turned a blind eye to their plagiarisation. Nonetheless, for 260 years, Leonardo's work was largely a secret, and not rediscovered again until the late eighteenth century. One can only speculate what impact Leonardo might have had on the subsequent development of anatomy if his portfolio of work had been published as intended. According to the medical historian Charles Singer (1956), anatomical illustration would have been advanced by centuries.

Leonardo's search for the soul

The first attempts of Leonardo to draw the brain date from about 1487 to 1493 and they show his fascination with the Cell Doctrine and, in particular, the location of the sensus communis. This term had been introduced by Aristotle who referred to it as the place in the body where all the senses converged – or rather that faculty of the psyche that allows the external world to be perceived. Aristotle, of course, had placed the sensus communis in the heart. In contrast, Leonardo, like most of his contemporaries, believed it existed in the brain. However, Leonardo had another reason to be interested in the sensus communis for he regarded it as the site which housed the soul. Although this idea opposed the view of thinkers such as Plato and St Augustine who believed the soul existed throughout the body, Leonardo rejected this theory by pointing out that if it was true, there would be no need for a place where all the senses converge.[11] Leonardo was also well aware that many of the medieval scholars, including Avicenna, had placed the *sensus communis* in the anterior ventricles, but Leonardo was not to accept this traditional view either. Indeed, in one of his earliest anatomical drawings of the head (circa 1490) entitled The Layers of the Head Compared to an Onion,[12] Leonardo

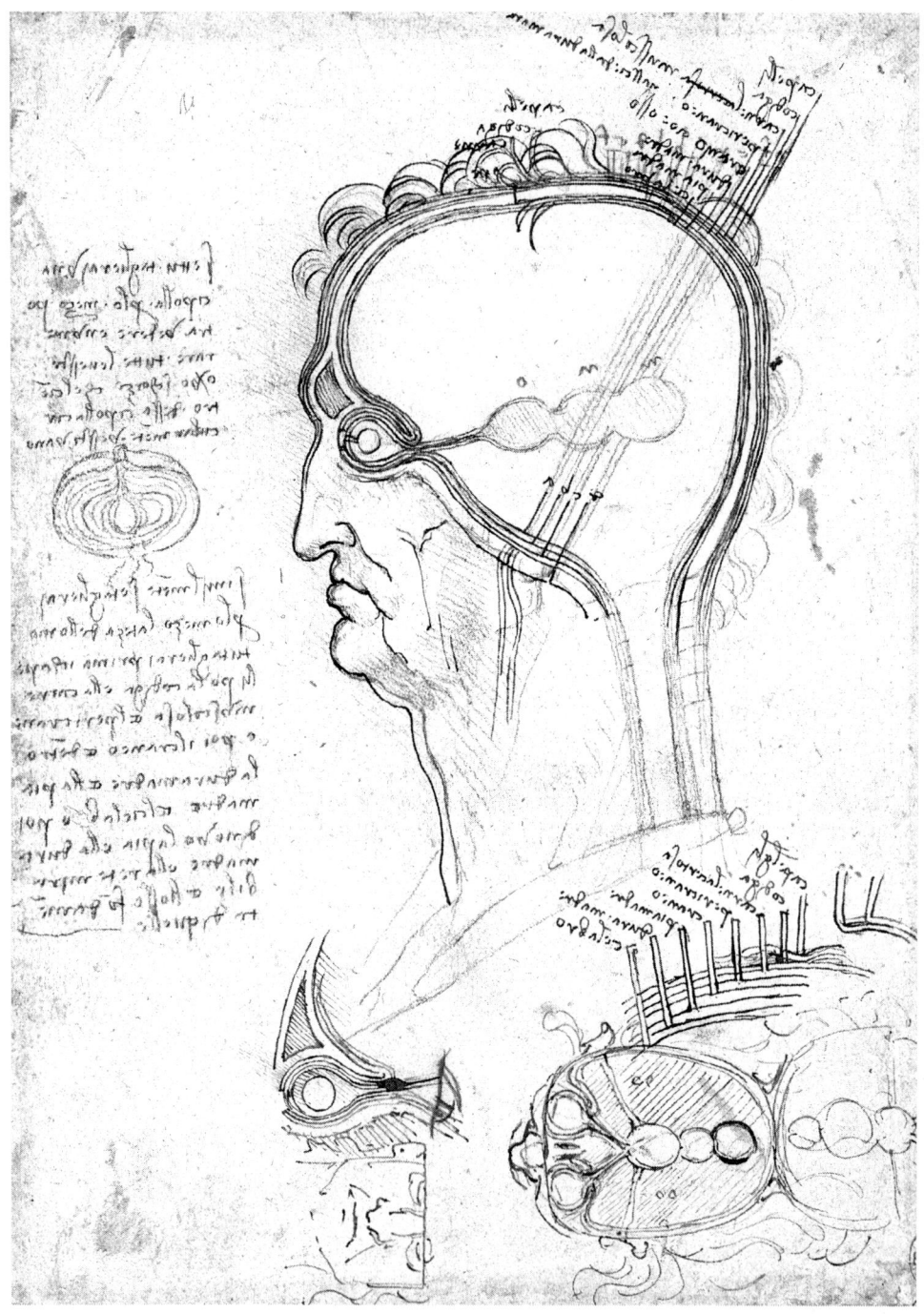

FIGURE 3.7 The Layers of the Head Compared to an Onion (*c*. 1490) depicting the layers of the skull and the ventricles as three interconnecting cells.

Source: Supplied by the Royal Collection Trust/© Her Majesty Queen Elizabeth II 2014.

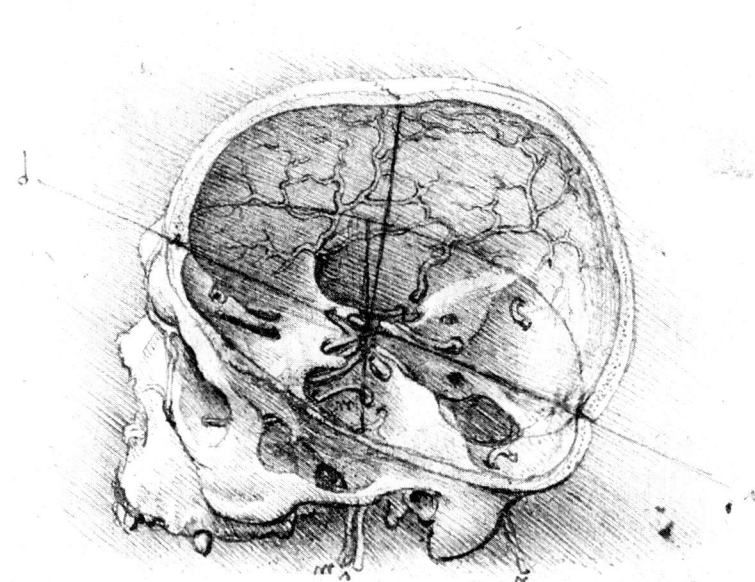

FIGURE 3.8 Drawing of the human skull and brain (*c.* 1489), which also shows the site where Leonardo believed the sensus communis was located.

Source: Supplied by the Royal Collection Trust/© Her Majesty Queen Elizabeth II 2014.

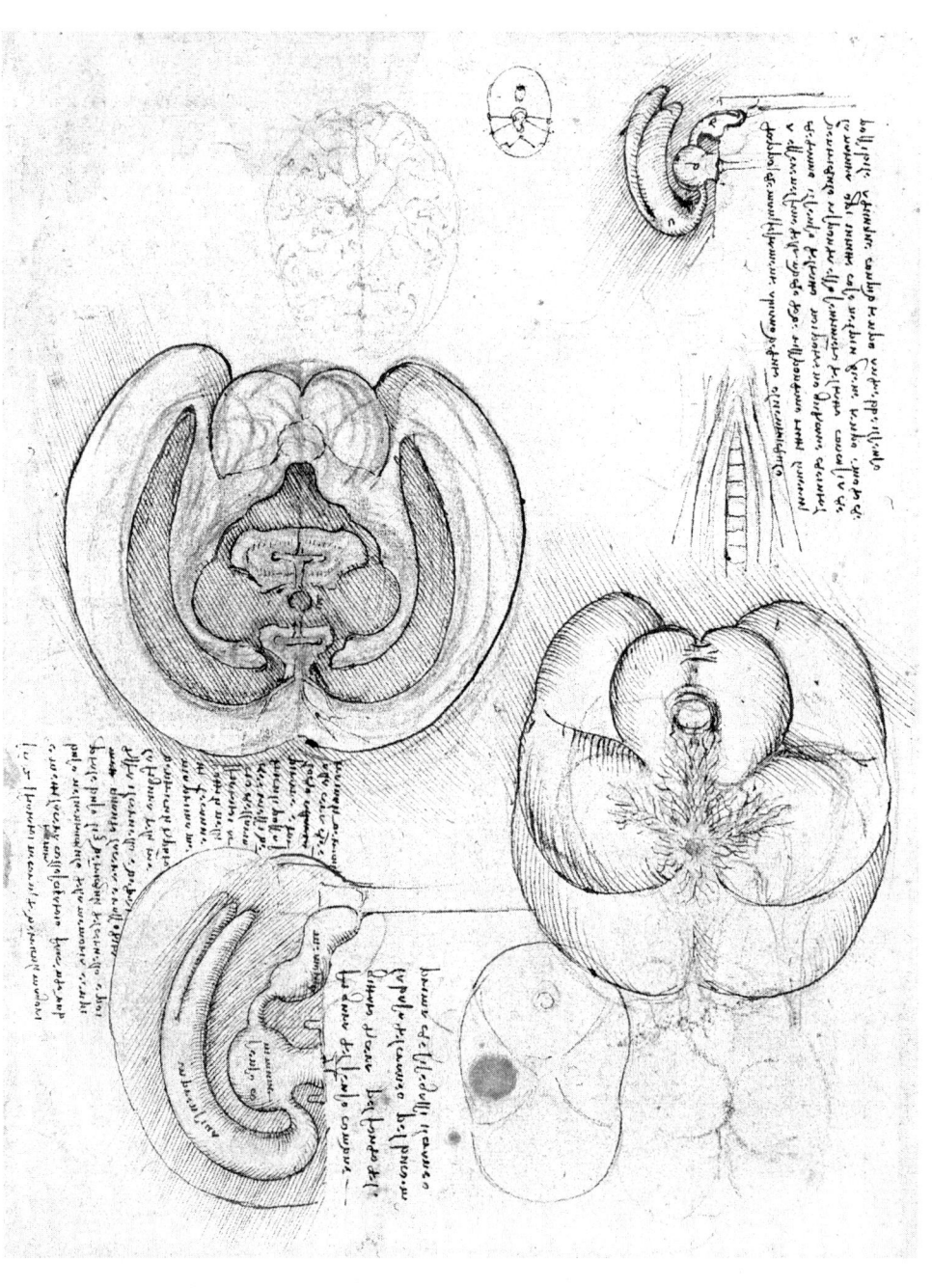

FIGURE 3.9 Although these drawings appear to provide a more realistic sketch of the human brain and ventricles by Leonardo (c. 1508) it is actually based on a cast from injecting molten wax into the ox brain.

Source: Supplied by the Royal Collection Trust/© Her Majesty Queen Elizabeth II 2014.

draws the ventricular system as three cells, and introduces the term *imprensiva* (i.e. where visual and auditory impressions arrive) to describe the anterior ventricles. In fact, Leonardo is the only anatomist to have ever used this term – a fact that confirms his great originality of thinking, but this is not the fabled sensus communis, for Leonardo now locates this faculty in the middle cell. Described by Leonardo as 'the eye without external light', the middle ventricle is now regarded as the place responsible for making sense of the visual images received from the first ventricle. From his notes it can be seen that Leonardo also gave this site the important responsibility of *comocio* (thought) and *volonto* (will). The third cell, or ventricle, was the site of memory, or the 'monitor' of the sensus communis.

In another drawing of the human skull and brain from around the same time, Leonardo leaves us in no doubt where he believed the sensus communis was located: he pinpoints it with a series of intersecting upright and diagonal lines superimposed on top of the anterior portion of the middle ventricle, at a point that marks the proportional centre of the skull (see Figure 3.6). It is also interesting to note that some of the cranial nerves are also shown as projecting to, or leaving from, the sensus communis.

Despite being opposed to animal cruelty (it is said that Leonardo would sometimes go to the local market to buy birds in order to set them free) and possibly a vegetarian, Leonardo nonetheless was prepared to undertake a few experiments on animals during the early phase of his anatomical studies (*c.* 1487). These too were primarily concerned to confirm the importance of the brain as the site of the soul's location. In one study, Leonardo took a frog and punctured its upper spinal cord – an act that immediately killed the animal.[13] From this, Leonardo reasoned that the brain was the site providing the essential vital energy for life, and by severing the upper spinal cord its power was stopped from reaching the rest of the body. However, this life force did not extend to the heart for in another experiment he found a frog could be kept alive for some hours without this organ. In fact, this may not have surprised Leonardo since a similar phenomenon had actually been observed by Aristotle in the tortoise.

Visualising the ventricles

Although the ventricles had been known from the earliest of times, and were accurately described by Erasistratus around 300 BC in Alexandria, one must remember ancient investigators had no effective way of visualising their complex shape after the brain was sliced open. This was partly because there were no methods at their disposal to harden tissue, allowing the serial slicing of tissue or firm dissection. Consequently, simple knife cuts only provided a minimal amount of information concerning their layout – a situation that encouraged medieval illustrators to portray the ventricles as simple cells with no effort at anatomical accuracy. However, around 1508, Leonardo initiated the second phase of his brain studies and worked out an ingenious way of solving this problem: he made a cast of the ventricles by injecting molten wax into the middle ventricle of an ox with a syringe. After making two vent holes in the horns of the anterior ventricles to allow fluid and air to escape, and allowing enough time for the wax to harden, Leonardo cut away the tissue of the brain to expose the injected material. This was the first time in the history of anatomy that a solidifying medium had been used to expose an internal bodily structure, and it revealed the ventricular system in unsurpassed detail. Indeed, by using this technique, Leonardo was able to draw (or rather imagine) the interior of the brain in a highly naturalistic way. While his illustrations are not without error, for the third ventricle is highly distorted, and he failed to fill the lateral ventricles completely, they do have a good resemblance to their true shape and connectivity. Interestingly, Leonardo's illustrations of the ventricles appear to be a composite taken from both an ox and human brain. According to the historian Charles Gross (1998), the drawing presented on page 68 (Figure 3.9) shows a collection of blood vessels at the base of the brain known as the *rete mirable* – a structure only found in the ox. However, the location of the cerebellum and general form of the ventricles are much closer to those found in a human brain.

The first printed anatomical text with illustrations

Although Leonardo's depiction of the ventricles was sadly never to be widely known, the times were changing in anatomical illustration, with one of the most important figures in this respect being the Italian Jacopo Berengario da Carpi. A man said to have engaged in violent quarrels and assaults, and even convicted of a robbery, Berengario was appointed Lecturer in Surgery at Bologna in 1502. He would later achieve fame as a doctor for his use of mercury in the treatment of syphilis – a disease then prevalent among the priests of Rome. However, Berengario was also a devoted anatomist and in 1521 published the fruits of his labours entitled *Commentaria super anatomia Mundini*.[14] Although the title suggests it is an abbreviated version of a much greater text by Mondino (who had also worked at Bologna over 200 years before him), the *Commentaria* is actually an original work, based on Berengario's own observations during the course of his dissection. A thick quarto[15] of over 1000 pages, the *Commentaria* is also the first illustrated anatomical textbook to be printed with a detailed series of 21 full-page woodcut text illustrations specially designed to accompany the text. Furthermore, it is generally recognised to be the first work since the time of Galen to introduce a considerable amount of new anatomical information based upon personal investigation and observation.

While the anatomical illustrations produced from the woodcuts are rather small, with most of the page on which they were printed being filled with text, they nevertheless represent a great advance on what had been produced before. The book also boasted an impressive front page which included an image of a dissection scene. A year later Berengario would publish a much shorted version entitled *Isagoge breves* providing a concise description of the anatomy of the human body and a manual of dissection. It can be considered the first ever textbook written specifically for students. Later, Berengario published a revised edition of this book which contained several new woodcuts, including two of the brain. For the record, these provide the first published view of the brain's ventricles.

If we forget Leonardo, then Berengario provides the first original thinking about the brain since antiquity. For example, he recognises the choroid plexus as composed of arteries and veins, and explains the pineal gland's protrusion into the posterior ventricle as a director of animal spirit flow. Berengario also reports a discovery that cast doubt on the greatest authority of them all in the sixteenth century, namely Galen. The claim was startling for despite having 'dissected many heads', Berengario was unable to find any evidence of the large network of blood vessels at the base of the brain in humans. This had been called the *rete mirable* by Galen, and it formed an important component of his physiological system since it was the place where animal spirits in the blood were transformed into psychic pneuma before entering the brain. In fact, Berengario was so sure of his observations he went so far to say Galen must have imagined the rete mirable on the strength of his own opinion. This was

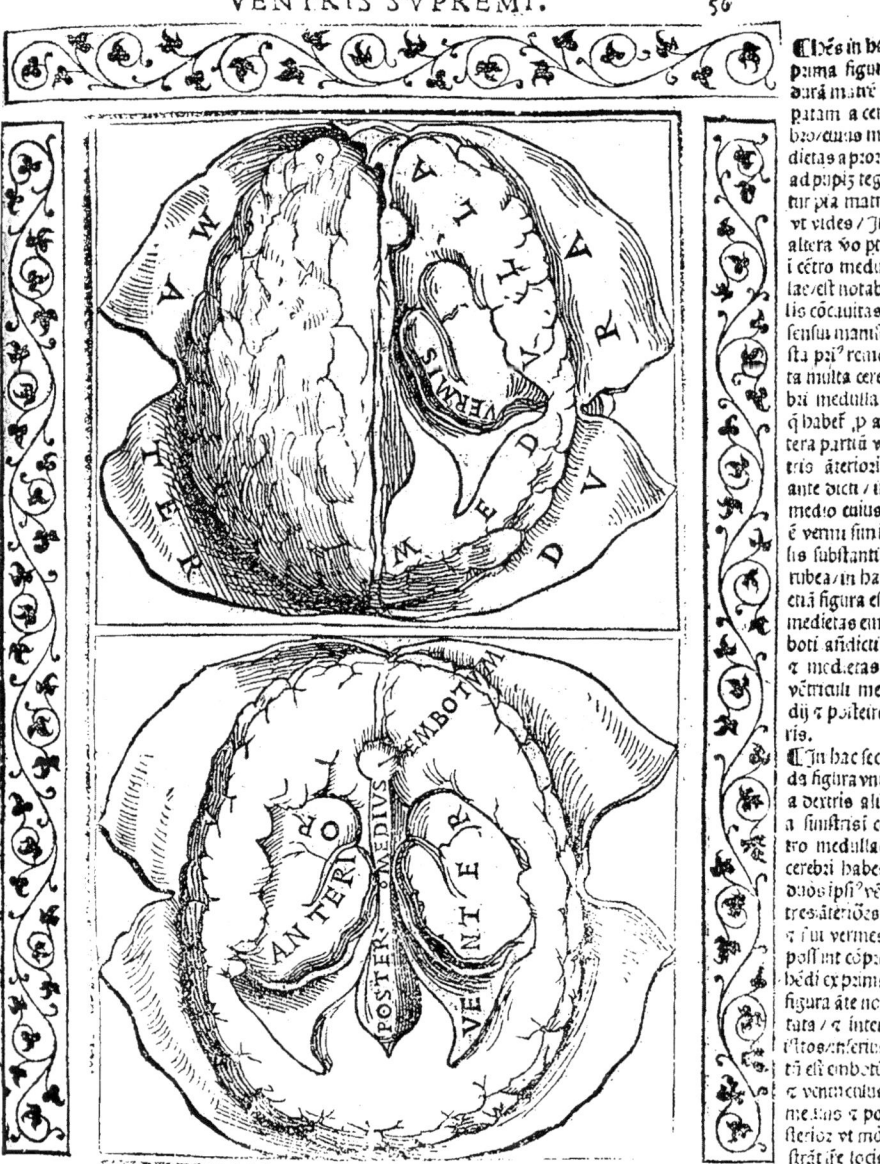

FIGURE 3.10 Woodcut illustrations made by Jacopo Berengario da Carpi (1523). The first shows the brain opened from above to reveal one of the anterior ventricles. The second reveals both anterior ventricles along with a midline structure called the embotum – possibly the exit for the pituitary gland to remove waste from the brain.

Source: Wellcome Library, London.

the first serious criticism of Galen since Roman times, although it did not appear to generate much controversy, perhaps because Galenic physiology at the time was considered sacrosanct and beyond reproach.

The reawakening: Andreas Vesalius

If there is one year marking the point when the medieval world ended and the modern era began, it is surely 1543. In this year, two books were published, within one week of each other that significantly weakened the stranglehold of dogma and traditional thought which had lasted for nearly 2000 years. The first was *De revolutionibus orbium coelestium* (*On the Revolutions of the Heavenly Spheres*) by Nicolaus Copernicus, a work questioning the authority of Aristotle, Ptolemy and Holy scripture by denying the Earth lay at the centre of the universe. Instead, the Earth was described as rotating on its axis once a day and circling around the sun every year – a heretical and profoundly radical idea for the times. Its repercussions would lead to the Italian scientist Bruno being burned at the stake and Galileo under threat of torture and death forced to renounce his geocentric beliefs. Although *De Revolutionibus* was placed by the Catholic Church on its index of prohibited books, it dealt a devastating blow to scholasticism and religious authority, changing the way in which man viewed himself in the grander scheme of things. A similar impact, this time in anatomy and physiology was provided by the second book, *De humani corporis Fabrica* (*On the Structure of the Human Body*) by Andreas Vesalius. Again it dealt a decisive break with the entrenched ideas of the past, for the *Fabrica* (as it became more widely known) would lead to a revision of concepts concerning the structure and function of the human body. Ideas that had been taught and accepted without question since the time of Galen were now questioned. Even more shocking, it proved Galen had made many errors in his anatomical work.

Vesalius is believed to have been born in Brussels on the last day of December in 1514. The son of a doctor and apothecary with royal patronage to Emperor Charles V, Vesalius once reminisced he was so intent on learning anatomy as a boy, that it led him to perform dissection on mice, moles, cats and dogs caught in the fields near his home.[16] He also lived close to a site where public hangings took place and it is likely the young Vesalius witnessed executions, along with the corpses left to hang as a deterrent to others. First embarking on an arts course at the Flemish University of Louvain, Vesalius moved to Paris at the age of 19 to study medicine. Here he became a keen collector of human bones, sometimes taking specimens illegally from public cemeteries – an experience he claimed enabled him to recognise any bone in the body by touch alone. However, due to a conflict between Emperor Charles and France, Vesalius would have to leave Paris, eventually completing his medical degree at the University of Padua in Italy. He must have been an exceptional student for the day after his graduation, the 23 year old Vesalius was appointed lecturer in surgery and anatomy on a beginner's salary of 40 florins. This was a remarkable achievement as Padua was the greatest medical university of its day, famed for its demonstrations of human dissection, which were held in a purposefully built theatre. Vesalius soon became part of its folklore, taking scalpel in hand himself rather than reading from a prepared text, and often using several cadavers at the same time in various states of dissection, to illustrate his anatomical teaching.

During his time in Paris, Vesalius had been formally taught dissection by the highly respected professor Jacques Dubois, better known by his Latin name Jacobus Sylvius.[17] Apparently the first in France to teach anatomy from a human corpse, his lectures were attended by hundreds of students. Sylvius was also the first to give separate names to many of the body's muscles (before him, the muscles had simply been referred to by numbers), and he applied a similar naming scheme to blood vessels, coining terms that are still used today including *jugular* and *subclavian*. But Sylvius was to be remembered less fondly for his blind support of Galen – a devotion that would later bring him into conflict with Vesalius. The seeds of the problem first arose in Paris when Vesalius realised the jawbone only had one part, and not two as Galen had taught. Although these types of disparity were not unknown to Sylvius, he dismissed them by arguing certain aspects of human morphology had changed since Galen's time. Nor was Sylvius alone in his admiration of Galen. The sixteenth century was a period when Galen's works formed the main basis of all anatomical education, and many believed progress beyond him was impossible. Vesalius may well have accepted this in Paris, but as his research continued at Padua, he began to observe further discrepancies – most notably the lack of the rete mirable at the base of the brain in man. But Vesalius was still not yet prepared to speak out against the authority of Galen. This reluctance is apparent in one of his first published works, *Tabulae anatomicae* (1538) which includes a series of six anatomical plates of the human skeleton and vascular system. In addition to the rete mirable which Vesalius depicts, the text also includes other Galenic errors, including a five-lobed liver and the venous system arising from the liver.

However, this reverence would not last. Two years later, Vesalius was asked to translate Galen's *On Anatomical Procedures* from its original Greek into Latin. This proved to be a decisive moment for upon reading the work more closely, Vesalius became suspicious of Galen's understanding of anatomy which included his knowledge of the human skeleton. It was then with great astonishment Vesalius realised that Galen had never dissected a human body. Instead, he had extrapolated all his human anatomy from operations performed on monkeys, and in some cases cattle and pigs. While this is not a serious indictment of Galen, for human dissection in the Greek and Roman world was prohibited at the time, it nonetheless invalidated much of his teaching. This was a stunning revelation, which called into question a huge corpus of work that had been revered for over 1400 years. After getting over the shock of his own stupidity for not recognising the truth earlier, Vesalius vowed to correct all of Galen's mistakes. This undertaking would result in the *Fabrica* – one of the most magnificent collections of anatomical drawings ever published.

De humani corporis Fabrica

The *Fabrica* has been described as one of the most daring and beautiful adventures ever undertaken by the human spirit which combines scientific exposition and art in a manner unprecedented in the sixteenth century and rarely equalled since. Consisting of seven books, the work attempts a complete anatomical and physiological study of every part of the human body, beginning first with the bones and skeleton, and then proceeding to the muscles, vascular system, nervous system, abdominal organs, thorax, heart and brain. More importantly, the *Fabrica* is based on Vesalius' own observation and dissection of human bodies, which he had largely undertaken at Padua.[18] More controversially, in doing this, Vesalius was not prepared to blindly follow the conventional wisdom laid down by Galen. In fact, Vesalius reports in the *Fabrica* that he had found over 200 instances where Galen's human anatomy was in error. Thus, Vesalius was not only encouraging anatomists to start exploring the inner workings of

FIGURE 3.11 The front page of *De humani corporis Fabrica* by Vesalius published in 1563.

Source: Wellcome Library, London.

the human body for themselves, but he himself had discovered a number of radically new anatomical facts about the human body that simply could not be ignored.

It has been said by the historian Vivian Nutton that there are few books so famous and yet so unfamiliar as the *Fabrica*, not least because it has never yet been fully translated into a modern Western language. Moreover, to read to the end of this large book is an arduous task in itself. Indeed, the first edition of the *Fabrica* is a huge tome (43 × 21 centimetres) of more than 700 folio pages, written in Latin, which contains 83 anatomical plates made up of 420 separate illustrations, along with hundreds of the little ornate drawings (e.g. cherubs) accompanying the text. Although this would have made the book prohibitively expensive for most buyers, the revolutionary message it conveys was enough to achieve its fame. Realising that his rejection of Galenic anatomy would convince few sceptics unless he could substantiate his own observations with clear proof, Vesalius went to great lengths to employ the best artists and wood-block cutters of his day. Curiously, the artists responsible for illustrating the *Fabrica* are not acknowledged by Vesalius and their identities have been the subject of much debate. It was once believed the figures were drawn by the great Venetian artist and painter Titan, although whether he could have produced such a huge mass of work is now regarded as unlikely. Thus, it is more probable that rough sketches made by Vesalius were given to a team of artists, who were asked to elaborate on them – perhaps overseen by the German-born Italian painter Jan Stephen van Calcar. These were then transformed into detailed woodcuts providing anatomical illustrations that generally took over a complete page, and of a quality that far surpassed anything that had previously been seen.

If there was ever doubt about the quality of the illustrations inside the *Fabrica* then one only has to examine the book's magnificent title piece. This is rightfully famous for its depiction of a crowded scene of public anatomy taking place in Padua. Vesalius is at the forefront, dissecting the exposed abdominal organs of a female cadaver while two barbers sit underneath the dissecting table, one of whom is sharpening a razor. There is a mixed audience of some 80 persons, arranged in tiers, including several members of the clergy, nuns, bearded patriarchs, assistants and students. At the centre of the scene a human skeleton stands erect in bold relief, while a naked male figure looks on from the side. There is also an artist sketching the proceedings believed to be Jan Stephen van Calcar. A monkey is shown in the bottom left-hand corner, a dog to the right, and cherubs hold aloft the Vesalian family coat of arms depicting three weasels. The work also gives Vesalius' title of Professor at Padua, and his gratitude to his Imperial Majesty and the Venetian State. Today, nearly 500 years later, the exquisite detail of the picture produced by an engraved woodcut cannot fail to impress. Interestingly, some copies of the book are also known to be bound in human skin – which apparently looks like leather.

In many ways, the *Fabrica* is a damning rebuttal of Galenic anatomy. Indeed, Vesalius does not hold back in his criticism from the beginning by pointing out in the preface that Galen only worked with animals and was 'deceived' by monkeys. Yet, it is far too simple to see the book as a total rejection of Galenism. In fact, much of the *Fabrica* borrows heavily from Galen whom Vesalius describes as the Prince of Physicians. Vesalius also acknowledges Galen when he asks his readers not to consider him 'disloyal and too little respectful of authority toward that author of all good things'. Thus, rather than being a new beginning, the *Fabrica* is a synthesis that integrates old with the new. Even so, the great authority of Galen had been undermined. This was the case when Vesalius pointed out that the *rete mirable* did not exist in humans – a momentous observation for it invalidated much of Galen's physiology concerning the manufacture of the psychic pneuma. However, there were many other embarrassing errors. Vesalius showed that the liver was not the source of the body's veins as ancient tradition insisted, and he challenged Galen's view that blood can pass between the right and left chambers of the heart through tiny holes. Vesalius even disproved religious scripture by showing that men and women each have 24 ribs. Bizarrely, it was still generally believed at the time that men had one rib less than women because of the tale in Genesis that God took a rib from Adam to make Eve. However, in pointing out these errors, the real achievement of the *Fabrica* was not to diminish Galen (or the Bible): it was rather to point the way forward for the free investigation of nature and the pursuit of truth.

The *Fabrica*'s depiction of the brain

The chapters of the *Fabrica* of most interest to the neuroscientist are the fourth which concerns the nervous system, and the seventh with its 15 portrayals of the human brain. The first illustration of the fourth chapter is typical of the book's general style. This shows a detailed and life-like drawing of the brain visualised from below with a number of structures clearly delineated (see Figure 3.12a). These include the stumps of the cranial nerves (Vesalius follows Galen and divides them into seven pairs) along with the cerebellum (B), the upper part of the brainstem or dorsal medulla (D) and the crossing of the optic nerves (H). The first illustration of the seventh chapter is similarly impressive (Figure 3.12b) showing a head with the roof of the skull removed to leave the brain exposed to observation. It depicts the tough fibrous outer layer of the brain or the dura (A and B) along with the middle meningeal artery passing over its surface (D). Another blood vessel called the superior sagittal sinus (C) can also be seen running from front to back over the top of the brain. In a second plate this sinus is shown to be connected with the cerebral veins.

A striking feature of the *Fabrica* is the way it makes a series of horizontal slices through the skull with the brain left *in situ*. This way of cutting the brain from top to bottom is not only effective in exposing the ventricles at different levels, but helps to reveal a number of internal structures. For example, in Figure 3.12c, the two anterior ventricles are clearly exposed (L and M) along with the brain's white matter (E, F, G and H), grey matter (D) and portions of the corpus callosum (I) which Vesalius correctly recognises as connecting the two hemispheres. A section cut below this level (Figure 3.12d) also shows the cerebellum, cerebral aqueduct (connecting the middle and posterior ventricles) and pineal gland. However, some structures in the *Fabrica* are not mentioned in the text, but can nevertheless be clearly seen in the illustrations.

These include the first ever depictions of an egg-shaped mass of grey matter located in the centre of the brain now called the thalamus; a striped group of nuclei buried under the cerebral hemispheres called the striatum; and a description of the midbrain including the colliculi. Vesalius also outlined a number of pathways including ones passing to the cerebellum and a band running on the surface of the thalamus (stria terminalis).

Vesalius' ideas concerning the functioning of the brain were tied to the prevailing pneumatic beliefs of his day and consequently less innovative. Although scathing of the Cell Doctrine which attributed the ventricles with different intellectual faculties, Vesalius admitted he was unable to form any opinion as to how the brain performed such activities as imagination, reasoning and memory. He also doubted whether an understanding of their function, or what he called the 'Reigning Soul', could ever be achieved through anatomical

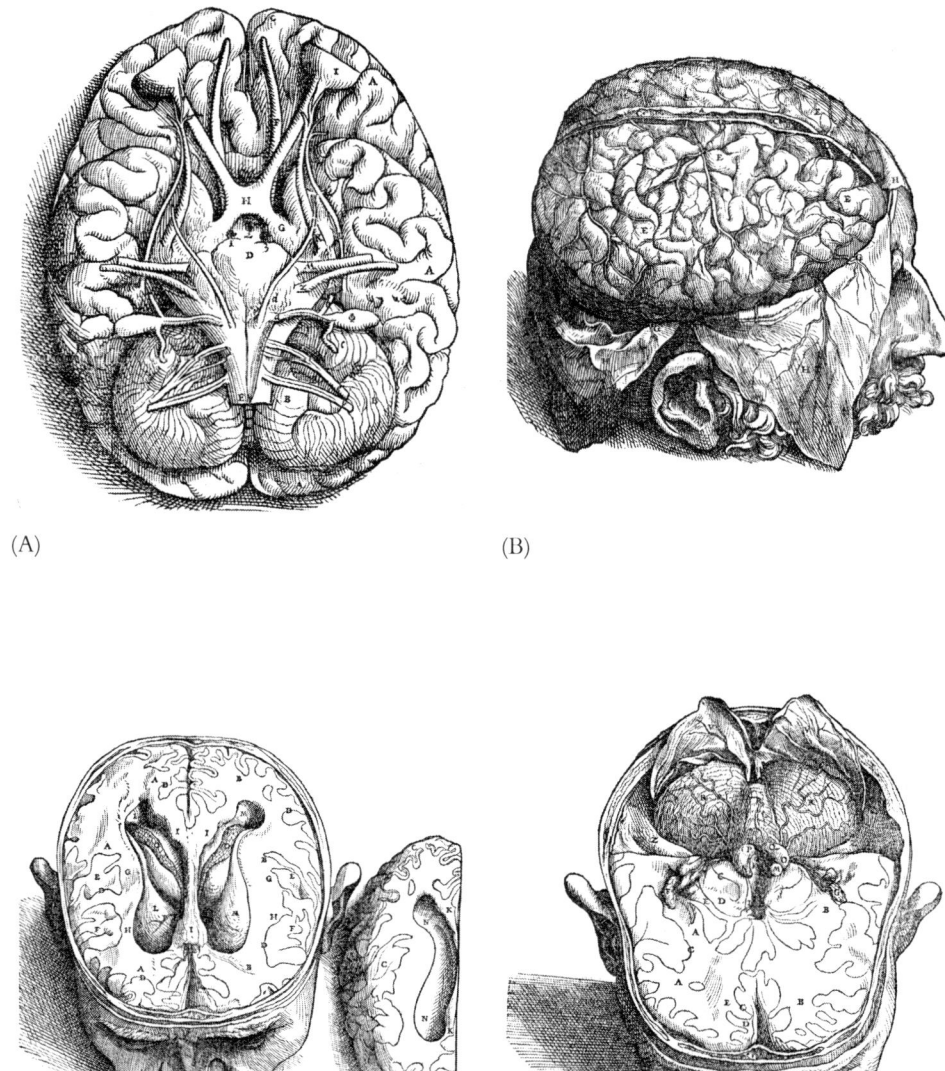

FIGURE 3.12 A series of four diagrams from the *Fabrica*: (A) the base of the brain from which the brainstem and cranial nerves can be seen, (B) the surface of the dura with some of its blood vessels, (C) the two anterior ventricles, (D) the upper surface of the cerebellum with cerebral aqueduct in the brain's centre. The pineal gland (P) is just to the side of this duct.

Source: Wellcome Library, London.

investigations alone. Noticing the human brain was similar in appearance to those of other animals with no special cavity or part, Vesalius was prepared, however, to explain its greater intelligence to its larger size. However, with no new perspective to guide him, Vesalius had to accept the basic tenets of Galen's physiology, despite disproving the existence of the rete mirable. This can be seen by his description of the ventricles as 'cavities and spaces in which the inhaled air, added to the vital spirit from the heart is . . . transformed into animal spirit.' Vesalius also believed that 'most delicate' animal spirits were distributed to the nerves where they served 'as the busy attendants of the brain'. Despite this, Vesalius admitted he could find no evidence to show that the nerves were hollow.

The impact of the *Fabrica*

The publication of the *Fabrica* rendered all previous anatomical works out of date, broke the dogma of ancient tradition, and set new objective standards for further progress. It might have been expected, therefore, that the book would be widely embraced. However, the *Fabrica* provoked mixed reactions, many of them unfavourable, especially from those who were not prepared to relinquish their long held beliefs in Galenism. The most virulent attack on Vesalius came from his former teacher in Paris, Jacobus Sylvius, who immediately called on him to renounce his heretical statements, confess his mistakes and declare publicly his faults. When these demands were ignored, Sylvius accused Vesalius of madness, and pleaded for royal intervention by asking the King that 'Vesalius be heavily punished and in every way restrained lest by his pestilent breath he poison the rest of Europe.' Several years later, in 1551, Sylvius continued his tirade by writing a book[19] whose aim was to expose 'the error-ridden filth' of that 'insolent and ignorant slanderer who has treasonably attacked his teachers with violent mendacity.' These criticisms appear to have had an effect on Vesalius, for within a year of *Fabrica's* publication, seemingly in a fit of depression, he burned the book's manuscripts and left Padua to become personal physician to the Holy Roman Emperor (Charles V) in Spain. Now concerned with his King's ailing health and travelling extensively, Vesalius would make no more original contributions to anatomy except correcting the second edition of *Fabrica* in 1555. He was just 30 years old. Despite its great expense, the *Fabrica* had sold reasonably well. An even greater success was its abridged version entitled the *Epitome* (first published in the same year as the *Fabrica*) which was written for students. A mere six chapters long with nine illustrations, printed on poorer paper but of larger size to make the details clearer, the *Epitome* was translated into French and German where it was widely read.

In 1562, after 11 years at the royal court, Vesalius resigned his post to undertake a pilgrimage to Jerusalem. The reasons for this decision are shrouded in mystery, with one source stating that Vesalius was forced to leave Europe after being accused of murder, following a botched autopsy on a noblewoman who was found to be still alive – an event attracting the interest of the Spanish Inquisition.[20] There are other tales, however, suggesting Vesalius was simply weary of his court service in Spain. Whatever the truth of the matter, Vesalius appears to have arrived in the Holy Land, only to be invited back to Padua to take up the position of professor, following the death of his successor Fallopius.[21] Tragically, he never completed the journey for his ship would be wrecked off the island of Zante (now Zákinthos) during a storm. Again, much confusion surrounds this event. One account holds that Vesalius was washed ashore only to be overcome by fever where he died alone in a vile and impoverished inn, whereas another states he died closer to the main town of Zante where a German honoured

FIGURE 3.13 Woodcut portrait of Andreas Vesalius (1514–1564) taken from the *Fabrica* of 1543.
Source: Wellcome Library, London.

him by erecting a monument in his memory. Although the second account may be more plausible, the monument has never been found. This may be due to the high incidence of earthquakes on Zante, one of which in 1893 was known to be particularly destructive since it left no building standing. Sadly, we will probably never know the final resting place of the founder of modern human anatomy.

Notes

1. This was encouraged by Mohammed who said, 'He who leaves his home in search of knowledge walks in the path of God.'
2. Historians disagree how much was known. One view is that a few scholars knew the *Timaeus* of Plato, a few logical excerpts from Aristotle, and the works of Isadore of Seville. However, these had relatively little impact on the intellectual life of the West.
3. Not to be confused by the more famous Posidonius (also confusingly of Apamea) who lived in the second century BC and was proclaimed in the Greco-Roman world as the greatest polymath of his age.
4. There is still some ambiguity about what Nemesius actually meant to say and the interested reader is referred to Green, C.D. (2003) *Journal of the History of the Behavioural Sciences*, 39, 131–142.
5. The original Latin word *'universitas'* refers to a group of persons associated with a society, although it would come to mean a guild of scholars. The earliest university is generally recognised to be Bologna, which was founded in 1088. Other early universities include Paris (1150), Oxford (1167) and Cambridge (1209).
6. A term first used by Aristotle as the place where all the senses come together.
7. This is far from certain. If true, then one would expect to find at least some evidence of their existence in ancient sources.
8. Leonardo also provides the first known clear illustration of a human fetus in utero.
9. However, Leonardo often tried to keep much of his work secret. To do this he sometimes wrote from right to left in an inverted form that could only be read by using a mirror, while punctuating his comments with a number of cryptic terms and pictures.
10. Leonardo also reports that it was easier to dissect the dry thin body of an old person than a young, fat and succulent one.
11. Or as he expressed it: 'It seems that the soul resides in this organ . . . called common sense. It is not spread throughout the body as many have thought but is entirely one part because if it were all pervading and the same in every part, there would be need to make the organs of the senses converge. The common sense is the seat of the soul.'
12. On the lower part of the page, the drawing also shows the head in transverse section, as if dissected at the level of the eyes.
13. It appears he got the idea after reading how an elephant could be slaughtered by driving a sharp spike into the back of its neck.
14. Commentary with very many additions on the anatomy of Mondino published with his original elegant text.
15. A quarto is a book printed on a sheet of paper folded twice to produce four leaves (or eight pages), each leaf a *fourth* the size of the original sheet printed.
16. It has been suggested that his boyhood interest in anatomy may help explain the drawings of the little fat cherubs that adorn the front pages of the *Fabrica*'s chapters.
17. One of the most common eponyms in neuroscience is *Sylvius*. The two most common uses of this term today is *Sylvian fissure*, which refers to the sulcus separating the temporal and parietal/frontal lobes, and the *Sylvian aqueduct*, which connects the third ventricle with the fourth. Neither of these, however, can be attributed to Jacobus Sylvius.
18. In 1543, Vesalius would perform his most famous public dissection in Basle on a notorious criminal called Jakob Karrer von Gebweiler who was beheaded for murder. With the help of a surgeon, Vesalius reassembled the bones of the skeleton. It is now on display at the Vesalianum Museum in Basle and believed to be the oldest anatomical specimen in the world.
19. *A Refutation of the Slanders of a Madman Against the Writings of Hippocrates and Galen.*

20 This story is found in a letter, written in Paris, 1556, by someone called Hubertus Languetus to Kaspar Peucer.
21 Again the truth of this story is uncertain as no documentation exists in the Paduan archives to confirm it.

Bibliography

Ashley-Montagu, M.F. (1955) Vesalius and the Galenists. *The Scientific Monthly*, 80, 230–239.
Bay, N.S.Y and Bay, B.H. (2010) Da Vinci's anatomy. *Journal of Morphological Science*, 27, 11–13.
Calkins, C.M., Franciosi, J.P. and Kolesari, G.L. (1999) Human anatomical science and illustration. *Clinical Anatomy*, 12, 120–129.
Castiglioni, A. (1935) The medical school at Padua and the renaissance of medicine. *Annals of Medical History*, 7, 214–227.
Castiglioni, A. (1943) Andreas Vesalius: Professor at the medical school of Padua. *Bulletin of the new Academy of Medicine*, 19, 766–777
Cavalcanti, D.D. (2009) Anatomy, technology, art and culture: towards a realistic perspective of the brain. *Neurosurgical Focus*, 27.
Choulant, L. (1920) *History and Bibliography of Anatomical Illustration*. University of Chicago Press: Chicago, IL.
Clark, K. (1958) *Leonardo da Vinci*. Penguin: Harmondsworth.
Clarke, C.D. (1949) *Illustration: Its Technique and Application to the Sciences*. Standard Arts Press: Butler, MD.
Clarke, E. (1962) The early history of the cerebral ventricles. *Transactions and Studies of the College of Physicians of Philadelphia*, 30, 85–89.
Clarke, E. and Dewhurst, K. (1972) *An Illustrated History of Brain Function*. University of California Press: Berkeley, CA.
Da Vinci, L. (1983) *Leonardo Da Vinci: Leonardo on the Human Body* (C.D O'Malley and J.B. de C.M. Saunders, Trans). Dover: New York.
De Gutierrez-Mahoney, C.G. and Schechter, M.M. (1972) The myth of the rete mirable in man. *Neuroradiology*, 4, 141–158.
De Lint, J.G. (1926) *Atlas of the History of Medicine. I. Anatomy*. H.K. Lewis: London.
Di Ieva, A., Tschabitscher, M., Prada, F., Gaetani, P., Aimar, E., Pisano, P., Levi, D., Nicassio, N., Serra, S., Tancioni, F., Arosio, M., Rodriguez y Baena, R., Di Ieva, A., Tschabitscher, M., Prada, F., Gaetani, P., Aimar, E., Pisano, P., Levi, D., Nicassio, N., Serra, S., Tancioni, F., Arosio, M. and Rodriguez y Baena, R. (2007) The neuroanatomical plates of Guido Vigevano. *Neurosurgical Focus*, 23 (1), E15.
Del Maestro, R.F. (1998) Leonardo da Vinci: the search for the soul. *Journal of Neurosurgery*, 89, 874–887.
Donaldson, I.M.L. (2008) Jacopo Berengario da Carpi: the first anatomy book with a complete series of illustrations. *Journal of the Royal College of Physicians*, 38, 375.
French, R.K. (1978) The thorax in history 3. Beginning of the Middle Ages. *Thorax*, 33, 295–306.
Green, C.D. (2003) Where did the ventricular localisation of mental faculties come from? *Journal of History of the Behavioral Sciences*, 39, 131–142.
Green, C.D. and Groff, P.R. (2003) *Early Psychological Thought: Ancient Accounts of Mind and Soul*. Praeger: London.
Gross, C.G. (1998) *Brain, Vision, Memory*. MIT Press: Cambridge, MA.
Gumpert, M. (1948) Vesalius: discoverer of the human body. *Scientific American*, 178, 24–31.
Herrlinger, R. (1970) *History of Medical Illustration*. Pitman Medical and Scientific: The Netherlands.
Infusino, M.H., Win, D. and O'Neill, Y.V. (1995) Modino's book and the human body. *Vesalius*, 1, 71–76.
Ione, A. (2010) Visual images and neurobiological illustration. In Finger, S., Boller, F. and Tyler, K. L. (eds) *Handbook of Clinical Neurology*, vol. 95. Elsevier: Amsterdam.
Joffe, S.N. (2009) *Andreas Vesalius: The Making, The madman, and The Myth*. Persona Books: Bloomington, IN.

Keele, K.D. (1964) Leonardo Da Vinci's influence on Renaissance anatomy. *Medical History*, 8, 360–370.
Kemp, S. (1996) *Cognitive Psychology in the Middle Ages*. Greenwood Press: London.
Lassek, A.M. (1958) *Human Dissection: Its Drama and Struggle*. Charles C. Thomas: Springfield, IL.
Lind, L.J. (1959) *Jacopo Berengario Da Capri. A Short Introduction to Anatomy (Isagogae Brevis)*. University of Chicago Press: Chicago, IL.
Linden, D.E. (2002) Five hundred years of brain images. *Archives of Neurology*, 59, 308–313.
Loechel, W.E. (1960) The history of medical illustration. *Bulletin of the Medical Library Association*, 48, 168–171.
Locy, W.A. (1911) Anatomical illustration before Vesalius. *Journal of Morphology*, 22, 945–988.
Manzoni, T. (1998) The cerebral ventricles, the animal spirits and the dawn of brain localisation of function. *Archives Italiennes de Biologie*, 136, 103–152.
Marenbon, J. (2007) *Medieval Philosophy: An Historical and Philosophical Introduction*. Routledge: London.
Nutton, V. (2003) Historical introduction. In Vesalius, A., Garrison, D. and Hast, M. *On the Fabric of the Human Body: An annotated translation of the 1543 and 1555 editions of Andrea Vesalius' De Humani Corporis Fabrica*. Karger: Basel.
Olry, R. (1997) Medieval Neuroanatomy: the text of Mondino dei Luzzi and the plates of Guido da Vigevano. *Journal of the History of Neurosciences*, 6, 113–123.
Pagel, W. (1958) Medieval and renaissance contributions to knowledge of the brain and its functions. In F.M.L. Poynter (ed.) *The Brain and its Functions*. Blackwell: Oxford.
Pevsner, J. (2002) Leonardo da Vinci's contributions to neuroscience. *Trends in Neurosciences*, 25, 217–220.
Quin, C.E. (1994) The soul and pneuma in the function of the nervous system after Galen. *Proceedings of the Royal Society of Medicine*, 87, 393–395.
Randall, J.H. (1953) The place of Leonardo Da Vinci in the emergence of modern science. *Journal of Historical Ideas*, 14, 191–202.
Russell, G.A. (2010) After Galen: late antiquity and the Islamic world. In Finger, S., Boller, F. and Tyler, K. L. (eds) *Handbook of Clinical Neurology*, vol. 95. Elsevier: Amsterdam.
Shanks, N.J. and Al-Kalai, D. (1984) Arabian medicine in the Middle Ages. *Journal of the Royal Society of Medicine*, 77, 60–65.
Simeone, F.A. (1984) Andreas Vesalius: anatomist, surgeon, court palatine and pilgrim. *American Journal of Surgery*, 147, 432–440.
Singer, C. (1956) Brain dissection before Vesalius. *Journal of History and Medicine*, 11, 261–274.
Singer, C (1957) *A Short History of Anatomy and Physiology from the Greeks to Harvey*. Dover: New York.
Singer, C. and Ashworth Underwood, E. (1962) *A Short History of Medicine*. Clarendon Press: Oxford.
Sironi, V.A. (2011) The mechanics of the brain. *Progress in Neuroscience*, 1, 15–26.
Smith, C.U.M., Frixione, E., Finger, S., and Clower, W. (2012) *The Animal Spirit Doctrine and the Origins of Neuropsychology*. Oxford University Press: New York.
Sudhoff, K. (1908) *Ein Beitrag zur Geschichte der Anatomie der mittelater speziell der anatomischen Graphik*. J.A. Barth: Leipzig.
Sudhoff, K. (1914) Graphische Darstellungen innerer Korperogane. *Arch. Gesch. Med.*, 7, 367–372.
Tan, S.Y. and Yeow, M.E. (2003) Medicine in stamps: Andreas Vesalius (1514–1564): father of modern anatomy. *Singapore Medicine*, 44, 229–230.
Tascioglu, A.O. and Tascioglu, A.B. (2005) Ventricular anatomy: illustrations and concepts from antiquity to renaissance. *Neuroanatomy*, 4, 57–63.
Vesalius, A. (1543) *Vesalius on the Human Brain (Being a Translation of a Section of his Fabrica of 1543)*. Welcome Historical Medical Museum. Oxford University Press: Oxford.
Vesalius, A. (1950) *The Illustrations from the Works of Andreas Vesalius* (J.B. de C.M. Saunders and C.D. O'Malley Trans). Dover: New York.
Whitaker, H. (2007) Was medieval cell doctrine more modern than we thought? In H. Cohen and B. Stemmer (eds) *Consciousness and Cognition*. Academic Press: London.
White, M. (2000) *Leonardo: The First Scientist*. Abacus: London.

4

SEARCHING FOR THE GHOST IN THE MACHINE

Cogito ergo sum ('I think, therefore I am')

Descartes

The brain is the source of sense and motion. It is the storehouse of thought and memory as well. Yet no diligent contemplation of its structure will tell us how so coarse a substance (a mere pulp, and that not over-nicely wrought) should subserve so noble an end.

Thomas Willis

Summary

The late sixteenth and early seventeenth centuries were a period of extraordinary intellectual ferment with far-reaching changes in the advancement of science. It also provided a decisive break with the past and ushered in the beginnings of the modern world, as old teachings were replaced by fresh knowledge obtained through experimentation and empirical endeavour. Accompanying this development was the beginnings of the industrial society, which would lead to further technological progress. One can point to many reasons for these changes, although underpinning it was a philosophical scepticism, as personified by Descartes, which laid a new foundation for the discovery of truth. Certainly in the decades after Descartes, many new thinkers questioned old authority and used scientific investigation to understand a wide range of phenomena. Within a century the new zeitgeist would achieve its most spectacular success when Newtonian physics was able to construct a sound mechanical and mathematical description of the universe, which remains viable in some respects today. However, similar concepts were also applied to biological processes. Again, Descartes was instrumental in this development for he had invented the 'reflex' – an idea that could explain how the body, and brain, acted in a mechanical way without an animating spirit to guide it. Nonetheless, Descartes was still living in a theological age, and despite the increasing emphasis on experimentation, most physiologists were not prepared to completely exorcise the soul from physiological functioning. This can be most readily seen in the work of Thomas Willis, whose *Cerebri anatome* published in 1664, provided the most thorough anatomical and physiological account of the brain, based on much practical innovation. Yet, the book was dedicated to the Archbishop of Canterbury, for the motivation behind this great work was to learn more about the soul. Thus, the seventeenth century was a time when two opposing views emerged concerning the brain: the mechanists believing the actions of the body were understandable through the laws of physics and chemistry; and the vitalists arguing for a unique soul-like force that infused the body with motion and life. This debate would continue well into the eighteenth century and its arguments became polarised by the disagreement between Albrecht von Haller who advocated the mechanical concepts of sensibility and irritability, and Robert Whytt who believed in a soul-like sentient principle. In short, this was a period when very different views existed concerning how the brain and nervous system worked.

Descartes: a new foundation for science

René Descartes (1596–1650) is recognised by many as the first truly modern thinker whose rationalism greatly contributed to the demise of the intellectual assumptions underlying the Scholastic-Aristotelian philosophy of his times. Indeed, his profound scepticism in relation to all traditional knowledge would enable him to replace old ideas with a new more objective search for the truth, and promote a very different mechanistic and scientific view of the world. Born in the small French town of La Haye, some 35 miles south of Tours in central France,[1] Descartes was the son of a prosperous lawyer who bequeathed him a generous annual income for life. At the age of ten, Descartes was sent to the celebrated Jesuit College of La Flèche, where he acquired a mastery of Latin and mathematics. Because of his delicate health, Descartes was allowed to spend mornings in bed, meditating and reading – a habit he is said to have continued for the rest of his life. Small in stature, with a large head, and always dressed in black, Descartes left La Flèche at 19 to study law at University of Poitiers, graduating in 1616. Keen to broaden his knowledge by travelling, which in those days could be achieved through voluntary military service, Descartes joined the army of Prince Maurice of Nassau, and a little later the Bavarian Military Service. Both of these forces were engaged in Europe's Thirty Years War. It was during his Bavarian service, while 'shut up in a stove-heated room' in a remote location on the Danube in 1619, that Descartes experienced a series of three dreams. These he interpreted to be a divine revelation telling him to devote the rest of his life to pursuing truth and determining the reality of things. Descartes decided to begin with philosophy, since the principles of all other sciences must be derived from it.

Following his conversion, Descartes continued to travel throughout Europe. Although his activities over the next nine years are not well documented, it is known he lived some of the time in Paris, and devoted himself to scientific pursuits including optics and mathematics. He also started writing a work entitled *Rules*, which he later abandoned. By the end of 1628, tired of travelling, Descartes moved to the Netherlands where he had the freedom to devote himself to writing. Here he would spend nearly 20 years, living in 13 different towns, and 24 addresses only known to a few confidants. The fruits of this endeavour were a number of books that became famous throughout Europe. Descartes had planned the first to be *Le Monde* (The World) concerned with showing how mechanistic physics could explain the vast array of natural phenomena in the world without referring to vague Aristotelian forms and principles.² This was a radical proposal. Aristotle's work was encyclopaedic in range, based on logic rather than experiment, and widely considered as infallible. It formed the basis of much university education and was supported by the church. To question it was paramount to turning the world upside-down, but such daring views were not without precedent. For example, in Italy, Galileo had questioned Aristotle by showing that different weights fall to Earth at the same speed,³ and was using mathematics to describe various types of physical phenomena such as the pendulum's swing. However, as *Le Monde* neared completion in the early 1630s, Descartes heard Galileo had been arrested and condemned by the Inquisition for supporting the Copernican view that the Sun lay at the centre of the universe. Faced with a potentially similar fate, Descartes suspended the publication of *Le Monde*. It would not see the light of day until 1664, some 14 years after Descartes' death.

Consequently, the first book to be published by Descartes was *Discourse on Method* (1637), an attempt to provide a new foundation for scientific thought based on certain and irrefutable knowledge. This was still a bold step in the seventeenth century when questioning traditional doctrines could easily be construed as heretical. At the heart of his new philosophy lay Descartes' great distrust in what he had been taught, or even what he could observe from his senses. Indeed, he realised it was difficult to view the world from a new perspective free from any preconceived notions. Descartes illustrated this by asking how he could be sure there was not some evil demon intent on systematically fooling his mind with sensory tricks? Or, put another way, how could he be sure he was not living in an imaginary or dream world?⁴ Descartes' answer was simple but profound: the only thing he could be absolutely sure of was the power of his own thought. This would be most famously encapsulated in his expression *Je pense donc je suis* (I think, therefore I am) which later became better known as the Latin *Cogito; ergo sum*.⁵ It is the starting point of his philosophical system on which everything else is built. The phrase has become famous in Western philosophy, both for its scepticism and the impetus it gave to the development of new philosophical and scientific thought. The Discourse was followed by *Meditations of First Philosophy* in 1641, outlining a more detailed account of Descartes' thought for establishing scientific truth. It continues to be essential reading for philosophy students today.

Another central theme in the work of Descartes is his belief that mathematics has the power to discover irrefutable proofs. In fact, the Cartesian approach to science is not unlike Euclidean geometry which attempts to discover basic truths and then transform them into a greater synthesis. This was yet another radical development for the time, as mathematics was widely regarded as something that dealt with abstract things and not the real world. Nonetheless, his pursuit of mathematics led Descartes to invent co-ordinate geometry – a method in which a single point in space was expressed numerically enabling lines and curves to be written as algebraic equations. This breakthrough, which philosopher John Stuart Mill called 'the greatest single step ever made in the progress of the exact sciences' had many practical and theoretical implications.⁶ It also led to the invention of calculus by Newton and Leibniz, which proved to be an excellent tool for describing nature's basic laws. The implications of using mathematics to describe and understand the universe were profound. If one believed in a universe that could be characterised through numbers and equations, then it was a small step to view all physical phenomena as part of a larger mechanical system governed by mass and motion. This contrasted with the Aristotelian view where the heavenly bodies were alive and divine, and God dwelled at the circumference of the universe where apparently he produced motion by being loved!

Although Descartes set out to establish a new scientific method, he also wanted science and religion to be compatible – not least because he regarded himself as a Roman Catholic. Thus, in a number of his works, Descartes attempts to show there is no contradiction between faith and reason by proving God's existence. He did this through a number of arguments, including one where he stated God had to exist for any intention to deceive on his part would clearly be absurd. A more subtle version of this idea was also expressed in *Discourse* when Descartes implies the perfection of geometry shows God's involvement in mathematics. Yet, in opposition to the Christian teachings of the time, Descartes believed the workings of the universe were Godless. In his view, God created the mechanical universe, but left it to run, ticking away like a watch, without further intervention. In fact, Descartes explained the 'start-up' process of life by postulating that God had introduced a fixed quality of motion into the universe at the time of creation. The phenomenon of life could thus be explained by the redistribution of motion that occurred from the countless collisions between minute particles of matter. This was a dangerously atheist view and one religious authorities were not ready to embrace. Indeed, all of Descartes' works would be banned by the Catholic Church in the 1660s and placed on their Index of prohibited books (*Index librorum prohibitorum*).[7]

Cartesian dualism

If Descartes' mechanical conception of the physical world wasn't controversial enough, he also applied the idea to living things by making the startling claim that animals and other forms of life are machines (*automata*). From the earliest days of mankind, through the great civilisations of Egypt and Mesopotamia, to ancient Greece, the workings of the body had been explained by a spiritual or soul-like force which had been called the *psyche* by Aristotle. This was generally regarded as an invisible breath-like force that imparted life and imbrued the body with motion. In fact, Plato defined the soul as 'the motion which can set itself moving'. It was an idea maintained by Medieval and Renaissance thinkers – not least because there was no viable alternative. Thus, living things were believed to contain a special vital energy, separate from their physical parts, which animated them. To think otherwise would have been unimaginable. Incredibly, Descartes broke with this long-held entrenched idea. For him, living things (with the exception of humans) were automated devices, containing parts similar to a mechanism of a watch, which operated together to produce movement. In other words, there was no *psyche* inside the machine to work them.

In *L'homme*, a book published in 1662, 12 years after its author's death, Descartes explained how this revolutionary idea came to him as a young man when walking in the Royal Gardens of Saint-Germain-en-Laye near Paris. The gardens were famous for their mechanical statues, which acted in life-like ways whenever a person approached them – a reaction caused by pressure plates hidden under the ground that activated water jets to hydraulically activate the figures. This must have been quite a sight, for Descartes recounts the statures could play musical instruments, pronounce words, and even threaten the onlooker with a trident! Although others had also begun to consider mechanical explanations of the body, including William Harvey who famously described the human heart as a pump in 1628, Descartes took the idea and applied it as a general principle of life's physiology itself. It was a formative moment in the history of biology for it implied that an explanation of life was possible through material processes belong to the physical world, and not some mysterious vague spiritual force. If true, then life and its various biological processes were also understandable in terms of physics and chemistry, and above all by scientific investigation.

Of course, the nervous system and brain was no different from any other part of the body, and Descartes sought to find a physical explanation for its function. Like others before him, he was tied to the prevailing beliefs of his day, which in the case of the nervous system had been understood through the action of *animal spirits* stored in the ventricles and flowing through the nerves. However, instead of regarding this as a weightless, invisible and self-moving force, Descartes now altered its composition by regarding it as a subtle fluid composed of particles that were distilled in the blood. In other words, the *animal spirits* now became a physical

FIGURE 4.1 Portrait of René Descartes (1596–1650).

Source: Wellcome Library, London.

entity (i.e. a chemical liquid). He also believed the brain received two types of spirituous particle: large ones which provided it with nourishment, and small ones that were used up as nervous energy – the latter being stored in the brain's ventricles. Thus, Descartes had tried to demystify the *animal spirits* by regarding them as being nothing more than a special class of volatile chemical particle. It was the first ever materialistic theory of the nervous system.

However, Descartes was not prepared to go the whole way with his materialistic theory which if true would have made him a mere machine. Indeed, right at the heart of Descartes' philosophy is his unyielding conviction he is different, for he is capable of self-determining thought and action. In contemplating this issue in *Meditations* he asks a fundamental question: 'What is a thing that thinks'? And Descartes concludes by stating, 'It is a thing which doubts, understands, affirms, denies, wills, refuses, which also imagines and feels.' In other words, a thinking thing has a mind – which for Descartes is the same as the soul with the power of self-determination and consciousness. Thus, Descartes is forced to regard the mind as a substance which is different from the material form of the body. This view is famously known as Cartesian dualism and it rests on the assumption that two different types of substance exist in our world: the physical *(res extensa)* and the mental *(res cogitans)*. Physical substances are material, 'extended' in space, and include particles which can be observed and measured. Mental substances, in contrast, cannot be examined for they do not have any material basis or fixed location in space.

Cartesian dualism was a major advance for it provided a major new way of understanding the relationship between mind and body, which despite being some 400 years old, continues to retain some supporters today.[8] Although the concept of dualism was clearly inherent in the earlier philosophy of Plato, its formulation by Descartes is very different since he rejected the soul as the essential life force. Thus, in his view, the physiological operation of human beings, like other animals, is mechanical and belongs to the material world. Of course, we are also fundamentally different and what sets us apart from other animals according to Descartes is our immaterial soul, which allows us to reason and establish with certainty we exist. Thus, for Descartes, human beings have a mind free from the influence of physical mechanics, whereas animals are simply automatons controlled by events in the natural world. In fact, this belief would even lead Descartes to deny animals felt pain – believing that their cries were mere mechanically induced behaviours.

The first account of the reflex

Descartes' most enduring legacy to neuroscience, however, would not be his account of dualism, but his invention of the reflex concept. If the body is indeed a mechanism and capable of action without the will's intervention, then it follows that Descartes must provide an explanation of how this automation can occur. Although Descartes did not use the exact term, he came up with the reflex – a sequence of events in which a stimulus from the external world elicits an involuntary response from the body. This was an idea without precedent and Descartes elaborates on it in several of his books, with the most famous example given in *L'homme* where a person is shown drawing his foot abruptly away from fire (see Figure 4.2). To explain this withdrawal response, Descartes proposed a new theory of nerve function. Instead of assuming a nervous spiritual force moved the limb, Descartes argued that the nerves conveying the burning sensation to the brain were composed of taut 'delicate threads' which passed all the way to the ventricles. When the person's foot was burned, the sensation from

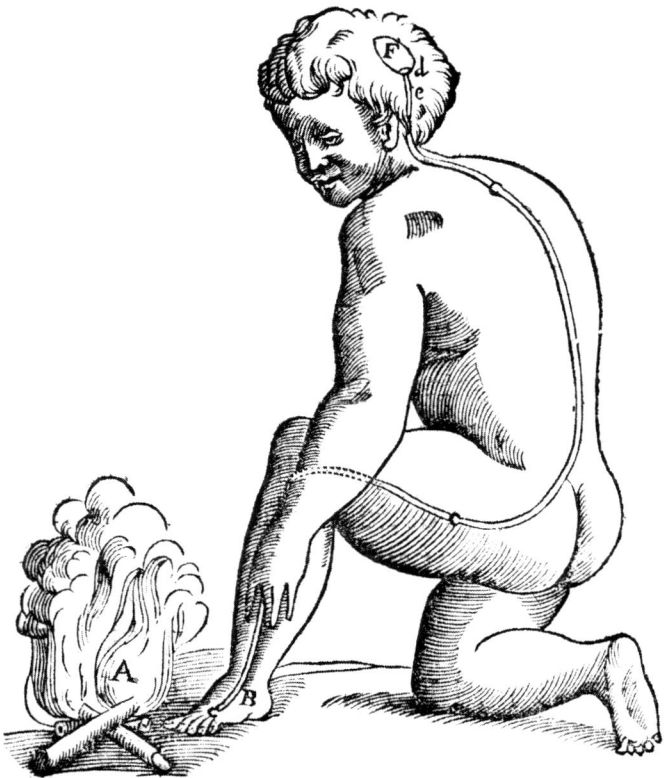

FIGURE 4.2 Descartes' illustration of the movement of 'animal spirits' in response to burning (from *L'homme* published in 1662). Note how the nerve impulse conveying the sense of fire is 'reflected' in the middle ventricle back along the nerve pathway to the foot.

Source: Wellcome Library, London.

the scalded skin pulled these threads tighter, opening tiny valves in the ventricular walls – a process Descartes likened to pulling on a cord at one end and causing a bell to ring at the other. The opening of these valves then caused a 'reflected' action involving the hydraulic release of *animal spirits* from the pressurised ventricles back into the nerves controlling the muscles of the foot. In turn, this caused the muscles to inflate leading to the foot's withdrawal.[9]

In some respects Descartes was extending an ancient theory of physiology first proposed by Galen, which viewed muscular movement occurring when *pneuma* was forced into the nerves from the ventricles of the brain. However, Descartes also introduced a new concept into the scheme by hypothesising the existence of valves in the ventricular walls which automatically caused a nervous message to be redirected back to generate movement. No self-governing spiritual force or *psyche* was needed to explain the behaviour. This may have been a simple advance over Galen, but it was a profound one as it offered a new approach by which to understand the actions of the nervous system. In fact, Descartes listed ten basic automated reflexes that he believed existed in ourselves and other animals. For the record these are (1) the digestion of food; (2) circulation of blood; (3) nourishment and growth; (4) respiration; (5) sleeping and waking; (6) sensory processes; (7) imagination; (8) memory;

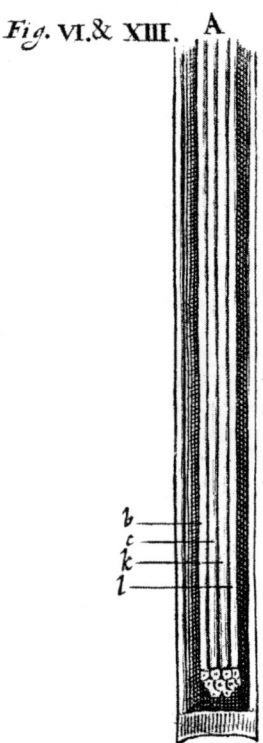

FIGURE 4.3 The internal structure of a nerve fibre as illustrated in *L'homme* (1662). The tubules labelled b to l are each described as containing a fine fibril that acts to convey sensations to the brain. These act as pulleys to open valves in the ventricles, which then allow animal spirits to pass down the outside of the nerve to the muscles.

Source: Wellcome Library, London.

(9) emotion, and (10) the movements of the body. These ideas were revolutionary. Reflexes not only implicated the nervous system in a far wider range of behaviours than previously supposed, but some such as imagination and memory were even involved in psychological processes traditionally associated with the mind. However, perhaps just as important, if all these reflexes were mechanical, then they were amenable to scientific investigation. These were powerful new concepts for the seventeenth century and they encouraged others to investigate the physiological workings of the body anew through experimental procedures. Thus, Descartes can be credited with establishing the modern scientific paradigm for examining the neurological basis of behaviour.

The site of mind–body interaction: the pineal gland

In formulating his account of dualism, Descartes had arrived at a position where the body was made up of physical matter, operating through reflexive action, while the mind was composed of an entirely different substance capable of its own self-determination. However, this raised a number of deep questions. For example, where does the soul exist? And, how does the 'soul-like' substance of the mind manage to interact with the material matter of the body? These are issues bedevilling all forms of dualism and Descartes was well aware of them. He first attempted to address the problem in *Passions of the Soul* where he argues there must be a specific site of interaction in the body between mind and body. Or as Descartes expressed it, there must be a place where the soul 'exercises its functions more particularly than in all the others.' Of course, this site had been regarded by all the great authorities as either located in the heart or brain. Descartes, however, proposed an alternative idea. For him, the soul's location was 'not the heart at all' or the 'whole brain'. Instead, it was 'a certain very small gland situated in the middle of the brain's substance . . .'. The structure that Descartes is referring to is the pineal gland.

The pineal gland was an anatomical structure with an ancient history. A smallish white structure which protrudes from the walls of the middle ventricle, it was first described by Galen who compared its shape to a pine cone. However, it is also clear from Galen that he was not the first to observe it, since others before him (perhaps Erasistratus) had attributed it with regulating the flow of *pneuma* through the ventricles. However, Galen disagreed, attributing little importance to the pineal gland, other than providing a support for the brain's blood vessels. Descartes would have been aware of Galen's writings, but he took a different view, more in line with his predecessors. For him the pineal gland occupied a strategic place within the ventricles, and stood out as the only singular organ in the brain, unlike all others which were paired. This was significant for Descartes who believed the soul was unitary and could not be divided into different parts. In addition, Descartes thought there had to be a place in the brain where the two images from the eyes came together to form a single sensation. Again, the pineal gland seemed to offer this site as he believed it was suspended in the middle ventricle.[10] Thus, it occupied an ideal position to detect animal spirits passing through the ventricular system, which were hypothesised to make small indentations on its surface.

But how does the pineal gland carry out the commands of the soul? To answer this, Descartes argued that the soul is able to cause fine and very rapid movements of the pineal gland – a response deflecting the flow of *animal spirits* towards the hollow nerves extending to the muscles. In this way, the pineal gland, could 'tilt' or 'pour' the *animal spirits* into the appropriate nerves. The idea was clearly reminiscent of the pipes controlling the hydraulic statues in the Royal Gardens. Sadly, Descartes' expertise in anatomy was limited. He was not only mistaken in his anatomical placement of the pineal gland at the centre of the middle ventricle, but by the seventeenth century it was also known the ventricles contained a watery substance and not an air-like one. The most damning indictment against the pineal theory, however, came from the Danish theologian Nicolas Steno who felt 'obliged to point out . . . the vast difference between Descartes' imaginary machine and the real machine of the human body'. Steno would show that the pineal gland was better developed in animals than in man – a finding inconsistent with the Cartesian notion of them having no soul. It seems Descartes was not entirely content with his pineal theory either, for in later years he would admit to its limitations in correspondence with friends. However, Descartes can be forgiven: the puzzle of how the mind interacts with the body remains arguably the most intractable problem in philosophy and brain science today.

The early years of the scientific revolution

In 1649, Descartes was invited to Stockholm by Queen Christina to become her private tutor. Sadly, he would soon succumb to the freezing climate and early morning tutorials, which required him to be in the Queen's attendance at 5 am in the morning. He died of pneumonia in 1650 at the age of 53.[11] By this time his books were becoming famous and his philosophy widely known. To give some indication of Descartes' influence, it is said that in the Netherlands around half of the physiologists considered themselves to be Cartesian, although this figure was far less in France where his teachings were banned. Descartes had nonetheless encouraged a time of intellectual change, assisted by innovative thinkers such as Copernicus, Galileo, Harvey and Vesalius, whose discoveries were shaking the foundations of traditional thought. And above all, Descartes had encouraged others to be sceptical of old truths and pursue new ones. However, his influence was only one of many contributing to this change. The gauntlet of new academic enquiry would be taken up by others, leading to what became known as the scientific revolution, which bought new methods of thinking and experimental investigation to bear on a wide range of subjects including physics, astronomy, anatomy and chemistry. It would also transform the Western world in a remarkably short space of time.

Another feature of the early seventeenth century was the development of amateur science, whose beginnings can be traced to the endeavours of a few wealthy gentleman sufficiently independent and curious about the wonders of the world to fund their own research. As their pursuit of 'natural philosophy' began to take place, there was a need to bring these individuals together in academic societies where they could present their findings. One of the earliest was the Royal Society, which is now believed to be the oldest extant scientific association in the world. The origins of this fellowship lie with a group of men that gathered for regular meetings at Oxford and London during the 1640s, who were determined to establish knowledge through experiment rather than rely on traditional accounts. In this quest, the founders decided to embrace the teachings of Sir Francis Bacon, an English statesman and philosopher, who had provided a damning indictment of scholastical authority in his *Novum organum scientiarum* (*New Instrument of Science*) published in 1620. Sometimes called the father of empiricism (i.e. the view that direct experience is the only true source of knowledge), Bacon stressed the importance of establishing facts from evidence-based argument including experimentation. For this reason he was highly critical of subjects such as astrology and alchemy for their lack of objective support. This was very different from the more popular Aristotelian approach which attempted to find fundamental truths through reason or simple argument alone.[12]

The Royal Society would include many of Britain's greatest scientists including Robert Boyle, Robert Hooke and Sir Christopher Wren. Another famous member and president was Sir Isaac Newton, author of *Philosophiae naturalis principia mathematica* (often just called the *Principia*), which provided the first complete mathematical exposition of time, force and motion (i.e. the universal laws of nature). Perhaps more famously, after watching an apple fall from a tree, Newton reasoned that there must be an attractive force called gravity. From this he went on to show that the gravitational force between two large objects is proportional to the product of their masses and weakens with the square of the distance between them. These concepts replaced Cartesian physics and solved virtually all the theoretical issues of his day, while proving seminal in the modern development of physics and astronomy. If there was any doubt about the power of mathematics to explain the world in the seventeenth century, it was dispelled by Newton who showed his methods to be far superior to anything that had gone before. Published less than 40 years after Descartes' death, the *Principia* marks a point where the widespread acceptance of the modern scientific method finally occurs. Yet, for all his scientific genius, Newton was a deeply religious man who ultimately saw his scientific work as an attempt to understand the plans of his divine creator. As he explained: 'Gravity explains the motions of the planets, but it cannot explain who set the planets in motion. God governs all things and knows all that is or can be done.'

Thomas Willis: the father of modern neurology

One of the most distinguished members of the Royal Society during its formative years was the English physician Thomas Willis. Born only 25 years after the birth of Descartes (their lives would overlap), Willis was not a mechanist but an iatrochemist, who held that chemistry was the basis of human function. He was also a doctor with a brief and unconventional medical education, although this would not stop him from becoming Sedleian Professor of Natural Philosophy at Oxford. In fact, Willis would use this opportunity to embark on an ambitious project to 'open heads and study their contents'. The result was the first ever monograph devoted exclusively to the brain, spinal cord and peripheral nerves entitled *Cerebri anatome* (1664). This work, assisted by a group of highly skilled colleagues known as the Vertuosi, combined among other things anatomical knowledge, autopsies, clinical observation and experimentation. Not only was it the first interdisciplinary investigation of the brain, but the book also introduced the Greek word *neurolgia* later translated into English as 'neurology' by Samuel Pordage. *Cerebri anatome* abounds in so much new information, and presents such an enormous contrast with the vaguer efforts of the past, that it was to remain authoritative for the next two centuries, and go through 23 editions.

A staunch churchman and Royalist during the English Civil War, Willis enrolled at Oxford in 1637 to study medicine, which at the time normally required taking several degrees over the course of 14 years. However his formal medical education was curtailed to only six months when Charles I moved his court into Oxford in 1642, turning it into a garrison town.[13] Despite his limited training, Willis was conferred his Bachelor of Medicine in 1646, allowing him to set up a doctor's practice. However, his royalist sympathies meant that he had few privileges after the fall of the monarchy in 1645, with Willis having to work long and hard to establish his reputation. Often travelling to nearby villages to treat his patients, it is even said he worked for a time as a 'piss prophet' at Abingdon market. Nor was Willis an imposing figure, with one account describing him as having middle stature, dark red hair resembling a pig, and a stammer. However, his public standing was to be greatly enhanced in 1650 when he revived a woman called Anne Green who had been publicly hung for murdering her young baby. Left to hang for half an hour, the woman's body had been left in a surgeon's house for 'anatomisation' when Willis heard a strange noise coming from the coffin. Willis then bought the woman 'back to life' by tickling her throat with a feather to provoke coughing. Anne Green not only survived, but married again to bear three more children.

The mid seventeenth century was a time when infectious diseases were common, with outbreaks of smallpox, influenza, and meningitis regularly affecting the inhabitants of Oxford and its surroundings. As a young doctor Willis described these outbreaks and examined their pathology. It would lead him to provide the first ever account of typhoid fever. Willis was also struck by the ways in which the mind was affected in feverish conditions noting, for example, how people stricken with meningitis were so impaired in their intellect, that they took leave of their friends. Because Willis had treated many of his patients over the course of several years and gained their trust, some agreed to his request of an autopsy. In the case of meningitis, Willis found the brains were often coated with a thick bloody mess with widespread signs of inflammation. Willis also developed a chemical theory of fever, rejecting the old Galenic notion of an imbalance in the humours, by replacing it with one where delirium arose from excessive fermentation. This may sound old-fashioned today, but it basically attributed fever to a disordered composition of the blood. For Willis, this also meant it was a condition that could be best treated with chemical cures such as Peruvian bark – instead of the older practice of blood-letting.

Fortune was to change for Willis following the restoration of Charles II in 1660, when a few months later he was appointed Sedleian Professor of Natural Philosophy at Christ Church. Now nearly 40 years old, Willis used his new position to learn more about neuroanatomy. Although Willis may have been encouraged to do this through his frustration at the inadequacy of knowledge concerning brain anatomy from the traditional accounts of Aristotle and Galen, a deeper motive ultimately influenced him. Willis was a conscientious Anglican, deeply interested in understanding God's divine wisdom and the nature of the human soul. Assuming the brain and nervous system formed part of an ordered system constructed by God, Willis believed its closer examination would hold the key to understanding the animal soul, along with the uniquely human one which was rational and immortal. Or as Willis put it, the study of anatomy can 'unlock the secret places of Man's Mind and [to] look into the living and breathing Chapel of the Deity.' Setting about his task with great fervour, Willis enlisted the help of several other experts including Robert Boyle, Richard Lower and Christopher Wren. Willis was also keen to use all the corpses at his disposal, and later confessed he had become 'addicted . . . to the opening of heads.' Fortunately, the University of Oxford had a relaxed view of dissection, for it had been given permission by Charles I in 1636 to claim the body of anyone who was executed with 21 miles of Oxford for this specific purpose. Willis took full advantage of this situation, often performing his operations in the inns and houses where the dead bodies lay.

Cerebri anatome

Cerebri anatome: cui accessit nervorum descriptio et usus (Anatomy of the brain and the description and use of the nerves) was published in 1664, dedicated to Willis's benefactor and friend, the Archbishop of Canterbury Gilbert Sheldon. If the reader was still in doubt about the book's underlying pious spirit, *Cerebri anatome* is also written in Latin – the language of religious authority. Although strictly speaking it was not the first book to be concerned exclusively with the brain and nerves (the credit for this goes to a Dutchman called Jason Pratensis who published De Cerebris Morbis in 1549), *Cerebri anatome* is far more comprehensive, providing an unsurpassed account of the brain with its emphasis on new experimental findings and clinical insights. Indeed, as Willis proudly boasted in its preface: 'I have not trod the paths or footsteps of others, nor repeated what has been told before.' This was a bold proclamation for it must be remembered *Cerebri anatome* was written during a time when the teachings of Hippocrates, Aristotle and Galen still dominated medical teaching. Nonetheless, the times were changing, as demonstrated by the fact that many of the traditional Aristotelians had been removed from their posts during the Royalist occupation of Oxford. These circumstances, along with Willis's support from people in high places, undoubtedly gave him the freedom to follow his own inclinations free from restraint.

FIGURE 4.4 (A) Engraving of Thomas Willis when he was 45 years of age by the artist D. Loggan, which was first used in Pathologiae Cerebri (1667). (B) Title page of *Cerebri anatome* first published in 1664.

Source: Both images provided by the Wellcome Library, London.

The *Cerebri anatome* is highly innovative throughout its 150 folio pages or so, and this is in no small part due to the collaborators who assisted Willis. These included Christopher Wren who drew many of the illustrations (actually engraved on copper plate which provided far superior detail than wood cuts), Robert Boyle who found new ways of preserving and hardening the brain with alcohol,[14] and Richard Lower responsible for much of the dissection. Unlike most other anatomists who had left the brain encased in its cranium, Lower and Willis removed the brain from the skull – a procedure which allowed them to examine it more carefully with magnifying glasses. Consequently, the *anatome*'s 15 plates of illustrations were to show the brain and nervous system in unsurpassed detail. It also led to a plethora of new brain areas being identified, and following Samuel Pordage's 1681 translation, a number of new anglicised terms entering the vocabulary which are still in use today. Indeed, many students are probably unaware that terms such as *anterior commissure, corpus striatum, inferior olives* and *stria terminalis* derive from Willis. Other neuroanatomical terms derived from Pordage's translation include *lobe, hemisphere, peduncle* and *pyramid*.

Perhaps the most celebrated highlight of *Cerebri anatome* is its depiction of the main blood vessels at the brain's base. These, of course, had proven controversial in the past, with Galen confusing them with a similar complex found in oxen called the *rete mirable* – a mistake only rectified by Vesalius in 1543. Willis confirmed that Vesalius was indeed correct by comparing the blood vessels from a human and several types of animal. Interestingly, while the human brain did not contain a *rete mirable*, it did exhibit a large ring of arteries that received blood from the two carotid arteries which coursed up the front of the neck. This arterial circle had

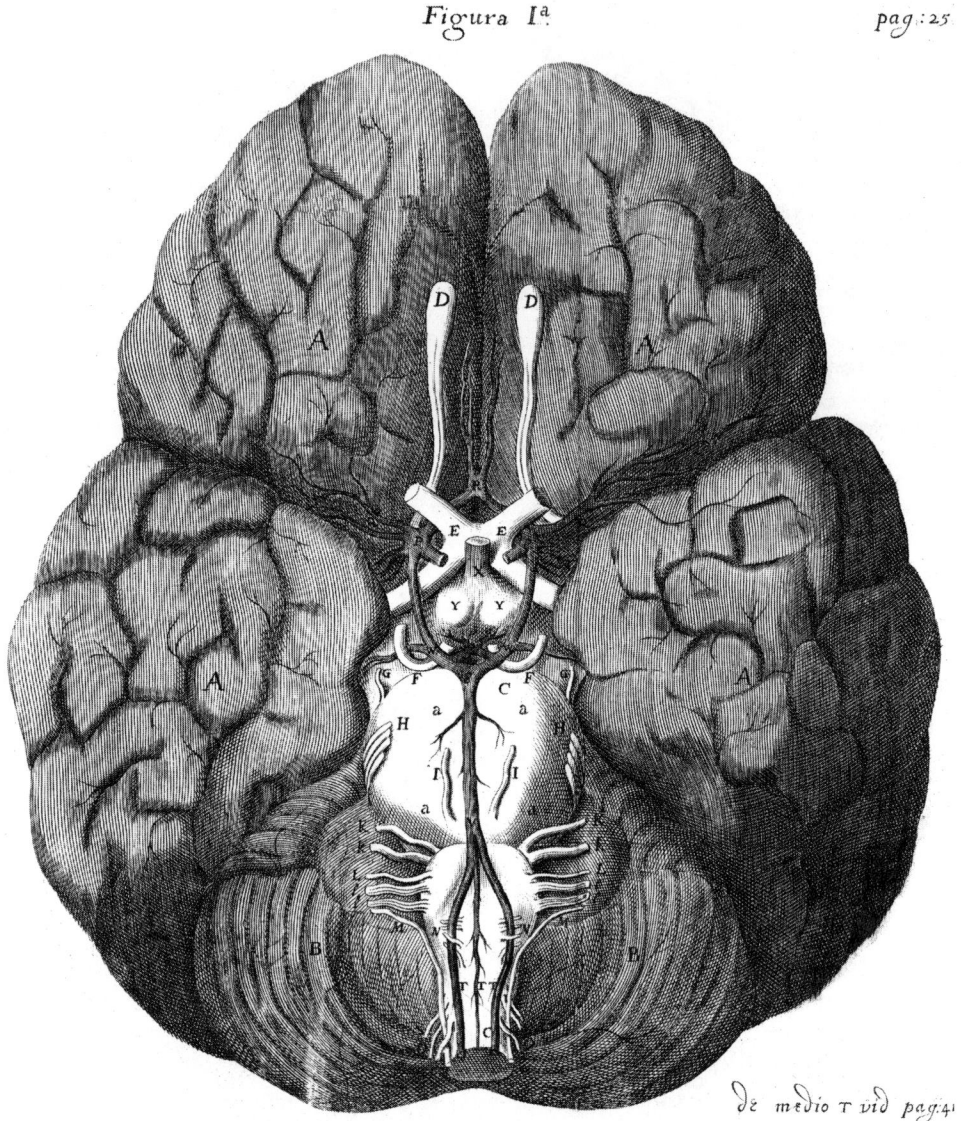

FIGURE 4.5 Figure from *Cerebri anatome* showing the base of the brain drawn by Christopher Wren. It illustrates the circle of Willis (E), the olfactory bulbs (D), the corpora quadrigemina (Y), and the cerebellum (B). The origin of ten pairs of cranial nerves can also be seen.

Source: Wellcome Library, London.

also been observed by Vesalius, but it was Willis who recognised its functional significance – reporting that he had come across an instance in one of his autopsies where an artery was blocked, but the individual enjoyed 'the free exercise of his mind and animal function.' From this, he realised the arteries were arranged in a ring for a very good reason: they ensured a constant supply of blood to the brain even when part of it was occluded. Today this structure is known as the Circle of Willis.[15]

Another impressive feature of *Cerebri anatome* was the use of new anatomical techniques which were used to trace the course and destination of blood vessels throughout the brain. This was a period when the blood supply to the brain was poorly understood. The circulation of the blood had only been established by William Harvey some 20 years previously, and cerebral haemorrhage was just beginning to be recognised as a cause of stroke by the German Johann Wepfer whose *Treatise on Apoplexy* first appeared in 1658. Despite this, there were many who still held onto the ancient idea that apoplexy arose from a blockage of animal spirits rather than blood. Willis provided convincing evidence against this idea, however, when he injected ink (using a syringe made from a quill and a dog's bladder) into the carotid arteries of a man who had died from cancer – one of the first times that this type of experimental procedure had been undertaken on a human brain. According to Willis the ink appeared as small dark spots on the outer cerebral cortex of the brain with 'the vessels creeping into every corner and secret place of the brain.' It also caused the brain to 'appear like a curious quilted ball'. There was little doubt following this demonstration that blood flowed into the brain from the arteries, which was then distributed throughout its tissues in a vast network of very fine vessels.

Cerebri anatome also provided researchers with a new outline of the brain's overall structure. Traditionally, since Aristotle, the brain had been divided into two regions: the cerebrum (cerebral cortex) and cerebellum (Willis refers to these as the *cerebra* and *cerebel* respectively).[16] However, in *Cerebri anatome*, Willis adds a third component called the corpora striata – so called because of its streaked grey and white appearance. Although this structure can be seen in the earlier work of Vesalius, he had not named it. Lying nestled under the cerebrum, Willis compared the corpora striata to a crossroads by Willis where sensory and motor spirits came together (see below). It should not be forgotten that about a third of *Cerebri anatome* also describes the peripheral nervous system, with this work being credited to Richard Lower, who 'indefatigably' followed the branches of the nerves to their 'most hidden places'. From his work, Willis was able to identify a total of nine cranial nerves, and provide the first account of the spinal accessory nerve controlling the neck muscles. Willis also named the vagus nerve (vagus being derived from the Latin 'to wander') which he showed had extensive projections throughout the upper body. He also gave the name '*solar plexus*' to the radiating nerve fibres of the abdomen as they reminded him of the sun's rays.

Localising functions in different brain areas

Cerebri anatome provided a new approach to understanding the brain in another important respect: Willis minimised the importance of the ventricles and placed mental functions in the brain's matter instead. While this was not entirely new as Galen had done something similar, Willis went much further and localised sensory, motor and intellectual functions to specific brain areas. Willis was particularly dismissive of the ventricles, regarding them as no more than empty watery spaces resulting from the folding of the brain. In fact, he thought they had no role other than to act as 'mere sinks' to collect waste material. And as evidence, Willis noticed a funnel-shaped passage leading out of the middle ventricle, which he believed passed excrement into the pituitary gland. This he referred to as the 'arse-hole' of the brain![17] Thus, the brain's *animal spirits* had to exist somewhere else, and Willis came up with an ingenious theory: he argued they were 'procreated' in its grey substance. Willis arrived at this idea when he examined the grey and white matter of the brain with a magnifying lens.

The grey matter, which is most conspicuous in the brain's outer cortical surfaces, appeared to have a rich supply of blood vessels, making it the ideal site in his view to produce *animal spirits*. In contrast, the white matter was much more fibrous, leading Willis to deduce it must provide the pathways by which the spirits reach the rest of the brain and spinal cord.

By focusing on the brain itself, Willis began to consider the functions of its three main regions: the *cerebri* (i.e. cerebral cortex); *cerebel* (i.e. the cerebellum) and *corpus striatum*. It led him to construct a surprisingly modern theory. The grey regions of the *cerebri* were for Willis the site responsible for intelligence and memory. As support, Willis noted that *cerebri* was larger in humans, and had many more 'folds and rollings' (i.e. gyri and fissures). He also correctly realised this enlarged the brain's surface area, especially in contrast to the flat and featureless cortical surface representative of birds and fish. The *cerebri* also contained the *corpus callosum*[18] which allowed animal spirits to 'travel from one hemisphere of the brain to the other, and return back again'. Willis also attributed this structure with fantasy and imagination. However, the *cerebel* located at the back of the brain, close to the neck, was different. This bulbous structure was large in both animals and man, and had a very similar pattern of folding. Willis knew the ancient authorities had sometimes attributed the posterior ventricle close to the *cerebel* with movement, and it appears his own clinical observations supported this general idea. Willis would come to recognise the *cerebel* as an organ where automated movements of the body were produced. He made this quite clear: 'The duty of the *cerebel* . . . is to supply animal spirits to certain nerves by which involuntary actions take place 'such as the heartbeat, normal respiration, concoction of aliment, production of chyme and many others.'

The most original aspect of Willis's localisation theory concerned the *corpus striatum*. For Willis, these two grey oval-shaped bodies 'streaked like ivory' occupied a pivotal position in the brain's architecture. Sitting on top of the upper reaches of the *medulla* – a Y-shaped structure which Willis referred to as the 'King's highway' in and out of the brain – the *corpus striatum* lay nestled under the folds of the *cerebri*. This allowed the *corpus striatum* to receive intellectual and behavioural commands from the *cerebri* lying above it. But just as important, Willis also came to regard it as a region which obtained impressions directly from the senses. To support his theory, Willis noted that the *striatum* of newly born blind puppies is not striped, indicating that visual experience is a necessary prerequisite for its development. In other words, the *corpus striatum* appeared to be the *sensus communis*, the fabled place first invented by Aristotle where all the senses merged. This theory was further strengthened by the *striatum's* location at the higher reaches of the King's highway – which made it an ideal site where willed action was executed, with its orders being channelled into nerve pathways passing down into the spinal cord. To back this idea, Willis also reported a case of striatal degeneration in one of his brain dissections where the individual had suffered from paralysis of the limbs.

Willis's abandonment of the ventricular or Cell Doctrine in favour of a theory favouring localised brain function was a major step forward in the history of the brain sciences. The theory was also innovative in other ways. For example, Willis believed the *cerebri* could exert executive control over the activities of the 'involuntary' *cerebel* – as occurs when one holds one's breath. Thus, Willis was also the first to give the brain a hierarchy of different functions. Moreover, while Willis was not a Cartesian, he embraced the reflex concept, and attributed the *cerebri* and *cerebel* with different types of response. For example, 'A gentle titillation of the skin' which caused someone to rub their skin was believed to be a reflected action involving the *cerebri*, whereas a painful stimulus leading to a quick pulse and heavy breathing was mediated by the *cerebel*. Willis's view of brain function was therefore both dynamic and integrative – concepts that are very much part of the modern lexicon of brain research today.

The rational soul, animal spirits and nervous activity

Willis's concept of the soul is one following in the tradition of Aristotle and Galen. That is, he attributes it with having three parts. Its most basic component was the *flamma vitalis*, which arises in the blood to generate life-giving heat. In the second part were the *animal spirits*, created when blood flowed over the surface of the brain, which then became stored as a liquid in the grey matter. Both the *flamma vitalis* and *animal spirits* were believed to be chemical in nature, existing in man and beast alike, and responsible for performing basic biological functions such as sensation and motion. However, the third type of spirit proposed by Willis was very different. This was the immortal and immaterial *rational soul*, found only in humans, capable of reason and free-will. The *rational soul* also had executive power to move the *animal spirits*. However, unlike his predecessors, Willis was the first to place the 'primary seat' of this higher spirit in the cerebral cortex (*cerebri*) – a conclusion he arrived at because of its greater structural complexity in man. The cerebral cortex also contained the *corpus callosum*, which Willis believed received sensory impressions and was responsible for the power of imagination.

Whereas Descartes had used mechanical principles involving threads and hydraulic forces to explain nervous function, Willis was an iatrochemist who relied on using chemical principles. However, this type of chemistry was very different from the type we understand today. Willis believed, for example, that all bodies are composed of five particles which in order of importance were spirit, sulphur, salt, water and earth. When these particles became mixed, 'fermentations' arose generating heat and movement. In turn, these reactions could explain all kinds of biochemical processes including the circulation of the blood, fevers, and digestion. To explain nervous activity, Willis argued that the nerves contained two types of liquid. The first was a 'nerve juice' created in the brain and channelled into the spinal cord where it 'irrigated' the nervous system. The second liquid contained *animal spirits* whose 'subtle corpuscles' were highly volatile which moved quickly through the less mobile nervous juice. Their motion was always set in motion by the rational soul located in the brain, likened to a breath moving upon waters. The destination of these *animal spirits* were the muscles where they combined with a spirit-like substance to create a reaction 'like the explosion of gunpowder.' In turn, this caused an effervescent air-like substance to be produced, making the muscles swell and produce movement. Despite this, Willis admitted he could find no evidence for the existence of hollow nerves to support his theory. In fact, Willis had been one of the first to use the microscope to scrutinise the nervous system, and from what he could make out, nerve fibres were compact. They were like sugar cane – a solid yet porous material not unlike a sponge.

The microscopic world revealed

The seventeenth century was also a time when microscopes were increasingly utilised to examine the fine structure of biological material. Although 'magnifying glasses' had been used in Roman times, and spectacles invented as early as the thirteenth century, the compound microscope was a later invention – possibly first developed in 1590 by two Dutch lens makers called Zaccharias and Hans Janssen. It would gradually lead to the availability of the first rudimentary microscopes with focusing devices by the time Thomas Willis was beginning his investigations at Oxford. In fact, one year after *Cerebri anatome*, the first book devoted to microscopic illustration entitled *Micrographia* (1665) was published by his colleague Robert Hooke. Using a microscope with a magnification of about ×50, *Micrographia* caused a sensation with its spectacular copperplate engravings. These included a tip of a needle resembling a craggy mountain, a detailed illustration of a flea, and a large picture (nearly two foot across) of a louse clinging on to a human hair. However, its most lasting claim to fame is Hooke's description of a slither of cork revealing thousands of tiny holes, which he thought resembled little rooms or 'cells'.[19] Little could Hooke have realised these holes were actually the remnants of the building blocks of life in the once living cork tree. Nor could have Hooke ever imagined that biologists would use his term to refer to the basic unit of both plant and animal tissue.

Considering his close association with Willis, it is curious Hooke did not report any inspection of nervous tissue in *Micrographia*. However, others were to do so, including the Italian Marcelo Malpighi, who examined brain matter in his 1665 treatise *De cerebro*. His observations appeared to show that the outer grey regions of the cerebral cortex contained a mass of tiny oval-shaped glands which extended like tiny intestines into the white matter. From this, Malpighi reasoned that these glands gave rise to long fine channels filled with a liquid that he assumed were 'vital spirits'. These observations have led some experts to claim that Malpighi can be credited as being the first person to observe nerve cells. However, others argue he was only describing blood vessels. Sadly, we will never know for certain what Malpighi actually saw in the brain, but we can be more definite about some of his observations.

For example, he described the structure of the lungs and its network of capillaries, allowing him to work out how air was able to enter the blood system. And, he was to find capillaries elsewhere in the body, a discovery leading him to realise the arterial and venous blood supplies are connected. In this way Malpighi completed the missing link in Harvey's theory of circulation. For these, and many other discoveries, Malpighi is often acclaimed as the founder of microscopic anatomy.

However, the greatest microscopist of all came from a more humble background. He was a draper from Delft in Holland called Anton van Leeuwenhoek with no formal scientific training and only able to speak Dutch. Around 1668, aged 36, Leeuwenhoek discovered a new way of making lenses, or rather glass spheres, which were far superior to those used by others. Closely guarding the secrets of his technique, Leeuwenhoek added to the mystery by claiming he was able to grind a lens from a grain of sand (which is almost certainly not true). Nonetheless, his lenses had a far greater magnification than any other at the time, and nobody would come close to matching them for another 200 years. Although the compound microscopes in the seventeenth century were rarely able to produce a magnification greater than about ×50, Leeuwenhoek constructed a more powerful instrument by placing the object to be examined on a fixed pin and using a tiny single lens clamped between two metal plates. The specimen was then bought into focus by adjusting a screw on the plates. While it was a simple device and needed to be held close to the eye in good light, it provided a magnification of up to ×200. It is possible that Leeuwenhoek even possessed some microscopes, which were significantly more powerful.

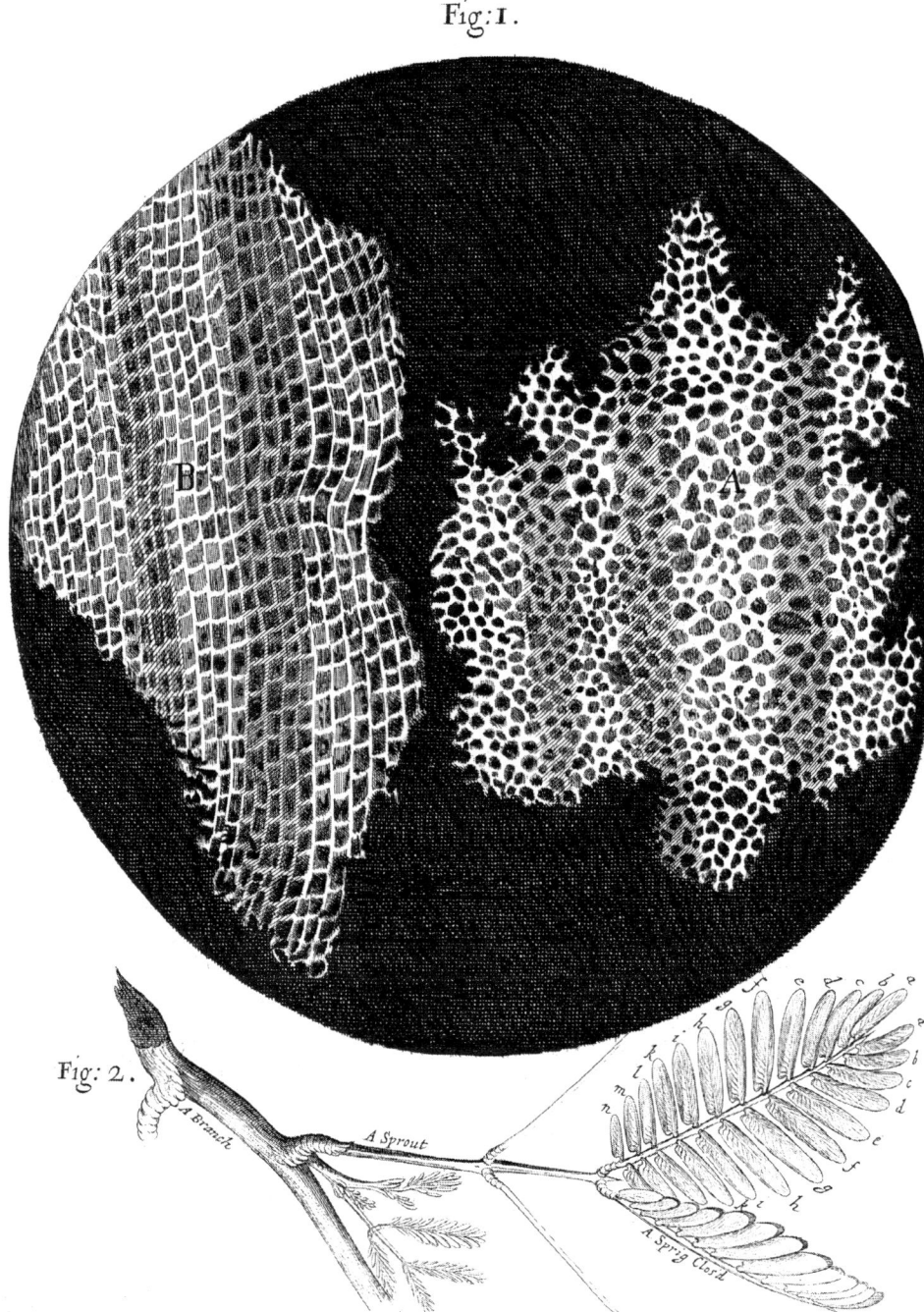

FIGURE 4.6 A slither of cork showing hundreds of little rooms or 'cells' as illustrated in *Micrographia* (1665). The word 'cell' would become an enduring one in the history of biology.

Source: Wellcome Library, London.

In 1673, Leeuwenhoek began reporting his findings to the Royal Society who published the letters in their journals. This correspondence would continue for over 50 years and result in the publication of 375 papers. Among these were a number of notable discoveries. These include the first description of single-celled organisms which he referred to as animalcules (some of which he found in the scrapings from his teeth), and red blood cells whose diameter he correctly gauged. Leeuwenhoek also showed muscles were made from fibres composed from thousands of finer filaments. Most memorably, he found that human sperm contained large numbers of little animalcules, less than a millionth the size of a coarse piece of sand, with thin undulating transparent tails. Leeuwenhoek also examined nervous tissue and sent a sectioned cow's optic nerve to the Royal Society in 1674. Like Willis, he was unable to show it was hollow. Rather, it appeared to be solid and contain many soft globules. Ten years later in 1684, Leeuwenhoek focused on the cortical area of turkey brain. Again his findings revealed nothing new as they confirmed Malpighi's own observations of rather featureless oily globules.

A test of the animal spirits theory

In the seventieth century all explanations of nervous function relied on the concept of animal spirits – essentially a type of liquid which flowed through nerve pathways into the muscles. For some theorists, such as Descartes, this caused them to expand, much like the pumping up of a balloon to produce movement. However, the seventeenth century was also a scientific age, and some investigators following Descartes realised this was an idea amenable to experimental investigation. One such person was the Cambridge physician Francis Glisson. Most famous for writing the first book exclusively dedicated to the liver (*Anatomia hepatis* in 1659) Glisson was particularly intrigued by the liver's relationship with the gall bladder, which seemingly discharged bile on its own accord without any nervous input or stimulus. This led him to propose the gland's muscles must have their own intrinsic energy which he called 'irritability'. However, this concept led to some further interesting predictions. For example, if irritability was found in the gall bladder, then why not other muscles of the body? More intriguingly, if irritability was intrinsic to the muscle, then Glisson realised it should not necessarily require the hydraulic inflow of animal spirits to invoke its action. In other words, the theory of irritability predicted that no muscle inflation should occur.

To test this hypothesis, Glisson placed a man's arm into an enclosed water container and asked him to open and close his fists. If the inflation theory was correct this action should have caused a fluctuating increase in the level of water. However, the volume of water remained unchanged or even fell slightly, forcing Glisson to conclude that muscles are 'shortened by an intrinsic vital movement,' and they 'have no need of an abundant afflux of spirits, either animal or vital'. In other words, it appeared to disprove the notion of animal spirits. As an alternative Glisson proposed that irritability was an inherent force in all biological tissue and not just muscle. Thus, in his view, all parts of the body moved in response to 'irritation'. This was an extremely powerful new idea for the times since it also helped explain why animals such as worms, eels and snakes were still capable of exhibiting movement long after they had been sliced into parts. Moreover, irritability could also explain why the heart continued beating after being removed from the body.

A similar conclusion was reached in the same year (1662) by a young Dutchman and skilled microscopist called Jan Swammerdam from the University of Leiden. In addition to his many anatomical discoveries, including the presence of eggs in the female ovary, the existence of valves in the lymphatic vessels, and the influx of blood as a cause of penile erection, Swammerdam undertook some of the first ever experiments in neurophysiology. He did this by using a nerve-muscle preparation taken from the frog's leg. That is, he took a muscle from the thigh with nerve stump attached, and then stimulated the nerve to produce muscular contraction or twitching. Now familiar to students of biology the world over, the technique has been described as one of the most important experimental protocols of all time. Swammerdam's choice of a frog was also fortunate for another reason: he found the animal still exhibited reflexive bodily movements when deprived of its heart, but not when it was decapitated – an injury which caused all movement to cease. On first sight this appeared to show the brain as the source of all movement. Then, Swammerdam discovered something else: if he placed a scalpel against the nerve stump of an exposed muscle, even in a decapitated frog, the muscle twitched. Swammerdam had therefore shown muscular contraction could occur without any connection between muscle and brain. It was yet another finding contradicting the traditional notion of 'animal spirits' flowing through the nerves to produce movement.

Swammerdam also performed another classic experiment to test the theory of animal spirits. He took a thigh muscle from a frog (with nerve attached) and immersed it in an airtight tube containing a small amount of water in its tip. Muscular movement could then be judged by an observable rising air bubble. The logic was simple: if movement was due to the increased flow of animal spirits into the muscle, then it should have raised the volume of the water. In fact, this was basically a more sophisticated version of Glisson's experiment. However, when Swammerdam stimulated the nerve with a fine wire, no bubble movement occurred. In other words, the muscle contraction did not increase the water's volume. This was yet further evidence against the idea of animal spirits and Swammerdam was unable to provide a viable explanation, other than to concede that an 'irritation' of the nerve was sufficient to produce muscular motion. A similar conclusion was also reached by Swammerdam's Danish colleague Niels Stensen (often shortened to Steno) who had described the heart as a muscle in 1667. Steno went beyond Swammerdam by showing that muscles did not change their overall volume when contracting, but they did change shape (the muscle bulged at one end).

Albrecht von Haller: irritability and sensibility

The concept of irritability was to be more fully developed by the Swiss doctor Albrecht von Haller. One of the most influential medical figures during the eighteenth century, Haller was a renowned polymath who wrote more than 1200 scientific works relating to almost every branch of human knowledge.[20] This included two encyclopaedias, four treatises on anatomy, seven on botany, five on bibliography, one book of poems, two theological works and four historical novels. However, his greatest work is generally regarded as his massive *Elementa physiologiae corporis humani* (*Elements of Physiology*) published in eight volumes between 1757 and 1766. This was so thorough that it led the nineteenth century physiologist Magendie to complain every time he thought of something new, he found Haller had done it already. Haller also authored *Primae lineae physiologiae* (*First Lines of Physiology*) published in 1747, which became a standard student text in universities. A Professor of Medicine at the newly founded University of Göttingen in Germany for much of his career (17 years), Haller was also renowned as an experimentalist who attempted to substantiate his claims with empirical evidence, which also led him to perform animal experiments on a large scale.

Although Haller was a devout Christian who believed in a free and immortal human soul, he was a mechanist in his physiological thinking. Consequently, Haller was strongly opposed to the traditional idea of the soul governing or closely integrated with the working of the body. In fact, he was more influenced by Newton, believing that the human body operated according to the same fundamental physical and natural laws which governed the universe. And, in this respect, Haller chose to adopt the idea of irritability as a basic physiological force, which was purely mechanical, and operated independently of the soul. He outlined his position in a monograph entitled *On the Irritable and Sensible Parts of the Body* in 1752, which simply defined irritability as that part of the body which becomes shorter upon being touched. Hence, unlike Glisson who thought irritability was a characteristic of all bodily tissues, Haller believed it only occurred in muscle. The book also gave numerous examples of how muscle tissue retained the power of movement for some time after it had been dissected from the body. For Haller this was proof that muscle fibres contained a physical property (which he also called the *vis insita*) that reacted with great force to a nerve impulse or other type of stimulation.

However, it was also apparent to Haller that the nervous system exhibited a different type of physiological property. Although nerves did not move when stimulated, they obviously were active in some way. Haller called this force 'sensibility' and proposed it could convey sensory information to the brain where the soul was located, or trigger the contraction of muscle. Haller also referred to this force as the *vis nervosa* – essentially a new way of describing what had previously been referred to as animal spirits. Haller was intent, however, to point out there was nothing mysterious about this force. In fact, he described it as nothing more than a thin, invisible, tasteless, odourless and very mobile liquid. In other words, the *vis nervosa* was similar to that proposed by Descartes: a volatile hydraulic force belonging to the physical world. Once motioned by the will located in the brain, it flowed to the muscles which were particularly sensitive or easily 'irritated' by its effects.

The distinction between muscular irritability and nervous sensibility was an important step forward over the traditional view of animal spirits. It helped show the body was not animated by some type of soul-like force which had remained in vogue since the time of Aristotle, but a reactive mechanical system that responded to stimulation. This not only helped to modernise theories concerning the working of the nervous system, but it also provided an idea that could be experimentally tested. Put simply, one could examine which parts of the body contracted upon stimulation (i.e. show irritability), or caused the animal distress (i.e. show sensibility). In fact, Haller would perform this type of procedure in nearly 200 experiments on a variety of animals. After exposing a particular part of the body, Haller would investigate the region with various forms of stimulation. This included pricking with needles and the application of chemical irritants using splinters of wood. An animal crying out in pain demonstrated an area with sensibility, whereas contraction or movement provided evidence of 'irritability'. This work, published in 1752, perhaps not surprisingly, attracted considerable criticism for its barbarism. Even Haller had to admit he felt a 'cruelty for which I felt such reluctance, as could only be overcome by the desire to the benefit of mankind.'

Haller also used similar procedures on the brain. This work revealed the outer grey surface of the cerebral hemispheres and cerebellum was insensitive to stimulation, producing neither expressions of pain or signs of movement. In contrast, irritation of the white matter caused the animal to squeal out in distress and frequently led to violent convulsions. These effects were obtained from many areas of the brain including the thalamus and medulla. From this, Haller concluded the brain's white matter provided the place where sensory impressions were

FIGURE 4.7 An engraving of Albrecht von Haller (1708–1777) taken from the first volume of his *Elements of Physiology* (1747). It is interesting to note that the many images of Haller invariably show him wearing a wig.

Source: Wellcome Library, London.

appreciated and movement generated. However, because he was unable to discern any clear differences in function between all the regions of white matter, Haller came to regard the medulla (from where the cranial nerves and spinal cord arose) as the most likely site for the *sensus communis*. Or as he explained it: 'the seat of the mind must be where the nerve first begins its formation or origin.' Haller was aware this contradicted Willis who had localised different functions to the cerebral cortex, striatum and cerebellum. However, when Haller examined these areas closely with various types of stimulation, he could not find anything

106 Searching for the ghost in the machine

that supported localisation of function. Assuming his experimental method was better, Haller blamed the 'errors' made by Willis on his propensity for 'substituting analogy' instead of rigorous experimental observation.

Robert Whytt and the sentient principle

In spite of his comprehensive research, Haller did not succeed in removing animal spirits or vitalist concepts from physiology. Indeed, an opposing school of thought was also highly influential during this time, which followed the Aristotelian–Galenic tradition by contending all living matter contained a mysterious and non-material soul-like force which animated the

ROBERT WHYTT, 1714-1766,
Professor of Medicine in the University of Edinburgh, 1747-1766.
Photograph by Crawford of the portrait "after Belucci" in the Royal College of Physicians, Edinburgh.

FIGURE 4.8 Robert Whytt (1714–1766) who believed in a non-material force called the *sentient principle* which governed the operation of the body.

Source: Wellcome Library, London.

body. It was a view held by a number of prominent physiologists during the eighteenth century, although Haller's sternest critic would turn out to be the Scotsman Robert Whytt. A professor at the University of Edinburgh, Whytt had studied medicine in London, Paris and Leiden, and was to become the King's personal physician. Whytt was also a staunch vitalist who published a strong defence of his views in 1751.[21] This described the possibility of a mechanical physiological system producing heat, heart-beat, or other movement essential to life as 'a notion . . . too low and absurd to be embraced by any but the most minute philosophers.' Instead, he believed that a non-material force existed, called the *sentient principle*, which acted throughout the brain and nerves to give the body life and self-motion. Although he did not deny that muscle had the mechanical ability to contract, Whytt believed it was unable to do so unless set into motion by the 'active principle' flowing through the nerves.

Haller was well aware of Whytt's criticism, and he robustly defended his theory of irritability and sensibility in return. It would prove to be the start of a longstanding conflict between the two men until Whytt's death in 1766. Whytt's strongest rebuttal, however, came in 1755 when he described his own experiments in a treatise aimed at discrediting Haller's work.[22] Like Haller, Whytt accepted there were many instances where decapitated animals, or isolated body parts, moved and twitched for some time after death. This, of course, had been taken as evidence by Haller to prove the existence of irritability. But Whytt had a different explanation: he held these actions occurred because small amounts of the sentient principle remained in the muscles after death.

Importantly, Whytt supported his argument with two other discoveries. The first came when he pinched a nerve, or stimulated it by heat, to produce a muscle contraction. He found that the muscle always produced a more vigorous response if its nerve remained attached to the spinal cord. For Whytt, this only made sense if the sentient principle existed in the nervous system as his theory maintained. The second discovery came when he pithed (i.e. destroyed) a frog's brain. Although this initially caused the frog's body to be flaccid, Whytt found the animal would slowly return to a sitting posture and even exhibit certain reflexes. For example, when the foot of the brainless frog was pinched, it was able to retract the leg away from the source of stimulation. According to Whytt these responses did not support Haller's notion of irritability. But they did support the idea of a sentient principle still existing in the spinal cord. In fact, this finding led Whytt to go on to show that only a small part of the spinal cord was needed for these reflexes to occur. Whytt could not have known it, but this was the first clear demonstration of what we now recognise as a spinal reflex – a simple automated movement controlled by the spinal cord without any intervention from the brain.

The debate between Haller and Whytt was one which greatly polarised the physiologists of the eighteenth century. The key issue was easily defined: was the body and its various organs an automatic machine fully capable of executing and controlling all its operations as first proposed by Descartes? Or was there some indefinable soul-like or animating principle at work as first proposed by the ancient Greeks? This was a question with profound implications for understanding the nervous system and brain. However, unknown to Haller and Whytt, the question was soon to take on a very different dimension with the discovery of intrinsic animal electricity by the Italian Luigi Galvani in the 1790s. Although this initially appeared to support the idea of an animating non-material force generated by the brain and flowing through the nervous system, an understanding of electricity would soon come to replace Whytt's sentient principle, and Haller's formulation of irritability and sensibility. It is to this story we now turn in the next chapter.

Notes

1. Now called La Haye-Descartes.
2. Among Aristotle's beliefs which shaped scientific thought up until the seventeenth century was the theory of the four elements (fire, water, earth and air) which was used to explain many physical phenomena. He also believed in the 'Unmoved Mover', an omnipotent God-like entity existing in the void beyond the stars that set into motion all matter. This maintained the movement of the heavens, keeping the Sun and planets travelling around the Earth.
3. Whether Galileo actually performed this visionary experiment from the leaning tower of Pisa is not known.
4. This problem is known today as the brain in the vat scenario. It is often illustrated by imagining a disembodied person whose isolated brain lies in a life-supporting vat. In theory, if the brain is wired to a supercomputer that stimulated its sensory nerve pathways in such a way to mimic reality, then the mind would live in an imaginary world. The idea has been explored in many ways including the movie The Matrix.
5. This expression first appears in his *Principles of Philosophy* (1644).
6. For example, it enabled the course of any object in space to be predicted given enough knowledge of its mass, speed and trajectory.
7. This list remained in use by the Catholic church until 1966.
8. A belief in Cartesian dualism has been held by some major figures in neuroscience including Charles Sherrington, John Carew Eccles, and Karl Popper. Another supporter of dualism is Roger Sperry, famous for his split-brain research.
9. It is important to note that Descartes believed that a given nerve pathway had both sensory and motor functions. Or, to be more specific, the 'threads' conveying the sensory impulses to the brain were thought by Descartes to be located in the inner pipes of the nerve fibre, and those sending information to the muscles in the outer pipes (see Figure 4.3).
10. The pineal gland is actually attached to the posterior wall of the middle ventricle.
11. Descartes was buried in Stockholm, although his body would be exhumed some 17 years later in 1667 and sent to Paris where it was interred. However, at his reburial, two parts of his body were missing: his skull and an index finger. The skull (allegedly that of Descartes) was later bought in 1821 at an auction in Sweden by a chemist called Johann Jakob Berzélius who donated it to the French Government. It is now kept at the Musée de l'Homme in Paris – apparently exhibited between a bust of prehistoric man and retired footballer Lilian Thuram.
12. In 1624 Bacon also wrote *New Atlantis*, a utopian novel, in which he described a scientific academy called 'Salomon's House' who trained scientists to conduct experiments and then apply the findings to the benefit of society. This is essentially the same idea as a modern science institute.
13. From the King's arrival in 1642 until its surrender in 1646, Oxford was the royalist capital of England, housing not only the King and his court, but also the central law-courts, the exchequer, Parliament, and a mint.
14. This is an important development. The brain is soft and easily damaged. It also deteriorates quickly. By perfusing the brain with alcohol, Willis and his colleagues were able to dissect it more effectively and examine it for longer.
15. Interestingly, the Circle of Willis shows considerable anatomical variation. In fact, the classic anatomy of the circle is only seen in about 35 per cent of cases.
16. In Willis's time, the understanding of these two main regions of the brain was very different from how we conceive them today. For example, the cerebrum included many subcortical centres that joined the two hemispheres, whereas the cerebellum included parts of the underlying brainstem and medulla which also ascended deep into the brain reaching as far as the corpus callosum and thalamus.
17. The idea of the ventricles acting as a 'sink' is essentially correct for they do act to remove metabolic waste material – although the main route of excretion is through the cerebro spinal fluid in the spinal cord.
18. It is important to note that Willis used the term *corpus callosum* to refer to all the medial cerebral white matter and not the central band of fibres, which connects the two hemispheres as it is designated today.
19. Hooke calculated the number of cells in a cubic inch to be 1,259,712,000.
20. Haller was also a member of the Royal Society and personal physician to George II.
21. Essay on the vital and other involuntary motions of animals (1751).
22. Observations on *The Sensibility and Irritability of the Parts of Men* (1755).

Bibliography

Akert, K. and Hammond, M.P. (1962) Emanuel Swedenborg (1688–1772) and his contribution to neurology. *Medical History*, 6, 255–266.
Allen, I.M. (1960) Seventeenth century neurology. *New Zealand Medical Journal*, 59, 241–246.
Arikha, N. (2006) Form and function in the early enlightenment. *Perspectives in Science*, 14, 153–188.
Bakkum, B.W. (2011) A Historical lesson from Franciscus Sylvius and Jacobus Sylvius. *Journal of Chiropractic Humanities*, 18, 94–98.
Bloch, H. (1988) Francis Glisson MD (1597–1677) The Glissonian irritability phenomenon and its roots, links and confirmation. *Southern Medical Journal*, 81, 1443–1436.
Cobb, M. (2002) Exorcizing the animal spirits: Jan Swammerdam on nerve function. *Nature Reviews: Neuroscience*, 3, 395–400.
Cottingham, J. (1986) *Descartes*. Blackwell: London.
Cottingham, J. (ed.) (1992) *The Cambridge Companion to Descartes*. Cambridge University Press: Cambridge.
Crombie, A.C. (1959) Descartes. *Scientific American*, 201, 160–173.
Cunningham, A. (2002) The pen and the sword: recovering the disciplinary identity of physiology and anatomy before 1800. I: Old physiology – the pen. *Studies in History and Philosophy of Biological and Biological Sciences*, 33, 631–665.
Cunningham, A. (2003) The pen and the sword: recovering the disciplinary identity of physiology and anatomy before 1800. II: Old anatomy – the sword. *Studies in History and Philosophy of Biological and Biological Sciences*, 34, 51–76.
Dewhurst, K. (1982) Thomas Willis and the foundations of British neurology. In Clifford Rose, F. and Bynum, W.F. (eds) *Historical Aspects of the Neurosciences*. Raven Press: New York.
Donaldson, I.M.L. (2009) The Treatise of Man (De Homme) by Rene Descartes. *Journal of the Royal College of Physicians of Edinburgh*, 39, 375–376.
Donaldson, I.M.L. (2010) Cerebri anatome: Thomas Willis and his circle. *Journal of the Royal College of Physicians of Edinburgh*, 40, 277–279.
Eadie, M.J. (2000) Robert Whytt and the pupils. *Journal of Clinical Neuroscience*, 7, 295–297.
Eadie, M.J. (2003) A pathology of animal spirits – the clinical neurology of Thomas Willis (1621–1675) Part I – background and disorders of intrinsically normal animal spirit. *Journal of Clinical Neuroscience*, 10, 14–29.
Eadie, M.J. (2003) A pathology of animal spirits – the clinical neurology of Thomas Willis (1621–1675) Part II – disorders of intrinsically abnormal animals spirits. *Journal of Clinical Neuroscience*, 10, 146–157.
Feindel, W. (1962) Thomas Willis (1621–1675) – the founder of neurology. *Canadian Medical Association Journal*, 87, 289–296.
French, R. (1969) *Robert Whytt, the Soul and Medicine*. Wellcome: London.
Frixone, E. (2006) Albrecht von Haller (1808–1777). *Journal of Neurology*, 253, 265–266.
Goodfield, G.J. (1960) *The Growth of Scientific Physiology*. Hutchinson: London.
Gross, C.G. (1997) Emanuel Swedenborg: a neuroscientist before his time. *The Neuroscientist*, 3, 142–147.
Haigh, E. (1984) Xavier Bichat and the medical theory of the eighteenth century. *Medical History Supplement no 4.*, pp. 47–65. London Wellcome Institute for the History of Medicine.
Hall, T.S. (trans) (1972) *Treatise of Man by Descartes*. Harvard University Press: Cambridge, MA.
Henry, J. (2008) *The Scientific Revolution and the Origins of Modern Science*. Palgrave-Macmillan: London.
Hierons, R. (1967) Willis's contributions to clinical medicine and neurology. *Journal of Neurological Science*, 4, 1–13.
Home, R.W. (1970) Electricity and the nervous fluid. *Journal of the History of Biology*, 3, 235–251.
Hughes, J.T. (2000) Thomas Willis (1621–1675). *Journal of Neurology*, 247, 151–152.
Isler, H. (2010) The development of neurology and the neurological sciences in the seventeenth century. In Finger, S., Boller, F. and Tyler, K. L. (eds) *Handbook of Clinical Neurology*, vol. 95. Elsevier: Amsterdam.

Jaynes, J. (1970) The problem of animate motion in the seventeenth century. *Journal of the History of Ideas*, 31, 219–234.

Lindskog, G.E. (1978) Albrecht von Haller: A bicentennial memoir. *Connecticut Medicine*, 42, 49–57.

McHenry, L.C. (1981) A history of strokes. *International Journal of Neurology*, 15, 314–326.

Martensen, R.L. (2004) *The Brain Takes Shape: An Early History*. Oxford University Press: Oxford.

Mazzolini, R.G. (1991) Schemes and models of the thinking machine (1662–1762). In Corsi, P. (ed.) *The Enchanted Loom: Chapters in the History of Neuroscience*. Oxford University Press: London.

Meyer, A. and Hierons, R. (1964) A note on Thomas Will' views on the corpus striatum and the internal capsule. *Journal of the Neurological Sciences*, 1, 547–554.

Molnar, Z. (2004) Thomas Willis (1621–1675) the founder of clinical neuroscience. *Nature Reviews: Neuroscience*, 5, 329–335.

Ochs, S. (2004) *A History of Nerve Function*. Cambridge University Press: Cambridge.

O'Connor, J.P.B. (2003) Thomas Willis and the background to Cerebri Anatome. *Journal of the Royal Society of Medicine*, 96, 139–143.

Pearn, J. (2002) A curious experiment: the paradigm switch from observation and speculation to experimentation, in the understanding of neuromuscular function and disease. *Neuromuscular Disorders*, 12, 600–607.

Porter, R. (2003) *Flesh in the Age of Reason*. Penguin: London.

Reynolds, E.H. (2005) Vis attractiva and vis nervosa. *Journal of Neurology, Neurosurgery and Psychiatry*, 76, 1711–1712.

Rousseau, G.S. (ed.) (1990) *The Languages of the Psyche: Mind and Body in Enlightenment Thought*. University of California Press: Los Angeles, CA.

Schwedenberg, T.H. (1960) The Swedenborg Manuscripts. *Archives of Neurology*, 2, 407–409.

Singer, C. (1957) *A Short History of Anatomy from the Greeks to Harvey*. Dover: New York.

Sloan, P.R. (1977) Descartes, the sceptics, and the rejection of vitalism in seventeenth-century physiology. *Studies in History and Philosophical Science*, 8, 1–28.

Spillane, J.D. (1981) *The Doctrine of the Nerves: Chapters in the History of Neurology*. Oxford University Press: Oxford.

Smith, C.U.M. (1998) Descartes' pineal neuropsychology. *Brain and Cognition*, 36, 57–72.

Smith, C.U.M. (2010) Understanding the nervous system in the eighteenth century. In Finger, S., Boller, F. and Tyler, K.L. (eds) *Handbook of Clinical Neurology*, Vol. 95. Elsevier: Amsterdam.

Swash, M. (2008) The innervation of muscle and the neuron theory. *Neuromuscular Disorders*, 18, 426–430.

Temkin, O. (1964) The classical roots of Glisson's doctrine of irritation. *Bulletin of the History of Medicine*, 38, 297–328.

Tubbs, R.S., Loukas, M., Shoja, M.M., Apaydin, N., Ardalan, M.R., Shokouhi, G. and Oakes, W.J. (2008) Constanzo Varolio (Constantius Varolius 1543–1575) and the pons Varolli. *Neurosurgery*, 62, 734–737.

Wallace, W. (2003) The vibrating nerve impulse in Newton, Willis and Gassendi: First steps in a mechanical theory of communication. *Brain and Cognition*, 51, 66–94.

Whitaker, H., Smith, C.U.M. and Finger, S. (eds) *Brain, Mind and Medicine: Essays in Eighteenth Century Neuroscience*. Spinger: New York.

Williams, A.N. (2003) Thomas Willis's practice of paediatric neurology and neurodisability. *Journal of the History of the Neurosciences*, 12, 350–367.

Willis, T. (1971) *The Anatomy of the Brain*. Tuckahoe, New York: USV Pharmaceutical Corporation.

Zago, S. and Meraviglia, M.V. (2009) Constanzo Varolio (1543–1575). *Journal of Neurology*, 256, 1195–1196.

Zimmer, C. (2004) *Soul Made Flesh: Thomas Willis, the English Civil War and the Mapping of the Mind*. William Heinemann: London.

5

A NEW LIFE FORCE

Animal electricity

In early September, at twilight, we placed . . . the frogs prepared in the usual manner horizontally over the railing. Their spinal cords were pierced by iron hooks, from which they were suspended. The hooks touched the iron bar. And, lo and behold, the frogs began to display spontaneous, irregular and frequent movements.

Luigi Galvani

If I do not greatly deceive myself, I have succeeded in realising . . . the hundred years dream of physicists and physiologists, to wit the identity of the nervous principle with electricity.

Emil Du Bois-Reymond

Summary

From the time of antiquity, thinking about the brain and nervous system had been dominated by the idea of a soul-like force, coursing through nerve pathways to convey sensation or produce movement. Even when we get to the mechanised ideas of Descartes in the seventeenth century, nerve physiology was still dependent on a nervous fluid containing animal spirits – an idea originating from the times of the ancient Greeks. However, the nature of this liquid, hypothesised by many to be derived from the blood and stored in the ventricles of the brain, was never fully elucidated. While the concept of animal spirits began to be questioned in the seventeenth century, and even replaced by irritability and sensibility by Haller in the eighteenth, it was only firmly rejected when investigators began to consider more seriously the nature of electricity. The origin of this new shift in thinking lay with the Italian Luigi Galvani in 1791 who believed he had discovered a form of electricity intrinsic to all living organisms. He also saw it as a non-material force with vital properties. However, at the time, many others disagreed, including fellow Italian Alessandro Volta, who thought instead that Galvani's electricity arose from an artefact of his experimental procedures. The dispute was to become one of the most fruitful in the history of science with Volta's discovery of the voltaic pile (i.e. the world's first battery) in 1800 seemingly disproving the existence of animal electricity. This situation would remain until the galvanometer was invented in the 1820s, which in the hands of certain researchers such as Carlo Matteucci, managed to confirm that nerves and muscle did indeed contain some type of intrinsic electricity as Galvani had maintained. Some 20 years later, with significantly improved instrumentation, the German Emil Du Bois-Reymond, not only showed that this force was likely to be physical in nature, but was transmitted along the surface of the fibre as a 'wave of relative negativity'. Since this was the signal that also caused the muscle to contract, Du Bois-Reymond called it the action current (later to be called the action potential). In effect, Du Bois-Reymond had discovered the nerve impulse. At last, after more than 2000 years, the notion of a spiritual type force governing the body was replaced by one involving electrical signals. The recognition of the nerve impulse as an electrical event was revolutionary, and arguably no other advance in neuroscience has provided such a definitive break with the past. But the crowning achievement of this period came in 1850 when Herman von Helmholtz measured the speed of this impulse at about 90 feet per second (or around 60 miles per hour). This was a finite speed, much slower than previously thought, which indicated that nerve function would one day be fully explainable in terms of biophysical and biochemical events.

The early history of electricity

Electrical energy has been known since ancient times. There are references to the use of electrical fish for numbing pain in Egyptian hieroglyphics, and around 585 BC the father of Greek philosophy Thales is said to have discovered the power of iron ore (magnetite) to attract other pieces of ore – a force he attributed to a soul-like psyche. This power fascinated many investigators down the ages, although the phenomenon was not called 'electricity' until 1600 when English physician William Gilbert derived the term from the Greek elektron meaning amber. He did this after finding amber attracted light objects such as hairs and dust if vigorously rubbed with a cloth.[1] Later, in the 1730s, Stephen Gray working at Charterhouse in London found that static electricity, produced by rubbing a glass cylinder, could be conducted over long distances (i.e. some 800 feet) by using wet twine. He also managed to pass a current through air across a room to 'magically' ignite alcoholic vapours. Most dramatically, Gray found this force passed through the human body. In one famous demonstration premiered in London in 1730, Gray suspended a young boy above the ground with silk ropes and generated an electric charge close to his feet, while placing a piece of metal foil near his nose. As the glass tube was rubbed with a cloth to generate electricity, the foil stuck to the boy's face with 'much vigour'. Not content with this, Grey also amused his audience by managing to draw a cracking blue spark from the boy's nose!

In order to generate electricity, Gray had used a simple 'friction' machine where a handle was turned to rub glass on a lump of sulphur. However, a more effective way of producing an electrical charge was to occur in 1745 when the Leyden jar was invented, by German Ewald Georg von Kleist, and Dutch physicist Pieter van Musschenbroek. This apparatus was essentially a glass jar half filled with water. But more importantly, the jar's inner surface was lined with metal foil, as was its outer surface which was grounded. By applying an electrical charge to the inner foil, the Leyden jar allowed electricity to be stored, and then released on demand by connecting the two surfaces with a metal conductor. This apparatus was also capable of producing some very powerful shocks, typically in the form of a spark, which at their most intense could knock a grown man off his feet.[2] Moreover, the amount of stored electricity was significantly increased by connecting a number of Leyden jars together. This would lead to some impressive public spectacles including one by the French abbot Jean Antoine Nollet in the presence of King Louis XV at Versailles in 1746, when a line of 180 soldiers were asked to hold hands. Much to the King's delight, all jumped en masse when subjected to an electric shock supplied by a large battery of jars.

When the Leyden jar was invented, the nature of electricity was poorly understood, with most people believing it was an artefact produced by friction. Electricity was also considered to be little more than a mysterious force, useful for parlour games or tricks, with little practical benefit. However, these views were to be radically revised with the work of Benjamin Franklin who was also a signatory to the American Declaration of Independence (1776). Franklin not only provided a plausible explanation for how the Leyden jar worked by recognising that positive and negative charges are built up in its inner and outer coatings respectively, which then form a current when the two surfaces are joined, but he also recognised electricity as a natural force. Indeed, Franklin achieved this in a famous experiment, allegedly in 1752, when he is said to have attached a silk string from a Leyden jar to a kite, which was then sent up into a thundercloud from a Philadelphia church spire. This increased the charge in the Leyden jar, thereby proving that its spark was the same energy produced by lightening. Whether Franklin ever performed this experiment is not certain, although it was performed by others – in some cases with lethal effects. One fatality was the German physicist Georg Wilhelm Richmann who was electrocuted in St Petersburg when attempting to replicate Franklin's experiment in 1753. It is said following his death, a red spot could be seen on Richmann's forehead and his shoes were blown apart!

Electricity was clearly a natural force of immense power with the power to kill. Equally, especially after the invention of the Leyden jar, it became increasingly recognised as something with healing properties. The administration of electric shocks was also capable of exerting a number of physiological effects including a quickening of the pulse, stimulating glandular secretions and producing violent muscular contractions. Consequently, the medical use of electricity became highly fashionable during the eighteenth century, being simple and cheap to administer, and widely seen as a panacea able to treat ailments of every conceivable kind. However, the ability of electricity to exert physiological effects also raised questions about its relationship with nervous activity. In fact, electricity had been hypothesised as the substance flowing through nerves as early as the 1720s when the clergyman Stephen Hales asked whether the force carried by the nerves may not 'act along their surfaces like electrical powers.' However, this idea was rejected by many physiologists. The leading critic was Haller who considered the possibility in 1762. Although he accepted an electrical force was 'very powerful and fit for motion', Haller dismissed it as the medium for nervous conductance since he believed it would quickly dissipate through the organs and tissues of the body. He also thought electricity would be unable to pass from a nerve to a particular muscle without spreading to other muscles. Thus, for Haller, the nervous force had to be a different kind of 'fine' and 'subtle' fluid, which was capable of reacting with an irritable force in muscle.

Luigi Galvani and the discovery of animal electricity

Although electricity had become recognised as an important and powerful natural force by the end of the eighteenth century, it seemed implausible to many that such energy could be created by the human body, let alone flow through the nervous system. However, these views had to be seriously reconsidered with the work of Luigi Galvani. Born in the famous university city of Bologna in 1737, it appears that Galvani had initially sought a career in the priesthood, but was persuaded by his father to study medicine instead. Graduating with degrees in both medicine and philosophy on the same day, Galvani pursued a successful dissertation on the formation of bones, before being appointed lecturer in anatomy. He would hold several academic appointments, including one of curator in a museum of anatomical waxes, rising to Professor of Anatomy at the University of Bologna in 1775. It was in this post, when demonstrating the electrical properties of frog legs to a group of students, that Galvani witnessed an event in the late 1770s which changed the course of scientific history.

Galvani had arranged a practical demonstration where the legs of a decapitated frog hung freely from a segment of spinal column. He did this to expose the nerve stumps of the leg, thereby allowing them to be stimulated with an electric current, or some other means, which caused the attached muscles to twitch. This was not a new experimental procedure. In fact, this frog muscle–nerve preparation had been pioneered by Jan Swammerdam a century earlier (see previous chapter). However, this time, Galvani saw something which was unexpected and inexplicable. This event occurred when a student serendipitously induced a powerful and convulsive kick of the legs by touching a nerve stump with a metal scalpel, at the precise moment a spark from a nearby Leyden jar was drawn. A scalpel by itself, coming into contact with the nerve, should not have produced such a vigorous muscle contraction, and the reaction only made sense to Galvani if the metal blade was somehow 'picking up' electricity through the atmosphere from the jar and transmitting it to the muscle. This accidental event would become the start of a ten-year crusade of experimentation for Galvani in which he attempted to understand the cause of this mystifying atmospheric form of electrical response.

Some of Galvani's earliest experiments, around 1780, set out to confirm whether the proposed nervous fluid (Haller's theories were then much in vogue) could convey an electrical force. He soon discovered it could. But curiously, Galvani found that muscular contractions were stronger if an electrical charge generated by a friction machine, stimulated the nerve stump rather than the muscle directly. This result was opposite to the one predicted by Haller's notion of irritability, which placed the force responsible for contraction intrinsic to the muscles. Even more puzzling, Galvani noticed strong muscle contractions could be produced by extremely weak electrical stimuli applied to the nerves, including a current generated from a sparkless Leyden jar that had almost been emptied or fully discharged. Clearly, nerves were particularly sensitive to the effects of electrical stimulation, and conducted this type of energy far more efficiently and forcefully than other tissues.

Galvani also set about confirming what he had first observed in his practical demonstration. That is, he produced strong contractions of the frog muscle–nerve preparation, when the nerve was touched by a scalpel and sparks were simultaneously drawn from the electric machine across the room. Having heard of Franklin's famous kite experiment, showing that electricity could pass through the atmosphere, Galvani not unreasonably wondered if thunderstorms could also similarly induce muscular contractions. To test this, he decapitated frogs, pierced them with a brass hook through the spinal cord, and hung them on the iron railings around

FIGURE 5.1 Few pictures of Luigi Galvani (1737–1798) exist. This is a print (lithograph) of Galvani that probably dates from around the early nineteenth century.

Source: Wellcome Library, London.

his house (the history of electrophysiology may have turned out very differently had he a wooden fence). A wire with one end touching the nerve was then fixed to an aerial attached to the house, while another wire touched the frog's muscles and dipped into the water of his well to act as earth. As he expected, during a thunderstorm, Galvani observed the legs twitch as if stimulated by an electrical charge.

However, Galvani was also meticulous enough to perform the same experiment on a calm day. To his surprise, he obtained the same effects – a puzzling result for there should have been no electricity in the atmosphere. In order to examine this effect further, Galvani set up the same type of arrangement in the confines of his laboratory. Here, he placed a decapitated frog on an iron plate and staked a brass hook through its spinal cord, and at this precise moment, the leg muscles twitched. Atmospheric electricity was not involved after all. In fact, Galvani had inadvertently made one of the greatest discoveries of all time: he had created bimetallic electricity, produced when two dissimilar metals come together. Keen to explore the new phenomenon further, Galvani examined different combinations of metals, which he found gave contractions, but of different intensities, depending on the ones used. Poor conductors, however, did not cause any reaction. Galvani also extended these observations by using metallic arcs – essentially a conductor made from two metals bent into a 'U' shape with one end attached to hooks in the spinal cord and the other to the muscles in the leg. These studies, again, showed the muscle contracted vigorously when they formed part of a circuit involving the two metals.

Galvani was now faced with trying to explain the phenomena he had been observing. The crucial question was simple: where was the electricity responsible for producing the muscular contractions coming from? Up until 1786, Galvani appears to have believed it was a result of 'metal electricity'. However, he later changed his mind and decided that the electricity must dwell inside the nerves themselves. He called this new force 'animal electricity' (animalis electricitas) and argued the two metals creating the circuit were prompting its circular flow. The concept of animal electricity and its implications for understanding nervous activity and movement were so revolutionary that Galvani continued his experiments for several more years before daring to publish them. His work would eventually appear in his 53 page *De viribus electricitatis in motu musculari commentarius* (*Commentary on the Effects of Electricity on Muscular Motion*) in 1791. First written in Latin, it was subsequently translated into Italian, German and French. It would make Galvani famous and convince many at the time that the secret of the 'nerve fluids' had been identified.

De viribus electricitatis not only gave a thorough description of Galvani's experiments, but also gave an account of his physiological theories concerning animal electricity. In this respect,

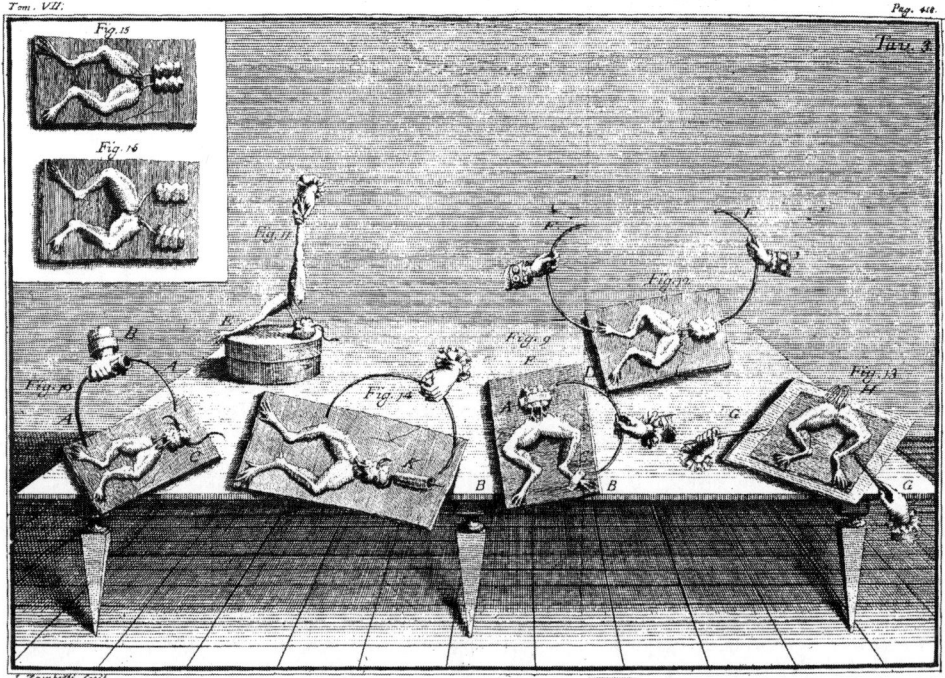

FIGURE 5.2 Plate 3 taken from Galvani's *De viribus electricitatis* (1791), which shows some of the methods for making an arc for passage of an electrical current along a nerve to produce muscular contraction.

Source: Wellcome Library, London.

Galvani was convinced that the energy was very different from the 'natural electricity' generated by different metals. In fact, he viewed it as a non-material life force which gave the body its animating power. Thus, Galvani's animal electricity was a newer variation of the traditional *animal spirits* concept. Galvani also believed this energy originated in the brain and flowed into the muscles through the nerves. Moreover, instead of attributing the muscle with the power of irritability, Galvani likened each individual muscle fibre to a small Leyden jar with its exterior and internal surfaces setting up opposite charges. This made the muscle fibre a store of two different kinds of electricity, putting it in a state of high tension or disequilibrium. Thus, the muscle fibre 'discharged' if a nervous fluid was infused into them. In other words, a nervous impulse was hypothesised to connect the two surfaces of the muscle fibre, in much the same way as a conductor connected the two stores of energy in the Leyden jar.

A rejection of animal electricity: Alessandro Volta

De viribus electricitatis caused great excitement and wonder, first in Italy and then in Europe, as investigators sought to replicate and verify Galvani's work. It has been said that frogs were decimated in large quantities to fulfil the demand, and the repeatable nature of Galvani's experiments meant they were even performed in fashionable ladies' saloons where they provided a good spectacle to all. Galvani had also sent abstracts of his work to fellow researchers, one of whom was the aristocrat and physicist Alessandro Volta at the nearby University of Pavia. He was a man of high scientific repute, who despite a lack of a university education, had dedicated his academic life to the study of physics, chemistry and above all electricity. It would lead him to invent a number of new instruments for electrical research, including the 'electrophorus' in 1775, which allowed positive and negative charges to be stored in metal plates; the condenser which made faint electric charges detectable; and the electrometer for measuring electrical tension. In addition, Volta is credited with discovering the gas methane in 1776. All of these inventions had helped Volta achieve international recognition, including membership of the Royal Society, by the time he received a personal copy of *De viribus* from Galvani in 1792.

At first, like many of his contemporaries, Volta was impressed with the quality of Galvani's research. Indeed, after replicating and confirming many of the experiments, Volta reported he had 'converted himself' to animal electricity and 'changed from incredulity to fanaticism.' However, upon continuing his investigations, Volta changed his mind. He first began to have his doubts when he found that a frog's muscle could be made to contract by bringing two different metals together on the surface of its nerve stump without either of them touching the muscle. From Galvani's perspective this should have simply meant a flow of electricity inside the nerve between the points of the two metals. However, this discovery showed muscles could be made to contract even when they were not apparently part of an electrical circuit, but how could two dissimilar metals produce such an effect? To find out, Volta undertook an unusual investigation by deciding to 'taste' the electricity. He did this by placing two metal strips, one lead and one silver, some distance apart on his tongue – a procedure causing him to experience a mild shock and sour taste. These results led Volta to reach a very different conclusion from Galvani. His sensations were not originating from a flow of intrinsic animal electricity, but by a steady electrical current produced by the two metals. In other words, the electricity was an artefact which was being generated outside the body.

The disagreement between Volta and Galvani was to become public during the last decade of the eighteenth century when both men attracted rival camps of supporters and detractors. One person to take an active interest in the debate was the German explorer and adventurer Alexander von Humboldt, whom Charles Darwin once called 'the greatest scientist traveller who ever lived.' Humboldt first learned about the controversy during a visit to Vienna in 1792, and over the next five years would undertake hundreds of experiments, some of which he performed on his travels, and pondered during his long hours of solitude.[3] Humboldt published his findings in a two-volume book in 1797 which made a convincing claim that both Galvani and Volta had been correct to some extent. In his view, the nervous system clearly produced a substance that made the muscles contract, and while the metal electrodes increased this effect, they did not generate it in the first place. Thus, Galvani had discovered two separate phenomena: bimetallic electricity and intrinsic animal electricity. In his quest for the truth, Humboldt was also not reluctant to experiment upon himself. Indeed, it is said that one day, after finding a dead bird in his garden, he placed a zinc blade in its beak and one of silver in its rectum to induce an electric shock – a procedure that made the bird flap its wings and walk. Not content with this, Humboldt performed the same test on himself with somewhat unpleasant consequences!

While it is known that Galvani and Volta had a cordial relationship, and often exchanged letters, the situation was not helped by the different personalities of the two protagonists. Volta was keen to publicise his work though publications and scientific meetings, whereas Galvani was in declining health and reluctant to be in the spotlight. There were also political forces at work. In 1796, just five years after the publication of *De viribus*, Bologna came under French control, and Galvani refusing to swear allegiance to Napoleon, was dismissed from his university post. Although later reinstated, Galvani died soon after at the age of 61 in 1798. In contrast, Volta supported the new regime and was even on friendly terms with Napoleon. He would later achieve political influence by becoming the Rector of Pavia University, and being elected as a senator of Lombardy.

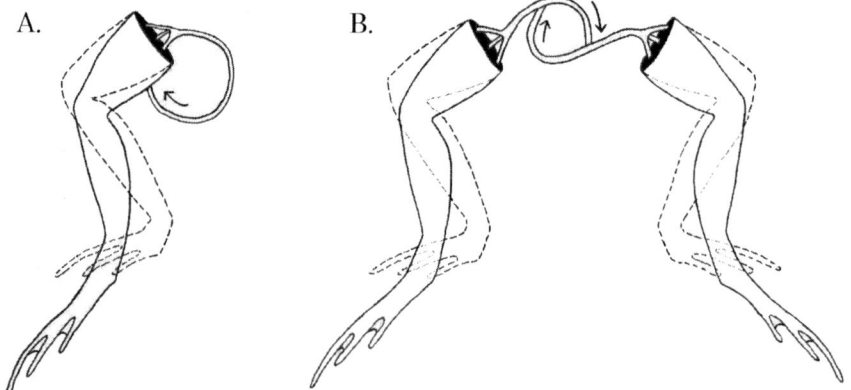

FIGURE 5.3 Two simple demonstrations (1794 and 1797 respectively) used to show the existence of animal electricity. In both instances muscular contraction occurs without any metallic involvement. (A) When an exposed nerve touches the muscle, the leg contracts. (B) When the right sciatic nerve touches the intact left sciatic nerve, then both legs contract.

Source: Both images are drawn by Charlotte Caswell.

FIGURE 5.4 Alessandro Volta (1745–1826) discovered the electric battery (voltaic cell) which for the first time allowed a constant electric current to be produced. This portrait appears to show Volta's approval of the Napoleonic regime.

Source: Wellcome Library, London.

With Galvani unable, or refusing, to engage in the academic dispute, it was left to his supporters to continue the argument for animal electricity. Nonetheless, Galvani is believed to have published one anonymous paper in 1794 that described an experiment in which a frog's muscle was cut open and touched with the freshly severed end of a spinal cord taken from another animal. This manipulation produced a contraction of the muscle. Since no external power source was involved other than the spinal cord, Galvani concluded the contraction must have been caused by intrinsic animal electricity. A second experiment by his colleague Eusebio Valli in 1797 provided further support when he bought the sciatic nerves of two dissected legs together – a procedure causing them both to vigorously contract (see Figure 5.3). However, for Volta, such experiments only reinforced his view that any two dissimilar substances could create electricity, although this effect was stronger for some materials, such as metals, than others.

Volta invents the battery

Despite the contribution of Humboldt to settle the matter, Volta maintained animal electricity lay not within the nerves or muscles of the frog leg, but in the metal strips used to complete the circuit. He was also not prepared to change his mind, and set about undertaking further experiments by combining different types of metal to determine what produced the most intense electrical shocks. It would lead to one of the greatest inventions of any age in 1800 when Volta discovered he could produce a strong electric current by placing a series of alternating copper and zinc disks on top of each other separated with cardboard layers soaked in acid. Called a voltaic pile, a steady electrical current was produced when a wire connected the top of the apparatus with the bottom. Remarkably, this contraption did not have to be recharged through a friction generator. Thus, Volta had invented the first ever battery which was able to provide a source of continuous electrical energy – a simple apparatus that would transform the world for evermore.[4] Before this invention, the study of electricity had been confined to the investigation of its static form. Now, scientists were able to work with a constant current that could be switched on and off at will.

It soon became clear that this form of electric current had many physiological effects, and was capable of producing strong convulsive muscle contractions. Furthermore, the alternating metal disks in the voltaic pile closely resembled the stack of muscle plates in the electric organ of the ray, and other fishes capable of generating shocks. Yet, Volta was more convinced than ever his battery disproved the existence of Galvani's animal electricity. For Volta, there was no need to postulate a non-material vitalistic electrical force, when it was obvious that physical electricity existed everywhere in nature. Moreover, his battery lent further support to the idea that animal electricity was an artefact. In fact, by the time the battery was invented, Galvani had been dead two years, and with Volta's greater scientific prestige, scientific opinion began to swing against animal electricity. In reality, however, the issue remained unresolved for no instrument was sensitive enough to record the tiny currents from muscle and nerve required to adequately test the theory. Because of this technological limitation, the question of who was correct concerning animal electricity would not be fully resolved for another 50 years or so. Nonetheless, history was to show that both Galvani and Volta were correct in certain respects.

Animating the dead: Giovanni Aldini

Following Galvani's death there was little progress in experimental research on animal electricity. Although this was partly due to Volta's explanation of it as arising from an external source, there was also a growing belief that any account of nervous activity must also incorporate physical and chemical principles. Nonetheless, electricity continued to be regarded with wonder and awe, and widely recognised as having a wide range of medical uses – a practice that became known as galvanism (a term generously coined by Volta).[5] One person, however, determined to keep the theory of animal spirits alive was Galvani's nephew Giovanni Aldini. A professor of physics at Bologna, Aldini had often helped his uncle with experiments, and taken much of the responsibility for corresponding with Volta. The founder of the world's first Galvanic society, Aldini also continued to insist that animal electricity was the vital force responsible for life itself. He also performed a number of experiments to show that frog legs would twitch when their nerves touched exposed muscle tissue without any metallic contact.

A new life force: animal electricity 121

FIGURE 5.5 Giovanni Aldini (1762–1834) who founded the world's first Galvanic society and noted for his demonstrations involving human corpses.

Source: Wellcome Library, London.

However, Aldini was to become most famous in the eighteenth century, not for his belief in animal electricity, but for his theatrical demonstrations in which dead bodies were 'reanimated' in life-like fashion with electrical stimulation. Aldini initiated these experiments in 1798 when he applied an electrical current to different areas of a freshly slaughtered ox brain. It produced grotesque changes of facial expression accompanied by eye and lip movements. Determined to extend his work to understanding 'the human animal machine', Aldini sought to experiment with human cadavers – a wish granted in 1802 when he was given permission to stimulate the brains of three criminals beheaded by an executioner's axe close to the law courts in Bologna. Performed in front of a public audience, Aldini's electrodes elicited grimaces, bodily contortions, and abrupt spasmodic movements of the limbs – effects that persisted up to three hours after death. In one instance, the arm of a corpse is said to have slowly raised itself eight inches from the table, and maintained its position even when a considerable weight was placed in the hand. When Aldini pushed his electrodes down between the cerebral hemispheres to reach the corpus callosum, he found the current induced a variety of facial expressions similar to those observed in the ox. He even tried to bring the men back to life by stimulating their hearts – but was disappointed when his attempts at resuscitation failed.

The great publicity surrounding these demonstrations encouraged Aldini in the autumn of 1802 to undertake a short tour of Europe. Although this was primarily to convince the scientific community of the medical uses of galvanism, the shows also involved electrifying human and animal bodies. The tour was not a simple journey as this was the time of the Napoleonic Wars. Nonetheless, Aldini visited Paris, and then England where he gave exhibitions in Oxford and London. These were public spectacles with one of the attendees being the Prince of Wales, who it is said, was amused by the proceedings. However, Aldini's most famous demonstration took place in front of the Royal College of Surgeons in London in 1803, which used the corpse of a 26-year-old man called George Foster who had just been hung at Newgate for drowning his wife and child. Assisted by two colleagues and generating electricity from a huge voltaic pile composed of 80 plates of zinc and copper, Aldini prodded various parts of the body with a long electrode. The first dramatic effect occurred when the mouth and ear were touched, causing the jaw to quiver, the face muscles to contort and an eye to open. Another startling response was produced when the electric current was applied to the rectum – a procedure making the body convulse so intensely that some members of the audience are alleged to have thought the corpse was coming back to life.[6] These events were even reported in *The Times* a few days later, although perhaps as more of a curiosity as they were sandwiched between accounts of pickpockets, and a fire in a carpenter's house in Hackney. Aldini was particularly pleased with his work for he later wrote the experiment 'surpassed our most sanguine expectations.' He also seemed to think Foster was close to being brought back to life, commenting that 'vitality might perhaps have been restored, if many circumstances had not rendered it impossible.'

Aldini may have been a flamboyant showman not afraid to court controversy, but his motives were ultimately scientific and therapeutic ones. Indeed, an important objective of his attempts to reanimate the dead was to see if electrical shocks could be used to resuscitate individuals who had been drowned or asphyxiated. In this respect he can be considered a pioneer of various electrotherapies that are still in use today, including cardiac electrical stimulation for the treatment of heart failure, and administering shocks to the brain for the relief of mental illnesses such as depression. Interestingly, Aldini devoted the last years of his academic career to matters more closely related to physics than medicine or biology. He tried to improve the construction and illumination of lighthouses, and invented novel procedures to improve the lighting of streets and buildings which included Milan's famous La Scala. He also undertook important work on the use of asbestos fibres to improve fire-fighting. In recognition of his work Aldini was made a knight of the Napoleonic Order of the Iron Cross and a councillor of the state of Milan. After his death in 1834, Aldini also left a considerable sum of money to found a school of physics and chemistry in Bologna.

The inspiration for Frankenstein

To some observers, Aldini's experiments appeared to show bringing the dead back to life was a real possibility. There were even some scientists who agreed. One such researcher was a professor of medicine at the University of Halle in Germany called Karl August Weinhold who published an account of his work in 1817.[7] A book of 116 short chapters, one of which is entitled *Observations on Seven Beheaded Criminals*, its main aim was to 'show' how bimetallic electricity could be used to bring life back to a corpse. In one experiment, Weinhold removed the spinal cord from a decapitated kitten and replaced it with a mixture of zinc and silver, since he believed it produced electricity in much the same way as a battery. According to Weinhold, this indeed caused the kitten's heart to start beating again along with movement of some muscles. Spurred on by his 'success', he repeated the procedure in another kitten whose brain was removed. Again, the animal was reinvigorated with Weinhold reporting it raised its head, opened its eyes and attempted to crawl 'with obvious effort' before sinking down exhausted. These observations are probably exaggerated and without merit since a dead body, even if showing signs of movement, does not prove it is alive as Aldini had showed. Nor were these responses ever verified by others outside his laboratory. Weinhold also appears to have been equally extreme and controversial in his personal views too. Said to be an abrasive man who did not care what others thought of him, Weinhold advocated population control in the poor by a compulsory surgical operation known as infibulation which entailed sewing the foreskin shut in order to eliminate childbirth.

Nonetheless the potential power of electricity to reanimate life fuelled the imagination of many thinkers. One person to be inspired was a 19-year-old author called Mary Shelley who used the idea to write *Frankenstein* – first published anonymously in 1818. The story of how Shelley came to write her novel is well known. After secretly marrying the writer Percy Shelley, the couple eloped to the shores of Lake Geneva where they stayed with the poet Lord Byron and a small circle of friends. Frustrated by an unseasonally wet summer, the group talked for long hours into the night, during which time they came up with a competition to see who could write the best horror story. Mary Shelley's contribution was her famous novel which tells the tale of a Bavarian doctor called Victor Frankenstein, aided by a hunchbacked assistant, who constructs a human body from bones and body parts collected from hospitals and dissecting rooms. By infusing the corpse with a 'spark of being' (delivered by a strike of lightning in later accounts) Frankenstein creates a being in the likeness of man which turns out to be a monster with disastrous consequences. The story was an immediate success. In the preface of the third edition of 1831, Shelley acknowledged she had been influenced to write her story by certain 'physiological writers of Germany'. There has been much speculation over the identity of these men, and it is conceivable Karl Weinhold was one of them.

The invention and use of the galvanometer

The question of animal electricity remained unsolved in the first quarter of the nineteenth century because no instruments were sensitive enough to detect the faint flow of electrical current within biological tissue. However, in 1820, the Danish physicist Hans Christian Oersted accidentally discovered a compass needle would deflect from magnetic north in response to the flow of electricity along a wire. This turned out to be a profound observation. Not only was electricity and magnetism related in some way (Oersted had, in effect, discovered electromagnetism), but it led the German Johann Schweigger in the same year to invent the galvanometer. Named in honour of Galvani, this instrument was able to measure the strength and direction of an electrical current. The first galvanometers were little more than a coil of copper wire surrounding a suspended magnetic needle or compass. In the absence of any current the needle pointed north. However, when electricity flowed through the copper wire the needle changed direction and, importantly, the stronger the current the greater the deflection. Schweigger also increased the sensitivity of the galvanometer by using multiple loops of wire to surround the needle. Unfortunately, this benefit was offset by the Earth's magnetic force which acted as a drag on the needle, thereby making fine measurements of the current unreliable. This problem was solved, however, by the Italian Leopold Nobili who invented the astatic galvanometer in 1825. By employing a double coil wound into a figure eight shape, and suspending a magnetic needle with their poles reversed in each of the two openings, Nobili overcame the background force of the Earth's magnetic field.

Nobili used his instrument to measure the flow of electrical current in a nerve dissected from a frog's leg for the very first time in 1827. He did this by modifying an experiment first undertaken by Galvani. In short, he placed a skinned frog's leg in a container of salt water, and then immersed the spinal lumbar plexus with sciatic nerve attached into another beaker.[8] Thus, the leg and spinal cord were completely separated. Nobili knew if he connected the two containers with a piece of moist cotton, the isolated frog's leg would then begin to twitch. However, instead of using cotton, Nobili made the connection by using the wires of his galvanometer. This procedure revealed the presence of a small but steady current (or what he termed the 'frog current') passing from the cut muscle to the spinal cord which lasted several hours. Unfortunately, the influence of Volta was still strong at the time, and Nobili did not believe he was observing an intrinsic electrical force. Rather, he thought the effect arose through the unequal cooling of nerve and muscle induced by his procedure. Nonetheless, we now know it was the first ever demonstration of an electrical current passing through a nerve – or what Galvani would have recognised as animal electricity.

The confirmation of this electrical current as a true biological one, however, would be established by Carlo Matteucci, a colleague of Nobili, who became Professor of Physics at Pisa in 1840. In one crucial experiment, Matteucci used what is known as the 'galvanoscopic frog' preparation. Here, the skinned leg of a freshly killed frog was placed in a glass tube and its exposed sciatic nerve allowed to hang freely out of the tube's open end. Matteucci knew if two metals, or separate substances in a different electrical state, were bought together on the nerve stump, the frog's leg would contract. However, the crucial question for Matteucci was whether this response could be produced by biological material alone. To test this, he placed muscle tissue on the nerve at two different points. By doing this, Matteucci came across a curious effect: providing he touched the nerve using one muscle with an intact surface, and another with a cut or exposed surface, then a current was produced causing the leg to contract. It was clearly an intrinsic form of animal electricity – albeit a puzzling one as the 'nerve current' depended on connecting two different types of muscle surface (intact and injured). Nonetheless, it was genuine as Matteucci replicated the experiment by pushing one wire from a galvanometer deep into a muscle, and then laying the other wire on the muscle's surface. This procedure caused a small current to pass through the galvanometer (from the outer surface of the muscle to its interior). Today, this is now recognised as an injury current.[9] No 'muscle current' was produced, however, if the electrodes were both placed on the intact surface, or injured interior.

There could be no doubt that the currents produced by Matteucci's 'galvanoscopic frog' preparation were being generated by the muscles themselves. Also, Matteucci provided another demonstration of this effect in 1843 by inventing a 'biological pile'. Here he took a series of sectioned skinned frog thighs arranged in serial order, and placed their intact surfaces in contact with the cut surface of their neighbours. It was, in principle, not unlike Volta's voltaic pile of dissimilar metals – except this time the battery only contained biological material. Using this preparation, Matteucci found the current recorded by the galvanometer increased steeply in proportion with the number of thighs. This was a particularly important finding for it also showed the electrical current was not an artefact of the galvanometer.[10] It was the strongest evidence yet for the existence of animal electricity.

Another seminal experiment was performed by Matteucci in 1842. He detached a frog's leg and exposed its sciatic nerve. He then lay an intact thigh muscle of a second frog onto this nerve pathway, and made it twitch with either mechanical or chemical stimulation. Matteucci found this immediately caused the recipient leg muscle to simultaneously contract. In other words, the twitching thigh muscle was generating enough energy through its action on the sciatic nerve to cause the recipient leg muscle to move. It was, in effect, the first experimental demonstration of a nerve impulse! Unfortunately, Matteucci did not recognise it as such. In fact, he even came to deny that the first muscle generated electricity. This is because Matteucci also found if he induced a prolonged contraction of the first muscle (a procedure known as tetanus), the contraction of the second muscle disappeared. Even more puzzling for Matteuci was the observation that the tetanus caused a decrease in his galvanometer reading which apparently showed the current had become weakened.[11] Unable to explain these findings, Matteucci believed some force other than electricity must be responsible for the neural transmission. Nonetheless, this experiment is widely seen as a landmark one in the historical development of electrophysiology.

Differing views: Johannes Müller and Emil Du Bois-Reymond

In 1844, Matteucci was awarded the Copley Medal by the Royal Society of London in appreciation for his research into bioelectricity. The society also described his frog pile as one of the most important discoveries of the age. Yet, despite the evidence supporting the existence of electricity in nerve and muscle, many investigators clung onto the idea of a mysterious spiritual 'vital force' flowing through the nervous system. One of the most influential advocates of this view was Johannes Müller at the University of Berlin who had created the world's first institute for experimental physiology along with a dynamic group of researchers. Müller was also renowned for his monumental *Handbuch der Physiologie* (*Handbook of Physiology*), containing nearly a million words, published in two volumes between 1834 and 1840. However, Müller was convinced that nervous energy was not electrical. He seems to have arrived at this position, in part, when he physically damaged (mashed) a nerve fibre, and showed that while electrical energy was still capable of passing through the damaged zone, mechanical stimulation of the nerve produced no muscle twitching.

Müller's alternative explanation was his theory of 'specific nerve energies'. This proposed each type of sensory nerve pathway had its own unique vital quality. So, for example, the optic nerves were in Müller's view only capable of conveying the sensation of light, whereas the auditory nerves only transmitted sound etc. How they did this he could not say, but to support his theory, Müller pointed out no matter how one stimulated the eye (e.g. with light,

FIGURE 5.6 Emil Du Bois-Reymond (1818–1896) was a German physiologist recognised as the father of electrophysiology, who is perhaps most famous for showing that the nerve impulse travels along the nerve fibre as a 'wave of relative negativity'.

Source: Wellcome Library, London.

touch, electricity, etc) all led to visual sensations. Thus, for Müller, it was inconceivable that all the different nerve pathways relaying sensory information were using the same type of nervous force (i.e. electricity) for their transmission. In other words, they were all unique in some way. However, there was a fundamental problem with this idea, and it was expressed by one of Müller's young research students called Emil Du Bois-Reymond: if true, then if we were able to cut and cross the visual and auditory nerves, we would hear with our eyes and see with our ears!

Despite his vitalistic views, Müller obtained a copy of Matteucci's work in 1840, and asked Du Bois-Reymond to investigate its claims. It was to be a pivotal moment in the young researcher's professional life for he spent the next 40 years devoted to the subject. At first, Du Bois-Reymond was frustrated by the limitations of the electrical recording equipment available to him. Although the invention of the galvanometer in 1820 had been a major breakthrough, it was slow reacting to the presence of current. Even worse, it gave imprecise measurements when electrical artefacts were generated from its electrodes coming into contact with biological tissue. But with great single-mindedness of purpose, Du Bois-Reymond set about solving these problems by building the most sensitive galvanometer of his day. The apparatus, which he called the multiplikator, took many months to build. Constructed from exceptionally light moving parts, it magnified the current using a coil of 24,160 turns, made from a very fine wire over a mile and a quarter in length. Du Bois-Reymond also greatly improved the sensitivity of the electrodes by making them non-polarizable (i.e. unable to generate electric currents). He achieved this by constructing them from glass vessels containing layers of filter-paper soaked in saline. While sensitive enough to measure the weak currents of muscle and nerve, they produced little source of electrical error.

In 1842, Du Bois-Reymond set about replicating many of Matteucci's experiments on muscle and nerve with his new apparatus. One of the first things he did was to confirm the existence of a 'injury current' which flowed from an outer surface of an intact muscle to a cut interior of another. Du Bois-Reymond also correctly deduced that this current arose because of two reasons: (1) the inside of the muscle was negatively charged compared to its outside; and (2) the inside of the muscle had a lower electrical potential. Both of these factors Du Bois-Reymond realised would cause positively charged electrical flow to pass into the cut muscle from its surface.[12] However, when Du Bois-Reymond produced a contraction in the intact muscle, the nature of this resting current changed. It now caused a new electric current to pass through his multiplikator on its way to the second injured muscle. He called this 'muscle current' or muskelstrom. Curiously, Du Bois-Reymond noticed this always caused his multiplikator's needle to deflect from positive to negative – an effect he referred to as 'negative variation'. In other words, a twitching muscle caused a negative electrical flow. Or put another way, the outer surface of the muscle was now becoming more negative compared to the inside.

Du Bois-Reymond also found this negative variation became more pronounced when he tetanised (i.e. repeatedly stimulated) a muscle with a chemical or series of electric shocks. Unlike Matteucci who had been puzzled by the apparent lack of a muscle current following tetanus applied to a muscle (his equipment was not sensitive enough to detect negative variation), Du Bois-Reymond realised a tetanised muscle actually produced a succession of short contractions with each one of these giving rise to a weak nerve current. Moreover, since they followed one another very rapidly, they produced a summated effect on the muscle – an effect increasing the negative variation. Thus Du Bois-Reymond had shown that a muscle contraction produced an electrical current, and tetanus produced an even stronger one.

The discovery of the action potential

In 1843, Du Bois-Reymond made a fundamental discovery that essentially changed the nature of neuroscience. He did this by applying an electric shock or 'pinch' to an exposed sciatic nerve stump, and then recording the changes in the current with an external electrode placed on the nerve. His multiplikator revealed that this produced a decrease in electrical potential which travelled along the surface of the nerve fibre as a 'wave of relative negativity'. Du Bois-Reymond, had of course, detected a similar electrical phenomenon (negative variation) in muscle. However, by using electrodes to record from the surface of a nerve fibre, he found the same energy moved as a pulse. More importantly, Du Bois-Reymond showed that this wave of negative variation provided the signal which made the muscle contract. Du Bois-Reymond called this phenomenon the action current (later to be called the action potential). It was the first ever demonstration of the nerve impulse, and Du Bois-Reymond was in no doubt about the great significance of what he had discovered. He triumphantly wrote:

If I do not greatly deceive myself, I have succeeded in realising . . . the hundred years dream of physicists and physiologists, to wit the identity of the nervous principle with electricity.

Du Bois-Reymond had not only discovered the nerve impulse, but shown it was electrical in nature – an important turning point in our understanding of the brain and nervous system. However, explaining how this force came about was a more difficult proposition. In an attempt to do so, he posited nerve fibres contain specialised spherical 'electric molecules' with negatively charged poles and a positive equator region, the idea being that these acted like miniature batteries to produce a current. This theory was soon shown to be untenable by his student Ludimar Hermann. Although embarrassed at his mistake, Du Bois-Reymond's pioneering of electrophysiological achievements could not be taken away from him. Moreover, he was convinced that the nerve impulse, along with all other physiological functions, were the result of physical and chemical events, belonging to the material world. In fact, he expressed his belief in 1842, when just 24, by making an oath pledging his support for the doctrine of materialism over vitalism – a vow also signed by several others in Müller's laboratory.[13] Perhaps more than any other figure in the nineteenth century, Du Bois-Reymond can be regarded as the first physiologist whose research laid the foundations for modern bioelectricity. His work was summarised in *Researches on Animal Electricity* whose first two volumes (containing over 1400 pages) were published in 1848 and 1849. A third volume appeared in 1884. Du Bois-Reymond would boldly claim in its preface that he had disproved the antiquated notion of mysterious 'life forces' by identifying the nervous principle with the material and scientifically observable force of electricity.

Hermann von Helmholtz

Hermann von Helmholtz was one of the greatest minds of the nineteenth century who made a large number of remarkable contributions to a variety of scientific disciplines, including physics, physiology, psychology and optics. He also was the first to accurately measure the speed of the nerve impulse in 1850. Born into a humble but scholarly Potsdam family, Helmholtz suffered delicate health as a child which confined him to home for the first seven years of his life. Nonetheless, his father gave him a good schooling, teaching him several languages and introducing him to philosophy. Although his family were unable to afford a university education for their son, Helmholtz managed to study medicine in Berlin through a government scheme that required him to serve as a surgeon in the Prussian army for several years after training. However, his peacetime military duties were few, and Helmholtz even had time to construct a small laboratory in his barracks. This work enabled him to join various academic circles in Berlin, leading to an acquaintance with Müller's laboratory. Here, Helmholtz would complete his doctoral thesis on the microscopic nerve structure of invertebrates in 1842 at the age of just 21. This work showed how nerve cells located in the spinal cord gave rise to long spindly fibres – a key discovery pivotal in the development of the neuron theory (see Chapter 7). A year later, Helmholtz graduated in medicine to begin his eight-year military obligation.

One of Helmholtz's main interests during his early period was the origin of animal heat. From the time of Aristotle onwards, this had been widely thought to be generated from a vital force, and for most physiologists it still held the key to understanding life. However, in

1847, Helmholtz showed there was nothing mysterious about the formation of animal heat when he proved that energy in a biological system is never created or lost, but only transformed from one form to another. Helmholtz also described this principle, known as the conservation of energy, in mathematical terms. It was an important advance for it showed the heat of the body could be explained solely in terms of chemical reactions. It further became clear that biological energy is derived from food which in turn gets its energy from the sun. The implication was clear: there was no reason to exclude the study of life, or indeed anything else, from the pursuit of science. It was yet another rejection of vitalism in favour of physics and chemistry.

Helmholtz also undertook pioneering work on the sensory systems of the body. For example, in 1851, he invented the opthalmoscope enabling him to view the living human retina for the first time. Four years later, the ophthalmometer was added to his inventions which allowed the curvature of the eye to be examined. His research in vision and optics, published in a definitive three volume book *Handbuch der physiologischen optik* (*Handbook of Physiological Optics*) established the modern foundations of optometry between 1856 and 1866. His work on vision also led him to correctly deduce the eye only contains three types of colour detector (red, blue and green) allowing investigators to understand colour blindness and the phenomenon of colour after-images. In acoustics he also correctly explained how certain structures in the cochlea of the ear resonate at certain frequencies to enable pitch and tone to be heard. Despite this, Helmholtz was aware the perception of many sensory events is a psychological process and not just a neural one. Also, among many other achievements, Helmholtz played a significant role in helping his student Heirich Hertz confirm James Clark Maxwell's theoretical prediction of radio waves. For these, and numerous other achievements, Helmholtz is widely regarded as the most important figure in German science during the late nineteenth century – pivotal in providing a radical break with traditional physics, and paving the way for later developments in the twentieth century.

Measuring the speed of the nerve impulse

After fulfilling military commitments, Helmholtz took up an appointment at the University of Königsberg in 1849 where he remained for six years. Du Bois-Reymond's discovery of a 'wave of negative variation' travelling along the nerve fibre capable of inducing a muscle contraction raised an intriguing possibility: if this was indeed the nervous message, then it should be possible to measure its velocity. At the time, however, this was widely regarded as an impossibility since the speed of electricity was widely believed to be too fast, or perhaps even instantaneous. Despite this belief, a number of researchers had attempted to estimate it. In fact, Müller's *Handbook* gave three estimates: the first by Haller judged it at 9000 feet per minute; the second by French physician Sauvages put it at 32,400 feet per minute; and the third source who is simply described as 'another physiologist' apparently estimated it at 57,600 million feet per second (if this value was true then it would be 60 times faster than light!). It appears Müller thought it conceivable that the nervous impulse was similar to the speed of light, and unlikely to ever be measured accurately. Yet, Helmholtz would prove him wrong in 1850.

Helmholtz set about trying to measure the speed of the nerve impulse by constructing an instrument called a chronograph which was capable of accurately measuring small fractions of a second. By connecting this timing mechanism to a galvanometer, Helmholtz established

FIGURE 5.7 Hermann von Helmholtz (1821–1894) who among his many scientific achievements measured the velocity of the nervous impulse in 1850.

Source: Wellcome Library, London.

a precise way of recording the time it took for the galvanometer's needle to be deflected by an electric current. Just as important, to increase the accuracy, Helmholtz added a switch to his apparatus that stopped the deflection of the needle in its track on the immediate onset of a muscle twitch. This switch was so precise that it could measure reaction time in terms of milliseconds (thousandths of a second).

To measure the speed of the nerve impulse, Helmholtz first isolated a motor (sciatic) nerve from the leg of a frog still connected to its gastrocnemius muscle. This nerve was roughly 40 mm in length. Helmholtz then began measuring the difference in time between the point of electrical stimulation applied to the nerve fibre and the start of muscular contraction. By stimulating the nerve at different points along its length and recording the twitch onset, Helmholtz was able to calculate the speed of the impulse. Although the time differences were minute, roughly in the region of 1.3 milliseconds, he was nevertheless able to estimate the nerve velocity to be about 90 feet per second, or around 60 miles per hour. This is a speed not only much slower than light, or the flow of electricity down a cable, but about one-tenth slower than the speed of sound.[14]

However, this slow velocity provided further evidence, if any were needed, that a physical–chemical process was involved in producing the nerve impulse. Furthermore, it was very different from the flow of electricity down a wire. Helmholtz's discovery had profound implications since it showed that an *active biological process* was responsible for the nervous message. Helmholtz referred to this nerve impulse as an *action potential* – a term still remaining in common use today.

In another set of experiments, Helmholtz examined the speed of the nerve impulse in humans. He did this by asking subjects to push a button when they felt various parts of their body touched by a weak electrical current. By using a simple method of subtraction, that is by comparing the difference in reported time between the two stimuli (e.g. touching the toe or thigh, or shoulder and wrist), Helmholtz managed to measure the rate of sensory nerve conduction to the brain. He found it to be between 165 and 330 feet per second – a figure roughly twice as fast than occurs in the frog. These subtraction methods also enabled Helmholtz to judge how long it took the brain to make a decision, or as he put it 'the process of perceiving and willing'. Although this figure varied greatly, he estimated the decision time to be around 0.1 second. With this transition from frogs to humans, Helmholtz's new reaction time methodology now provided a way for other investigators to begin measuring the speed of mental activities. It was, in effect, the beginnings of experimental psychology.

The measurement of the nervous impulse by Helmholtz in 1850 took place less than 60 years after the discovery of animal electricity by Galvani in 1791. It had been an extraordinary period of time for scientific progress into understanding electricity and the nervous system. The work of Galvani was a major turning point in overturning the antiquated notion of the nervous system containing hollow tubes through which *pneuma* or animal spirits flowed. However, it would require the combined efforts of many other researchers including Matteucci, Du Bois-Reymond and Helmholtz to build on this work, and show all living organisms were not powered by a spiritual 'vital force'. Instead, they were governed by physical and chemical processes that were understandable through scientific investigation. Du Bois-Reymond was to make this point most clearly in his famous oath: 'No forces other than the common physical–chemical ones are active within the organism.' Galvani could never have realised when he first observed a frog leg twitch in response to a nearby spark from a Leyden jar, that his theory of animal electricity would be so pivotal in laying down the foundations of modern electrophysiology and neuroscience. Although there was still much to learn about the nerve impulse, the scientific road ahead was now open for these discoveries to be made.

Notes

1. We now know that Gilbert was describing static electricity, which is generated by the build-up of negatively charged electrons within a material, which in turn creates a force by attracting positively charged particles.
2. It is said that van Musschenbroek tested the apparatus on his student Andreas Cunaeus and subjected him to such a powerful shock that he refused to partake in any more investigations.
3. Among his many achievements were his travels in Latin America and the mapping of the Orinico River.
4. First communicated to the Royal Society of London on 20 March 1800.
5. We still talk of being 'galvanised' into action.
6. It is said that a Mr Pass, who was beadle of the Surgeon's Company, was so alarmed during the demonstration that he died of fright soon after his return home.
7. *Experiments on Life and its Primary Forces through the Use of Experimental Physiology* (1817)
8. The sciatic nerve runs from the lower leg and innervates the leg.
9. Although the 'injury current' is a genuine electrical phenomenon, and could be said to support the idea of animal electricity, it is also an artefact of the experimental procedure. This is because the

injured or interior part of the muscle is negative with respect to the intact surface. Thus, Matteucci was providing a path for the flow of current.
10 At the time investigators were well aware that whenever a metal was put into contact with biological tissue a small current was produced. Thus the 'electrophysiologist' of the first part of the early nineteenth century was faced with an apparently unsolvable puzzle: how to demonstrate a true biological current when metal electrodes had to be used to detect it. Matteucci's frog pile overcame this objection.
11 Matteucci was correct. Tetanus does cause the nerve current to weaken. We now know tetanus induces sustained muscular contraction by producing a rapid series of nerve impulses.
12 It is a fundamental law of physics that a positive charge is attracted to a negative one (and vice versa). Similarly, a high potential (or charge) will be attracted to a lower one.
13 The oath read: No other forces than the common physical–chemical ones are active within the organism. In those cases which cannot at the time be explained by these forces one has either to find the specific way of form of their action by means of the physical mathematical method, or to assume new forces equal in dignity to the physical-chemical forces inherent in matter, reducible to the force of attraction and repulsion.
14 Helmholtz was fairly accurate in his estimation. We now know that the speed of nerve conduction varies depending on the type of nerve fibre with the impulse being faster in large diameter myelinated axons. For example, the fastest neurons can conduct action potentials at up to 200 miles per hour, with the slowest speed being around 50 miles per hour.

Bibliography

Baigrie, B.S. (2001) Bioelectricity and the origins of physiology. *Optics and Photonics News*. October, 42–45.
Barbara, J.G. and Clarac, F. (2011) Historical concepts on the relations between nerves and muscles. *Brain Research*, 1409, 3–22.
Bertucci, P. (2007) Sparks in the dark: the attraction of electricity in the eighteenth century. *Endeavour*, 31, 88–93.
Bischof, M. (1995) Vitalistic and mechanistic concepts in the history of bioelectromagnetics. In Beloussov, L.V. and Popp, F.A. (eds) *Biophotonics – Non Equilibrium and Coherent Systems in Biophysics, Biology and Biotechnology*. Bioinform Services: Moscow.
Brazier, M.A.B. (1960) The historical development of neuropsychology. In Field, J., Magoun, H.W. and Hall, V.E. (eds) *Handbook of Physiology*. Waverly: Baltimore, MD.
Bresadola, M. (1998) Medicine and science in the life of Luigi Galvani (1737–1798). *Brain Research Bulletin*, 46, 367–380.
Bresadola, M. (2008) Animal electricity at the end of the eighteenth century: the many facets of a great scientific controversy. *Journal of the History of Neurosciences*, 17, 8–32.
Cajavilca, C., Varon, J. and Sternbach, G.L. (2009) Luigi Galvani and the foundations of electrophysiology. *Resuscitation*, 80, 159–162.
Clower, W.T. (1998) The transition from animal spirits to animal electricity: a neuroscience paradigm shift. *Journal of the History of Neurosciences*, 7, 201–218.
Darrigol, O. (2003) Number and measure: Hermann von Helmholtz at the crossroads of mathematics, physics and psychology. *Studies in History and Philosophy of Science*, 34, 515–573.
De Kerk, G.J.M. (1979) Mechanism and vitalism: a history of the controversy. *Acta Biotheoretica*, 28, 1–10.
Dorsman, C. and Crommelin, C.A. (1957) The invention of the Leyden jar. *Janus*, 46, 275–280.
Dougan, A. (2008) *Raising the Dead: The Men who Created Frankenstein*. Birlinn: Edinburgh.
Durgin, W.A. (1912) *Electricity: Its History and Development*. McClurg: Chicago, IL.
Fara, P. (2002) *An Entertainment for Angels: Electricity in the Enlightenment*. Icon Books: Cambridge.
Finger, S. and Law, M.B. (1998) Karl August Weinhold and his 'science' in the era of Mary Shelley's *Frankenstein*: experiments on electricity and the restoration of life. *Journal of the History of Medicine and Allied Sciences*, 53, 161–180.

Finger, S. and Wade, N. (2002) The neuroscience of Helmholtz and the theories of Johannes Mueller Part 1: nerve cell structure, vitalism and the nerve impulse. *Journal of the History of Neurosciences*, 11, 136–155.

Finger, S. and Piccolino, M. (2011) *The Shocking Story of Electric Fishes*. Oxford University Press: New York.

Finkelstein, G. (2006) Emil du Bois-Reymond vs Ludimar Hermann. *C.R. Biologies*, 329, 340–347.

Glickstein, M. (2014) *Neuroscience: A Historical Introduction*. MIT Press: Cambridge, MA.

Helferich, G. (2004) *Humboldt's Cosmos*. Gotham Books: New York.

Hoff, H.E. (1936) Galvani and the pre-Galvanian electrophysiologists. *Annals of Science*, 1, 157–172.

Holmes, F.L. (1993) The old martyr of science: the frog in experimental physiology. *Journal of the History of Biology*, 26, 311–328.

Home, R.W. (1970) Electricity and the nervous fluid. *Journal of the History of Biology*, 3, 235–251.

Home, R.W. (2002) Fluids and forces in eighteenth-century electricity. *Endeavour*, 26, 55–59.

Kettenmann, H. (1997) Alexander von Humboldt and the concept of animal electricity. *Trends in Neurosciences*, 6, 239–242.

Kipnis, N. (1987) Luigi Galvani and the debate on animal electricity. *Annals of Science*, 44, 107–142.

Locy, W.A. (1930) *Biology and its Makers*. Henry Holt & Co.: New York.

McComas, A.J. (2011) *Galvani's Spark: The Story of the Nerve Impulse*. Oxford University Press: Oxford.

Mauro, A. (1969) The role of the voltaic pile in the Galvani-Volta controversy concerning animal vs. metallic electricity. *Journal of the History of Medicine*, April, 140–150.

Meulders, M. (2010) *Helmholtz: From Enlightenment to Neuroscience*. MIT Press: Cambridge, MA.

Morus, I.R. (1998) Galvanic cultures: electricity and life in the early nineteenth century. *Endeavour*, 22, 7–11.

Morus, I.R. (2011) *Shocking Bodies: Life, Death and Electricity in Victorian England*. The History Press: Stroud, Glos.

Moruzzi, G. (1996) The electrophysiological work of Carlo Matteucci. *Brain Research Bulletin*, 40, 69–91.

Parent, A. (2004) Giovanni Aldini: from animal electricity to human brain stimulation. *The Canadian Journal of Neurological Sciences*, 31, 576–584.

Pera, M. (1992) *The Ambiguous Frog: The Galvani-Volta controversy on Animals Electricity*. Princeton University Press: Princeton, NJ.

Piccolino, M. (1997) Luigi Galvani and animal electricity: two centuries after the foundation of electrophysiology. *Trends in Neurosciences*, 20, 443–448.

Piccolino, M. (1998) Animal electricity and the birth of electrophysiology: The legacy of Luigi Galvani. *Brain Research Bulletin*, 5, 381–407.

Piccolino, M. (2006) Luigi Galvani's path to animal electricity. *C.R. Biologies*, 329, 303–318.

Piccolino, M. and Wade, N.J. (2011) Carlo Matteucci (1811–1868), the 'frogs pile' and the Risorgimento of electrophysiology. *Cortex*, Sept, 1–2.

Prosser, C.L. and Curtis, B.A. and Esmail, M. (2009) *A History of Nerve, Muscle and Synapse Physiology*. Stipes Pub: IL.

Rothschuh, K.E. (1973) *History of Physiology*. Krieger: New York.

Schmidgen, H. (2002) Of frogs and men: the origins of psychophysiological time experiments, 1850–1865. *Endeavour*, 26, 142–148.

Schuetze, S.M. (1983) The discovery of the action potential. *Trends in Neurosciences*, 6, 164–168.

Sleigh, C. (1998) Life, death and Galvanism. *Studies in History, Philosophy and Biomedical Science*, 29, 219–248.

Steinbach, H.B. (1950) Animal Electricity. *Scientific American*, 183, 40–43.

Walker, W.C. (1937) Animal electricity before Galvani. *Annals of Science*, 2, 84–113.

Westheimer, G. (1983) Hermann Helmholtz and origins of sensory physiology. *Trends in Neurosciences*, 6, 5–9.

Whitaker, H., Smith C.U.M. and Finger, S. (eds) *Brain, Mind and Medicine: Essays in Eighteenth Century Neuroscience*. Springer: New York.

6

THE RISE AND FALL OF PHRENOLOGY

As the skull takes its shape from the brain, the surface of the skull can be read as an accurate index of psychological aptitudes and tendencies.

Franz Joseph Gall

The writings of Drs Gall and Spurzheim have not added one fact to the stock of our knowledge respecting either the structures or the functions of man . . . Such is the trash, the despicable trumpery, which the two men, calling themselves scientific inquirers have the impudence gravely to present to the physiologists of the nineteenth century, as specimens of reasoning and induction.

Dr John Gordon

Summary

In 1798, the German born Franz Joseph Gall proposed a new 'science' of the human mind, which became more widely known as phrenology – a term popularised by his one-time colleague and later rival Johan Spurzheim. This was a radical departure from previous ideas since the new doctrine proposed the brain's cortical surface was made up of a number of different regions with specific mental or temperamental faculties. The implications of this theory were also an affront to many as it assumed the underlying physiology of the brain provided the seat of one's character and behaviour. Today, this view is not seen as unreasonable. However, in the late eighteenth century it was a concept which radically went beyond the philosophical and religious teachings of the day. In particular, it seemed to deny that a self-determining spiritual entity, or mind, governed the morals and responsibilities of the individual. Although Gall never achieved the scientific recognition he sought for his work, and perhaps in some respects deserved, phrenology was to develop into the most popular fad of the nineteenth century, capturing the imagination of many who saw it as a means of self-improvement and a vehicle for social and educational reform. Today, it is regarded as little more than a form of character divination, undertaken by feeling the bumps of the head, somewhat akin to astrology and palmistry. Even in Gall's day, phrenology was ridiculed by many investigators. Yet, phrenology was to prove an important step forward in the history of neuroscience, especially the biological study of the mind. For one thing, phrenology's rise in popularity, despite the protestations of the Church, quickly led to a change in the academic climate where the soul was exorcised from brain functioning, thereby making it more amenable to objective and scientific investigation. Phrenology also helped establish psychology as a biological science, encouraging a more naturalistic approach to the study of behaviour, and paving the way for evolutionist theories that saw man as part of the animal kingdom. However, perhaps most important of all, was phrenology's insistence that mental faculties could be localised to discrete areas of the brain. While this idea was rejected for much of the nineteenth century by brain researchers, including Pierre Flourens who was widely regarded as the greatest authority of them all, Gall was eventually proven correct – at least in some respects. Despite its many failings, phrenology represents a point in the historical development of neuroscience where a significant break with the dogma and assumptions of the past was made. Consequently, it was crucially important in formulating a more modern way of understanding the brain.

Franz Joseph Gall: the founder of phrenology

Around the same time as Galvani and Volta were formulating their theories on the nature of animal electricity in Italy, a different type of revolution concerning the relationship between brain and mind was taking place in Austria, with the emergence of phrenology. The founder of this movement was Franz Joseph Gall whose great break with the past concerned the brain's cerebral cortex, which he believed contained localised areas with specific intellectual or temperamental faculties. This was a theory which directly opposed the Christian belief prevalent in the late eighteenth century that people were spiritual entities with unitary souls. However, Gall's intention was more than simply describing the faculties of the mind: he also believed this knowledge could be used to construct an entirely brand new science of human nature with implications for the individual and society. Gall initially called his system 'schädellehre' (the doctrine of the skull), and later 'craniology' (the science of the head) and 'organology'. However, much to his disapproval, the system become known as 'phrenology' (derived from the Greek *phrēn* meaning 'mind' and *logos* meaning 'knowledge'). While distancing himself from the more popularist phrenological developments which were to arise in the nineteenth century, it nevertheless remains that Gall was the inventor of phrenology's main principles and methodology, which held the skull was moulded by the shape of the underlying brain. Hence, a large cerebral organ was associated with a cranial protuberance, making it possible to measure people's character by examining the shape of the skull. This also marked a departure from previous theories of self for it recognised the brain's physiology as providing the seat of all mental functions, including traits such as love, moral behaviour and spiritual character. It was a materialistic theory which flatly contradicted the theological teachings of the time which viewed the soul, and not the brain, as the agent freely responsible for all personal choice.

Gall was described in one biography as a complex man who had three passions: science, gardening, and women.[1] The son of an aristocratic Italian wool merchant, born in the small village of Tiefenbrunn located in southwest Germany, Gall was encouraged to enter the Roman Catholic priesthood, but chose to pursue medicine instead. Beginning his studies at Strasbourg where he learned dissection and anatomy from the French naturalist Johann Hermann, Gall completed his training in 1785 under Maximilian Stoll at the University of Vienna, who was then the most famous Austrian physician of his day. This training held Gall in good stead for he was offered the post of personal physician to the Holy Roman Emperor Franz II.[2] However, even at this early stage of his career, Gall was intent in pursuing research into the subject of physiognomy – the belief that a person's intelligence and temperament could be determined by the physical appearance of their face. This was a popular subject in the late eighteenth century,[3] and Gall later confessed he had become drawn to it as a nine-year-old schoolboy when he noted a classmate with bulging eyes who had a remarkable ability to memorise pages of text. Later, at university, Gall came across other students with similar characteristics – an observation leading him to believe the ability for remembering passages of verbal material lay in the frontal region of the brain behind the eyes. This discovery, it seems, was the first step leading Gall to consider the possibility that other psychological functions might be localisable in the same manner.

Gall began exploring these ideas more systematically in the early 1790s, as a doctor with access to a large psychiatric hospital where he had free reign to examine the facial and cranial features of its patients. It soon led Gall to extend his pursuit to other individuals with various gifts including writers, artists and statesman, as well as those with criminal tendencies such as thieves and murderers. He was also not adverse to using unconventional practices to broaden his knowledge, for it is said Gall enticed many 'street urchins' into his home with cake and brandy so he could learn more about the lower classes. Nonetheless, Gall's motive was an honourable one: he sought to correlate character traits with cranial prominences. Gall was also interested in examining the skulls of the deceased, and if given the opportunity, comparing them with the brain removed at autopsy. In this respect, Gall was particularly interested in individuals who had shown marked bravery, cunning or cold-blooded murder. This was an interest which also required the help of the Viennese Minister of Police, and by 1802 Gall had amassed a collection of over 300 skulls. It was an obsession that did not go unnoticed in Vienna where it is said that some of its citizens were so alarmed they wrote clauses in their wills prohibiting the use of their skulls for Gall's research.

By the end of the eighteenth century, Gall had established the basic tenets of his craniology, which stressed how the mind's moral and cognitive abilities were physiologically determined and governed by localised regions of the cerebral cortex. Although he appeared hesitant to publish his ideas, Gall did lay out the fundamental basis of his system in 1798 when one of his letters was published in the main literary journal of the Holy Roman Empire *Der neue Teutsche*. However, in the main, Gall presented his findings in public lectures, some given in his own home, which he often illustrated with examples of brain dissection. One person to attend these lectures was Johann Casper Spurzheim, later to become Gall's most important disciple and collaborator. While the lectures proved highly popular in the social circles of Vienna, with some taking it upon themselves to publish their own pamphlets and notices, they also attracted the disapproving attention of the authorities. This was not only because they were seen as subversive to religious and state authority, but rumours abounded that Gall's talks frequently included sexual themes which encouraged impropriety – a situation perhaps not helped by his reputation for womanising. Consequently, Gall was told to stop his lectures in 1801 by the Emperor Francis who accused him of going against 'the first principles of

FIGURE 6.1 Franz Joseph Gall (1757–1828) the founder of craniology which assumed the mind's moral and cognitive abilities were physiologically determined and governed by localised regions in the cerebral cortex.

Source: Wellcome Library, London.

morality and religion'.[4] Two weeks later, Gall was formally ordered to stop lecturing and prohibited from publishing his work by the Austrian government.

There is little doubt this situation gave Gall's craniology a considerable amount of publicity. Indeed, it is said that twice as many pamphlets on his system were published in 1802 than in the year before. Encouraged by this interest, Gall decided to leave Vienna in 1805 at the age of 47, to undertake a lecture tour of central Europe. Accompanying him was an entourage including his 'attendant and dissectionist' Spurzheim, a servant, a wax modeller, two monkeys and the best part of his skull and cast collection. The tour was intended to be both an entertainment show and a scholarly enterprise, with Gall arguing he had arranged his lectures so they were of interest to everyone. However, his critics accused him of 'greed' and 'charlatanism'. Whatever the truth of the matter, Gall became one of the most famous men in Europe. The tour would be so successful that it continued for over two years, with visits to over fifty cities in Germany, France, Switzerland, Holland and Denmark – while all the time steering clear of the distractions posed by the Napoleonic wars. Gall was also feted by the rich and famous, although the unorthodox practice of disseminating his research discoveries through a sensational lecture tour did not impress the more conservative and critical academic authorities. Nonetheless, the debates and controversies surrounding craniology were widely reported in magazines and periodicals of the day where they generated much curiosity. For the first time in its history, the subject of the brain was attracting a popular interest.

Gall in Paris

Following the tour, Gall settled in Paris, and except for a brief trip to England in 1823, lived there for the rest of his life where he was relatively free to lecture and publish his work. He arrived in 1807 as an internationally famous figure and only planned to stay for one year. However, since Gall spoke French fluently, he soon established himself as one of the capital's most popular physicians with a high class clientele which included ten ambassadors. Gall also sought to gain recognition for his work by joining the main French academic societies of the time, although in this regard he was less successful. There were many factors working against Gall, not least Napoleon's known antipathy towards him,[5] and perhaps the Emperor's general dislike of Germans. There were other reasons also, not least the social and religious implications surrounding craniology which ran against the prevailing orthodoxy of the times. Consequently, Gall would find his attempts to become part of the respected scientific community in France thwarted. Although Gall and Spurzheim had submitted a report of their much admired and less controversial anatomical work (see below) to the Institut de France just four months after arriving in Paris, they were denied membership. This would have given Gall a commanding position in world science. The reason for the rejection has been subject to much speculation, although there were clearly political motives at work. Later, in 1821, long after Napoleon had been exiled, Gall would again be refused membership of the Académie des Sciences (the Institut's successor) with the same outcome, despite having become a naturalised French citizen.

These setbacks, however, did not deter Gall's determination to pursue his interest in craniology. Soon after settling in Paris, he began writing what turned out to be a massive four volume work entitled *Anatomie et physiologie du système nerveux*[6] published between 1810 and 1819 with the first two books co-authored with Spurzheim (1810 and 1812). Written in French, the book included an atlas of over 100 plates which made it very expensive (1000 francs) and a limited circulation. Although receiving praise in some quarters, the book attracted much criticism. In response to this censure, which included an accusation of atheism (a charge he denied by pointing out the brain had an innate faculty of religion), Gall published a revised and much cheaper six-volume edition entitled *Sur les fonctions du cerveau*[7] between 1822 and 1825. This work, aimed more at the general public, also omitted the name of Spurzheim. Interestingly, there would be no English translations of this book until *On the Functions of the Brain* appeared in 1835. Even then, the book was only published in America. One reason for this was undoubtedly due to the fact that Spurzheim's phrenology had, by now, far superseded Gall's craniology in general popularity.

The neuroanatomy of Gall and Spurzheim

In the controversy surrounding Gall, it is often overlooked that the first volume of his work with Spurzheim is concerned entirely with neuroanatomy. Furthermore, the two men would make a number of significant contributions to the field. Indeed, one of the book's main innovations was a novel form of dissection. Instead of slicing the brain in sections as most others had done with a knife, Gall and Spurzheim hardened it in alcohol, and then took it apart from the bottom upwards, teasing away its structures with their fingers – a technique allowing the brain's fibre bundles to be observed and followed. While the method was not entirely original, going back to the sixteenth century Italian anatomist Arcangelo

Piccolhommi,[8] Gall and Spurzheim used it to make several new discoveries. For example, they were to realise that the brain's white matter was made up of fibrous material, unlike the grey which had a different composition. Although Thomas Willis has made a similar claim in his *Cerebri anatome* (1664), Gall and Spurzheim went much further. They showed the white matter contained two types of fibre: (1) those projecting considerable distances to other brain regions or the spinal cord; and (2) those fibres confined to the cerebral cortex or passing through its commissures between the two hemispheres. Today, these are known as 'projection' and 'association' fibres respectively. It was an important step forward in understanding how the different parts of the brain were connected and organised.

By pulling away and examining the divergent fibres, Gall and Spurzheim made another fundamental discovery: they demonstrated that the white matter of the *medulla oblongata* (located in the brainstem) gave rise to a 'bundle of threads' that ascended upwards into the forebrain and fanned out into all regions of the *cerebral cortex* to reach its surface. In addition, the thalamus which sits centrally in the brain gave rise to similar threads. Importantly, the distribution of its fibres was not uniform, but patchy, innervating some cortical areas more intensely than others. This provided strong anatomical support for the idea that the two cerebral hemispheres contained many different and localised nervous centres. In other words, there was no single centre in the cortex where everything converged. Today, we know that Gall and Spurzheim were actually describing a system called the *reticular activating system* which is involved in maintaining the brain's arousal. Nonetheless, at the time, these fibres provided anatomical justification for their theory of craniology.

In addition, Gall and Spurzheim also realised that the cerebral cortex not only received fibres from the medulla, but was the origin of fibrous pathways travelling down into the *medulla*. In fact, these fibres passed through a pyramidal-shaped ridge called the *pyramidal decussation* before passing to the spinal cord.[9] It had been known since the work of Giovanni Morgagni in the sixteenth century that injuries to one side of the brain produced paralysis and weakness to the opposite side of the body. Later, in 1709, the Italian doctor Domenico Mistichelli speculated this might occur because descending pathways crossed in the pyramidal region of the brain. Gall and Spurzheim provided support for this idea by showing the fibres of the right pyramid terminating in the left side of the spinal marrow, and vice versa. Another of Gall and Spurzheim's neuroanatomical achievements were to trace the origin of the cranial nerves. In doing this, they were to show that the trigeminal nerve was not merely attached to the pons as had previously been thought, but extended to a structure called the *inferior olive* in the medulla.

Of course, Gall and Spurzheim's most important contribution to neuroanatomy concerned the organisation of the cerebral cortex. Keen to elucidate its structure, they not only provided the most detailed and accurate illustrations of its surface features, but recognised that the shape of its gyri and foldings were consistent across individuals. They also recognised that the foldings of the cortex, so prominent in man, helped to greatly increase its surface area. Perhaps, just as important, by describing the cortex as a multitude of different organs they also changed thinking about its function. To put their achievements in some perspective, prior to their investigations it was widely believed the cortex was merely a protective rind (*cortex* means 'rind' in Latin), a glandular structure (the early microscopists saw globules in the cortex), or a vascular structure made up of small blood vessels. Gall and Spurzheim had helped change this type of thinking, and the German anatomist Johann Reil was to write: 'In Gall's anatomic demonstrations of the brain I have seen more than I thought possible for one man to discover in the course of a long life time'. Even Pierre Flourens who proved to be the most vociferous critic of phrenology, and more than any other responsible for its ridicule, confessed an admiration for Gall's anatomical work.

Exorcising the soul from the brain

Gall's '*schädellehre*' provided such a radically new way of understanding human nature, that it appeared to many onlookers to have come from nowhere. However, Gall's materialistic organology was not without its antecedents. Indeed, it must be remembered Gall developed his theories at the height of the Enlightenment when alternative views were more acceptable. One person to influence Gall was the Swiss naturalist Charles Bonnet who suggested the brain might be an 'assemblage of different organs' in his *Analytical Essay on the Faculties of the Soul* of 1760. Bonnet came to this view after his 89-year old grandfather, almost blind from cataracts, began to experience an extraordinary range of visual hallucinations. With remarkable insight, Bonnet identified the hallucinations as arising from the 'irritation of the fibres of the visual organ'. It was an explanation that presumed his grandfather had a 'over active' visual region of the brain. In other words, the brain was not a unitary structure after all. Another influence on Gall was the German philosopher and Lutheran pastor Johann Gottfried Herder who argued in the 1790s that the traditional division between mind and body was misconceived. For him, the soul of humans, properly understood, is 'physiology at every step'. He also believed that every function in the body must have a controlling organ in the brain. Interestingly, unknown to Gall, there was also yet another important proponent of the localisation view. This was the Swedish theologian Emanuel Swedenborg, who founded the Swedenborgian church in 1784. Swedenborg had earlier turned his attention to investigating the brain, and realised that the cerebral hemispheres contained regions with different functions. He appears to have done this largely from observations of individuals who had suffered strokes. Sadly, this work would lay forgotten in the royal Swedish Archives for over a century until it was re-discovered in 1868.

Nevertheless, the general consensus at the end of the eighteenth century was that the cerebral cortex acted as a unified whole – a belief bolstered by the work of Haller who had been unable to show any differentiation of cortical function in his stimulation experiments. However, Gall advocated the opposite view, by proposing the mind is not one functioning unit, but a collection of different parts. He also argued the person's aptitudes and personality originated from distinct structures from within the cerebral cortex. Gall most clearly expressed his thinking in this matter in the forward of *Anatomie* where he lays down the four basic assumptions of his system. According to Gall these were: (1) the moral and intellectual faculties are innate; (2) their exercise or manifestation depends on brain organisation; (3) the brain is the organ of all propensities, sentiments and faculties; (4) each propensity, sentiment and faculty is governed by a specific organ. Although it may not be immediately apparent, Gall was proposing that the mind arises from the material or physical operation of the brain. If this was true, then it seemed to many to reject the idea of humans having a divine and spiritual soul with choice and freedom. Taken to its logical conclusion, it therefore reduced man to the status of an animal without responsibility. Although Gall denied he was an atheist, he nonetheless did little about refuting this objection, apart from claiming that the soul depended on the 'material organs' of the brain. Craniology was therefore the first theory of the brain to exorcise the vitalistic soul from its functioning.

Feel the bumps, know the man

Although the personal divination of head-reading was only one aspect of Gall's organology, it was the one that attracted the most attention, and it became the best remembered aspect of phrenology. Because Gall believed the size of a particular brain region had a direct effect on determining the shape of the skull, it followed therefore that a careful mapping of its protrusions and depressions would reveal the nature of the person's individuality. As already mentioned, Gall established his first faculty as a schoolboy when he noticed a classmate with bulging eyes with a good rote memory. It was a striking feature, which Gall would later confirm as a university student when he noticed other students with the same characteristics. Gall would use this basic method to discover other organs, and by the time he came to write his *Anatomie*, he had extended the number of faculties to 27. Although honest enough to admit he had not discovered them all (his organs only covered about two-thirds of the cranial surface) Gall divided his faculties into two groups: those shared by humans and animals, and those uniquely human (see Figure 6.2). The partitioning of the mind into smaller parts was controversial enough, but making human comparisons with animal behaviour was even more provoking – especially as this was some 50 years before Darwin's *On the Origin of Species*. Of these faculties, the instinct for reproduction, or sexual desire, which Gall placed in the cerebellum, was the most contentious. It was the first faculty to be numbered in his classification, and Gall left no doubt that he regarded it as the most important. In fact, he devoted 120 pages to the subject of sexual desire in his *Anatomie*, which was roughly six times more than he spent on any other faculty of the mind.

Gall claimed to have first discovered the faculty of sexual desire when a woman patient who was highly promiscuous collapsed into his arms. As she fell, Gall noted the back of her neck was unusually thick, which led him to suspect her cerebellum was well developed. To confirm this, Gall set about examining the neck in other individuals with strong erotic inclinations. He also sought further supporting evidence, which led him to point out that males have bigger necks, which reflected their stronger sexual instinct, and the sexual foreplay of human and beast alike frequently involves attention to the neck area. Not shy to report from his own experience, Gall also confirms the nape of the neck is an erogenous zone. To bolster his claims further, Gall resorted to even more dubious arguments. These include the assertion that only animals with cerebella mate by sexual union, and this brain area swelled up and became hot as they became sexually active! Gall also stated that castration early in life affected the size of the cerebellum, although it had little effect once the animal reached maturity.

Other faculties of the mind were justified in much the same way. For instance, Gall localised 'destructiveness' to a cranial region just above the ear by observing it was the largest part of the cranium in carnivores. He also found it to be prominent in a professional executioner. Similarly, he identified the organ of 'cunning' (situated above the temple region) as particularly well-developed in a group of thieves, although it was noticeably flat in a group of errand-boys who had been invited to Gall's house because of their honesty and integrity. Although Gall's preferred method was to examine human skulls, he also compared the crania of different species, and was particularly interested in pets with unusual abilities – including a 'lost' dog that had been able to find its way home from a distant place. Yet, incomprehensibly, Gall took little interest in individuals where brain damage was present, or what he referred to as 'accidents of nature', as in the case of stroke or cerebral injury. Had Gall examined such individuals, it is likely his craniology would have enjoyed far more respect. Nor did Gall show any interest in experimentation.

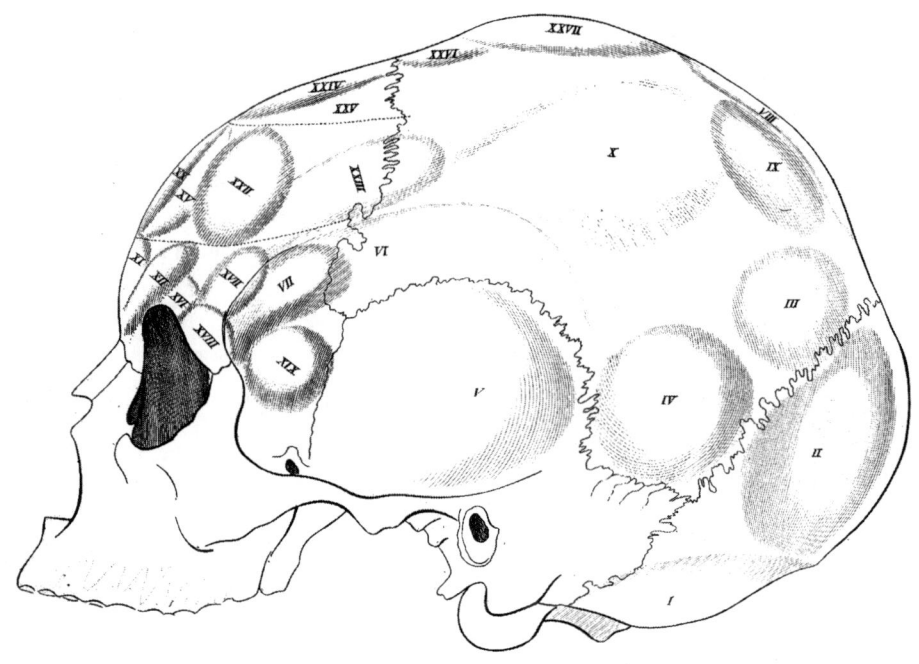

Faculties shared by humans and animals		*Directly human*	
1	Reproductive instinct	20	Wisdom
2	Love of one's offspring	21	Sense of metaphysics
3	Affection or friendship	22	Satire and wit
4	Instinct of self-defence or courage	23	Poetic talent
5	Destructiveness or tendency to murder	24	Kindness
6	Cunning	25	Mimicry
7	Desire to possess things	26	Religious sentiment
8	Pride	27	Firmness of purpose
9	Vanity or ambition		
10	Circumspection		
11	Memory for facts and things		
12	Sense of place		
13	Memory for people		
14	Memory for words		
15	Sense of language		
16	Sense of colour		
17	Gift of music		
18	Sense of numbers		
19	Mechanical or architectural sense		

FIGURE 6.2 A plate taken from Gall's *Anatomie et physiologie du système nerveux*, followed by a table listing his faculties as proposed in 1812.

Source: the plate is from the Wellcome Library, London.

Craniology and social reform

To make matters worse for its critics, craniology was the first theory in the history of the brain with wider implications for social, educational and even political change. Indeed, from an early stage in his thinking, Gall was interested in the causes of criminality and would go on to use his theories to advocate the reform of the penal system. Controversially, because he regarded the moral and intellectual tendencies of an individual to be innate, Gall argued for criminal behaviour to be considered as a disease of the brain, rather than a result of moral laxity. This belief also dictated treatment. Because the cerebral organs like other parts of the body can be strengthened by use and weakened by disuse, the purpose of treatment, therefore for Gall, was to identify the 'faulty organs' of the brain, and then attempt a re-education of the individual. Although Gall accepted punishment was a necessary deterrent for making criminals suppress their natural inclinations, his critics nonetheless pointed out that if humans have little responsibility for their actions, then their deviant behaviour could not be regarded as sinful or evil. In other words, it made criminal behaviour more acceptable or even tolerable. This was not a position many in authority were prepared to accept.

Craniology also had potentially important ramifications for educational reform. In fact, Gall believed that education, including moral and religious teaching, should be targeted to stimulate the unique human organs of the brain at the expense of the more primitive animal ones. Or as he put it: the 'faculties of an inferior order' can be made 'to obey those of a superior order'. Thus, Gall suggested the organologist could make an analysis of the strengths and weaknesses of an individual's character, and then recommend ways in which the 'good' faculties can be trained to overcome the effects of the 'bad' ones. Gall also regarded education as essential for all citizens, and made no distinctions between individuals of different races, stating that 'All men have the same brains, therefore the same faculties and tendencies . . . a Negro and a European stand on the same level of the scale of the animal kingdom.' However, for its opponents, craniology with its tolerance and promise of reactionary new solutions to the problems of education, was deeply amoral and troubling.

Spurzheim and the popularisation of phrenology

Johann Spurzheim was perhaps an unlikely collaborator for Gall, having studied Hebrew, divinity and philosophy at the German University of Trèves (Trier) before moving to Vienna in 1799. But upon meeting Gall in 1800, Spurzheim was so inspired by the new science of craniology that he began studying for his medical degree the same year. He also became Gall's personal assistant, a position requiring him to perform brain dissections in lectures and demonstrations. Spurzheim proved himself adept at this task and took over much of the responsibility for their anatomical research. However, over the course of their relationship, the two men came to hold different opinions about the nature and purpose of craniology, which would cause their partnership to end in 1813. The exact reasons for this split are not clear, although it appears that Spurzheim wanted to broaden the subject's appeal and extend its benefits to more people. The two men also disagreed on the nature of human nature with Gall believing in evilness as a fundamental propensity, whereas Spurzheim had a more positive and optimistic view of the individual. Thus, intent on promoting his new vision of craniology with its emphasis on self-improvement, Spurzheim broke with Gall halfway through the writing of *Anatomie* to prepare for a lecture tour of Britain in 1814. It appears the two men had

FIGURE 6.3 Johann Spurzheim (1776–1832) who helped popularise phrenology after collaborating with Gall.

Source: Wellcome Library, London.

originally planned to go to Britain together, but Spurzheim had other ideas. He studied English for six months without Gall's knowledge and then severed their relationship in 1813.[10]

Spurzheim would make a number of lecture tours throughout Britain between 1814 and 1817, which generated considerable interest, not least because none of Gall's works were available in English at the time. It was also during the course of these lectures, sometime around 1815, the naturalist Thomas Forster is said to have suggested the word 'phrenology' which Spurzheim adopted as his own.[11] Over the course of many years, phrenology was made distinct from craniology in a number of fundamental ways. Most significantly, Spurzheim changed its emphasis to understanding the personality traits and aptitudes of the common man. This had never much interested Gall who believed the association of cranial protrudences with mental faculties could only be conclusively found in a small number of individuals possessing particularly 'well-developed' organs. However, Spurzheim thought differently, emphasising it was possible to use his method to judge personality and provide a means for self-development in all individuals and not just a select few. To stress this point, Spurzheim compared the brain with a muscle, which could be exercised and strengthened through education. Spurzheim also extended the number of Gall's faculties from 27 to 35. In doing this, he omitted the ones he thought were inherently sinful or immoral such as the propensity for murder and thievery, but added three with religious associations. As Spurzheim would write, 'All faculties in themselves were good and given for a salutary end.' This clearly made his phrenology a lot more palatable than Gall's organology. Perhaps most important of all, Spurzheim also broadened the popular appeal of phrenology by replacing Gall's use of skull diagrams with a chart that mapped the faculties on a head of a living person.[12] With this innovation, one no longer had to be an expert in brain anatomy to understand phrenology.

Spurzheim was a prolific author and the first account of his new system appeared in his *The Physiognomical System of Drs Gall and Spurzheim* in 1815. It was essentially an abridged version of Gall's four volumed *Anatomie*, and the first comprehensive account of phrenology in English, with Spurzheim confident enough to exclaim: 'It is impossible to find any object of greater importance than this, or of more durable interest.' However, the controversial reputation of Gall's craniology had preceded it, and the book attracted criticism from certain quarters including a vitriolic attack from one of the most influential literary and political magazines of its day called *The Edinburgh Review*. The condemnation was written by an anatomist and member of the Royal College of Surgeons called Dr John Gordon, and his words show the contempt to which craniology and phrenology were held by some:

> The writings of Drs Gall and Spurzheim have not added one fact to the stock of our knowledge respecting either the structures or the functions of man: but consist of such a mixture of gross errors, extravagant absurdities, downright misstatements and unmeaning quotations from scripture, as can leave no doubt, we apprehend, in the minds of honest and intelligent men as to the real ignorance, the real hypocrisy, and the real empiricism of its authors . . . Such is the trash, the despicable trumpery, which the two men, calling themselves scientific inquirers have the impudence gravely to present to the physiologists of the nineteenth century, as specimens of reasoning and induction.

In response to this damning indictment, Spurzheim hastily made his way to Edinburgh in an attempt to restore his damaged reputation. He made a special effort to confront his adversaries, and undertook a series of lectures including one in Gordon's own lecture theatre where he is said to have given an impressive demonstration of his anatomical skills. Nor did this criticism halt the wider and more general interest in phrenology. If anything the exposure appears to have increased its popularity, which Spurzheim was keen to exploit by writing several more books. This included *The Physiognomical System* and *Examination of the Objections made in Britain against the Doctrines of Gall and Spurzheim* (1817) which used the word 'phrenology' for the first time. Although Spurzheim would spend much of his time in Paris after 1817 where he married, he returned to Britain in 1825 after his writings were ridiculed in France. By now, Spurzheim was keen to distance himself from craniology, boasting that he had rendered 'systematic and philosophical what had been in Gall's hands merely rude and detached facts.'

George Combe

One person to attend Spurzheim's lectures in Edinburgh and be converted to phrenology was the Scottish barrister George Combe, who perhaps more than any other figure managed to popularise the subject. In the absence of Spurzheim who had returned to Paris in 1817, Combe wrote *Essays on Phrenology* in 1819, and would continue to publish widely read accounts of the subject up until 1847. Said to be an opinionated egoist, Combe travelled extensively through Britain, Germany, and the United States on lecture tours. He also founded The

Edinburgh Phrenological Society in 1820, the first of more than 40 societies to be founded in Britain alone. Often attracting physicians, educators or lawyers with an interest in reforming society, these groups regarded themselves as academics who frequently published their own journals. However, Combe's most famous book was written in 1828 entitled *The Constitution of Man*. This expounded the importance of phrenological principles for understanding man's purpose in the world, although it also controversially denied the existence of a Christian God.

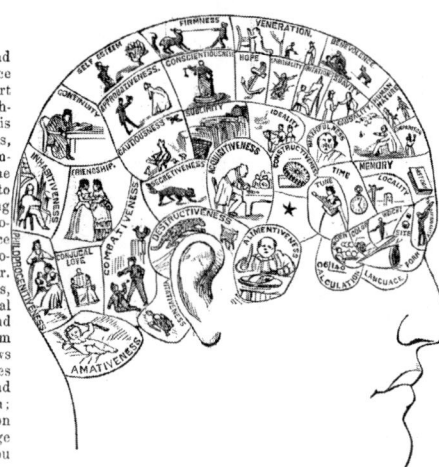

FIGURE 6.4 Leaflet advertising the practice of Professor Thomas Moores, a practical phrenologist of Leeds during the 1870s.

Source: Wellcome Library, London.

Nonetheless, it sold over 350,000 copies between 1828 and 1900, with only *The Bible* and *Pilgrim's Progress* selling more copies over this period. Combe also spent much of his time campaigning for social justice, visiting prisons and asylums, which provided phrenology with some degree of respectability. Although interest in phrenology was to show a decline in the 1830s, it remained fashionable enough for Combe to be summoned to Windsor in 1846 to advise on the education of the Prince of Wales, and examine the heads of his brother and sister.

What happened to phrenology?

After establishing itself in Britain, interest in phrenology spread to the United States where it reached its greatest popularity with both Spurzheim and Combe invited to give lecture tours. Spurzheim received a particularly rapturous welcome during his visits to New York and Boston in 1832, being made a honoured guest at Harvard and Yale Universities. This was a period, only 50 years or so after the American declaration of independence, which had promoted the importance of 'life, liberty and the pursuit of happiness'. Consequently, Spurzheim's lectures readily found an appreciative audience as they provided a seemingly scientific basis for the American belief in hard work and self-advancement. But after just six frenzied weeks of touring, perhaps exacerbated by his zeal for earning money and love of fame, Spurzheim suddenly fell ill and died. The cause according to one paper was 'over-exertion' although a more likely reason was typhoid. Spurzheim was only 56 and his death was widely mourned with his funeral taking place in Boston where it was attended by over three thousand people. One week after his death the Boston Medical Society would even go so far to state that they viewed 'the decease of Dr Spurzheim and the termination of his labours as a calamity to mankind'. Despite his untimely death, Spurzheim had agreed to have his brain, skull and heart removed and preserved in jars of alcohol, so they could be viewed by the public. It was reported that Spurzheim had a large 57 ounce (1616 gram) brain, which was some 10 ounces larger than normal. The skull is now housed in the Harvard Medical School, Boston, Massachusetts.

Some idea of the social and cultural impact phrenology had in America in the years following the death of Spurzheim can be gleaned from the way the Fowler brothers capitalised on the fad. Beginning in the 1830s, they turned phrenology into big business by touring, giving lectures and opening establishments where customers could pay a fee to get their head and features read. The Fowlers also marketed just about every type of phrenological apparatus, including heads with neatly numbered areas, and books for self analysis or prognostication. An indication of just how popular phrenology became can be gleaned by reading Hewett Watson's *Statistics on Phrenology* which records a figure of some 64,000 phrenological works and more than 15,000 plastic heads being sold in the year of 1836 alone. Phrenological exhibitions were also regular occurrences in the big cities, with their practitioners consulted on numerous issues from choosing employees, selecting marriage partners, or even diagnosing illness. This popularity was maintained until around 1845, although interest continued up until the end of the century, as shown by the fact *The American Phrenological Journal* was published from 1838 to 1911. References to phrenology also abound in nineteenth-century literature. For example, Charlotte Bronte's heroes tend to have high foreheads and wide set eyes in marked contrast to the villains, and Sherlock Holmes was known on occasion to use phrenological analysis in his criminal investigations. An interest in phrenology even led to

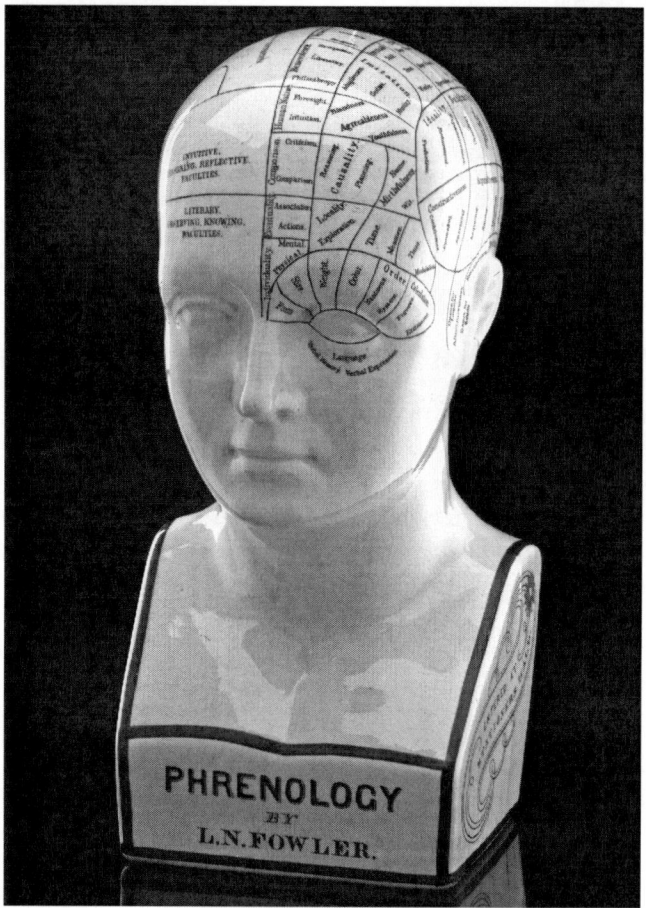

FIGURE 6.5 A phrenological bust designed by the Fowler Brothers, made in Staffordshire during the late nineteenth century.

Source: Wellcome Library, London.

the composer Haydn having his head stolen following his burial in Austria in 1809. Apparently, the thieves were interested in examining whether his 'bump of music' was 'fully developed' (it was). The skull was eventually returned to the rest of the bodily remains in 1954, completing a 145 year burial process.

Gall had died some four years before Spurzheim from a stroke in 1828 on his country estate Montrouge near Paris. He was 71. Despite his failing health, this had not stopped him marrying his younger mistress of some 13 years' standing when his first wife died in 1825. Shortly before his death, it is said Spurzheim had requested an audience with his mentor who accepted. However, in the opinion of his doctors, Gall was too weak to receive him. In life, Gall remained somewhat embittered, believing he had never gained the respect and recognition he deserved from the wider scientific community for his work. In death, the animosity continued. Despite his insistence he was not an atheist, Gall's works were placed on the Catholic list of prohibitive books (the *Index librorum prohibitorum*) in 1806, which meant that he was denied a Christian burial in consecrated ground. At his behest, Gall's cranium was added to his own collection of over 300 skulls and given to one of his doctors, called Fossati, to examine. His analysis revealed Gall's organs responsible for the 'preservation of the species' and 'friendship' were well-developed. The skull also showed a tendency for perseverance, indifference to others opinions, benevolence, wit, and a gift for inventing new ideas. His brain, which weighed 1358 grams (less than Spurzheim) was preserved in alcohol and given to a faithful disciple for safe-keeping. A collection of Gall's skulls and death masks are now on display at the Rollet Museum in Vienna.

Gall's legacy

There is little doubt Gall was sincere in his belief of craniology being a legitimate science of the mind, which helped cast away old speculative philosophical theories and merited serious academic interest. In many respects he has been proven correct. For one thing, his view that mental phenomena have a biological origin that can be 'discovered', is one virtually all modern day brain investigators would agree with. Similarly, his insistence the brain is not one functioning unit, but a collection of different autonomous parts, is another idea that is supported by today's research – including the most modern scanning techniques including fMRI. Considering these theories were introduced over 200 years ago, in an academic climate strongly opposed to such materialistic concepts, there are some grounds for regarding Gall as a true visionary who helped modernise our views of the brain. Yet, neither his theories, nor those of phrenology, ever received scientific credibility, and the reasons for this are not hard to understand. Despite its strengths, Gall's organology was also deeply flawed in many respects.

One of the most damning indictments of organology and phrenology is the assumption that a dominant or well-developed faculty of the mind can be revealed and assessed by feeling the surface contours of the skull. A related premise is Gall's belief that innate predispositions, or parts of the brain that were used a great deal, would become larger than neglected ones. Again, the consequence of this growth would be to bulge out the shape of the overlying cranium. Today, we know, as did researchers in the nineteenth century, that the shape of the brain has no bearing whatsoever on the shape of the skull. Nor is there any evidence that the human brain gets bigger with experience or learning. Thus, the main assumption of organology is simply not true. Even had there been some validity in the concept of brain growth affecting the overlying skull, Gall's attempts at cranial measurement was highly subjective and not amenable to quantification or independent verification. On top of this, the faculties of mind making up Gall's classification were somewhat curious and bizarre. Today, character traits such as 'carnivorous instinct' and 'tendency for murder' have no place in the lexicon of psychology.

Another problem concerning organology were the methods Gall used to delineate his psychological faculties. These were always highly subjective and not amenable to falsification. This can be seen, for example, when Gall attempted to support his theory of the *cerebellum* as an organ of sexual instinct after an amorous woman with a thick neck fell into his arms (see above). Instead of taking objective measurements of the neck in a group of patients with high sexual desire as most researchers would today, Gall sought only to confirm his theory through other naturalistic observations. In doing this, Gall overlooked observations that did not fit their theories – a situation making it impossible to obtain contradictory evidence. Hence, when Gall encountered a man with large projecting eyes, which was supposedly a sign of good rote memory, but with a poor ability to learn verbal material, he explained it away by arguing the large eyes were the result of rickets or hydrocephalus. Similarly, when a disagreeable person was found with a well-developed organ of benevolence, Gall accounted for it by claiming the activity of other organs counteracted its effects. However, the most famous example occurred when Spurzheim is said to have examined the skull belonging to Descartes. Finding it to be remarkably small in the anterior and superior regions of the forehead (allegedly the site where the faculty for rationality and reflection was located), Spurzheim boldly proposed that Descartes was not such a great thinker as had previously been supposed. A similar example is also said to have occurred when the French anatomist Francois Magendie kept the brain of the great mathematician Pierre Laplace. Inviting Spurzheim to examine the brain, Magendie replaced it with one that had belonged to an imbecile. Unaware of the switch, Spurzheim is reported to have spoken in glowing terms of its intellectual qualities.

Yet, the overall impact of Gall's organology on the development of brain research and psychology has to be seen as a positive one. For one thing, Gall was the first major figure to pursue a science of human behaviour, which made a decisive break from the teachings of the church, by replacing the soul with a materialistic view of the mind, dependent on the physical structure of the brain. This view is known as monism (in contrast to dualism) and is the position adopted by most neuroscientists today. In addition, Gall's belief in animal-like predispositions in humans, which at the time was highly shocking, would later receive support from Darwin and Freud. It also should not be forgotten that phrenology had a major impact on the development of academic psychology as it encouraged the development of better character typing, and the use of rating scales and inventories to measure personality.

There is perhaps no other figure that divides opinion as much as Franz Joseph Gall. The eminent British neurologist Macdonald Critchley summed up Gall's influence by referring to him as a: 'great though misguided and perhaps even slightly ridiculous figure in the rise of progressive science.' The historian Edwin Boring added, that while the theory of phrenology was wrong, it was 'just right enough to further scientific thought', but others give Gall greater praise. For example, Erna Lesky in an anthropology of Gall's writings published in 1979, claimed Gall was the father of the behavioural sciences; a forerunner of Darwin; a great criminal anthropologist; and the instigator of social reform. Although one can come to their own opinion over his greatness, there is little doubt Gall was an important figure in shifting the zeitgeist of his times. By introducing new concepts and questions that were revolutionary for their time, he helped change the academic climate, and provided an important bridge to developing new theories of brain function.

Lesioning the cerebral cortex: Pierre Flourens

Although Gall had shown no inclination to test his craniology with experimental methods, his theories would encourage others to do so, including most importantly Marie-Jean Pierre Flourens. Born in the Languedoc region of France, Flourens was a child prodigy who enrolled in the medical school at Montpellier when just 15 years old. Graduating four years later in 1813, with renowned surgical skills, Flourens moved to Paris to become the protégée of the highly respected naturalist and zoologist George Cuvier. This was a time when Gall was famous in Paris, and it appears Flourens flirted with phrenology, along with materialism, in his early years there. He even wrote a complimentary review of Gall and Spurzheim's first volumes of *Anatomie*, and attended some of Gall's lectures in 1815. However, by 1820, Flourens had

FIGURE 6.6 Pierre Flourens (1794–1867). French physiologist who pioneered the lesioning technique as a way of examining the brain's involvement in behaviour.

Source: National Library of Medicine.

turned against Gall, rejecting physiological materialism in favour of Cartesian dualism. The reasons for this decisive break are not clear, although the controversies surrounding Gall are likely to have played their part. In addition, dualism was by far the most popular view in French academic circles and supported by Cuvier. There is little doubt Flourens's adoption of this doctrine would have helped him gain more influential friends, thereby assisting his rise through the scientific establishment.

In the early 1820s, Flourens initiated experimental work examining the functions of the brain through the method of ablation – or what is more commonly known today as lesioning. This is essentially the technique where one attempts to assess the behavioural workings of the brain by cutting out its parts, and then examining the consequences of the operation for the animal. Although this method goes back to Galen, it had been re-introduced as an experimental procedure by the Italian Luigi Rolando in 1809 who used it to remove the *cerebellum* and *cerebral hemispheres* in a variety of animals – operations that had caused loss of movement and sleepiness respectively. Flourens's use of the method, however, would be far superior: lesioning a number of distinct areas, and improving the post-operative procedures which enabled his animals to fully recover from their surgery before testing. For this reason, Flourens is often credited with inventing the technique and introducing it as an important means of examining the relationship between brain and behaviour. The technique of lesioning was a considerable experimental advance over the 'pricking, pinching and compression' methods used by Haller and his followers in the years before.

Flourens presented his findings to the Académie des Sciences in 1822, and published them in *Recherches expérimentales* two years later.[13] Although this work was not specifically written with the aim of discrediting phrenology, its main findings provided a damning experimental criticism of it. Flourens described the effects of brain lesions, both big and small, on a wide variety of animals. Of particular interest were the cerebral hemispheres, which most concerned the phrenologists. If the principles of phrenology were correct, then one would have expected lesions placed in different cerebral locations to produce significant variations in behaviour. However, Flourens was unable to identify any change in behaviour from small lesions placed anywhere in the cerebral cortex. Nor were the deficits pronounced when he destroyed larger areas. When Flourens removed an entire cerebral hemisphere from a pigeon, the only impairment he could identify was a loss of vision in the opposite sided eye. In fact, the only way Flourens was able to produce a striking behavioural deficiency was to remove both hemispheres – an operation that caused a loss of almost all sensation, and an inability to instigate purposeful behaviour. A decerebrate pigeon, for instance, was not able to orient to a loud noise, or show any inclination to eat, unless food was placed into its beak. The bird also remained motionless and under-aroused as in a state of perpetual sleep. Despite this, Flourens found the pigeon could fly without difficulty when thrown high into the air.

Given these results, Flourens reasoned the cerebral hemispheres not only provided the site of perception, but were responsible for exercising mental functions such as memory, will and judgement. Moreover, because the site of the lesion was unimportant, this showed the cerebral cortex could not be divided into functional units, or faculties, as the phrenologists claimed. Flourens also found there was a critical cut-off point when large lesions were made to the brain. Thus, if a function such as vision was eventually lost, others such as hearing and movement would disappear at the same point too. This was convincing evidence that sensation, perception and volition were represented diffusely throughout all parts of the cerebral cortex and functioned as an integrated whole. Or as Flourens put it: 'All sensations, all perceptions, and all volition occupy concurrently the same seat in these organs. The faculty of sensation, perception and volition is then essentially one faculty.' In other words, no one area was more important than any other. All regions of the cerebral cortex had an equipotent role in thought and behaviour. It was also damning experimental evidence against the doctrine of phrenology.

Lesioning the cerebellum and medulla

Another area of the brain to be examined by Flourens was the cerebellum, which was the region where Gall had placed the faculty of sexual desire. Flourens lesioned this structure in a number of different animals and only found an impairment limited to action and movement. The deficit also depended on the amount removed. For example, when small slices were cut from the pig's cerebellum, the animal only exhibited a slight alteration in its stance and gait. However, larger lesions caused it to 'stagger in a drunken fashion', with complete removal of the cerebellum rendering the pig unable to move or stand erect. As with the cerebral cortex, however, no specific part of the cerebellum was critical for producing these deficits. Flourens also noticed that cerebellar lesions in pigs did not affect their sight, hearing or intelligence. Interestingly, pigeons with cerebellar damage would attempt to fly (unlike those with cerebral cortex lesions) but could not co-ordinate their movements to stay aloft for any length of time. Flourens concluded that the cerebellum is responsible for controlling and co-ordinating the movement and posture involved in walking, jumping, flying and standing. This was a very different function to the one linking it with sexual instinct as proposed by Gall.

The other brain region to come under Flourens' scrutiny was the *medulla oblongata*. In evolutionary terms, this is the oldest part of the brain, otherwise known as the brainstem, which is continuous with the spinal cord. Flourens found medulla lesions produced the most devastating effects of all, with ablation typically causing the death of the animal by stopping respiration. This finding was not a surprise for fellow French physiologist Jean-César Legallois had observed the same effect in 1812 when cutting the medulla at the level of the eighth cranial nerve. Nonetheless, Flourens continued his investigations in an attempt to narrow down the location of this region more precisely. It led him to conclude the region responsible for respiration was no bigger than a pin-head. In this pursuit, Flourens also located an area nearby involved in controlling heart rate. Although the medulla appeared to contain tiny localised areas with different functions, Flourens still maintained it served an integrated function, as it has a common overall purpose in keeping the organism alive. He would refer to the *medulla oblongata* as the *noeud vital* or 'vital knot' of the brain.

Flourens published a second and much revised edition of *Recherches expérimentales* in 1842. While this included a number of other achievements, including his discovery of the ear's semicircular canals in balance (first made in 1828), the main thesis of the book had not changed. Flourens had proven to his satisfaction the medulla oblongata was responsible for maintaining the vital processes for life; the cerebellum for producing movement; and the rest of the brain including the cerebral cortex for higher cognitive functions and sensory processing. He also called this last structure the 'brain proper' (or encephalon) and attributed it with the broad function of intelligence. Flourens also believed the encephalon was subject to an overall *action commune* (or common action) causing it to act as a whole with components that 'concur, consent and are in accord'. In other words, it was not divisible into a number of smaller mental organs. How this unifying process operated Flourens could not say, but he did point out this 'grand principle' was following in the tradition of Descartes who saw the brain as the instrument of a unitary soul. Flourens also noted his findings were in accord with Haller who had not been able to distinguish with his stimulation experiments any differentiation of sensory, motor or intellectual functions through its white matter.

Phrenology re-examined

In 1842, Flourens published a small book entitled *Examen de la phrénologie*, which made his views on phrenology known to the general public. Its second edition was translated into English a few years later as *Phrenology Examined*. By now, Flourens occupied an exalted position at the Collège de France, and secretary to the Académie des Sciences. Moreover, his reputation for sound experimental research meant his authority was held in the highest regard. *Examen de la phrénologie* was dedicated to Descartes, and written according to Flourens 'in opposition to a bad philosophy'. Although the book reported little of Flourens' own experimental work, it nevertheless provided a damning indictment of the idea that mental faculties could be localised in discrete areas of the brain. As he put it: 'Gall's whole doctrine is one series of errors which press upon each other cumulatively'. Yet, Flourens it seems, had some degree of respect for his German adversary admitting that he had been impressed with Gall's anatomical work. He even conceded that Gall had done much good in reinforcing the idea of the brain as the organ of mind. However, Flourens could find nothing good to say about Spurzheim, accusing him of stealing Gall's craniology and being a fool intent on confirming his own extreme beliefs.

Gall was aware of Flourens' experimental lesioning work during the 1820s, and with some justification criticised it as one that 'mutilates all the organs at one, weakens them all, and extirpates them all at the same time.' In effect, Gall was arguing that the lesioning approach was an inappropriate and crude technique by which to examine localisation. By then, Gall was a lone voice, and after his death there were few supporters left who could provide an adequate retort in his defence. The weight of evidence was in favour of Flourens, and *Examen de la phrénologie* provided a sound refutation of phrenology. Flourens had paved the way for a new physiological approach to examining the brain, with his assurance of mental functions such as speech, thought and memory being spread through the higher brain regions, supported by experimental evidence. Thus, few were prepared to challenge Flourens in his anti-localisation beliefs, not least for fear of being ridiculed as a phrenologist. Yet, as we shall see later, the exalted position of Flourens in this regard was a bar to further progress. The truth would turn out somewhat differently. It has been said that Gall had the right type of idea, but adopted the wrong method. In contrast, Flourens used a better experimental method, but undertook work on the wrong type of animal (i.e. those lacking a well-developed forebrain). Because of this situation, it would be many years before opinion started to shift away from Flourens, to the possibility of cortical localisation again.

Notes

1. Ackerknecht, E. and Vallois, H. (1956).
2. Franz II was the last Holy Roman Emperor ruling from 1792 to 1806.
3. The popularity of physiognomy can be judged by the fact that Johann Lavater's four volume *Physiognomical Fragments* went through 55 editions between 1775 and 1810.
4. Gall vigorously protested his innocence, saying among much else that young ladies rarely came to his lectures and were always chaperoned.
5. Napoleon is alleged to have described Gall as a 'German quack'.
6. The full title was: *Anatomie et physiologie du système nerveux en general, et du cerveau en paiticulier, avec des observations sur la possibilite de reconnautre plusieurs dispositions intellectuelles et morales de l'/wmme el des animaux par la configuration de leurs têtes.*
7. The full title was: *Sur les fonctions du cerveau et sur celles de chacune de ses parties avec des obsevations sur la possibilité de reconnaitre les instincts, les penchans, les talens, ou les dispositions morales et intellectuelles des hommes et des animaux, par la configuration de leur cerveau et de leur tête.*
8. Arcangelo Piccolhommi (1526–1605) introduced the term 'cerebrum' for the cerebral cortex and 'medulla' for the white matter.
9. This pathway is now known as the *corticospinal tract* and is involved in producing voluntary movement.
10. Although he departed Paris in 1813, Spurzheim did not arrive in England until March 1814 due to a brief stop in Vienna to complete his medical degree.
11. It has also been claimed that the term 'phrenology' was invented by the celebrated American physician Benjamin Rush in 1805 (Clark and Jacyna 1987).
12. It is perhaps most telling, in private letters to his future wife, Spurzheim admits that he was most interested in making money and achieving fame for his system.
13. *Recherches expérimentales sur les propriétés et les fonctions du système nerveux.*

Bibliography

Ackerknecht, E.H. (1958) Contributions of Gall and the phrenologists to knowledge of brain function. In Poynter, F.M.L. (ed.) *The Brain and its Functions: An Anglo-American Symposium, London, 1957.* Blackwell: Oxford.

Ackerrknecht E.H. and Vallois, H.V. (1956) *Joseph Franz Gall, Inventor of Phrenology and his Collection.* University of Wisconsin Press: Madison WI.

Boring, E. (1929) *A History of Experimental Psychology.* Appleton-Century: New York.

Burrell, B. (2004) *Postcards from the Brain Museum.* Broadway Books: New York.

Clarke, E. and Jacyna, L.S. (1987) *Nineteenth Century origins of Neuroscientific Concepts.* University of California Press: Berkeley, CA.

Combe, G. (1827) *The Constitution of Man, Considered in Relation to External Objects.* Neil: Edinburgh.

Critchley, M. (1965) Neurology's debt to F.J. Gall (1758–1828). *British Medical Journal*, II, 775–781.

Davis, J.D. (1955) *Phrenology Fad and Science.* Yale University Press: New Haven, CT.

Fancher, R.E (1979) *Pioneers of Psychology.* Norton: New York.

Flourens, P. (1824) Experimental researchers on the properties and functions of the nervous system in vertebrate animals. Trans Dennis, W. (1948) *Readings in the History of Experimental Psychology.*

Flourens, P. (1846) *Phrenology Examined.* (Trans Meigs, C de L) Hogan and Thompson: Philadelphia, PA.

Fulton, J. (1927) The early phrenological societies. *Boston Medical and Surgical Journal*, 196, 398–400.

Goodwin, C.J. (2008) *A History of Modern Psychology.* John Wiley: New Jersey, NJ.

Greenblatt, S.H. (1995) Phrenology in the science and culture of the nineteenth century. *Neurosurgery*, 37, 790–804.

Hedderly, F. (1970) *Phrenology: A Study of Mind.* Fowler: London.

Hoff, T. L. (1992) Gall' psychophysiological concept of function: the rise and decline of 'internal essence'. *Brain and Cognition*, 20, 378–398.

Hothersall, D. (2004) *History of Psychology.* McGraw-Hill: New York.

Hunt, M. (1993) *The Story of Psychology.* Anchor: New York.

Kannard, B. (2009) *45 True Tales of Disturbing the Dead.* Grave Distractions Press: Nashville, TN.

Klein, D.B. (1970) *A History of Scientific Psychology.* Routledge and Kegan: London.

Lesch, J.H. (1984) *Science and Medicine in France: The Emergence of Experimental Physiology, 1790–1855.* Harvard University Press: Boston, MA.

Lesky, E. (1979) *Writings of Franz Joseph Gall.* Hans Huber: Bern.

Marshall, J.C. and Gurd, J. (1994) Franz Joseph Gall: genius or charlatan? *Journal of Neurolinguistics*, 8, 289–293.

Marshall, J.C. and Gurd, J. (1995/6) Johann Gasper Spurzheim: quack or thomist? *Journal of Neurolinguistics*, 9, 297–299.

Neuburger, M. (1981) *The Experimental Development of Experimental Brain and Spinal Cord Physiology Before Flourens.* John Hopkins University Press: Baltimore, MD.

Olmsted, J.M.D. (1953) Pierre Flourens. In Underwood, E.A. (ed.) *Science, Medicine and History II.* Oxford University Press: London.

Pearce, J.M.S. (2009) Marie-Jean-Pierre Flourens (1794–1867) and cortical localization. *European Neurology*, 61, 311–314.

Rawlings, C.E. and Rossitch, E. (1994) Franz Joseph Gall and his contribution to neuroanatomy with emphasis on the brainstem. *Surgical Neurology*, 42, 272–275.

Rezende-Cunha, F. and de Oliveira-Souza, R. (2011) The pyramidal syndrome and the pyramidal tract. *Arq Neuropsiquitr*, 69, 836–837.

Shortland, M. (1987) Courting the cerebellum: early organological and phrenological views of sexuality. *The British Journal for the history of Science*, 20, 173–199.

Simpson, D. (2005) Phrenology and the neurosciences: Contributions of F.J. Gall and J.G. Spurzheim. *The Australian and New Zealand Journal of Surgery*, 75, 475–482.

Swazey, J.P. (1970) Action propre and action commune: the localization of cerebral function. *Journal of the History of Biology*, 3, 213–234.
Temkin, O. (1947) Gall and the phrenological movement. *Bulletin of the History of Medicine*, 21, 275–321.
Temkin, O. (1953) Remarks on the neurology of Gall and Spurzheim. In Underwood, E.A. (ed.) *Science, Medicine and History II*. Oxford University Press: London.
Throesch, B. (2011) *The Schadellehre, Phrenology and Popular Science*. Honours Thesis: University of Michigan.
Tizard, B. (1959) Theories of brain localization from Flourens to Lashley. *Medical History*, 3, 132–145.
Van Wyhe, J. (2002) The authority of human nature: the *schädellehre* of Franz Joseph Gall. *The British Journal for the History of Science*, 35, 17–42.
Van Wyhe, J. (2004) *Phrenology and the Origins of Victorian Scientific Naturalism*. Ashgate: Aldershot.
Van Wyhe, J. (2004) Was phrenology a reform science?: towards a new generalisation for phrenology. *History of Science*, 42, 313–331.
Walsh, A.A. (1972) The American tour of Dr Spurzheim. *Journal of the History of Medicine*, 27, 187–205.
Wells, F.C. and Crowe, T. (2004) Leonardo da Vinci as a paradigm for modern clinical research. *The Journal of Thoracic and Cardiovascular Surgery*, 127, 929–944.
Young, R.M. (1968) The functions of the brain: Gall to Ferrier (1808–1886). *Isis*, 59, 251–268.
Young, R.M. (1970) *Mind, Brain and Adaptation in the Nineteenth Century*. Oxford University Press: Oxford.
Zola-Morgan, S. (1995) Localization of brain function: The legacy of Franz Joseph Gall (1758–1828). *Annual Review of Neuroscience*, 18, 359–383.

7

THE NERVE CELL LAID BARE

I must declare that when the neuron theory made, by almost unanimous approval, its triumphant entrance on the scientific scene, I found myself unable to follow the current of opinion.

Camillo Golgi

The brain is a world consisting of a number of unexplored continents and great stretches of unknown territory.

Santiago Ramón y Cajal

Summary

At the beginning of the nineteenth century, the fine composition of the brain was essentially unknown. Although the use of the microscope by early pioneers such as Hooke and Leeuwenhoek in the 1600s had allowed some investigators to observe what appeared to be tiny glands in the brain, the magnification of these instruments was far too weak and aberrant for nervous tissue to be seen in any further detail. Consequently, neuro-anatomists did not know what nerves looked like, or how the brain was structurally organised. This situation would be rectified by the end of the nineteenth century. An important advance was the invention of the achromatic lenses in the 1820s, creating microscopes capable of magnifying up to a thousand times with clear resolution. It was a magnification that, in theory, could expose the nervous system. Another important breakthrough was the cell theory in the late 1830s, which showed animal and plant tissue to be composed of independent, self-governing, cells. Despite this, the structure of the brain resisted visualisation. At first the new microscopes could do little better than view it as an indiscriminate mass of granulated corpuscles, and others denied the brain was composed of cells. A new method of visualising brain tissue was needed, and this occurred in 1873 when the Italian Camillo Golgi discovered the silver impregnation stain. Instead of staining nerve cells indiscriminately, it only highlighted a few individual cells in any given piece of brain tissue – a characteristic causing them to stand out in bold black relief and be fully scrutinised with microscopy. However, the person who became most famous for doing this, was the Spaniard Santiago Ramón y Cajal. His improvement of the Golgi method helped reveal the nerve cell in its entirety, allowing him to determine its various individual components. From his observations, Cajal was also able to deduce the direction of flow of information through neural networks by realising that the nerve cell collected information from its dendrites, which it passed through the cell body into the axon. Another of Cajal's great achievements took place in the 1890s when he recognised that nerve cells are separated by tiny junctions (called synapses by Charles Sherrington). It helped show that the brain was composed of individual cells, much like any other part of the body, which became known as the 'neuron doctrine'. Although this theory was disputed by Golgi who argued the brain and nervous system was joined together in an elaborate reticulum, Cajal would be proved correct. The establishment of the neuron doctrine, along with the synapse, is one of the great moments when modern neuroscience can be said to have been born.

The microscope at the turn of the nineteenth century

During the seventeenth century, the development of the microscope had opened up a brand new world of exciting new discoveries (see Chapter 4). Robert Hooke, for example, in 1665, wrote *Micrographia*, which created a sensation by revealing many common objects in unforeseen detail, and Anton van Leeuwenhoek had developed his own lenses and microscopes to observe a variety of curious life forms including blood cells, spermatozoa, and bacteria. He referred to these as 'animalcules' giving the first indication of the cellular basis of life. The microscope also allowed Marcelo Malpighi to identify capillaries in the lungs, which completed the link between arteries and veins, thereby confirming Harvey's circulation of blood in the body. However, as important as these findings were, the instrument's power of magnification was generally no better than ×50, and this remained the upper limit for the next 150 years. Moreover, the early microscopes suffered from chromatic aberrations, which produced rings of colour surrounding the focused object, along with other flaws that distorted visual images at higher magnifications. Not surprisingly, this led to many spurious observations, and rendered microscopic investigation so contentious, that by the early nineteenth century the microscope had fallen into disrepute. Indeed, the greatly respected French biologist Francois-Xavier Bichat, widely regarded as the founding father of histology and pathology, was to insist the only observations to trust were those made by the naked eye. By the beginning of the nineteenth century, the majority of physiologists agreed with him.

Further progress would have to wait until 1826 when the English wine merchant and amateur scientist Joseph Jackson Lister (father of the famous surgeon Joseph Lister) built a microscope that contained achromatic lenses. Although the first achromatic lens is believed to have been invented by an English barrister called Chester Moore Hall around 1730, he kept his invention a secret and was also seemingly unaware of its importance. Consequently, the achromatic lens was only known to a few opticians up until the nineteenth century. Lister built his microscope by combining two or more achromatic lenses of different glass, which he found greatly minimised optical sources of error. The visual acuity of this apparatus was later increased by immersing the focusing lens in oil. The results were microscopes with a 600-fold magnification and a resolution getting close to the limits imposed by light. However, other developments taking place around this time were also essential in allowing these instruments to reach their full potential in biological investigations. New ways of hardening and preserving tissue by using alcohol and other fixatives, for example, allowed specimens to be kept longer without deterioration. Mounting media such as Canada balsam enabled permanent slides to be created. And, new methods of staining with dyes such as carmine and haematoxylin added colour to different components of biological tissue making them easier to visualise. Together, these were exciting new developments in microscopy enabling biologists, at last, to have confidence in what they were observing.

The cradle of histology: Jan Purkinje

One of the first researchers to use an achromatic microscope in a biological context was Jan Evangelista Purkinje. Born in 1787, the son of an estate manager from Bohemia (now part of the Czech Republic), Purkinje was to become one of the most famous scientists of his time. Initially matriculating from theological school with the intention of ordaining as a priest, Purkinje changed his mind and turned to studying philosophy and medicine at the University of Prague instead. Here he graduated in 1819 with a thesis on the subjective effects of vision. In the same year he also discovered an unusual optical illusion, now known as the Purkinje shift, which occurs when a blue object appears brighter than a red one in decreasing light, although both appear to be of equal brightness in normal bright light.[1] Following his graduation, Purkinje worked in the anatomical institute of the university, before accepting a position as Professor of Physiology and Pathology at the University of Breslau (now Wroclaw in Poland) in 1823. This was a great achievement for Breslau was a prestigious and highly patriotic Prussian university, created in 1811, only rivalled by Berlin. Moreover, Purkinje's Czech nationality was a considerable disadvantage even though his appointment had been recommended by the great German philosopher and politician Johann Goethe. Nonetheless, Purkinje took full advantage of his new opportunity by becoming a Prussian citizen and establishing himself as a member of the German intellectual and political elite.[2]

Two years after his appointment, Purkinje asked his employers for an achromatic microscope – then a highly expensive piece of equipment. Seven years later, in 1832, his wish was granted when he received a purpose built microscope built by the best maker in Vienna (Simon Plössl). The instrument had cost 220 gulden: a sum that was equal to an average man's salary. While waiting for his microscope to arrive, Purkinje had continued his researches by using hand-held magnifying lenses to examine biological material – work leading to the discovery of the 'germinal vesicle' (the nucleus) in the eggs of birds. Another of his pastimes involved overdosing himself with various drugs and recording their visual and behavioural effects.[3] But the arrival of the achromatic microscope opened up many new possibilities, and despite being forced to start his investigations at home, due to a lack of space at the university, Purkinje set about earnestly examining the microanatomy of animal and plant tissues. In doing this, he was helped by his devoted students who, it is said, gathered around the instrument 'like a pack of hungry wolves'.

Purkinje's new microscope soon led to some notable discoveries including an account of sweat glands in 1833, the first observation of ciliary motion in embryonic cells in 1834, and the recognition of the protein-digesting power of pancreatic extracts in 1836. As a reward for his dedicated work, Purkinje was presented with a second high class microscope (built by Pistor and Schiek of Berlin) in 1836, but perhaps more importantly, he received his own university building for research and lecturing in 1839. It was, in effect, the world's first institute dedicated to microscopical research. Now known as the cradle of histology for its many discoveries, it also housed the world's first practical microtome – a bladed instrument allowing wafer thin sections of fixed size to be cut for microscopic analysis. This instrument, first constructed under Purkinje's leadership in 1841, was a major advance, for prior to its invention, microscopists had been forced to manually prepare tissue by using razors. The microtome now allowed reproducible clean cuts to be made of such thinness that light could shine through the tissue allowing its structure to be better visualised. This would prove to be particularly valuable in the examination of nervous material.

The first depiction of a nerve cell

Armed with his new microscope, Purkinje turned his attention to the brain in the mid 1830s, and enlisted one of his favourite students, Gabriel Valentin, to assist him. Purkinje had first become interested in the nervous system during 1829 when he began to examine the large nerve bundles, or ganglion, of the spinal cord by soaking them in a potassium solution, and then prising them apart with needles. This work led him to realise that the ganglia were actually composed of many smaller fibres. Although Purkinje was not the first to discover this (Leeuwenhoek had shown that nerve fibres were composed of finer tubes, which were later called 'nervous cylinders') it nonetheless encouraged him to extend his studies to the brain. The results of this endeavour were published in 1836 under Valentin's name, although it is clear that Purkinje contributed much of the work. While the histological preparation of the material was simple, involving little more than washing slices of tissue in water and potassium, Valentin managed to microscopically examine parts of the *cerebral cortex* and *cerebellum* from several animals including humans. His main finding showed that the brain was composed of tiny globules. Again, this was not new as similar structures had been seen by others including Leeuwenhoek and Malpighi. However, Valentin went further by drawing what he observed. Thus, he enjoys the distinction of being the first person to provide a detailed depiction of an individual brain cell (see Figure 7.1).

Velentin's illustration shows the globules of the brain to be flask-shaped with a fluid-like substance full of tiny granules in their interior. The globule also had a distinct outer membrane with a small dark corpuscle at its centre, which he recognised as a nucleus. There was also the beginning of a tail-like appendage that Valentin describes as a fibre. Despite this, Valentin did not believe the globule and its appendage was joined or continuous, for he had come across instances (or so he believed) where each was entirely surrounded by its own membrane. Consequently, Valentin came to regard the globule and fibre as separate entities. In other words, he reasoned they only gave the appearance of being joined together as they were closely juxtaposed. Nonetheless, he had the foresight to realise the entire nervous system, from brain to spinal cord, was comprised of just these two components. This was a significant step forward as it revealed the nervous system to be composed of vast numbers of tiny simple units. The issue of whether the cell body and its fibre were physically joined, however, would be one to be hotly debated for many years to come. In this respect, Valentin was to be proven wrong.

A year later, in 1837, Purkinje presented a more thorough account of his research to a scientific congress in Germany, describing the cellular composition of a number of brain areas, including descriptions of cells taken from the *substantia nigra, locus coeruleus, inferior olive, thalamus,*

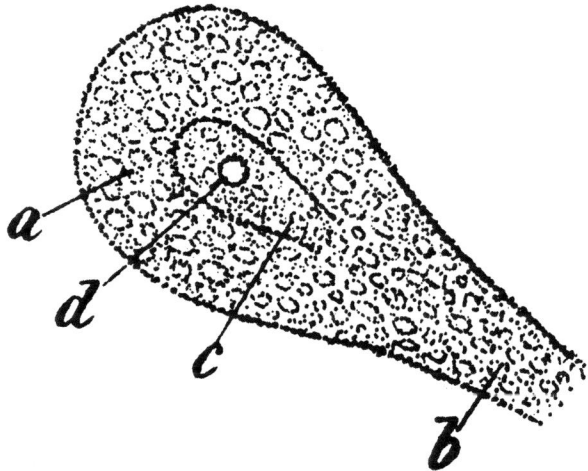

FIGURE 7.1 The first depiction of a nerve cell (or globule) by Gabriel Valentin in 1836, which is most likely a Purkinje cell taken from the cerebellum. (a) cell body; (b) appendage or the beginnings of the axon; (c) central region of the cell; (d) nucleus.

Source: Plate VII from Valentin 1836.

and *hippocampus*. This work was a seminal moment in the history of neuroscience, for it showed that different brain regions have different anatomical compositions. Purkinje's work was especially detailed regarding the structure of the cerebellum, which had several layers, with the outermost one containing a neatly organised row of large globules or 'flask-shaped ganglionic bodies'. Today, these are called Purkinje cells and recognised as some of the largest neurons in the brain. These globules also revealed the beginnings of elongated appendages, which Purkinje referred to as 'processes'. In fact, Purkinje discerned two different types of process: the first with small tails which entered the cerebellum's grey matter near its outer surface, and a second type located on the other side of the nerve body disappearing into the underlying white layer. Although there is debate over what Purkinje was describing, some commentators have speculated the first were *dendrites*,[4] and the second were nerve fibres (or *axons*). Purkinje also agreed with Valentin in recognising the nervous system as composed of both globules and fibres, but was more cautious in assessing their relationship by admitting 'nothing definite could be ascertained about their connection.'

Despite not committing himself to their connection, Purkinje had some interesting ideas on how the globule and nerve fibre might interact. In his view, the globules of the brain were confined to its grey matter where they acted as 'energy generators'. In contrast, the nerve fibres provided the 'power cables'. This theory was quite revolutionary considering telegraphy was in its infancy and the galvanometer was only just beginning to confirm the existence of electrical currents in nerve and muscle. Purkinje also proposed the nerve fibres formed long loops, descending from the brain to innervate a bodily organ such as a muscle, before returning back to the brain in a closed pathway. In this way, the electric fluid, or power, inside the nerve fibre was thought to be continuously circulating around the body. While this theory is known to be wrong today, it corresponds to a surprisingly modern idea: namely that brain cells act like electrical power generators and their fibres as telegraph lines.

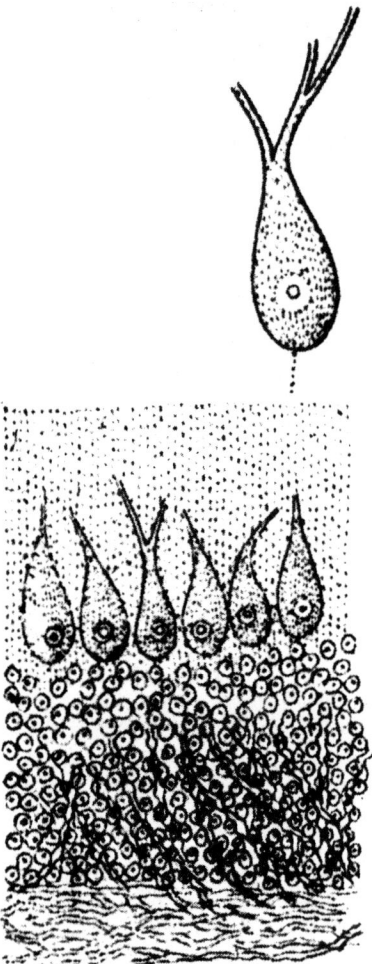

FIGURE 7.2 The cellular structure of the cerebellum presented by Purkinje. Showing what could be discerned before the Golgi silver technique was invented, it highlights from top to bottom, the large corpuscles (or Purkinje cells), a mass of smaller granular cells, and the beginning of nerve fibres.

Source: Plate II from Purkinje 1837.

In contrast to many great pioneers intent on promoting their reputation, Purkinje had an aversion to writing and publicising his own work. Indeed, one source described him as being unassuming, self-effacing and naïve, accompanied by a disinterest in personal glory or personal gain.[5] This is also reflected in the fact that many of Purkinje's discoveries are reported in the doctoral theses of his students, or brief reports and obscure lectures. In some instances, Purkinje did not even bother to put his own name on work he was supervising – thereby giving undue recognition to his collaborators. This is believed to have occurred, for example, in the case of a doctoral student called J.F. Rosenthal who used the term *achsencylinder* ('axis cylinder') in relation to the nerve fibre in 1839. This term became widely used during the nineteenth century although it is almost certainly coined by Purkinje. It would develop into the more

PURKINJE
OIL PAINTING BY PETER MAIXNER

FIGURE 7.3 Jan Evangelista Purkinje (1787–1868) who established the University of Breslau as the 'cradle of histology' where he, and his students, made many important discoveries.
Source: Wellcome Library, London.

modern term 'axon' by Kolliker around 1896. Another term to come from Purkinje's laboratory was 'protoplasm' which now refers to the watery substance inside cells. In addition, at least 18 eponyms are named after Purkinje, including Purkinje fibres found in the inner ventricular walls of the heart, and Purkinje vesicles found in the nuclear portion of an ovum

A landmark in modern biology: the cell theory

Robert Hooke could never have realised, when examining a thin slice of dried cork under his microscope in 1655, the word he would use to describe the chambers, namely 'cell' (from the Latin *cella* meaning a small room or cubicle) was to have such an enduring impact. Nor could he have known these holes were the residual skeletons of the most fundamental and simplest unit of life (i.e. the living cell) of which all plants and animals are made. In fact, Hooke believed the cells formed in the cork were simply the pipes or channels through which the natural juices of the plant flowed. And, perhaps because of this, Hooke never reported seeing any 'living cells' under his microscope.[6] However, six years after the publication of Hooke's *Micrographia*, the Royal Society received two manuscripts, which showed plants were composed of cellular units. The first came from a Warwickshire vicar called Nehemiah Grew who described plant material as composed from 'a huge mass of small cells or rigid bubbles.' The second, sent from the Italian Marcello Malpighi, reported much the same thing, except he preferred the Latin terms *utriculus* (meaning a small bottle) and *sacculae* (meaning a small bag). Although there has been debate over the priority of this discovery, it need not concern us here. More importantly, by the late seventeenth century, biologists had come to recognise the tissues of plants, at least, were largely composed of large numbers of very tiny small chambers or cells.

The realisation that animal tissue was also composed of cells took far longer. Although Leeuwenhoek reported seeing tiny particles in the blood (the first description of red blood cells) in 1673, along with other types of 'animalcule' in the years after, it occurred to nobody that these corpuscles had some homology with animal tissue. This was hardly surprising. Plant cells have large regular rectangular shapes with visible walled membranes – unlike animal cells, which are smaller and more irregularly structured. Consequently, no animal tissue examined at the time presented a comparable appearance to plant material. It was natural, therefore, to suppose animals were organised in some fundamentally different way. This obviously led to much speculation. Many agreed with Leeuwenhoek who held all animal substances were composed of vaguely defined globules, although by the eighteenth century the idea of tissue being made from fibres had become a popular alternative. And, there were other possibilities. The Italian Felice Fontana, for example, believed all animal tissue was composed of twisted cylinders (structures he observed in nerves, tendons and muscles), whereas Purkinje increased the complexity of microanatomy to three components: fluids, corpuscles and fibres.

A further development would have to wait until 1833 when the Scottish botanist Robert Brown noticed all plant cells contain a small dark nucleus. Although a nucleus had been seen many times before, Brown recognised it as a feature of *all* plant cells. If this was the case, then clearly the nucleus served an important and universal function. This soon led the German botanist Matthias Schleiden in 1838 to propose the nucleus was responsible for determining the cell's growth and intracellular activities. In other words, the nucleus acted as the command centre of the cell. From this, it was a small step to reach the more important conclusion that all plants are composed entirely of individual cells and their products. However, the situation was less clear regarding animal tissue. Later in the same year, Schleiden was having dinner with his close friend Theodor Schwann.[7] As the two men were discussing the role of the nucleus in plant cells, Schwann realised he had also seen the nucleus in embryonic animal cells. It was a vital connection for in that moment he recognised plant and animal cells shared much in common. Furthermore, if the nucleus governed the development of plant cells, then why not animal cells? Schwann outlined this new thinking in a 1839 book[8] now regarded as the starting point for two of the most important ideas in biology: (1) the cell is the basic building block of all living things; and (2) all living organisms, whether plant or animal, are composed of cells. This is now known as cell theory, and is one of the four central pillars on which modern biology is built, along with Darwin's theory of evolution; Mendel's laws of inheritance; and Crick and Watson's discovery of DNA.

The nerve cell begins to take shape: Robert Remak

Although few have heard of his name, Robert Remak was one of the most remarkable microscopists of the nineteenth century. A devout Jew born in the Polish city of Posnan, Remak would be barred from university positions for much of his life due to the anti-Semitism of his times. Consequently, he was forced to pursue his research by taking unpaid positions and supporting himself through his medical practice. Studying medicine under Johannes Müller in Berlin, where he would spend the rest of his life, Remak first obtained a compound microscope in 1836 while working towards his doctorate. He then used it to examine the different stages of neural growth in the embryonic rabbit. This proved to be an extremely good choice. As Remak watched the 'thin-walled tubes' of the peripheral nerve fibres develop, he noticed some were surrounded with a light-coloured sheath, whereas others lacked this outer coat. We now know this sheath is a fatty substance called myelin, which covers the axon. Thus, Remak was distinguishing for the first time myelinated and non-myelinated fibres. During this work, Remak also noted small gaps, or 'interruptions' in the sheathed fibres – an observation that he thought was an optical artefact. Remak was wrong. These gaps do exist in the myelin and are now known as the nodes of Ranvier, named after Louis-Antoine Ranvier who formally recognised them in 1871 – some 35 years after Remak.

Two years later, in 1838, Remak made another pivotal discovery. While observing an 'organic fibre' grow out from the spinal cord during embryonic development, he saw it emerge from a globule (or cell body). In other words, the globule and fibre were not separate entities as Valentin had claimed, but were one and the same unit. This was a discovery of enormous significance for several reasons. For one thing, it introduced the concept of one fibre per nerve cell. Also, for the first time, a nerve fibre could be conceived as arising from a cell body in the spinal cord, and running outwards into the periphery to innervate structures such as muscles. However, of even greater significance, it indicated the globule and fibre must act together as a functional unit in their own right. In other words, it was the first ever indication of the nerve cell's true existence. However, this finding was initially regarded with considerable scepticism – not least, because it ran counter to the prevailing views of the day. This situation was not helped by Remak's criticism of Valentin whom he accused of having poor microscopic technique. And, to make matters worse, Remak argued the interior of a nerve fibre was composed of a solid substance, which contradicted the great Purkinje who held that the fibre contained some form of electrical fluid.

Remak's great contribution to neuroanatomy, however, would be eclipsed by a discovery made in the late 1840s regarding the formation of life itself. Forced to finance his own research in Berlin, Remak discovered red blood cells in the chick embryo arose from a process of binary fission (the splitting of the cell into two). This was a vital contribution to cell theory in itself, but Remak went on to show this was true for all cells, including the fertilised egg (or ovum), which divided to give rise to leaf-like cell layers. In turn, these rolled up into tubes to eventually form the organs of the body. In short, Remak was the first biologist to show virtually every cell in the animal body arises through the binary fission of pre-existing cells beginning with a fertilised egg. This discovery should have made Remak one of the most famed, if not the most famous, biologists of all time. However, again, his ideas were

Robert Remak (1815–65).

FIGURE 7.4 Robert Remak (1815–1865) the first anatomist to fully describe cell division and show all animal cells derive from already existing ones. He is also noteworthy for discovering myelin and being the first to propose nerve fibres extend from the cell body.

Source: Wellcome Library, London.

poorly received. In fact, binary fission was not accepted by other biologists for another 15 years, until Rudolf Virchow in 1855 realising it might be correct, published the theory as his own. Even today, Virchow is normally credited with the discovery, while the crucial contributions of Remak are largely forgotten. It may well be that Virchow took advantage of the fact Remak was an unpopular figure in Berlin without a formal university position – a situation only rectified in 1859 when Remak's reputation had increased sufficiently to get him a salaried academic appointment at the age of 44. However, Remak saw his predicament differently as shown when writing to one of his few friends, Alexander von Humboldt, where he described his life as one 'frustrated by religious and political prejudice.'

Nerve cells are incorporated into cellular theory: Albert von Kölliker

The idea that cells form the basic unit of all living things gained general acceptance during the late 1830s. However, for many, the nervous system was an exception to the rule. Nerves were not only known to be long and spindly structures, which made them unlike any other cell in the body, but the fact that they were responsible for conveying information rapidly suggested they were joined together in some way rather than composed of separate units. To make matters worse, the components of the nervous system were very tiny and structurally complex, making them difficult to visualise even with the most powerful microscopes. It was not surprising, therefore, that two schools of thought arose concerning the structure of the nervous system. Some agreed with Remak who thought nerves were composed of cells with a fibre attached to a 'nucleated globule'. Others sided with Valentin who believed that the fibre and cell body were different entities.

One person to turn his attention to the structure of nervous tissue was the eminent Swiss anatomist Albert von Kölliker. A medical student of Zürich University, Kölliker made his greatest discovery as a young research student in 1841, when demonstrating spermatozoa and ova were both cells. This finding greatly helped to clarify the true nature of reproduction. Four years later, in 1845, in his medical thesis, Kölliker showed the division of the nucleus always preceded the splitting of the cell into two, thus confirming the embryo always arose from a single egg. These discoveries were essential contributions to cell theory, bringing Kölliker widespread recognition, and leading him to a professorship at the University of Würzburg in Bavaria in 1847 – a post he would hold for 55 years. Recognised as a tireless researcher and author of several hundred research papers on a diverse range of topics,[9] Kölliker's opinions were widely respected by his peers. His books include *Handbuch der gewebelehre des menschen* (*Handbook of Human Histology*) first published in 1852 and which went through many editions,[10] and the three-volumed *Mikroskopische anatomie* (*Microscopic Anatomy*) in 1854.

Kölliker first began examining nervous tissue in the early 1840s when still a student at Zürich by inspecting the embryonic development of the spinal cord. Like Remak, he confirmed most nerve fibres originated from a cell body located in the spinal cord. Or, as he put it: 'thin nerve fibres take their origin from ganglionic corpuscles.' In 1849, Kölliker published a number of drawings to show what he had observed. These surpassed all others by revealing the nerve fibre as a 'dark-bordered' thread-like appendage (or 'pale process') extending from a larger nucleated cell body. Today, we recognise Kölliker was describing the initial segment of the axon and its myelinated covering. These drawings were later incorporated into Kölliker's *Handbuch*, adding to the bulk of evidence supporting the idea of the nerve unit as a cell. Interestingly, while Kölliker never claimed priority for showing the continuation of the nerve fibre with its cell body, he made no mention of Remak either.[11] Perhaps it was because Kölliker provided illustrations to back up his claims, or he depicted many other types of nerve cell in the various editions of his *Handbuch*. Whatever the reason, it is Kölliker and not Remak who is normally given credit for being the first to make a convincing case for the existence of independent nerve cells.

By the 1850s there was considerable evidence supporting the idea of the 'nerve unit' being some type of cell – not least because it had an outer membrane and contained intracellular fluid with a granular-like composition. However, Kölliker was aware nerve cells exhibited certain structural features, which made them unique. One was its fibre or 'axis cylinder'. Although Kölliker believed this formed part of the cell, he was unable to convince everyone,

FIGURE 7.5 (A) Albert von Kölliker (1817–1905) who made many great contributions to the field of histology, including evidence supporting the concept of the nerve cells as units giving rise to fibres.

Source: Wellcome Library, London.

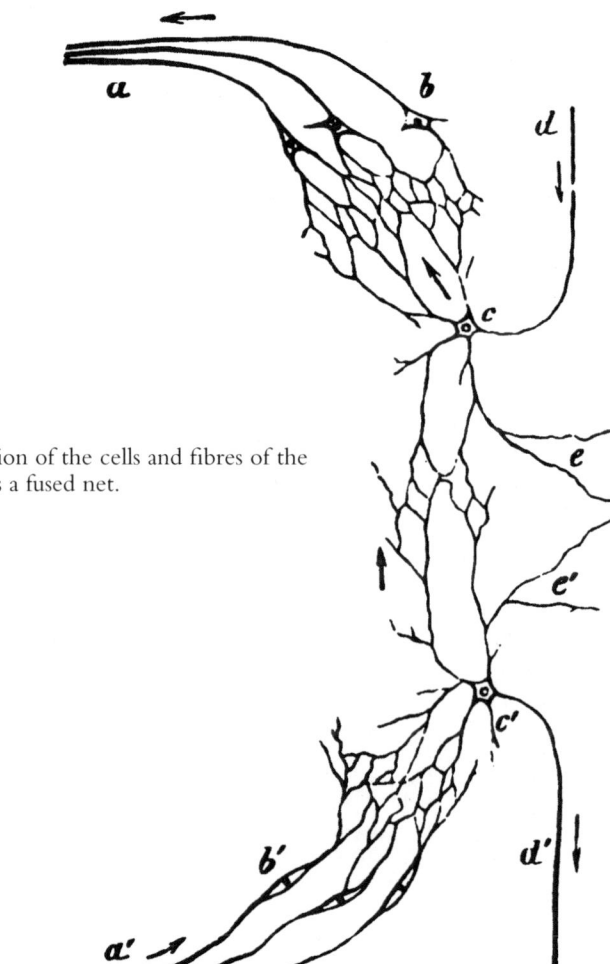

FIGURE 7.5 (B) Kölliker's illustration of the cells and fibres of the spinal cord, which shows them as a fused net.

Source: from Kölliker 1867.

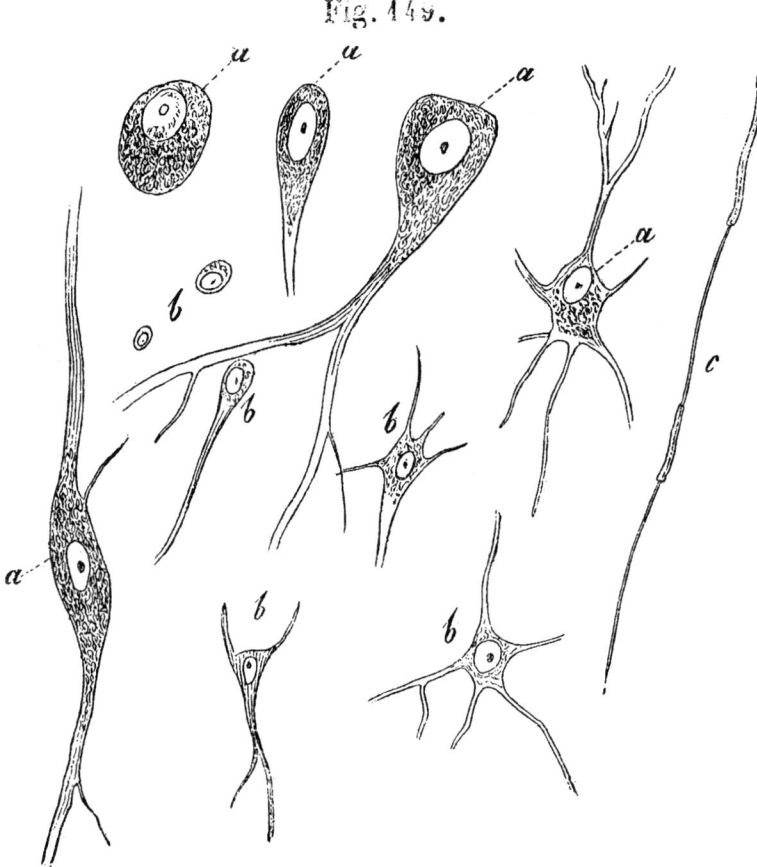

FIGURE 7.6 Cells taken from the grey layer of the human cerebral cortex as illustrated in Kölliker's *Handbuch der Gewebelehre des Menschen* published in 1853.

Source: Wellcome Library, London.

including the supporters of Valentin who maintained it was a secondary process distinct from the cell's body. The issue was further complicated by the discovery that the myelin surrounding the axon cylinder did not derive from the nerve cell itself, but from a separate cellular structure lying outside it (see page 164 and note 7). Most crucially, Kölliker was also unable to judge whether the terminal branches of the nerve axon fused with other cells, or remained separate from them. This had now become a crucial issue. All other cells in the body were known to be independent self-governing units and physically separate from one another. Clearly if nerve fibres fused with one another, as many believed, this raised serious questions about whether they were really 'cells'. The fact that suspected nerve cells in the brain were too small and intricate for them to be observed compounded the problem. Kölliker doubted whether the truth would ever be known, despairingly writing in 1853: 'it is in the highest degree probable that in many places it will be altogether impossible to demonstrate the origin of fibres from nerve cells.'

The isolated nerve cell: Otto Deiters

Kölliker had provided the first realistic illustrations of nerve cells in 1853. However, these were significantly improved upon during the 1860s, mainly as a result of an accidental discovery made by a German professor at the University of Erlangen called Joseph von Gerlach. A leading anatomist of his time, Gerlach sought new ways of staining tissue so its fine structure could be better visualised. In 1854, Gerlach began to use a dye called carmine, a pigment obtained from certain insects including cochineal, which stained various tissues red. At first, Gerlach was unable to get satisfactory results with nervous material, but in 1858 he happened to pour a solution of carmine over a section of the cerebellum previously treated with an application of ammonium carminate. Leaving the preparation overnight in his laboratory,

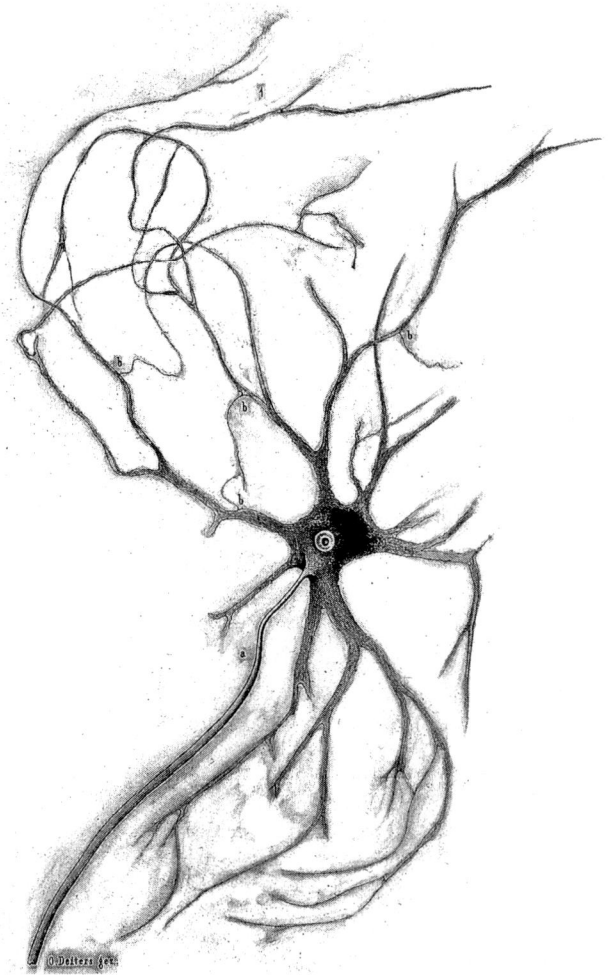

FIGURE 7.7 An illustration of a carmine stained nerve cell by Otto Deiters (magnified some 300–400 times) taken from the bovine grey matter of the anterior horn.

Source: Wellcome Library, London.

Gerlach returned the next day to find the stain had produced a clear differentiation of nerve fibres and cells. Although Gerlach was not the first to utilise stains to examine nervous tissue, or even the first to use carmine, his discovery is generally recognised as representing the modern founding of neurological staining. Gerlach also disseminated his stains and methods to other investigators.

One person to take advantage of this new technique was a young researcher called Otto Deiters at the University of Bonn. Dieters first started using carmine around 1860 when he attempted to stain nerve fibres projecting from the anterior horn of bovine spinal cord.[12] After successfully staining these fibres, Deiters was then able to use fine needles to tease out and isolate the individual 'nerve units'. While extremely delicate and painstaking work, once extracted from their surrounding tissue, the nerve units could be magnified some 300–400 times and their features scrutinised. This was the first time individual nerve cells had been isolated for microscopic analysis. More importantly, it clearly showed them with a cellular body, joined to a long, single, 'axis cylinder', which was sheathed with myelin for most of its length. Interestingly, Deiters also found that the cell body was also covered with many smaller protrusions, which he referred to as 'protoplasmic prolongations'. These are today called dendrites. Thus, Deiters can be recognised as the first neuroanatomist to clearly differentiate between dendrites and the axon.

Deiters also used staining methods to observe a number of brain areas including the *pontine area*, *inferior olive*, and *periventricular grey substance*. Remarkably, he was even able to tease out, and draw, a star-shaped cell from the brain now known as an astrocyte.[13] If nothing else, Deiters had shown the brain was a composite of different cells. He also examined the nuclei and fibre tracts of the lower brainstem (*medulla*) by using an extensive series of stained carmine sections cut in a variety of planes. This work supported the idea that the *medulla* is a vastly more complex modification of the spinal cord. Sadly, Deiters was never to see his work in print, for he tragically died of typhus at the age of 29, leaving behind an unfinished manuscript. Fortunately, his book would be edited posthumously and published in 1865 through the endeavours of colleague Max Schultz.[14] It immediately became celebrated for its exquisite illustrations of isolated nerve cells, which were considered so beautiful by some that Deiters was even compared to the great romantic poet Keats. His book also mentions for the first time two structures located in the brainstem: the *reticular formation* (formatio reticularis) and the *lateral vestibular nucleus* (now called the nucleus of Deiters).

It might have been expected from his drawings that Deiters would have supported the group of researchers led by Kölliker who supported the theory of the independent nerve cell. However, Deiters is normally recognised as a reticulist. This is the idea that fibres of the nervous system are joined together by anastomosis, much like the connection between an artery and vein. Although Deiters admitted he had never once encountered a physical joining of nerve cells, the situation was complicated by the protoplasmic protrusions – for they appeared to be connected to a second, more numerous and diffuse system of processes. Although these protrusions were too small to allow close observation, Deiters could not deny that they might 'ramify' and 'unite' together in a complex network.[15] They also appeared to be ideally located for interconnecting the populations of nerve cells in a given region. This was not a characteristic typical of a cell. As regards the nerve fibre (axon) Deiters was unable to make a judgement for he had been unable to follow them far enough to their endings. In fact, this was to pose a significant problem for neuroanatomists for many more years to come.

However, the most influential advocate of reticulum theory was to be Joseph von Gerlach. After spending many years perfecting a variety of different staining techniques, and using them in a number of preparations including invertebrates, Gerlach arrived at a position where he was convinced the nervous system was composed of a delicate but extremely complex network of 'axis fibres', which fused with the protoplasmic prolongations. Indeed, Gerlach found this pattern of connectivity in both the spinal cord and cerebral cortex. Publishing his findings in 1872, Gerlach's arguments were so convincing that most neuroanatomists fully accepted the concept of anastamosis. Kölliker too seems to have been persuaded for he had already endorsed a similar idea in the fifth edition of his *Handbuch*, published in 1867, by providing an illustration of how segments of the spinal cord might integrate sensory input and motor output fibres through large nets of joined branched extensions in its grey matter.

The silver impregnation stain: Camillo Golgi

The use of carmine, along with a variety of other stains, had gone some way to revealing the microscopic structure of the nerve cell. However, many fundamental questions remained unanswered. This was particularly true for the brain where the carmine stain did little more than indiscriminately reveal masses of poorly defined 'granules' with an occasional nucleus or small rim of cytoplasm being highlighted. What was needed was a more selective staining method capable of identifying all the various components of the nerve cell – especially one capable of following the axon along its entire length. Remarkably, against all the odds, such a method would be discovered in 1873 by the Italian physician Camillo Golgi.

Born in the historic north-western Italian city of Bresica in 1843, Golgi was the son of a country physician. Studying medicine at the nearby University of Pavia, Golgi graduated at the age of 22 after writing a thesis on the pathological causes of mental illness. Following this, Golgi was employed in the university's psychiatric institute before taking up a directorship in 1872 at a hospital 'for incurables' located in Abbiategrasso, near Milan. This unlikely place was to be the location of Golgi's famous discovery. Isolated from the scientific community, with no facilities for performing histological work, Golgi set up a simple laboratory in the kitchen of his small hospital apartment. And, it was here, one evening in 1873, he chanced upon a new method for staining nerve cells. Golgi never publicly revealed how he came across the procedure, and the only clues we have come from the Swedish anatomist Gustaf Retzius who described the discovery in 1933 (some 60 years after the event). According to Retzius, Golgi was attempting to impregnate the pia mater of the brain with a solution of silver nitrate[16] when he accidentally spilt some of the solution on to tissue hardened with potassium dichromate. It produced a chemical reaction causing the material to turn black. However, when Golgi examined this tissue more closely under a microscope, to his surprise, he realised only a small percentage of the nerve unit had been stained.[17] Far from being a disadvantage, this meant the nerve components stood out in black bold relief against a silvery-yellow background. Golgi must have gazed in disbelief, for his stain revealed with unprecedented clarity the fine structure of nerve cells and their fibres down to the most minute protoplasmic processes.

In 1873, Golgi published a short account of his new stain in an Italian medical journal, which also included a detailed description of a completely stained nerve cell taken from the cerebral cortex. However, as Golgi was relatively unknown outside Italy, this paper did not attract a great deal of wider attention. Undaunted, Golgi set about an intensive study of the

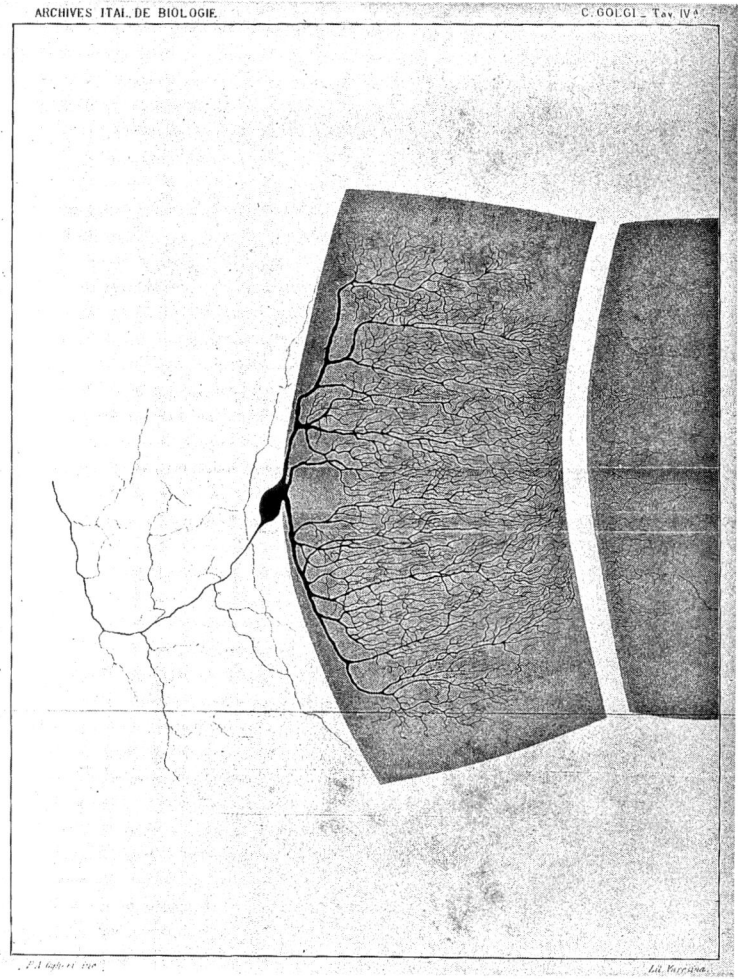

FIGURE 7.8 A human Purkinje cell (type 1) stained by Golgi (1883).
Source: Wellcome Library, London.

brain and nervous system with his new technique, resulting in clear depictions of material taken from the cerebellum (1874), olfactory bulb (1875) and hippocampus (1883). He also gave a thorough account of the spinal cord in 1881. Together, this work was beginning to reveal the fine anatomical structure of the nervous system in unsurpassed detail. Golgi's endeavours also identified previously unseen features of individual nerve parts. For example, he showed for the first time that axons were not always singular fibres, but in some cases they split into smaller branches. This led Golgi to distinguish two types of axon: those having a myelin sheath with few branches (type 1); and those without a myelin sheath which repeatedly branched along their length (type 2). Similarly, Golgi demonstrated that the protoplasmic prolongations were more complex than had previously been imagined. They invariably had intricate tree-like patterns, which in some cases formed second, third and fourth orders of

FIGURE 7.9 Camillo Golgi (1843–1926) the inventor of the silver impregnation method of staining in 1873. This stained a small proportion of nerve cells black enabling them to be clearly observed in their entirety under a microscope.
Source: Wellcome Library, London.

branches from a dendritic trunk. In this respect, the suspicions of Deiters regarding the finer division of the protoplasmic prolongations had proven correct.

Golgi also speculated about the functional significance of his observations. By carefully examining hundreds of slides, taken from many different areas of the nervous system, Golgi came to believe axons (i.e. the nerve fibre) always fused with other axons. To be more specific, he proposed the type 1 nerve fibres merged with the anastomosing branches of type 2 nerve fibres. Thus, according to Golgi, the grey matter of the brain and spinal cord contained a 'diffuse nervous network' (or what he called a *rete nervosa diffusa*) built from a huge continuous reticulum of interlocking fibres. Put another way, the nervous system was essentially a meshed net of axon cylinders. However, in contrast to the view of Gerlach, Golgi did not believe the axon cylinders merged with the protoplasmic prolongations located on the body of the nerve cell. Instead, he believed the protoplasmic prolongations were 'free-endings' associated with blood vessels, where they most likely served a nutritive role in maintaining the nerve.[18]

Objections to the nerve net theory

In 1876, Golgi was appointed professor of histology at Pavia University, and although would depart for a short time, returned in 1881 to take up the chair of general pathology. Golgi had won a prize for his research the previous year from the Lombardy Academy of Science and seen his work published in several publications later to be combined in a lavishly illustrated book (1885).[19] Yet, Golgi's discoveries were not attracting much attention outside Italy. This situation was partly rectified in 1882 when the *Archives Italiennes de Biologie* was founded, which promoted Italian research abroad and contained translations in other languages. Even so, it was not until 1885 when the psychiatrist Eugen Bleuler demonstrated the silver impregnation technique at a conference in Zurich that Golgi's research began to attract international interest. Two years later, it was championed by Kölliker who gave an enthusiastic account of its use to a scientific conference in Würzburg. He was so impressed that he crowed in no uncertain terms: 'We know of no other procedure which reveals with such perfection the nerve cells of the grey matter and the neuroglial elements.'

However, as Golgi's staining method began to gain recognition, his theories concerning the network structure of the nervous system came under increasing criticism. One of the first investigators to question the idea of an anastomosed nervous network was the Swiss anatomist Wilhelm His at the University of Leipzig. His had followed the development of the human embryo by examining tissue taken from aborted foetuses at various stages of their growth. During this work, His discovered that the spinal cord and its nerves arose from a primitive type of cell called a neuropile. However, as His watched the axon fibre grow from the neuropile, followed by the emergence of the cell body and its dendrites, at no point could he see the axons fuse together as Golgi insisted. Summing up his observations in a brief paper published in 1886, His rejected reticulum theory and concluded nerve cells are like all other cells of the body. That is, they remain independent and self-contained entities throughout their life. For the record, His is also responsible, in 1889, for replacing the term 'protoplasmic prolongation' with '*dendrite*' (the word being derived from the Greek 'dendron' meaning 'tree').

Evidence supporting the individual nerve cell was also provided by another Swiss researcher called August Forel.[20] In 1879, Forel had taken up a professorship at the University of Zürich where he began to examine the origin of the cranial nerves arising from the brainstem. In order to do this, Forel chose to use a technique called Wallerian degeneration.[21] It was a method that involved lesioning a given nerve pathway near its ending, and then following the 'backward' degeneration that occurred to its fibre – a process that began some three to four days after the lesion. In fact, Forel first undertook the technique by cutting the nerves projecting to the tongue and then tracing the degenerating fibres back into the brain. He then used the technique to lesion other cranial nerves. To his surprise, all of these lesions only produced cell degeneration in a small and highly localised area of the brainstem. This finding contradicted reticulum theory, which predicted a much greater spread of degeneration. At first, Forel could not explain his findings, but while on holiday in the mountains of his native Switzerland, he realised they occurred because the neural degeneration always remained confined to the cell whose fibre had been cut. In other words, it went against the reticulum concept and provided further support for cell theory.[22]

Unknown to other neuroanatomists during this period, evidence against the reticular theory was also being compiled by a young Norwegian marine biologist called Fridtjof Nansen.

A native of Oslo, Nansen had been appointed curator of the Zoological Museum in Bergen, which housed marine collections taken from Norwegian Arctic surveys. Keen to use this unique opportunity to undertake research for his Ph.D. thesis, Nansen decided to examine the nervous organisation of simple marine invertebrates, including worms, crustaceans and molluscs, using the Golgi stain. His decision proved to be a wise one as these marine creatures have simple nervous systems with relatively large neurons, allowing their anatomy to be observed in detail. Nansen's thesis, published in 1887, was to include over 100 illustrations of nerve cells. Crucially, it revealed them as separate units with no evidence of axonal or dendritic fusing. Unfortunately, Nansen's work would have little immediate impact. Two days after submitting his thesis, he began an expedition across the interior of Greenland, followed by a three-year voyage through the ice flows of northern Russia. Such exploits would help Nansen be recognised as one of the founders of Polar exploration. Although later employed at the University of Oslo, he never pursued anatomical research again. Nansen turned to politics and won the Nobel Peace prize in 1922 for his endeavours to help war refugees. It is likely that his anatomical work would have been forgotten if it were not for the efforts of German Wilhelm Waldeyer who referred to it in a paper published in 1891 (see below).

The founder of modern neuroscience: Santiago Ramón y Cajal

If there is one person above all others who can be considered the father of modern neuroscience it is surely Santiago Ramón y Cajal. Recognised by many as the finest neuroanatomist of the nineteenth century, Cajal provided the experimental facts through his groundbreaking tissue staining techniques to disprove reticulum theory. In its place he replaced it with the neuron doctrine whose basic tenet holds that the nervous system is composed of individual cells. However, Cajal's achievements were far greater than this, with his discoveries and insights leading to an unprecedented understanding of nerve structures and transmission. Among these was his recognition of the synapse, and the realisation that the flow of information though the nerve cell passes from the dendrites to soma (cell body) to axon. Yet, the early portends for the young Cajal were not promising. The son of a struggling country doctor, born in 1858, in the small village of Petilla de Aragon in north eastern Spain, Cajal was a headstrong and rebellious youth who was expelled from school and frequently punished for bad behaviour.[23] First apprenticed as a shoe-maker and then a barber, Cajal dreamed of becoming a painter (his art skills were later used to great effect in his anatomical work). Fortunately, his interest turned to anatomy when his father taught him about the human skeleton using bones unearthed from a local graveyard. Later reminiscing, Cajal wrote: 'Henceforth, I saw in the cadaver, not death . . . but the marvellous workmanship of life.' It would also encourage him to undertake a bachelor's degree with preparatory medical training from the University of Zaragoza. Graduating in 1873, Cajal extended his medical expertise by serving in Cuba (then under Spanish rule) as a military doctor. However, he would spend less than two years in Cuba before contracting malaria and tuberculosis. In a state of frail health, Cajal returned home. By then it was evident to him that being a physician was not going to be his vocation.

Towards the end of 1875, Cajal began work as an auxiliary professor at the University of Zaragoza – a position also allowing him to pursue his doctorate. Teaching himself the basics of microscopy, Cajal used his skills to study the pathogenesis of inflammation – obtaining his

FIGURE 7.10 Photographic portrait of Santiago Ramón y Cajal (1852–1934), the founder of modern neuroscience who did more than any other figure to establish the neuron doctrine and the synaptic organisation of the nervous system.

Source: Wellcome Library, London.

Ph.D. just two years later. In the same year, Cajal visited Madrid, where he spotted a modest French Verick microscope for sale with a magnification of ×800. Using every peseta he had saved from Cuba, Cajal purchased the instrument and took it back to Zaragoza where he set up a laboratory in his home.

Still in frail health, Cajal would suffer a pulmonary haemorrhage in 1878 in the aftermath of his tuberculosis. Nonetheless, his home-based investigations, some of it examining how nerve fibres terminated in muscle, enabled Cajal to take up an academic position at the University of Valencia in 1884. With its great historical and architecture splendour, Cajal enthusiastically embraced his new city. However, one year after his arrival, an outbreak of cholera broke out, killing thousands of people. Cajal turned to finding a treatment, leading him to set up a culture to grow the bacterium, which he then used in a vaccine to provide inoculation. The civic authorities were so grateful for this endeavour they awarded Cajal a 'magnificent' Zeiss microscope in 1885. This model was far superior to his Verick, which he described as 'a rickety door bolt' in comparison. For Cajal, there could have been no greater gift.

Two years later, in 1887, at the age of 35, Cajal took up a professorship at the University of Barcelona. However, arguably, a more crucial event was to occur in the same year when Cajal returned to Madrid. In his travels, Cajal was invited to the laboratory of psychiatrist Don Luis Simarro who had just returned from a conference in Paris, where he was shown samples of nervous material stained by Golgi's silver impregnation method. Cajal was dumbfounded by what he saw, writing later in his biography that 'One look was enough' to reveal that the 'Savant of Pavia' had made the nerve cells stand out in black with 'unsurpassable clarity' even down to their 'finest branchlets'. It was an inspirational moment for Cajal who returned to Barcelona determined to learn the technique. His first attempts, however, proved disappointing. The Golgi stain already had a reputation for being capricious, and it more often than not produced disappointing results, despite its occasional success.

Cajal set about improving the technique. He initially did this by staining the tissue twice (the 'double impregnation' technique), which produced a more intense colouring reaction. Although the process was slow and laborious, it had its rewards: the finer processes of the nerve cell were now clearer to see. Moreover, the axons could be followed for much further. It was a marked improvement over Golgi's technique where the axon fibre often appeared thin, faint, and prone to disappearance. Cajal also realised the Golgi method was poor at highlighting axons with a myelinated sheath.[24] He overcame this problem, however, by using foetal and immature nervous tissue, taken before myelination had taken place. Again, this was a significant advance, for it allowed the neural connections to be visualised much more clearly. And, by examining the development of young animals at different ages, Cajal could see how nerve cells created the basic structure of the nervous system.

The nerve cell starts to give up its secrets

Once he had improved the reliability and intensity of the Golgi technique, Cajal engaged in a feverish burst of research activity where his 'ideas boiled up and jostled each other.' Despite his excitement, Cajal was worried about the difficulties of living in Spain, which was then on the periphery of the scientific world. Concerned his research would not be disseminated to scientists abroad, Cajal took the unusual step of publishing his own journal from his own funds.[25] This involved printing sixty copies of each issue, which Cajal sent to the most important neuroanatomists of Europe. The inaugural copy first appeared in 1888. Containing six articles all written by Cajal, the journal's format was highly innovative, helping to set new standards for this type of research publication with its detailed and unique illustrations. The first paper, for example, concerned the anatomy of the bird's cerebellum. Although he did not report anything fundamentally new about its structure, Cajal's genius lay in the way he depicted how the cerebellar components fitted together and were connected by various pathways. Thus, from his illustrations, the reader could see from a glance how the cerebellum was structured. In fact, the illustrations were actually a composite of many different drawings, based on many hours of meticulous microscopic examination, and expressed with some degree of artistic imagination from Cajal's own anatomical knowledge. Nonetheless, they captured

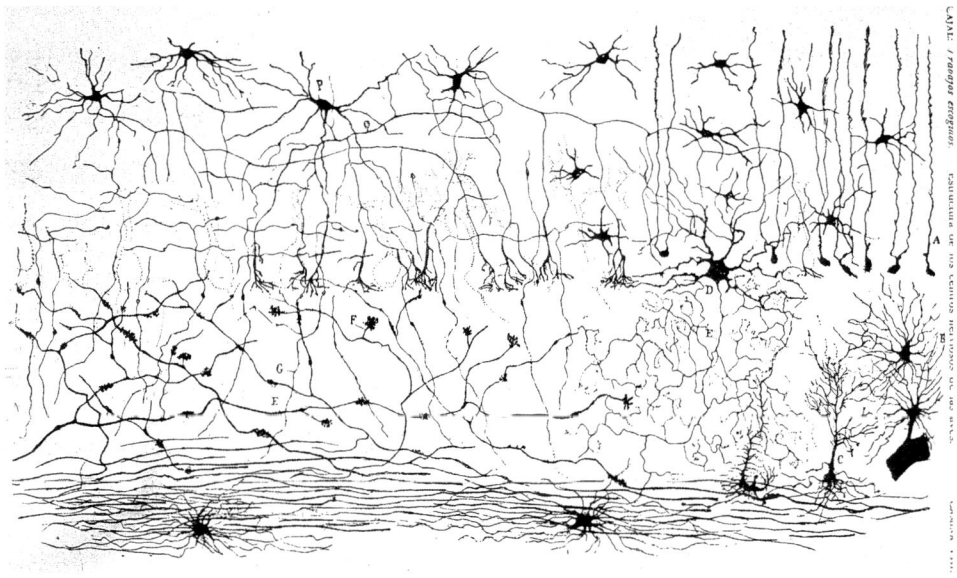

FIGURE 7.11 One of Cajal's earliest illustrations of the cerebellum (from the pigeon) using the Golgi stain (1888). It shows a number of cell types including the heavily branched stellate cells (D) and mossy fibres (E).

Source: Wellcome Library, London.

perfectly the structural complexity of the cerebellum in stunning detail. Even today, over 130 years later, Cajal's drawings of the nervous system have not been significantly improved upon.

When Cajal started his neuroanatomical research, he was like most scientists of his time, a reticularist, believing the nervous system was a continuous network of interconnected fibres. However, right from the beginning of his published work, Cajal reported being unable to find any evidence for Golgi's notion of axonal fusing. While accepting the dendrites end freely (as Golgi believed), Cajal's observations led him to conclude the same was also true for the axon fibres and their branches. In fact, with his characteristic forthrightness, Cajal asserted the nerve cell was 'an absolutely autonomous canon'. He also stated that communication between neurons was through contiguity and not continuity. This was a strong denouncement of Golgi's reticulum theory, which held the axons fused together. Another disagreement with Golgi also concerned Cajal's description of dendritic spines. These are extremely fine and small (micron-size) thorn-like projections on the dendrites. Although these tiny knobbly protrusions had been observed by Golgi, he had dismissed them as artefacts produced by the silver staining technique. However, Cajal correctly realised the dendritic spines were an extension of the main dendrites, which occurred on many large neurons throughout the brain, including the cerebellar Purkinje cells.

Following his initial publications of 1888, Cajal used his new improved Golgi method to examine other areas of the nervous system including the retina, tectum, cerebral cortex, olfactory bulb and spinal cord. It was a remarkable burst of research activity, which must have required Cajal to spend endless hours staining his material and obsessively peering down his microscope to draw what he was observing. Over the course of this work, Cajal also

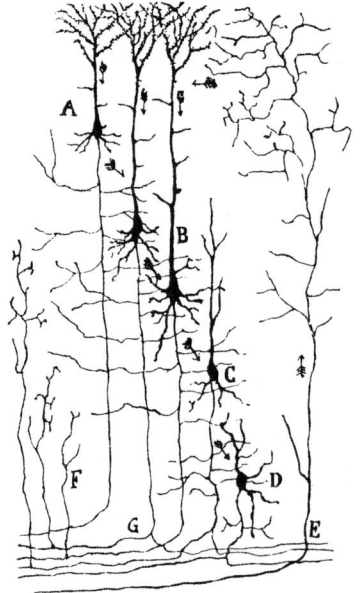

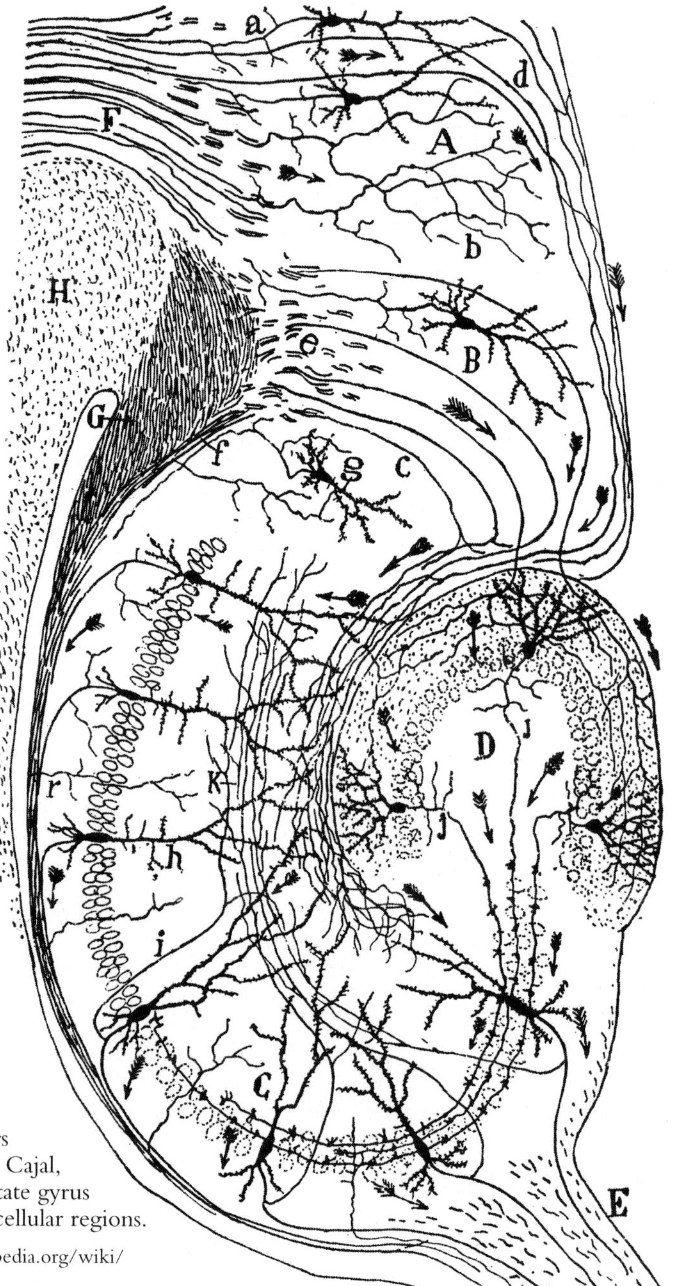

FIGURE 7.12
(A) Drawing showing the various layers of the cerebral cortex (From Ramón y Cajal, 1892, p. 471). (B) Drawing of the dentate gyrus of the hippocampus showing its main cellular regions.

Source: From Wikipedia Commons. en.wikipedia.org/wiki/File:CajalHippocampus.jpeg

came to realise that each area of the brain had a uniform and predictable structure, characterised by its own array of cell types and neural organisation. Thus, the anatomical fabric of the cerebellum was distinctly different from the cerebral cortex, and so on. This was yet further evidence that the brain was not an undifferentiated mass of nervous tissue as many reticulist theorists would have liked to believe.

Cajal also realised something else of fundamental importance. When examining the cerebellum, Cajal noticed the axons of the small granular cells always terminated on the dendrites of the Purkinje cells. And, as his research continued, Cajal realised a similar type of connectivity occurred in other brain areas. That is, the axon fibres always passed to the denser arrays of dendrites located on recipient cells. This was a crucial observation for it pointed to a new way of understanding how the nerve cell conveyed information. Golgi, of course, had dismissed the importance of dendrites by arguing they only served a nutritive role, leaving the axons to provide the main point of contact in his reticulum. He had even denied the cell body was involved in conveying the nerve impulse – a function he attributed solely to the axons. However, Cajal reached a different conclusion: reasoning if axons projected to dendrites, they must be sending nerve impulses to them. In turn, the cell (or soma) presumably collected this information before conveying an impulse down its own axon. This was a revolutionary idea. It implied the dendrites were the receptive part of the nerve cell, while the axon was the emitting and distribution apparatus. Although Cajal realised this theory was hypothetical, by the end of 1889, he had introduced this new dendritic-soma-axonal conduction theory in a Spanish journal with international readership. He called it the law of dynamic polarisation.

The neuron doctrine

Cajal had laid the foundations for a new understanding of the nervous system, yet his anatomical work was not attracting attention from other scientists. Consequently, in an act of some desperation, Cajal travelled to Berlin in 1889 to attend a congress frequented by some of the most eminent neuroanatomists of his day. Unable to speak German and only managing a little broken French, Cajal set up his microscopes and invited the delegates to examine his silver-impregnated slides. One person to view the work was Albert von Kölliker. Now in his 72nd year, and the most distinguished delegate at the conference, Kölliker had come around to believing the weight of evidence supported the idea of axonal fusion, but after viewing Cajal's slides he changed his mind. In fact, Kölliker was so enthused by what he saw, he invited Cajal to his 'luxurious hotel' and treated him to dinner. He also introduced Cajal to a number of distinguished guests who were soon keen to learn his technique. Within a year, a summary of Cajal's research findings were published in two prestigious German journals, with Kölliker confirming the independence of the nerve cell with his silver method in 1890. Kölliker even set about learning Spanish so he could read Cajal's earlier publications. Cajal had finally managed to impress the rest of the scientific world with his pioneering research.

In the years following the Berlin Congress, many others adopted Cajal's methodology to reveal the fine structure of the nervous system. There was also an increasing awareness that the earlier findings of His and Forel provided additional support for Cajal's theories. The evidence was so persuasive, and its implications profound, that the German anatomist Wilhelm Waldeyer felt compelled to write a six-part review article to support the nerve cell concept in 1891. Although Waldeyer had undertaken little of the research himself, his arguments made an authoritative case that nerve cells were the main structural, embryological and functional units of the nervous system. Waldeyer also pointed out nerve cells were similar to others in the body. In other words, they conformed to the tenants of cell theory. However, Waldeyer did make one original contribution in a way he probably could not have foreseen: he referred to nerve cells as neurons. It was an inspired choice of term for the new theory would become known as the neuron doctrine. Waldeyer's innovation has since been described as the quintessential example of the power of the word in science – a single term encapsulating an entire biological principle. It was endorsed by Kölliker in the sixth edition of his *Handbuch* (1896). By now, he was using much the same basic terminology as we use today, including the terms neuron (although Kölliker preferred neurondendron), dendrite and axon.

The neuron doctrine attracted vociferous criticism from the reticulists. Nonetheless, it gained in popularity during the last decade of the nineteenth century. However, the question concerning the direction of information flow through the nerve cell still continued to bother Cajal. He had deduced the flow of nerve impulses from dendrite to axon passing through the *nerve body* in 1889. Cajal began to question this theory when he discovered some cells existed in the nervous system that had two axons with a cell body located 'off the main track' (these are now called a *bipolar cells*). This led Cajal to modify his *dynamic polarisation* theory slightly in 1897. Now he simply stated the cell body does not always intervene directly in the conduction of the impulses, and in some instances the nervous current is able to go directly from the dendrites to the axon. Cajal called this new theory *axipetal polarisation* and presented it as a law at a medical congress in Valencia in 1891. Later Cajal would discover tiny strands within the nerve cell called *neurofibrils*, resembling tiny wires, that passed continuously from dendrite to axon, which provided further support for his theory.

The synapse is named

In a period barely spanning a decade, Cajal had made a number of fundamental discoveries concerning the structure and function of the nervous system. Perhaps most of all, he had helped establish the *neuron doctrine*, which recognised the nerve cell as the basic unit of the brain and spinal cord. However, this also led to a baffling and awkward question: how could information flow within an 'infinitely fragmented' nervous system, as opposed to a continuous neural reticulum? If the neuron doctrine was correct, then each nerve cell was a separate entity. In other words, the axon and dendrite did not fuse together as reticulum theory demanded (this was also sometimes called the continuity hypothesis). However, the alternative option, otherwise known as the contact hypothesis, was also problematical, since it remained to be known how the impulse travelled from one nerve cell to another across a tiny gap. In this respect, the Golgi technique could offer few clues. The strength of the Golgi stain had been its random selectivity. By only staining a few nerve cells in any tissue sample, it had made them highly visible for a microscopist to observe. This was also a weakness when it came to trying to establish their connections, because it was unlikely the stain would highlight two consecutive nerve cells. Even if this did occur, the visualisation of the contact was beyond the resolution of the light microscope. Thus, the Golgi method, even in the hands of Cajal, was not suitable for revealing how the axon endings terminated with other cells.

Nonetheless, during the 1890s, evidence began to favour the contact hypothesis. One source of support came from the *neuromuscular junction* – the site where the endings of the motor nerves extending from the spinal cord meet the skeletal muscles. The German anatomist Wilhelm 'Willy' Kühne had devoted a considerable part of his career to examining this contact point, and initially favoured the continuity hypothesis. However, in 1886 he changed his mind when improved microscopic techniques allowed him to see a clear gap between the motor fibres and muscle. Thus, the nerve endings did not fuse into the muscle, but fell just short of its surface. Kühne reported his findings to the Royal Society in 1888, speculating the nerve ending was open in some way, allowing its fluid to escape and cross the tiny gap, before reaching the muscle. Of course, this naturally led to the bigger question: could a similar process be taking place in other areas of the nervous system as well?

One of the approaches used by Cajal to address the contact question was to examine the embryological development of nerve cells. As Wilhelm His had already shown, the embryonic nerve fibre always arose from a more primitive cell called the neuropile. Cajal also carefully observed this process through his microscope, and discovered the growing nerve fibre always exhibited a curious swelling at its head, which he called a 'growth cone'. This bulbous part of the fibre also had its own unique form of 'amoeboid movement', resembling a battering ran that pushed aside all obstacles in its path in its quest to make contact with its target cell. Cajal was particularly struck by this growth in the chick's cerebellum, which showed the axons of the *climbing fibres* seemingly seeking out Purkinje cell bodies in purposeful fashion before tangling with their dendrites. In a sudden stroke of insight, Cajal realised this movement came about because the Purkinje cells produced a chemical luring the axons to their destination – a theory now known to be correct. As far as Cajal could make out, at no point in this growth process did the axon terminals fuse with the dendrites. Rather, they appeared to end bluntly when making contact at their destination.

In 1894, Cajal received a surprise invitation from the Royal Society of London to deliver the Croonian lecture for that year. The decision to invite Cajal for such a prestigious event had been supported by Charles Sherrington who himself was undertaking research on the spinal circuitry of reflex action – work which required him to carefully map the pathways of nerves and their connections. His attempts to trace the flow of information through the spinal cord had also convinced him that the nervous system was composed of individual nerve cells whose fibres made contact without fusing. Daunted by the prospect, Cajal travelled to London, giving his talk in French, while staying with the Sherrington family for the duration of his visit.[26] The two men were to remain great friends, and several years later Sherrington would recount how Cajal had insisted on converting the guest bedroom into a temporary laboratory for his microscopical research – which he then typically undertook by providing a running commentary on his observations with humorous anthropomorphic language.

It is not clear if the two men discussed the possibility of naming the contact point during this visit. However, soon after the Croonian lecture, Sherrington was invited to write a chapter in Sir Michael Foster's popular *Textbook of Physiology*. Sherrington accepted the task, but became frustrated by having no term for the junction he believed existed between nerve cells. At first coining the term *syndesm* (from the Greek meaning 'to clasp'), he changed it to *synapse* after the classics scholar Arthur Woollgar Verrall suggested it provided 'a better adjectival form'. The word would first appear in Foster's seventh edition of his textbook (1897) where the synapse was simply defined as a 'surface of separation'. Sherrington could only speculate how information crossed this junction. Nonetheless, he was certain synapses existed because he had discovered from his own research that the transmission of impulses along nerve fibres was much slower at their endings. Although definite visual proof of synapses in the nervous system was not fully confirmed until the invention of the electron microscope in the 1950s, the pharmacological evidence for their functional existence proved overwhelming long before. The question of whether synapses passed chemical or electrical messages, however, was to prove a more problematical issue, and one leading to the famous 'sparks versus soup' debate in the first half of the twentieth century (see Chapter 12).

Golgi and Cajal win the Nobel Prize

In 1906, Golgi and Cajal were told of their Nobel Prize in recognition of their work on the structure of the nervous system. This annual prize, then worth roughly 1 million American dollars in today's money, had been initiated in 1901 from a legacy provided by Alfred Nobel, to be conferred on those who had given 'the greatest benefit on mankind'. Golgi and Cajal were the first anatomists to receive the honour, and was the only time the two men ever met. Sadly, their friendship did not get off to a good start, when Cajal went to greet Golgi at the train station, but was largely ignored as the Italian was more concerned to find his hotel. Nor did things improve. The prize ceremony took place soon after, with both men required to give an acceptance speech on separate days. Golgi gave the first, surprising the distinguished audience by resolutely defending his reticularist position. He also began the talk by stating that the neuron theory was 'going out of favour', and then proceeded to spend much of his efforts at denouncing it. Shocked at Golgi's 'display of pride and self-worship', Cajal's feelings turned to anger as the criticism of the neuron doctrine continued. Although Cajal endured the ordeal, he was 'trembling with impatience' at being unable to correct 'so many odious errors and so many deliberate omissions.' The next day Cajal gave his talk, which provided a more measured account, which expressed his belief in nerve contiguity and not continuity. History has since shown that it was Cajal and not Golgi who was correct in regards to the neuron doctrine and many other aspects of nerve function.

In Golgi's defence it can be pointed out Cajal's position was not universally accepted in 1906, with reticulum theory still retaining a number of supporters. Moreover, the concept of a synapse was a cumbersome and implausible concept for many, leaving important questions about the nature of neural transmission unanswered. The reticulists also pointed out with some degree of justification, that if neurons only made contact, then the conduction of electrical impulses would be much slower. All of these objections cast doubt on the neuron doctrine, which was also slow to advance in the early years of the twentieth century. Consequently, the reticulum theory lingered on, helped by Golgi who continued to denounce neuron theory in his advancing years. It is even said that he refused any discussion of it among his students.

Regardless of who was ultimately correct, Golgi and Cajal were both great pioneers of neuroanatomy. In addition to his work on the nervous system, Golgi used his stain to discover the Golgi tendon (an organ found in muscles that detects stretching), and the Golgi apparatus (a structure involved in the manufacture and transport of protein and lipids within the cell). Golgi also undertook important research on malaria by elucidating the life cycle of the malarial agent plasmodium in red blood cells. Elected Rector of the University of Pavia, Golgi died at the age of 81 in 1926. Cajal was also highly prolific and summarised much of his work in the three-volumed *Histologie du système nerveux de l'homme et des vertébrés* (*Histology of the Nervous System of Man and Vertebrates*) published in 1911. Containing nearly 2000 pages and 887 illustrations, this book still can't fail to impress the reader today with its remarkable graphic intricacy and written detail of the spinal cord and brain. It still forms the cornerstone of our modern understanding of the nervous system's cellular organisation.[27] Cajal also authored other books including *Advice for a Young Investigator* (1916), and a colourful biography entitled *Recollections of my Life* (1917). He even wrote several fictional tales and popular scientific works intended for a broad public under the pseudonym of 'Doctor Bacteria'. Towards the end of his life, Cajal's fame and reputation was so great, the Spanish authorities tried to put his portrait on bank notes and postage stamps. Although he refused to permit this, he did agree to become a life senator in Madrid. Retiring from university work in 1922, Cajal continued writing until his death at the age of 83 in 1934. An obituary by the Italian neurologist Ernesto Lugaro eulogised that Cajal had contributed more to the knowledge of neuroscience than all the efforts of his fellow colleagues put together. There are few who would disagree even today.

Notes

1 Purkinje had first noticed the effect during one of his long walks when he noticed that red flowers looked more bluish-red as dawn approached.
2 It is interesting to note that Purkinje always used the German way of spelling his name ('Purkinje') until he returned to Prague in 1849 when he adopted the Czech spelling ('Purkinj_').
3 During this time Purkinje also identified nine different patterns of fingerprint, although made no mention of their value for personal identification.
4 The tree-like protrusions found all over the cell body whose function is to collect neural information from other nerve cells.
5 See Tan and Lin (2005).
6 Whether Hooke observed any plant cells is actually disputed. There is one plate in *Micrographia* that may show the underside of a stinging nettle leaf illustrated with cells – but this is far from certain. If he did observe them, then it is clear he did not realise their true significance. The interested reader is referred to *The Birth of the Cell* by Henry Harris (1999).
7 This work had led Schwann to discover that the myelin sheath surrounding nerve fibres is derived from a separate cell – now known as a Schwann cell.
8 *Mikroskopische Untersuchungen über die Übereinstimmung in der Struktur und dem Wachstum der Thiere und Pflanzen* (*Microscopic Researches into Accordance in the Structure and Growth of Animals and Plants*).
9 It is said by one source that Kölliker made important contributions to understanding the structure of smooth muscle, skeletal muscle, skin, bone, teeth, blood vessels and the viscera. He was also the first to realise that arteries have walls made of muscle, and that cells contain large numbers of fine granules which turned out to be mitochondria (the power source of the cell).
10 The sixth *Handbuch* would also be famous for introducing the term 'axon' in 1896.
11 This is most certainly unfair on Remak although Kölliker did recognise him as the discoverer of the axis cylinder (see Shepherd 1991).
12 These are now known to be motor neurons projecting to the muscles.
13 An astrocyte is a type of glial cell, which were first discovered by Rudolf Virchow in 1858 who likened them to 'glue' holding the nerve components together. Today, we know glial cells serve a number of important functions in maintaining the integrity and function of nerve cells.
14 *Untersuchungen über Gehirn und Rückenmark des Menschen und der Saugethiere* (*Investigations on the Brain and Spinal Cord of Man and Mammals*).
15 The 'second' set of fibres observed by Deiters have been the subject of much speculation. They could be secondary dendrites, although most investigators today believe they were the 'incoming' branches of axon endings from other nerve cells. In fact, Deiters even drew some of them with triangular bases – now known to be an important morphological feature of axon terminals.
16 Golgi may well have been attracted to this substance as it had just begun to be used in the production of photographic film.
17 For reasons that are still unknown, the method only stains some 1 to 5 per cent of cells.

18 We know today that some cells of the brain (now known as glial cells) do indeed have processes that terminate in the blood vessels – although these are not nerve cells. It is likely that Golgi observed and confused these two types of cell.
19 *Sulla fina anatomia degli organi centrali del sistema nervoso* (*On the Fine Anatomy of the Central Nervous System*).
20 While working in Munich, Forel had constructed a microtome big enough to slice the human brain, which had enabled him to undertake a widely acclaimed analysis of the *tegmental region* (an area of the upper brainstem) in 1877.
21 First undertaken by the Englishman Augustus Waller in 1850.
22 Forel's discovery in 1886 probably occurred before that of His in the same year. However, the paper by Forel was delayed in press for several months and did not appear until 1887 – some three months after the work by His was published. Consequently, His is normally given priority in this matter.
23 Including a time, aged just 11, when Cajal managed to make home-made explosive and blow a hole in the town's main gate!
24 This is because the sheath acts as a barrier to the silver nitrate entering the axon.
25 *Revista trimestral de histologia normal y patologica*.
26 There is some doubt as to whether this was the first time the two men met. Sherrington had gone to Spain in 1885 to help investigate the outbreak of cholera. The official Nobel Prize biography states that Sherrington and Cajal met during this period, although Sherrington later denied this in his eulogy of Cajal in 1934.
27 There is a recent English translation, which any serious student of neuroscience should make an effort to examine. See Swanson and Swanson (1995).

Bibliography

Akert, K. (1993) August Forel: cofounder of the neuron theory (1848–1931) *Brain Pathology*, 3, 425–430.
Anderson, C.G. and Anderson, B. (1993) Koelliker on Cajal. *International Journal of Neurosciences*, 70, 181–192.
Andres-Barquin, P.J. (2001) Ramon y Cajal: a century after the publication of his masterpiece. *Endeavour*, 25, 13–17.
Andres-Barquin, P.J. (2002) Santiago Ramon y Cajal and the Spanish school of neurology. *The Lancet*, 1, 445–452.
Barbara, J.G. (2006) The physiological construction of the neurone concept. *C.R. Biol*, 329, 437–449.
Bennett, M.R. (1999) The early history of the synapse: From Plato to Sherrington. *Brain Research Bulletin*, 50, 95–118.
Bentivoglio, M. (1996) 1896–1996: The centennial of the axon. *Brain Research Bulletin*, 41, 319–325.
Cannon, D. (1949) *Explorer of the Human Brain: The Life of Santiago y Cajal*. Henry Schuman: New York.
Clark, G. and Kasten, F.H. (1983) *History of Staining*. Williams and Wilkins: London.
Clarke, E. and O'Malley, C.D. (1968) *The Human Brain and Spinal Cord*. University of California Press: Berkley, CA.
DaFrano, C. (1926) Camillo Golgi (1843–1926). *Journal of Pathological Bacteriology*, 29, 500–514.
De Carlos, J.A. and Borrell, J. (2007) A historical reflection of the contributions of Cajal and Golgi to the foundations of neuroscience. *Brain Research Reviews*, 55, 8–16.
DeFelipe, J. (2002) Sesquicentenary of the birthday of Santiago Ramon y Cajal, the father of modern neuroscience. *Trends in Neurosciences*, 25, 481–484.
Edwards, J.S. and Huntford, R. (1998) Fridtjof Nansen: from the neuron to the north polar sea. *Endeavour*, 22, 76–80.
Fodstad, H. (2001) The neuron theory. *Stereotactic and Functional Neurosurgery*, 77, 20–24.
Golgi, C. (1906) The neuron doctrine – theory and facts. In *Nobel Lectures, Physiology and Medicine 1901–1921*. Elsevier: New York, 1967.
Grant, G. (2007) How the 1906 Nobel Prize in Physiology and Medicine was shared between Golgi and Cajal. *Brain Research Reviews*, 55, 490–498.
Guillery, R.W. (2005) Observations of synaptic structures: origins of the neuron doctrine and its current status. *Philosophical Transactions of the Royal Society. B*, 360, 1281–1307.

Harris, H. (1999) *The Birth of the Cell*. Yale University Press: New Haven, CT.
Haymaker, W. (1953) *The Founders of Neurology*. Thomas: Springfield, IL.
Jones, E.G. (1994) The neuron doctrine 1891. *Journal of the History of Neurosciences*, 3, 3–20.
Jones, E.G. (1999) Golgi, Cajal and the neurone doctrine. *Journal of the History of Neurosciences*, 8, 170–178.
Judas, M. and Sedmak, G. (2011) Purkyne's contribution to neuroscience and biology: Part 1. *Translational Neuroscience*, 2, 270–280.
Katz-Sidlow, R.J. (1998) The formulation of the neurone doctrine. *Archives of Neurology*, 55, 237–240.
Koeppen, A.H. (2004) Wallerian degeneration: history and clinical significance. *Journal of the Neurological Sciences*, 220, 115–117.
Kölliker, A. (1867) *Handbuch der Gewebelehre des Menschen*, 5th edn. Engelmann: Leipzig.
Lagunoff, D. (2002) A Polish Jewish scientist in nineteenth century Prussia: Robert Remak (1815–1865). *Science*, 298, 2331.
Llinás, R.R. (2003) The contribution of Santiago Ramón y Cajal to functional neuroscience. *Nature Reviews Neuroscience*, 4, 77–80.
Lopez-Munoz, F., Boya, J. and Alamo, C. (2006) Neuron theory, the cornerstone of neuroscience, on the centenary of the Nobel Prize award to Santiago Ramon y Cajal. *Brain Research Bulletin*, 70, 391–405.
Marshall, L.H. and Magoun, H.W. (1998) *Discoveries in the Human Brain*. Humana Press: Totowa, NJ.
Mazzarello, P. (1999) *The Hidden Structure: A Scientific Biography of Camillo Golgi*. Oxford University Press: Oxford.
Mazzarello, P. (1999) A unifying concept: the history of cell theory. *Nature Cell Biology*, 1, E13–E15.
Mazzarello, P. (2011) The rise and fall of Golgi's school. *Brain Research Reviews*, 66, 54–67.
Mazzarello, P., Garbarino, C. and Calligaro, A. (2009) How Camillo Golgi became 'the Golgi'. *FEBS Letters*, 583, 3732–3737.
Pannese, E. (1996) The black reaction. *Brain Research Bulletin*, 41, 343–349.
Pannese, E. (1999) The Golgi stain: invention, diffusion and impact on neurosciences. *Journal of the History of Neurosciences*, 8, 132–140.
Pedro, J.A.-H. (2001) Ramon y Cajal: a century after his masterpiece. *Endeavour*, 25, 13–17.
Pokorný, J. and Trojan, S. (2005) Purkinje's concept of the neuron. *Casopis Lekaru Ceskych*, 144, 659–662.
Purkyně, J.E. (1838) *Bericht über die Versammlung deutcher Naturforsher und Arzte in Prag in September, 1837*. Prague. Part 3, Section 5: A Anatomische Verhandlungen, pp. 177–180.
Ramón y Cajal, S. (1892) Elneuvo concepto de la histologia de los centros nerviosos. III. – Corteza gris del cerebro. *Rev. Ciencias. Méd. Barcelona*, 18, 457–476.
Ramón y Cajal, S. (1906) The structure and connexions of neurons. In *Nobel Lectures, Physiology and Medicine 1901–1921*. Elsevier: New York, 1967.
Ramón y Cajal, S. (1911) *Histologie du Systeme Nerveux de l'Homme et des Vertebretes*, Vols. 1 and 2. A. Maloine. Paris, 1911.
Ramón y Cajal, S. (1954) *Neuron Theory or Reticular Theory?* Instituto Ramon y Cajal: Madrid.
Ramón y Cajal, S. (1966) *Recollections of my Life*. MIT Press: Cambridge, MA.
Rapport, R. (2005) *Nerve Endings: The Discovery of the Synapse*. Norton: New York.
Shepherd, G.M. (1991) *Foundations of the Neuron Doctrine*. Oxford University Press: Oxford.
Shepherd, G.M., Greer, C.A., Mazzarello, P. and Sassoe-Pognetto, M. (2011) The first images of nerve cells: Golgi on the olfactory bulb 1875. *Brain Research Reviews*, 66, 92–105.
Smith, C.U.M. (1996) Sherrington's legacy: Evolution of the synapse concept, 1890s–1990s. *Journal of the history of Neurosciences*, 5, 43–55.
Smith, C.U.M. (1997) 1997: Centenary of the synapse. *Endeavour*, 21, 49–51.
Sotelo, C. (2003) Viewing the brain through the master hand of Ramon y Cajal. *Nature Reviews Neuroscience*, 4, 71–77.
Swanson, N. and Swanson, L.W. (1995) *Histology of the Nervous System of Man and Vertebrates by Ramón y Cajal*. vols 1 and 2. Oxford University Press: Oxford.
Tan, S.Y. and Lin, K.H. (2005) Johannes Evangelista Purkinje (1787–1869): 19th century's foremost phenomenologist. *Singapore Medical Journal*, 46(5), 208–209.

Valentin, G.C. (1836) Uber den Verlauf und die letzten Ende der Nerven. *Nova Acta Physico-medica Academiae Caesareae Leopoldnp-Carolinae Naturae Curiosorum*, 18, 51–240.
Venkatamani, P.V. (2010) Santiago Ramon y Cajal: Father of neurosciences. *Resonance*, November, 968–976.
Wade, N.J. and Brozek, J. (2001) *Purkinje's Vision: The Dawning of Neuroscience*. Lawrence Erlbaum: New York.
Winklemann, A. (2006) Wilhelm von Waldeyer-Hartz (1836–1921): An anatomist who left his mark. *Clinical Anatomy*, 20, 231–234.
Wolpert, L. (1996) The evolution of 'The Cell Theory'. *Current Biology*, 6, 225–228.

8

THE RETURN OF THE REFLEX

The initial cause of any human action lies outside the person.

Ivan Mikhailovich Sechenov

The brain is waking and with it the mind is returning. It is as if the Milky Way entered upon some cosmic dance. Swiftly the head mass becomes an enchanted loom where millions of flashing shuttles weave a dissolving pattern . . . a shifting harmony of subpatterns.

Charles S. Sherrington

Summary

The concept of a reflex, or an involuntary physiological response produced by an external stimulus, was first illustrated in *L'Homme* by Descartes in 1662. It was a pivotal movement in the history of physiology where the flesh and bone of bodily movement previously accountable in terms of a soul-like animating force, now became explainable through mechanics. Despite this, the concept of a reflex was slow to develop in the century after Descartes because vitalist beliefs were not so easily dismissed. This is seen, for example, in the work of the eighteenth century physiologist Robert Whytt who highlighted the importance of nerve pathways for certain types of reflexive movement, yet continued to insist on a spiritual force (or 'sentient principle') to explain the body's action. A more modern understanding of the reflex would have to wait until the anatomy of the spinal cord was better understood, and this advance took place through the endeavours of Charles Bell and Francois Magendie in the early part of the nineteenth century. Although the priority for their discoveries was disputed, the two men identified separate spinal pathways for the transmission of sensory and motor information. This would help enable the English physiologist Marshall Hall to elucidate the spinal and medullary mechanisms of a large range of complex reflexive behaviours, which he believed was free of any vitalist force. These concepts, however, reached their culmination with Charles Sherrington, whose research career spanning over 50 years, was to explain in great detail the integrative nature of the reflex. In particular, he was to emphasise the importance of both excitation and inhibition, and show how all levels of the nervous system, including the cerebral cortex, were responsible for reflexive actions. However, the reflex not only dominated physiological thinking. Following the work of Sechenov in the mid-nineteenth century, reflexes increasingly became recognised as important components in mental operations. They were also used to explain learning by Ivan Pavlov – an idea that was to have a major impact on the development of psychology. Karl Lashley, for one, would be inspired to search for the neural basis of conditioned reflexes in the brain, and while this proved more difficult than he imagined, his failure was to result in a new concept of reflexive action by student Donald Hebb. His theory proposed learning and memory involved 'circuits' of reflexive neural activity in the brain, whose electrical 'reverberations' becomes strengthened or weakened through changes taking place at the synapse. Today, Hebb's ideas remain highly influential in a number of fields relevant to brain research including neuroscience, psychology and artificial intelligence.

Early accounts of involuntary action

It has been known since the ancient Greeks that many complex bodily actions, including those of the heart, lung and stomach, can occur without any intervention of the will. However, such behaviours were not regarded as reflexive as is understood in the modern sense. Rather, they were considered to be under the control of an indwelling vital force (the *psyche* in Aristotelian terminology) governing the activities of the body. In fact, Galen even stated involuntary actions were illusory. In his view they were caused intentionally by the soul, but instantly forgotten, giving them the appearance of being autonomous. This belief also helped explain the problem of 'sympathy'. This was a concept going back to Hippocrates who argued the four humours of the body must be in harmony (or sympathy) for good health. However, by the time of Galen the problem had become one of trying to explain how the body could work in a co-ordinated fashion with 'sympathy' between its different parts. The difficulty arose, for example, when attempting to explain how a disturbance in one part of the body (say a movement of the uterus) could lead to an abnormality elsewhere (hysteria). Although Galen was aware many instances of sympathy had a nervous origin arising from the spinal cord, allowing animal spirits to travel from one organ to another, he also recognised other cases. These could occur, for example, from a connecting blood supply which unified different parts of the body (e.g. pregnancy and lactation in women), or even situations where vapours rose from the stomach to disturb the brain (e.g. gastrointestinal illness). Clearly, without a major shift in thinking regarding the operation of the body, especially its nervous system, the concept of sympathy would remain tied to such antiquated beliefs.

It would be another 1500 years or so, before Descartes formulated the concept of a reflex to explain involuntary movement. This is famously presented, for example, in *L'Homme* (1662) where Descartes accounts for a person pulling their hand away from a hot object, with reference to a nerve pathway passing to and from the brain (see Figure 4.2). It was a remarkable idea for its time and one of the most important advances in establishing the modern foundations of physiology. Even so, Descartes' new theory, which saw the behavioural response as independent from the volition of the soul, was not widely accepted. To make matters worse, the finer details of Descartes' theory was shown to be incorrect as he located the 'reflected' action in the walls of the brain's ventricles. However, others following Descartes would give more plausible accounts including Thomas Willis who recognised the cerebellum as an important site for controlling involuntary action. In addition to the cerebellum, Willis also attributed a similarly important role to the intercostal nerves (later to be called the sympathetic trunk) and the vagus nerve, which passed to many organs of the body including the heart and lungs. All of these systems not only arose from the posterior part of the brain, but Willis also recognised there was extensive interconnections, functional unity and 'sympathy' between them. Consequently, he came to see these nerve pathways as having integrated actions that did not necessarily involve the brain. However, unlike Descartes, Willis resorted to the idea of animal spirits flowing through nerves, and a soul extended throughout the whole body, where it could effect sympathetic action.

The reflex as the mediator of sympathy: Robert Whytt

The reflex concept would be more fully developed in the eighteenth century by Scotland's so-called first 'neurologist' Robert Whytt (see Chapter 4). Appointed as Professor of Medicine at Edinburgh in 1747, Whytt was a strong opponent of the Cartesian or mechanist view of physiology. Instead, Whytt believed the body's mechanical actions were subordinate to a spiritual entity capable of self-motion called the *sentient principle* – an animistic force diffused throughout the body where it controlled sympathy through its 'wonderful union' with the nervous system. Yet, despite having these views, Whytt discovered several types of involuntary action. The first of these appeared in an essay published in 1751 where Whytt discusses how the eye's pupil gets smaller in the presence of bright light.[1] This reaction had first interested Haller who thought it occurred because the iris (the muscle surrounding the pupil) became irritated, which caused it to contract without any nervous intervention. But Whytt came to a different conclusion: proposing it arose when light fell on the highly sensitive retina at the back of the eye. This produced an unpleasant sensation, which was communicated to the sentient principle in the optic nerve and brain. It responded by initiating the 'motions' necessary to lessen the offending cause, causing the pupil's contraction, so less light entered the eye. However, Whytt also noticed something else: the pupil's response to light occurred in both eyes – even if one of them was covered. This demonstrated *sympathy* between the eyes, and Whytt could only explain it by assuming the nerve pathway from one eye passed to the brain, where the sentient principle acted to evoke the response in the other eye. In fact, Whytt went one step further by identifying the *optic thalamus* as the site of this action after undertaking an autopsy in a child whose pupils had been unreactive to light and finding a cyst in this area of the brain.

Four years later in his *Physiological Essays* (1755), Whytt made another important discovery. It was well known to eighteenth century investigators that a decapitated frog retained a heart beat for several hours after death, and produced muscular movements long after the head had been severed from the spinal cord. In fact, the headless frog often exhibited a co-ordinated sitting posture which revealed a 'sympathy' between its different muscles. It was even capable of exhibiting various reflexes such as retracting a leg from noxious stimulation. For Haller, these responses simply meant that its muscles continued to be 'irritable' for some time after death. But Whytt disagreed. For him, this behaviour was due to residual amounts of the sentient principle remaining in the spinal cord following the frog's decapitation. Indeed, Whytt set about proving his theory by showing the headless frog lost its action and sympathy if the spine marrow was also destroyed with a hot wire. This encouraged Whytt to locate the sentient principle more precisely and, in doing so, he discovered only a small part of the spinal cord was necessary for the limb withdrawal to occur. It was the first clear demonstration of what we now recognise as a spinal reflex – a simple involuntary response, controlled by a small segment of the spinal cord, without any intervention of the brain.

Although Whytt never used the term 'reflex' he was contributing significantly to its greater understanding. And he further did this by specifically defining involuntary actions as those requiring a *stimulus* that took place without the 'power of the will'. In other words, he realised such motions were triggered by an external event (a stimulus) which led to an automatic response. Such reactions also required the intervention of the spinal cord, and while Whytt

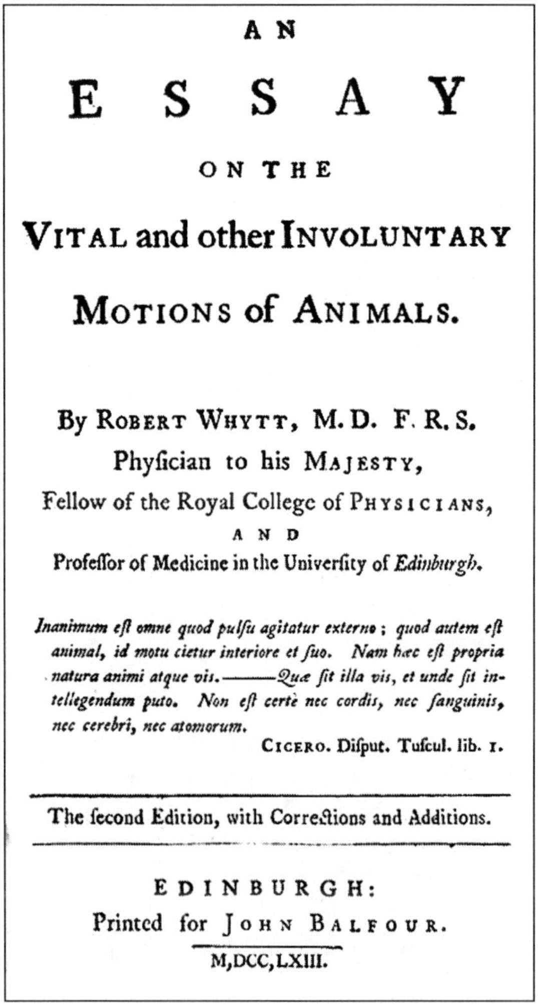

AN

ESSAY

ON THE

VITAL and other INVOLUNTARY

MOTIONS of ANIMALS.

By ROBERT WHYTT, M.D. F.R.S.
Physician to his MAJESTY,
Fellow of the Royal College of PHYSICIANS,
AND
Professor of Medicine in the University of *Edinburgh*.

Inanimum est omne quod pulsu agitatur externo; quod autem est animal, id motu cietur interiore et suo. Nam hæc est propria natura animi atque vis.———Quæ sit illa vis, et unde sit intellegendum puto. Non est certè nec cordis, nec sanguinis, nec cerebri, nec atomorum.
CICERO. Disput. Tuscul. lib. I.

The second Edition, with Corrections and Additions.

EDINBURGH:
Printed for JOHN BALFOUR.
M,DCC,LXIII.

FIGURE 8.1 Title page of 'An Essay on the Vital and Involuntary Motions of Animals' published by Robert Whytt in 1751, which provides some of the earliest discussion of reflex action.

believed only a 'small part' of the sentient principle was devoted to their activity, he nonetheless regarded them as essential to the body's overall integrated function (i.e. sympathy). The list of actions Whytt recognised as involuntary was also particularly impressive and included functions such as respiration and ejaculation.[2] These 'sympathetic' motions, however, stood in marked contrast with voluntary action, produced by the 'power of the will' located in the brain. This required no external stimulus for they were governed by the soul. In addition, Whytt recognised the existence of natural motions such as the heart-beat that again required no stimulus.

The Bell–Magendie law

Further developments in understanding the reflex would have to wait until the neural anatomy of the spinal cord was more fully elucidated. This occurred in the early part of the nineteenth century. Up until then, knowledge of the spinal cord had not been significantly improved upon since the time of Galen, who had described the spine growing out of the brain like the trunk of a tree, with paired nerves (right and left) forming branches which divided into thousand of twigs as they passed into the body.[3] He also demonstrated the effects of spinal transactions, which deprived the body of sensation and movement below the cut. Interestingly, Galen also recognised the spinal cord as a bundle of nerves that ran without interruption from brain to muscle (the motor part), or in the reverse direction (the sensory part). He arrived at this conclusion as a young doctor after examining a patient who had fallen from his chariot, and suffered a loss of sensation in his fingers without loss of motor control. This finding indicated the sensory pathways to the spinal cord were different from the motor ones going to the muscles. Later, he described the sensory nerves as 'soft', and the motor ones as 'hard'. Galen's differentiation of sensory and motor nerves, however, was seemingly forgotten by his successors. For example, Descartes believed individual nerve pathways conveyed both sensory and motor information, and this was endorsed by Haller who wrote: 'I know not a nerve which has sensation without also producing motion.'

The first experimental evidence distinguishing between sensory and motor nerves was provided by the Scot Sir Charles Bell. A man who had learned surgery under his older brother in Edinburgh, and who had made several enemies there, Bell moved to London in 1804 to increase his academic prospects. He would live in London for over 30 years before returning to Scotland as Professor of Surgery at Edinburgh in 1836. During this time Bell was to become a highly prolific medical author whose name is now associated with a form of facial paralysis (Bell's palsy). Bell also came to the recognition of the art world in 1806 after publishing *Essays on the Anatomy of Expression in Painting* – a stunning visual work combining anatomy with art which has since become a classic. It may be that his art skills were better than his surgical ones, for Bell would see military action at the Battle of Waterloo where it is said 90 per cent of his amputee patients died from the surgery.[4]

However, Bell's most important anatomical work was to involve the spinal cord. Around 1810, Bell appears to have had an important insight when he reasoned it was most likely that the most anterior regions of the spinal cord were connected with the frontal areas of the brain. Similarly, Bell thought it probable that the more posterior areas of the brain were associated with the rear of the spinal cord. If true, this had significant implications for the front part of the brain had traditionally been linked to sensation, while the posterior part (i.e. cerebellum) had been linked with involuntary reflexes. To test his idea, Bell stunned a rabbit, opened its spine, and pricked different parts of its spinal cord. Although this procedure produced little effect when stimulation was applied to the posterior part of the spinal cord, it caused the animal to convulse when made to its anterior parts. This was an unexpected finding indicating the frontal regions of the brain and spinal cord did not serve a sensory function after all. Rather excitation in this part of the spinal cord caused seizures in the musculature of the body; it therefore appeared to have a motor function.

Bell followed up these observations with further experiments. In particular, he turned his attention to the 31 pairs of nerves leaving the spinal cord on either side along its length. These ganglia were known as spinal nerves, and it had been established by the early nineteenth century that they were formed from two roots: those leaving the spinal cord anteriorly (sometimes called the ventral roots) and those posteriorly (the dorsal roots).[5] Bell began to examine the effects of cutting these roots in rabbits. He soon confirmed what he had previously discovered in his spinal pricking experiments. That is, cutting the anterior roots induced convulsions, while lesioning the posterior roots had little behavioural consequence. Bell published his findings in a privately circulated pamphlet of a hundred copies in 1811.[6] It clearly attributed the anterior roots of the spinal cord with a motor function, although Bell was much more vague about the posterior roots, which he claimed served the 'secret' operations of the body.

Bell's research on the spinal roots was not widely known. Curiously it was not even reported in his anatomical textbook (1816). In fact, the wider dissemination of the work would have to wait until presented by his brother-in-law Alexander Shaw to the Royal Society in Paris in 1822. One of the attendees listening to this talk was the noted physiologist Francois Magendie. Born into a poor family that had embraced the French Revolution, Magendie had risen through the chaotic times to become a doctor and then researcher at the Anatomical Institute of France. He had also achieved a reputation for performing experiments on himself and animals using arrow poisons from Java and Borneo – work leading to the discovery of strychnine in 1818. Known for his forthright and abrasive personality, which frequently engendered hostility from others, Magendie was also a notorious vivisectionist who was not averse to inflicting great cruelty on animals in experiments and public exhibitions. Such work had not gone unnoticed in England.[7] After listening to Shaw's talk, Magendie immediately set about undertaking his own studies by exposing the spinal cord in young puppies and cutting their roots using scissors. Magendie found when he cut the anterior roots it caused convulsions and paralysis. However, severing the posterior roots deprived the animal of sensation, including that of pain, below the level of the lesion. The implication was clear: the anterior roots were responsible for conveying information to the muscles (as Bell had suggested), whereas the posterior roots conveyed sensory input such as touch into the spinal cord.

Magendie published a brief communication of his discovery in 1822, and later backed it up with more substantial research. Although Magendie acknowledged Bell was the first to attribute the anterior roots with a motor function, he claimed priority for demonstrating the sensory nature of the posterior roots. It was a claim Bell would strongly dispute, leading to an acrimonious and longlasting rivalry between the two men. This situation was not helped by anti-French sentiment of the times, and the cruelty of Magendie's experiments. The conflict was to last until Bell's death in 1842. Today, most historians accept Magendie was justified in his claims, with Bell providing little evidence to indicate the sensory nature of the posterior roots.[8] However, regardless of who was right, the two men had between them established an important new neurological concept: that is, there are separate motor

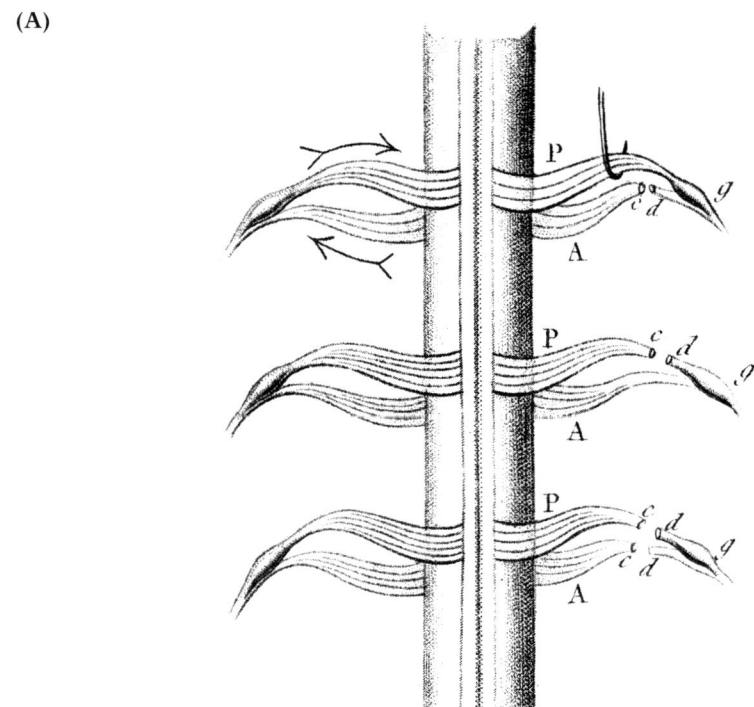

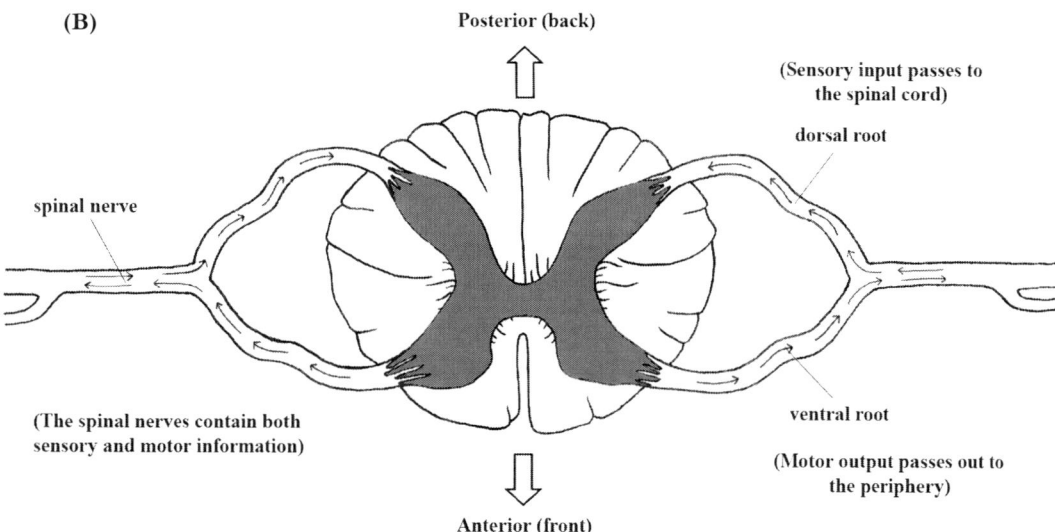

FIGURE 8.2 (A) A image of the spinal roots taken from an 1860 course of lecture notes from Charles Brown-Sequard (Source: Wellcome Library, London). **(B)** The spinal roots. The anterior (ventral) root is responsible for relaying "outgoing" motor fibres into the body, whereas the posterior (dorsal) roots convey sensory information into the spinal cord (Image drawn by Charlotte Caswell).

FIGURE 8.3 (A) Charles Bell (1774–1842) who was the first person to examine the behavioural effects of lesioning the spinal cord roots, which would lead him to recognise the anterior nerves served a motor function. (B) Francois Magendie (1783–1855) who identified the sensory function of the posterior spinal cord roots in 1822.

Source: Both images are from the Wellcome Library, London.

and sensory pathways in the spinal cord. In short, the motor fibres leave the spinal cord through its anterior roots, whereas sensory fibres enter the spinal cord by its posterior roots. This arrangement is now known as the Bell–Magendie law and it was to provide a crucial advance in understanding reflexive action.

The excito-motory reflex: Marshall Hall

Bell and Magendie avoided becoming entangled in the controversies surrounding the issue of whether the spinal cord contained a sentient principle capable of self motion. Nonetheless, the establishment of the Bell–Magendie law provided the anatomical foundations for a new understanding of the reflex, and the person generally credited with making this next step was Marshall Hall. Born in Basford near Nottingham in 1790, Hall graduated as a doctor of medicine from Edinburgh in 1812 and spent time studying in Paris, Göttingen and Berlin. After returning to Nottingham to work as a physician, Hall moved to London in 1826 where he continued practising medicine. A free spirit who never came to hold a professional position in a hospital or university, Hall undertook most of his research at home. This did not deter him, for he proved to be a tireless researcher who wrote over one hundred and fifty papers on a wide range of topics.[9] He also published 19 books. Hall's lack of formal academic advancement may have been due to a certain arrogance for it said he rarely acknowledged the work of others, which sometimes led to accusations of plagiarism. However, it is for his experimental work on the reflex in a wide variety of lower animals including frogs, lizards, eels, snakes and turtles that he would be most remembered – a devotion which he claimed in 1850 had taken up 25 years of his life, or 25,000 hours of his time.

Hall confessed he had first become interested in reflex action during the late 1820s, when observing a severed newt's tail that was still capable of twitching in response to a touching scalpel blade. Although this type of response had been reported by others, it fascinated Hall who set out to learn its mysteries. In one experiment, Hall severed the spinal cord of a snake between its second and third vertebrae – an operation causing paralysis below the level of the cut. However, Hall noticed the paralysed snake's body was not without motion for it showed vigorous writhing when stimulated. Again, this response was not unexpected. Similar reactions had been produced in frogs by Robert Whytt, who had attributed this motion to a soul-like force in the spinal cord. However, the times had changed and Hall arrived at a very different conclusion. For him there was no need to attribute this action to a spiritual force. Rather, it made more sense to explain the reflex as an 'excito-motory system' that received sensory input from the posterior roots, and integrated it within that particular segment of the spinal cord, to produce a co-ordinated response via the anterior roots. In other words, the spinal cord 'reflected' a stimulus into action through a purely mechanical action. Furthermore, according to Hall, such a response could take place without sensation in the brain or higher volition.

Hall first presented his ideas to the Royal Society in 1833.[10] He not only attributed the spinal cord with a mechanical action (or what he called a 'diastaltic' function), but extended

FIGURE 8.4 Marshall Hall (1790–1857) who rejected vitalist accounts of the reflex and replaced them with a mechanical explanation involving the nerves of the spinal cord and medulla.

Source: Wellcome Library, London.

it to include the medulla of the brain. By doing this, Hall was attempting to show that involuntary reflexes were more complex and involved in a far wider and diverse range of behaviours than had previously been supposed. Indeed, Hall had found though his experiments that reflexes not only governed the overall muscle tone of the animal, but controlled respiration, along with other automated functions such as sneezing, coughing and vomiting. Hall also realised that certain complex behaviours such as swimming were likely to involve chains of various reflexes, which required the integration of different sections of the spinal cord or medulla. In fact, he even went so far to say that disturbances of reflex mechanisms might explain various disorders such as asthma, chorea and epilepsy. In other words, neurological disease could have a reflexive basis. However, the situation above the level of the medulla was different with Hall believing the cerebral cortex to be the seat of voluntary movement and the conscious mind. Although Hall accepted reflexive action could be controlled by the higher centres of the brain, to produce even more complex and co-ordinated movement, the human mind was self-determining.

Psychic reflexes: Ivan Mikhailovich Sechenov

Hall had stopped short of attributing the higher regions of the human brain with reflexive action. For him, the regions laying above the medulla governed thought, voluntary action and free will, and were not reducible to physiological laws. In this respect, Hall was a tacit dualist who probably agreed with Flourens in denying localisation of function in the cerebral cortex. However, Hall had laid the foundations for others to extend the reflex concept to higher brain areas, and one person to do this was the German psychiatrist Wilhelm Griesinger. Referring to Hall's dualist theories as 'anachronistic' in 1843, Griesinger proposed reflexes were also at work in mental operations. He called these *psychische reflexaktion* and said they could occur both consciously or unconsciously. Another to make a similar claim was Thomas Laycock who became the first Englishman to hold the Chair of Medicine in Edinburgh in 1855. Laycock pointed out that higher areas of the brain, including the cerebral cortex, were no different from the rest of the nervous system. Consequently, Laycock believed the brain was subject to the same laws of reflex action as the spinal cord, except the reflexes became more complex and highly evolved. As evidence, Laycock pointed to several reflexes that he believed were cerebral in origin, including pathological laughing, and the fearful scream of a rabid patient when exposed to water. While Laycock believed the cerebral cortex acted as the organ of free will, he nevertheless thought there were many reflexes where the mind played no part, including instinctive behaviours, emotional responses, and even some acts of intelligence.

The person to become most associated with the idea of psychic reflexes, however, was the Russian physiologist Ivan Mikhailovich Sechenov. The youngest son of a nobleman and peasant mother, whose early career was spent as an engineer, Sechenov graduated with a medical degree from Moscow in 1856 at the age of 27. Furthering his studies abroad, Sechenov worked in several distinguished laboratories, including those of Müller and Helmholtz in Berlin. However, it was while working in Paris with Claude Bernard, during the early 1860s, that Sechenov made his most important discovery. Sechenov was examining the withdrawal of a frog's leg from an acid solution (i.e. a simple spinal reflex) when he realised he could inhibit this movement by stimulating the segment of the spinal cord just above the site where it was generated. This was an unexpected finding for up to this point all reflexes had been regarded as essentially excitatory – as occurs, for example, when a stimulus causes a muscle contraction. Examining this response further, Sechenov stimulated other areas of the nervous system including the medulla and thalamus of the brain. The results of this research showed stimulation not only suppressed reflex activities in the lower extremities of the body, but also autonomic responses such as blood flow. Sechenov had made a vital discovery: inhibition was a prerogative of a higher area of the nervous system over a lower one.

This simple discovery had profound implications. For one thing, Sechenov realised a stimulus need not elicit a reflexive action directly (as Hall had insisted), but could release a reflex from its inhibition. To give an example, Sechenov pointed out that a small grain of dust can elicit a sneeze. How a small stimulus produced such a violent response had long puzzled researchers. Now it was explainable by arguing the itch of the nose released the neural reflexes responsible for the sneeze from their inhibition. This also greatly increased the scope of what a reflex could be. Indeed, a reflex constrained by an 'inhibited end' need not only be a motor response, but a stored memory or some type of a thought. Taking this idea to its logical conclusion, Sechenov realised all human behaviour and mental activity could be understood in reflexive terms. In fact, it would lead him to a theoretical position where he would deny thoughts were the cause of voluntary behaviour. Instead, for Sechenov, thoughts were nothing more than psychic reflexes released from their inhibition by external stimulation. Although Sechenov did not deny the existence of consciousness, he believed there was no need to attribute it

FIGURE 8.5 Ivan Mikhailovich Sechenov (1829–1905) who developed the idea of psychic reflexes governing all thought and behaviour. He was also the first to realise complex patterns of reflexes could be held in check by a simple process of inhibition.

Source: National Library of Medicine.

with volition. Instead, it was a series of inconsequential mental events triggered by external events. In this way, Sechenov was able to reduce all mental activity, emotions and behaviour to the excitation or inhibition of reflexes. Sechenov published his ideas in his book *Reflexes of the Brain* in 1863. They were, however, highly controversial, especially in nineteenth century Russia, where his assertion that 'the initial cause of any human action lies outside the person' was regarded as anti-religious, immoral and dangerous to society. The Czarist regime even considered indicting him under the penal code.

Sechenov also made another important contribution to the concept of psychic reflexes: he proposed they were subject to modification by learning. Prior to Sechenov, reflexes had always been regarded as pre-wired neural circuits, which elicited the same invariant response throughout the animal's lifetime. However, Sechenov recognised reflexive action could be modified by experience. Taking this idea one step further, Sechenov came to view human development, from infant to adult, as a process in which inhibition gradually exerts greater control over more basic and innate reflexes through the process of learning. Indeed, for Sechenov, this learning process was the very thing that made humans special. Although it was very difficult to see the reflexive nature of behaviour in an adult, Sechenov argued if one could observe all external events from childbirth, then the true development of learning would be revealed. The idea that innate reflexes could be modified through experience would be taken up and formulated in more exact terms by his compatriot Ivan Pavlov.

The integrated reflex: Charles Scott Sherrington

Few would deny the greatest and most extensive analysis of the reflex was to be undertaken by Charles Scott Sherrington. His writings on the subject spanned nearly seventy years and resulted in over three hundred publications. It also led to his great book *The Integrative Action of the Nervous System* (1906), which synthesised many aspects of the fledgling neuron doctrine in an attempt to explain how the reflex operated as the simplest unit of nervous co-ordination. This work is now recognised as one of the founding books of modern neuroscience.[11] Although Sherrington's achievements were wide-ranging, they were principally to show three things: (1) to map out in great detail the neural circuitry underlying a variety of reflexes; (2) to show how activity in the nervous system requires both excitatory and inhibitory information; (3) to demonstrate how the unified individual depends on the integration of reflex activity involving every level of the nervous system. Perhaps above all, the real secret of *Integrative Action* lies in the fact it provided a set of concepts by which the nervous system could be understood. It should also not be forgotten that Sherrington is credited with coining the term 'synapse' which was also central to his understanding of the reflex.

According to his official biography, Sherrington was born in Islington, London, in 1857, the son of a country doctor who died during his infancy. However, it is almost certainly the case that Sherrington was actually the illegitimate son of the classical scholar and surgeon Caleb Rose, who only got around to marrying his mother in 1880. Sherrington spent much of his childhood in Great Yarmouth and Ipswich, where Rose worked as a physician, and from whom he appears to have developed an abiding love for literature − a passion often manifest in his later works. However, it was to medicine Sherrington first turned, starting medical studies at St Thomas's Hospital in London in 1876.[12] Three years later, Sherrington moved to Cambridge, joining the laboratory of Sir Michael Foster, who had just founded *The Journal of Physiology*, the first publication dedicated to physiological research. Not long after his arrival, Sherrington was given a unique opportunity after the Medical Congress of 1881 in London had witnessed a highly publicised dispute between David Ferrier and Friedrich Goltz concerning the behaviour of decorticated animals (see next chapter). One person to adjudicate on the matter was the Cambridge professor John Newport Langley. Asked to examine two brains, one provided by Ferrier (from a monkey) and one from Goltz (a dog), Langley enlisted the help of the young Sherrington. It was to lead to his first scientific paper describing the damage in Goltz's dog in 1884.

After completing his medical studies at St Thomas's, Sherrington moved to Strasbourg where he spent a year in Goltz's laboratory examining the effects of motor cortex damage. Around the same time, Sherrington was also asked by two British societies to investigate cholera outbreaks then taking place in Spain and Italy. After establishing a further expertise in infectious diseases, Sherrington was able to follow this up by working with the eminent pathologist Rudolf Virchow, and bacteriologist Robert Koch, in Berlin. It proved a highly productive time for Sherrington who was able to develop a diphtheria vaccine, which he was forced to use in 1894 to save the life of his seven-year-old nephew. Near the end of his life, Sherrington would proudly write to a friend, that the young boy had grown to be six feet in height, and distinguished himself by serving with distinction in the First World War.

In 1891, Sherrington took up a post in a London veterinary hospital, which also provided excellent research facilities called the Brown Institution. Later, Sherrington would admit he was unsure of his chief research interests at this time, but became focused on neurophysiology after being inspired by Walter Gaskell at Cambridge who spoke of the endless mysteries concerning the functions of the spinal cord. Consequently, he began to focus his attention on spinal reflexes. One of Sherrington's first concerns was the patella reflex, otherwise known as the knee jerk, which is initiated by tapping the tendon below the kneecap. This response, now used routinely by doctors to assess the health of the nervous system, had first been described by two Germans Carl Westphal and Wilhelm Erb, independently of each other, in 1875. However, they had arrived at different opinions about how the reflex occurred. Westphal believed it was a simple mechanical twitch of the quadriceps (the muscle running along the front of the thigh), which was transmitted from the force of the tapped tendon. In contrast, Erb considered it to be a genuine reflex, involving a sensory pathway entering the spinal cord from the tendon, that was then turned back to contract the quadriceps muscle by another nervous pathway.

Using monkeys as subjects, Sherrington set about understanding the patellar reflex, first by identifying the spinal roots mediating the response, and then cutting the nerves conveying information to and from the leg muscles. It proved Erb was correct: the reflex involved a nervous impulse from the tendon to the spinal cord, which was re-directed back out to the quadriceps. However, Sherrington found this reflex was much more complex than previously envisaged, since tapping the tendon *also* caused reduced muscle tension in the hamstrings that run along the back of the thigh. This was to be the first demonstration (1891) of something Sherrington repeatedly found as a consistent feature of all spinal reflexes: namely the reciprocal action of agonist and antagonistic muscles. From this, Sherrington would realise the notion of a simple or excitatory reflex was in reality 'a purely abstract conception'. Indeed, a reciprocal response is essential for reflexive action, including the knee jerk, because the reduced tension of the hamstrings is a necessary adjustment to allow the quadriceps to stretch. Sherrington also deduced this reflex must involve two motor neurons: one producing excitation of the quadriceps, and the other one causing inhibition of the hamstrings.

The discovery of proprioception

In 1895, Sherrington became Holt Professor of Physiology at the University of Liverpool, a position he held until moving to Oxford in 1913. Realising little further progress would be made into understanding the neuroanatomy of reflexes, unless the functional roles of the motor and sensory nerves were more fully established, he set out to examine the anterior and posterior roots. It was to involve a decade of painstaking research, assessing the behavioural consequences of individually severing each root, from the lumbar-sacral region of the spinal cord, in a variety of different animals. Sherrington's procedure typically involved anaesthetising the creature, and exposing their anterior (motor) spinal roots – which were then stimulated with a mild electric current. By doing this, Sherrington found most skeletal muscles received an intermingling of nerve fibres from several (usually two or three) roots leaving the spinal cord at different levels. Thus, Sherrington was able to establish the region of the body innervated by each spinal nerve and the movements it produced. Sherrington also used a similar method for mapping the areas of the body that contributed to each sensory nerve. For example, by noting reflexive action after pinching areas of skin and then cutting the dorsal roots above and below the ones he wanted to study, Sherrington was able to establish the territories of touch (or 'islands of sensibility') that fed into each root. Nobody else had before mapped the functions of the posterior and anterior roots in such detail.

During this research, Sherrington discovered something else of great importance. Previously, it had been assumed all nerves passing from spinal cord to muscle only produced movement. Furthermore, there was no reason to regard muscles as sensory organs since they did not appear to contain any receptors. Indeed, it was generally accepted that other tissues of the body, such as the skin and joints, were better candidates for detecting sense information. However, when Sherrington lesioned the anterior roots (an operation that should have destroyed all the motor nerves) he found nearly half of the fibres in the spinal nerve were unaffected. There could only be one explanation: these were sensory fibres from the muscles heading into the spinal cord. This raised several other questions – not least, where were the sense organs in the muscle located? Sherrington had a hunch. In 1860, the German August Weismann had discovered long bundles of fine fibrous threads embedded in muscle, called muscle spindles by Willy Kühne in 1863. Sherrington suspected these served as sensory organs. To test this idea, he lesioned the posterior (sensory) spinal roots, and followed the degeneration back to the muscle. His theory was confirmed: muscle spindles were sensory organs. Further examination revealed these structures detected the amount of stretch taking place in the muscle, and initiated a compensatory reflex contraction when the muscle was stretched by a heavy load. In 1906, Sherrington called this type of response 'proprioception'. He also viewed it as the sixth sense of the body, distinct from touch, which provided the nervous system with information regarding the position of the muscles and joints – information that is crucial for the control of posture.

An examination of more complex reflexes

Much animal behaviour, of course, is composed of highly complex and synchronised automated actions such as walking or running. Sherrington suspected these types of behaviour were also the result of integrated and sequenced chains of much simpler spinal reflexes, and set out to examine this possibility in decerebrate animals.[13] This operation eliminated the effects of higher brain areas, including the cerebral cortex, on reflex activity, enabling him to focus entirely on the involuntary spinal cord. Sherrington had first undertaken this procedure in 1896, in an attempt to avoid the troublesome use of anaesthetics, which depressed reflex activity. To his surprise, the operation made things worse, since it caused many muscles to become rigid and extended. Sherrington called this state 'decerebrate rigidity'. Sherrington also realised it occurred because the higher brain areas were no longer inhibiting the areas below. Consequently, the rigidity came about as muscles went into spasm from the 'unchecked' activity of their sensory fibres, which excited the motor nerve leading back to the muscles. Sherrington even managed to locate the descending source of this inhibition: the anterior-lateral descending tracts of the spinal cord.

Despite producing muscular rigidity, Sherrington was able to induce a number of complex behaviours in his decerebrate animals. For example, when Sherrington stimulated the left foreleg of a decerebrate cat, it stepped forward, while the corresponding left back leg moved backwards. Furthermore, as this happened, the two legs on the right side of the body exhibited opposing movements. Thus, Sherrington had discovered one of the basic reflexes involved in walking: that is, a reflex mediated entirely by the spinal cord involving the reciprocal interaction of several limbs. Sherrington also examined the scratch reflex, which he initially elicited by tickling a dog on its shoulder. Later, he would use a small pin inserted under the skin, called an 'artificial electrical flea', to stimulate the itch over various parts of the body. Careful observation of the scratch reflex showed it was composed of two components: the scratching motion and an adjustment for maintaining balance. The scratching motion was a reflex consisting of a series of alternating knee, ankle and hip movements, involving a total of nineteen muscles, which beat rhythmically five times a second. To maintain balance, the animal also had to bring into action a further seventeen muscles. Without this adjustment the animal would fall over. Remarkably, the decerebrate animal was able to follow the flea all over its body and scratch in the appropriate place. Although the external sensors in the skin initiated the order to start scratching, the reflex was also dependent on the proprioceptors located in the muscles to keep the animal upright.

Sherrington's legacy

In 1906, Sherrington published *The Integrative Action of the Nervous System*, based on a series of ten Silliman lectures, given at Yale College in New Haven, two years previously. Covering nearly two decades of research, it was an attempt to provide a modern synthesis of how the nervous system worked utilising concepts borrowed from the neuron doctrine. In particular, Sherrington's book emphasised the role of the synapse. Prior to Sherrington, the synapse had largely been considered as a hypothetical entity. However, Sherrington's careful examination of reflex activity had shown it to have important functional implications. In particular, he had revealed it to be the site where excitation and inhibition was exercised, which allowed reciprocal responses of different muscles to occur – a fundamental reflex mechanism. However, Sherrington also realised the synapse had another vitally important role: it provided the site where the 'decision making' over different competing reflexes was made. One place where synapses exerted this effect was on the motor neuron. The cell bodies of these neurons are situated in the grey matter of the spinal cord, and Sherrington recognised them as providing the 'final common pathway' to the muscles, but how do they know when to fire or not? Sherrington realised motor neurons must receive both excitatory and inhibitory synaptic input from many diverse sources, including other segmental areas of the spinal cord, and higher brain areas. In turn, all this information was 'added together' by the cell – not unlike working out an algebraic sum of its effects. This balance of excitatory and inhibitory input, therefore, determined whether the motor neuron would send an impulse to the muscle. This idea is called summation, and is now recognised as the fundamental mechanism by which nerve cells make their decisions.

But perhaps the most important feature of *Integrative Action* was the demonstration that simple reflexes are the fundamental building bricks of behaviour. Although a reflex can be defined simply as an innate automated response bought about by a specific stimulus, Sherrington also realised this idea was 'a convenient, if not probable fiction'. This is because all parts of the nervous system are intimately connected. Indeed, as Sherrington had shown with the patellar knee jerk, every contraction of muscle movement requires the relaxing action of an opposing muscle. Thus, all reflexes must involve more than one neural component. Extending this idea further, Sherrington used it to show how more complex behaviours arise from much larger chains of simple reflexes, sometimes requiring the integration of many different neural circuits, at higher levels of the spinal cord and brain. Sherrington also recognised that as one ascends through the nervous system, the neural anatomy of reflexes become more circuitous and the complexity of behaviour increases. Consequently, the most complex reflexes arose from the cerebral cortex, which Sherrington compared to a complicated switchboard with millions of on–off switches, similar to a great telephone exchange. Sherrington's legacy therefore was to show that much of our behaviour, at any given point in time, must involve vast numbers of nerve cells and synapses acting together in a beautifully organised, integrative, reflexive, way.

In 1913, Sherrington was invited to become Waynfleet Professor of Physiology at Oxford, where he would remain until his retirement in 1936 at the age of 78. By now he had been bestowed with many honours including a knighthood in 1922, and the Nobel Prize in 1932 (along with Edgar Douglas Adrian) for 'discoveries regarding the functions of neurons'. In addition, he had also written another classic book entitled *Mammalian Physiology: a Course of Practical Exercises*, this time for students. Much beloved by colleagues for his modest and friendly personality, Sherrington continued to write in his later years, turning his attention to many other subjects including philosophy and poetry. He also published his thoughts concerning the relationship between mind and brain in *Man and his Nature* published in 1940 when he was 83 years old. Somewhat surprisingly for a man who had dedicated his life to involuntary reflexes, Sherrington admitted to believing in a form of Cartesian dualism where the mind was not subject to physical laws – although it was somehow capable of interacting with, and directing, the brain's reflexive machinery. Famously, Sherrington would describe the relationship of brain and mind as being similar to an enchanted loom:

> The brain is waking and with it the mind is returning. It is as if the Milky Way entered upon some cosmic dance. Swiftly the head mass becomes an enchanted loom where millions of flashing shuttles weave a dissolving pattern . . . a shifting harmony of subpatterns.

This idea has proven to be an enduring image, inspiring countless generations of researchers to contemplate how the mind arises from the electrical activity of the brain. Sherrington also wrote *The Endeavour of Jean Fernel* (1946), a scholarly biography involving much Latin and

FIGURE 8.6 Charles Scott Sherrington (1859–1952). A man whose vast list of achievements over a long career are difficult to summarise briefly, although he is perhaps best remembered for his *The Integrative Action of the Nervous System* first published in 1906.

Source: Wellcome Library, London.

French translation of a sixteenth century French doctor who stressed the importance of establishing facts over traditional belief.[14] He continued to follow his academic interests until late in his life, despite being short-sighted and suffering from rheumatoid arthritis. Sherrington died from a sudden heart attack at the age of 94 in an Eastbourne nursing home.

Conditioned reflexes: Ivan Pavlov

Sherrington regarded the reflex as a response determined by a pre-wired neural circuit, fixed by the connections of the nervous system. However, the reflex would come to be seen very differently by the Russian physiologist Ivan Pavlov. Born in the small town of Ryazan, a hundred miles south east of Moscow, Pavlov was the oldest of 11 children (six of whom died in childhood) and destined for a career in the priesthood. However, he apparently changed his vocation after reading and being inspired by Sechenov's *Reflexes of the Brain*. Despite the strong opposition of his father who refused to support him financially, Pavlov enrolled as a student in natural sciences at St Petersburg in 1870. He soon distinguished himself by winning a gold medal for his work on the pancreas gland in 1875. After receiving his medical degree in 1879, and a doctorate in 1883, Pavlov undertook a period of research in the German universities of Breslau and Leipzig, where he concerned himself with cardiovascular and gastrointestinal physiology. Returning to St Petersburg in 1890, Pavlov secured two appointments: one becoming Director of Physiology at the Imperial Institute for Experimental

Medicine (a position he would hold for 45 years), and the second as Professor of Pharmacology at the Imperial Medical Academy. Pavlov would quickly gain an international reputation for his work at these institutions.

By all accounts Pavlov was a devoted, tireless and meticulous experimenter whose chief interests in St Petersburg concerned the physiology of circulation and digestion. Although the Imperial Institute for Experimental Medicine possessed Russia's most advanced facilities for research, it suffered from financial difficulties, and Pavlov was forced to work under relatively impoverished conditions compared to the rest of Europe. Nonetheless, he set about investigating the nervous control of secretion in digestion by inventing a surgical technique allowing him to implant small collecting vessels called fistulas into the stomach and intestines. Pavlov was now able to directly observe the processes of digestion in the intact and conscious animal, and to examine the chemical composition of their digestive fluids. It was a major advance over the work of previous investigators who had investigated the same phenomena in dying animals under anaesthesia. These innovative methods allowed Pavlov to open up a new window into the workings of the digestive system, resulting in a number of fundamental discoveries. This included explaining the role of the vagus nerve in the secretion of gastric juices and the brain's control over the pancreas gland. It would lead Pavlov to become the first Russian and physiologist to win a Nobel prize in 1904.[15]

It was during the early stages of this work, sometime around 1891, Pavlov turned his attention to the collection of saliva. This task proved somewhat more problematical than the collection of stomach juices, because saliva was readily elicited by the visual sight of food (or what Pavlov referred to as 'psychic excitation'). Of course, this was not entirely unexpected as the response formed the basis of the well known expression 'one's mouth is watering'. At first, it appears Pavlov regarded it as little more than a confounding variable in his attempt to accurately measure saliva. However, as the research continued, it became increasingly clear to Pavlov that his dogs were anticipating when they were about to be fed – a response showing they were *learning* something about their situation. Pavlov's interest in the psychic release of saliva was further aroused when one of his research students, Sigizmund Vul'fson, observed that dogs salivated with proportionally smaller amounts of saliva when the food was presented further away from the animal. Pavlov could not explain the learning of psychic reflexes, or the distance effect in terms of fixed pre-wired reflexes. Yet, clearly, the animal was adjusting to changes in its environment in some way. By 1902, Pavlov was so impressed with the potential of these new discoveries he ceased his work on digestion and concentrated on learning instead.

The most famous experiment associated with Pavlov is one where he is alleged to have rung a bell whenever a dog was presented with food. Initially in this situation, the food naturally elicits salivation, whereas the bell is neutral and does not produce any response. However, after repeated pairings of bell and food together over many trials, Pavlov found the dog would also come to secrete digestive juices in response to the bell alone. Thus, the dog had learned the association between the bell and food – a response that Pavlov referred to as a conditional reflex. In fact, this description is partly fiction: Pavlov used many stimuli to elicit conditioning in his dogs, including a metronome, buzzer, black-square and lamp flash, but apparently never a bell.[16] The origins of this seminal experiment are also obscure, and may have been first performed by one of Pavlov's students, Vasiliĭ N. Bol'dyrev in 1905.[17] Nonetheless, Pavlov was to use the method as his basic paradigm to explore conditioned reflexes over the subsequent years. His research would lead him to propose two types of reflex existed: inborn and learned.

(A)

(B)

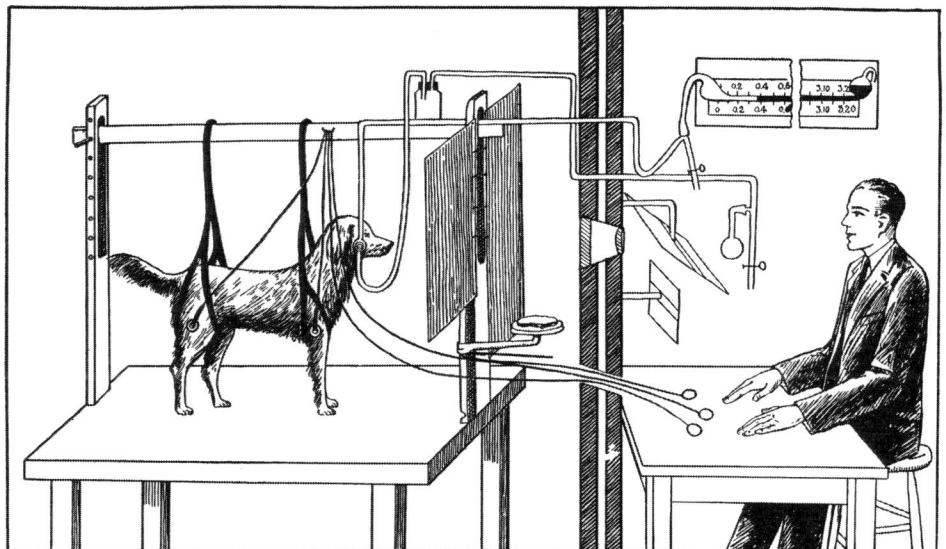

FIGURE 8.7 (A) The Russian physiologist Ivan Pavlov (1849–1936) who began examining conditioned reflexes around 1904 after his dogs began to salivate in anticipation of being fed. (B) Experimental set-up used by Pavlov to examine conditioning.

Source: Both images are from the Wellcome Library, London.

The former, such as salivating in response to the smell of food, was a result of pre-wired neural connections in the brain. The latter, such as salivation in response to the sound of a buzzer, were the result of new reflexive pathways being created in the cerebral cortex. Importantly, both types of reflex were regarded as mechanical events that did not depend in any way on mental factors or consciousness.

Pavlov's impact on psychology

The main reason why conditioned reflexes are so important is that Pavlov believed he had identified the fundamental unit of learning in animals and human beings. All learned behaviour, in his view, was nothing more than 'a long chain of conditioned reflexes' whose acquisition, maintenance and extinction were subject to universal laws. It was basically the same position taken by Sechenov in 1863, and attributed the cause of all behaviour to stimulus events taking place outside the organism. Viewed in this way, the person was therefore little more than a machine, operated by unconditioned and conditioned reflexes, with the mind 'looking on', but playing no role in initiating action or producing behaviour. These ideas were to become very influential, especially in the field of experimental psychology where they stimulated a new movement called behaviourism.

Pavlov's work on conditioned reflexes had first attracted the attention of American psychologists in 1909 through an article written by Robert Yerkes. This was a period when psychology was dominated by experimental methods that were widely regarded as unscientific by its critics. One person seeking a radically new approach to psychology was the Harvard professor John Broadus Watson. Frustrated by psychology's lack of rigour and objectivity, Watson published *Psychology as the Behaviorist Views It* in 1913, which sought to make the study of behaviour a natural science. By doing this, Watson was hopeful psychologists would turn away from studying the mind, and instead establish irrefutable laws of behaviour that could be directly observed and measured. At the time of writing this paper, however, Watson had been unaware of Pavlov's research. However, when finally introduced to the work in 1915 by colleague Karl Lashley, Watson immediately recognised it provided a model for his behaviourism. By focusing on conditioned reflexes, Watson believed he could finally erase mentalistic and subjective explanations from psychology, thereby transforming it into a genuine scientific discipline.

The behaviourist approach, sometimes known as stimulus–response psychology, became highly influential in the early part of the twentieth century, and emphasised the importance of the environment in shaping behaviour. Indeed, Watson himself contributed to this development in the winter of 1919–1920, with a young graduate research student Rosalie Rayner, when he used the principles of Pavlovian conditioning to elicit fear in a nine-month-old child who has since become known as Little Albert.[18] It is an experimental study that is also one of the most infamous in the history of psychology. Watson and Rayner acquainted Albert with a pet rat. After becoming accustomed to the animal, the experimenters then loudly clanged a metal bar with a hammer when the rat was in Albert's presence. Seven pairings of the rat and noise was sufficient to produce a conditioned fear response in Albert, with the child exhibiting a state of distress when given the animal. Refusing to play with the rat again, Albert also generalised his anxiety to other objects in the room including a rabbit, dog, a seal coat, cotton wool and a bearded Santa Claus mask. These responses endured over the subsequent month, at which point Albert's mother removed him from the study. Regrettably, at no point had Watson made any effort to remove his fear and confessed Albert's behaviour was 'likely to persist indefinitely'. Nor could psychologists provide any further help, for Albert's true identity was soon forgotten. In fact, it would remain unknown until 2009, when psychologist Hall P. Beck at Appalachian State University, identified Albert as Douglas Merritte who had died in 1925 from hydrocephalus when just six years old. Although Watson claimed Little Albert was in perfect health during the study, this now seems unlikely. The collaboration between Watson and Rayner also had repercussions in other ways. Watson had an affair with his younger assistant, leading to front-page news, a divorce, loss of his faculty position at John Hopkins University, and a new career in advertising.[19]

Despite the controversies, behaviourism had a profound influence on the course of psychology during the first half of the twentieth century, providing the primary paradigm by which to study behaviour from the 1920s to around 1950. For the behaviourist, complex behaviour was simply the sum of past learning experiences, and the individual modifiable into anybody the psychologist desired through conditioning. Indeed, Watson was to memorably express this belief in no uncertain terms in 1930:

> Give me a dozen healthy infants . . . and I'll guarantee to take any one at random and train him to become any type of specialist I might select – doctor, lawyer, artist, merchant-chief, and, yes, even beggar man and thief, regardless of his talents, penchants, tendencies, abilities, vocations and race of his ancestors.

Put into its broader context, this meant behaviourists could also act as social engineers, helping society to scientifically engineer individuals to fit their environment. These ideas would reach their fullest expression in the work of B.F. Skinner. The inventor of the operant box (essentially a controlled environment in which the animal could be conditioned by the process of reinforcement), Skinner provided an alternative to Pavlov's 'classical conditioning' where an existing behaviour is shaped by associating it with a new stimulus, by formulating operant conditioning in which behaviour is shaped by reward and extinguished by punishment. The potential benefits of Skinner's psychology were expressed in his utopian novel *Walden Two* (1948) which portrayed a small society where children are rigorously conditioned from birth by positive reinforcement for the benefit of all. Despite going beyond Sechenov and Pavlov, Skinner still continued to reject mentalistic explanations of behaviour, viewing free will as illusory and describing consciousness as something that only provided a running commentary on what was happening to the individual.

Searching for the engram: Karl Lashley

Pavlov not only provided a theoretical foundation for behaviourism, but inspired others to search for the neural correlates of conditioned reflexes – an endeavour that became most associated with the work of Karl Lashley. Graduating in zoology from West Virginia, and obtaining a Masters in Bacteriology from Pittsburgh, Lashley seems not to have been particularly interested in psychology until going to Johns Hopkins where he was taught by Watson in 1911. Although studying for his doctorate at the time, Lashley became a voluntary assistant to Watson, which resulted in a number of publications, including a contribution to his supervisor's book *Behavior* in 1914. However, Lashley had a more biological orientation than Watson, and it was cultivated further after his doctorate by working with the Washington surgeon Shepherd Ivory Franz. Best known for his work on brain injured patients, Franz also carried out lesion experiments in animals to assess the functions of the frontal lobes. Working with Franz provided Lashley with a unique opportunity to develop his surgical expertise – skills he realised could be used to examine the neural basis of Pavlovian conditioning. In fact, Pavlov had hypothesised that conditioned reflexes involved the formation of new neural connections in the cerebral cortex. Or, more specifically, they resulted in a new neural pathway being created that connected a sensory region (e.g. auditory cortex), with an area mediating responses (e.g. motor cortex). Lashley saw if this was the case, then it followed that lesions to the cerebral cortex should disrupt, or even abolish, the learning and retention of conditioned responses.

Lashley first set about attempting to locate the neural site of memory (or what he called the 'engram') at the University of Minnesota, where he had been made Assistant Professor in 1920. However, most of this work would be undertaken in Chicago after moving there in 1926. His main strategy involved training rats on visual discrimination or maze tasks, and then assessing whether lesioning various parts of the cerebral cortex disrupted subsequent performance. The logic was simple: if Lashley could find a site where a lesion impaired the ability of the animal to perform a well learned task, it would provide evidence for its anatomical location. However, the results were not as Lashley expected. For example, after training his rats to run through an eight-alley maze, Lashley found small lesions which removed about 20 per cent of the cortex had little adverse effect on retention. In fact, Lashley had to remove

FIGURE 8.8 Karl Lashley (1890–1958) who attempted to locate where memory engrams are stored in the brain. He would conclude that memory is stored throughout the cerebrum, with all parts of the cortex playing an equal role in its storage.

Source: With permission from Robert Boakes, 1984, *From Darwin to Behaviourism*.

around half of the cortex before he observed a significant maze deficit. Even then, the rats could relearn the task with subsequent training. Nor did it matter where Lashley placed the lesion. No site was more critical than any other for disrupting the learning.

On the basis of these experimental findings, Lashley devised two general laws about the localisation of the engram which he outlined in his 1929 book *Brain Mechanisms and Intelligence*: (1) memory is stored throughout the cerebral cortex – a principle he called 'mass action'; (2) all parts of the cortex play an equal role in the storage or memory – which he called 'equipotentiality'. It was as if the neural correlates of maze learning were stored everywhere or no place in particular – a finding not explainable in terms of Pavlov's theory. It also led Lashley to reject the assumptions of behaviourism with its simple emphasis on explaining all behaviour in terms of conditioned reflexes. Another problem with the conditioned reflex concept was that many types of skilled behaviour such as playing a musical instrument were simply too fast for them to be directed by a chain of stimulus response actions (a golf swing, or a tennis player attempting to return a high velocity serve are two other examples). Consequently, Lashley reasoned they had to be 'centrally supervised' in some other way. Thus, there must be a more complex representation in the brain, or 'schemata of action' which was able to generate behavioural responses independently of any stimulus. However, he was never able to provide a convincing explanation for the phenomena. After some 30 years of research, Lashley was to conclude in 1950, in a paper entitled *In Search of the Engram*, that despite all his attempts to localise the memory trace, he had not been able to do it.

Reflexes as cell assemblies: Donald Hebb

Although Lashley came to reject the idea of explaining behaviour in terms of conditioned reflexes, a new conception of reflexive action would be proposed by one of his students – the Canadian Donald Hebb. Born in the small fishing town of Chester, Nova Scotia to parents who were doctors, Hebb's ambition was to be a novelist. To this end, he obtained his Bachelor of Arts degree in 1925, and took a number of jobs before enrolling as a part-time psychology Masters student at Montreal's McGill University in 1928. Having to support himself by teaching, Hebb became head of an elementary school in a working class district of Montreal. This was not to last, however, as Hebb became bedridden for a year with a hip infection, and then tragically lost his wife of just 18 months in a car accident. Forced to recuperate, Hebb read Sherrington's *The Integrative Action of the Nervous System* and Pavlov's *Conditioned Reflexes*, which had both been published in English in 1927. It would motivate him to write a Masters thesis (from his bed) on the learning of spinal reflexes in the foetus in 1932. Following his recovery (he would always have a small limp from his infection), Hebb took up a doctorate position under Karl Lashley, then in Chicago, after a recommendation by Boris Babkin who had worked with Pavlov in St Petersburg. Here, Hebb examined the visual deficits of rats reared in the dark, before taking up another position, five years later with Wilder Penfield at the Montreal Neurological Institute, where he assessed patients who had undergone brain surgery for epilepsy. After another spell working with Lashley, this time at the Yerkes Laboratory in Florida, Hebb was appointed professor at McGill University in 1948.

While in Florida, Hebb had started work on his great book *The Organization of Behavior*. This was to be published in 1949. By now, Hebb was seeking a new way of understanding behaviour. Agreeing with Lashley that stimulus–response reflex theories had failed in their attempt to explain learning, he realised a different approach was needed. Consequently, he began to consider many aspects of behaviour that were held to be off limits to experimental psychologists at the time, including subjects such as attention, intelligence, vision and perception. This was not an aim fitting well within the zeitgeist of the times, which was dominated by B.F. Skinner, who denounced as 'physiologising' anyone who sought to explain behaviour in terms of mental processes or vague physiological processes. But Hebb was keen to develop a theory going beyond the laws of conditioning. Undaunted by the criticism of the behaviourists, Hebb set out to show how the structure and function of the brain, at least in theory, could go beyond stimulus–response physiology to give rise to mental activity and behaviour.

One of Hebb's most important innovations in *The Organization of Behavior* was to replace the reflex with a neural circuit containing a large assembly of cells. Instead of viewing the reflex as a one-way chain of neural events from stimulus to response, Hebb created a loop in the 'middle' which contained a large number of additional neurons.[20] The reflex arc was now circular and not linear. This concept had a number of important advantages over the simple reflex. For one thing, if set into motion, a cell assembly could produce 'reverberations' within the circuit, thereby providing a mechanism to maintain activity in the cerebral cortex long after the stimulus had terminated. This situation did not occur with a simple reflex that caused an automated and immediate response. However, this reveberatory activity gave the circuit another important feature: it allowed learning to take place since the neural network was capable of being strengthened with repeated sensory experience. So, for example, in the case of a baby learning to recognise its milk bottle – the initial sensory experience was hypothesised to establish the neural circuit, with its repeated stimulation resulting in learning. However, Hebb did not end there for he recognised the cell-assembly provided yet another crucial function. If it was excited by external sensory stimulation then it could be responsible for perception; but if active after the sensory stimulation has ceased, it provided the neural basis for imagery and thought.

But where in this neural circuit was the site responsible for the changes that allowed reverberatory activity to take place? For Hebb there was only one place: the synapse. In fact, he would come up with a memorable law, which explicitly stated the role of the synapse in creating reveberatory activity and learning that is now widely known as Hebb's rule:

> If an axon of cell A is near enough to excite a cell B and repeatedly or persistently takes part in firing it, some growth or metabolic change takes place in one or both cells, so that the efficiency of neural transmission between the two is increased.

Put simply, Hebb proposed continuous stimulation of a neuron in a circuit would lead to a structural change at its synapse, which then increased the likelihood of reveberatory activity being maintained. And, if reverberatory activity continued for long enough, some form of permanent memory was likely to result. This was an important theoretical development, perhaps not fully recognised at the time, which viewed the synapse as the crucial site of change in the nervous system. It also helped explain how memory could be stored in a network diffusely distributed throughout the brain. That is, once the synaptic connections had been strengthened in a cell assembly, they effectively acted as a 'stored memory', which could be elicited again by further stimulation of the circuit, or its excitation from other neurons or networks.

Although *The Organization of Behavior* was a ground breaking work, it was largely theoretical and dealt with issues which could not be experimentally verified at the time. Despite this, the book's arguments were persuasive to many, and it helped to play a role in the demise of behaviourism and the rise of cognitive psychology. Perhaps the most important support for Hebb's synaptic rule was to come in 1973 when British neurophysiologist Timothy Bliss and Norwegian Terje Lomo, working in London, discovered a phenomenon which they called long-term potentiation. Working with the perforant pathway which projects into the hippocampus, Bliss and Lomo found its stimulation caused longlasting voltage changes in its recipient cells. In some cases these occurred weeks after the stimulation – proving the synapses in the pathway had been strengthened by the initial electrical impulses. Since this discovery, long-term potentiation has been found in many other areas of the brain, and shown to be an important substrate of learning and memory. A synapse that is 'strengthened' as a result of learning, now commonly known as a Hebbian synapse, has also become an essential concept in neural network computer programming and artificial intelligence. Hebb could probably never have imagined his radical, but simple rule, would be such an integral part of brain research in the twenty-first century.

Notes

1. *Essay on the Vital and other Involuntary Motions of Animals.*
2. In addition to the already mentioned pupillary reflex to light, they included: (1) digestion; (2) coughing; (3) sneezing; (4) erection of the penis; (5) ejaculation; (6) genital responses in the female; (7) blushing; (8) bladder movements; (9) heart beat; and (10) respiration.
3. Neither Vesalius or Willis had focused much attention on the spinal cord itself – instead they had been more concerned with describing the cranial (direct from the brain) and sympathetic trunk (a ganglion that runs along the length of the spinal cord on either side).
4. The anatomist Robert Knox made the claim.
5. The spinal roots appear to have first been described by the Swiss anatomist Johann Jacob Huber (1707–1778).
6. Entitled 'Idea of a new anatomy of the brain', submitted for the observation of his friends.
7. After learning of one of Magendie's public dissections involving a greyhound that was apparently nailed down ear and paw, dissected, and then left overnight for further butchery, the Irish MP Richard Martin, is said to have been so incensed that he introduced his landmark bill banning animal cruelty in the United Kingdom in 1822. He also called Magendie a 'disgrace to Society.'
8. This is the view of Dr Paul Cranefield who in 1974 published a detailed treatise on the matter *The Way In and the Way Out. Francois Magendie, Charles Bell and the Roots of the Spinal Nerves*. Mount Kisco: Futura Publishing.
9. Some of the things Hall wrote about included bloodletting, diseases of women, slavery, and the problem of sewage disposal.
10. On the reflex function of the medulla oblongata and medulla spinalis.
11. The book has been compared in greatness with Newton's *Principia Mathematica*, and Harvey's *de Moto Cordis*. See Levine, D.N. (2007) Sherrington's 'The integrative action of the nervous system: A centennial appraisal'. *Journal of the Neurological Sciences*, 253, 1–6.
12. According to one biography (Cohen 1958), Sherrington had intended going to Cambridge, but a family financial problem meant that his two brothers (William and George) went ahead of him, with Charles having to enrol at St Thomas's.
13. Decerebration is the elimination of higher brain function in an animal by removing the cerebral cortex – normally undertaken by cutting across the brainstem.
14. Fernel is also credited with first using the terms 'physiology' and 'pathology'.
15. Although the Nobel Prize was awarded on the basis of Pavlov's work involving digestion, when he received his prize, contrary to custom before and since, his oration discussed the subject of conditioned reflexes rather than the work on digestion.
16. See Black, S.L. (2003) Pavlov's dogs: for whom the bell rarely tolled. *Current Biology*, 13, 426.
17. This is somewhat later than is often stated for Pavlov first described his early experiments on psychic conditioning to Western scientists at a conference in Madrid in 1903.

18 The 'Little Albert' experiment has also been subject to various degrees of distortion and mythologising in many works of psychological literature. The interested reader is referred to Harris, B. (1979) Whatever happened to Little Albert? *American Psychologist*, 34, 151–160.
19 Watson and Rayner married in 1921 and remained together until her death in 1935.
20 Hebb had actually got the idea of the reveberatory circuit from the American neurobiologist Rafael Lorente de No who had first formulated the idea in the 1930s.

Bibliography

Babkin, B.P. (1946) Sechenov and Pavlov. *The Russian Review*, 5, 24–35.
Bartlett, F.C. (1960) Karl Spencer Lashley: 1890–1958. *Biographical Memoirs of the Fellows of the Royal Society*, 5, 107–118.
Beach, F.A. (1961) Karl Spencer Lashley. *Biographical Memoirs of the National Academy of Sciences*, 35, 162–204.
Beritashvili, I.S. (1968) A modern interpretation of the mechanism of I.M. Sechenov's psychical reflex medium member. *Progress in Brain Research*, 22, 252–264.
Berkowitz, C. (2006) Disputed discovery: vivisection and experiment in the nineteenth century. *Endeavour*, 30, 98–102.
Boakes, R. (1984) *From Darwin to Behaviourism: Psychology and the Mind of Animals*. Cambridge University Press: Cambridge.
Brown, R.E. and Milner, P.M. (2003) The legacy of Donald O. Hebb: more than the Hebb synapse. *Nature Reviews Neuroscience*, 4, 1013–1019.
Bruce, D. (1986) Lashley's shift from bacteriology to neuropsychology, 1910–1917, and the influence of Jennings, Watson and Franz. *Journal of the History of Behavioral Sciences*, 22, 27–44.
Burke, R.E. (2007) Sir Charles Sherrington's The Integrative Action of the Nervous System: A centenary appreciation. *Brain*, 130, 887–894.
Carmicheal, L. (1926) Sir Charles Bell a contribution to history of physiological psychology. *Psychological Review*, 33, 188–203.
Cohen, Lord Birkenhead (1958) *Sherrington: Physiologist, Philosopher and Poet*. Liverpool University Press: Liverpool.
Cope, Z. (1959) *The Royal College of Surgeons of England: A History*. A Blond: London.
Corson, E.R. (1910) Sir Charles Bell: the man and his work. *Bulletin of the John Hopkins Hospital*, 21, 171–182.
Cranefield, P.F. (1974) *The Way In and the Way Out. Francois Magendie, Charles Bell and the Roots of the Spinal Nerves*. Futura: Mount Kisco, NY.
Eadie, M.J. (2000) Robert Whytt and the pupils. *Journal of the History of Neurosciences*, 7, 295–297.
Eadie, M.J. (2008) Marshall Hall, the reflex arc and epilepsy. *Journal of the Royal College Physicians of Edinburgh*, 38, 167–171.
Fearing, F. (1970) *Reflex Action: A Study in the History of Physiological Psychology*. MIT Press: Cambridge, MA.
Fentress, J.C. (1999) The organization of behavior: revisited. *Canadian Journal of Experimental Psychology*, 53, 8–19.
Fulton, J.F. (1947) Sherrington's impact on neurophysiology. *British Medical Journal*, 2, 807–811.
Gibson, W.C. (2002) Sir Charles Sherrington, OM, PRS (1857–1952). In: Rose, F.C. (ed.) *Twentieth Century Neurology: The British Contribution*. Imperial College Press: London.
Glimcher, P.W. (2003) *Decisions, Uncertainty, and the Brain*. MIT Press: Cambridge, MA.
Goodwin, C.J. (2008) *A History of Modern Psychology*. John Wiley: New York.
Gordon-Taylor, G. and Walls, E.W. (1958) *Sir Charles Bell: His Life and Times*. E&S Livingstone: London.
Granit, R. (1967) *Charles Scott Sherrington: An Appraisal*. Double Day & Co.: New York.
Gray, J.A. (1980) *Ivan Pavlov*. The Viking Press: New York.
Green, J.H.S. (1958) Marshall Hall (1790–1857): A biographical study. *Medical History*, 2, 120–133.

Grigoriev, A.I. and Grigorian, N.A. (2007) I.M. Sechenov: the patriarch of Russian physiology. *Journal of the History of the Neurosciences*, 16, 16–29.

Grimsley, D.L. and Windholtz, G. (2000) The neurophysiological aspects of Pavlov's theory of higher nervous activity. *Journal of the History of the Neurosciences*, 9, 152–163.

Hebb, D.O. (1949) *The Organisation of Behavior: A Neuropsychological Theory*. John Wiley: New York.

Hoff, H.E. and Kellaway, P. (1952) The early history of the reflex. *Journal of the History of Medicine and Allied Sciences*, 7, 211–249.

Hunt, M. (1993) *The Story of Psychology*. Random House: New York.

Kanunikov, I.E. (2004) Ivan Mikhailovich Sechenov – the outstanding Russian neurophysiologist and psychophysiologist. *Journal of Evolutionary Biochemistry and Physiology*, 40, 596–602.

Lashley, K. (1929) *Brain Mechanisms and Intelligence*. University of Chicago Press: Chicago, IL.

Lashley, K.S. (1950) In search of the engram. *Society of Experimental Biology No. 4: Physiological Mechanisms in Animal Behavior*, 454–482. Cambridge University Press.

Leff, A. (2003) Thomas Laycock and the romantic genesis of the cerebral reflex. *Advances in Clinical Neuroscience and Rehabilitation*, 3, 26–27.

Levine, D. N. (2007) Sherrington's 'The integrative action of the nervous system': A centennial appraisal. *Journal of the Neurological Sciences*, 253, 1–6.

Liddell, E.G. (1960) *The Discovery of Reflexes*. Oxford University Press: London.

Manuel, D.E. (1996) *Marshall Hall (1790–1857): Science and Medicine in Early Victorian Society*. Rudopi: Amsterdam.

Milner, P. (1993) The mind and Donald O. Hebb. *Scientific American*, January, 124–129.

Molnar, Z. and Brown, R.E. (2010) Insights into the life and work of Charles Sherrington. *Nature Reviews Neuroscience*, 11, 429–436.

Orbach, J. (1998) *The Neuropsychological Theories of Lashley and Hebb*. University Press of America: Lanham, MD.

Pare, W.P. (1990) Pavlov as a psychophysiological scientist. *Brain Research Bulletin*, 24, 643–649.

Pavlov, I.P. (1927) *Conditioned Reflexes*. Oxford University Press: Oxford.

Pearce, J.M.S. (1993) Sir Charles Bell. *Journal of the Royal Society of Medicine*, 86, 352–354.

Pearce, J.M.S. (1997) Marshall Hall and the concepts of reflex action. *Journal of Neurology, Neurosurgery and Psychiatry*, 62, 228.

Rousseau, G.S. (1990) *The Languages of the Psyche: Mind and Body in Enlightenment*. University of California Press: Berkeley, CA.

Samoilov, V.O. (2007) Ivan Petrovich Pavlov (1849–1936). *Journal of the History of the Neurosciences*, 16, 74–89.

Seung, H.S. (2000) Half a century of Hebb. *Nature Neuroscience*, 3, 1166.

Sherrington, C.S. (1906) *The Integrative Action of the Nervous System*. Scribner: New York.

Swazey, J.P. (1969) *Reflexes and Motor Integration: Sherrington's Concept of Integrative Action*. Harvard University Press: Boston, MA.

Tan, S.Y. and Graham, C. (2010) Medicine in stamps: Ivan Petrovich Pavlov (1849–1936). *Singapore Medical Journal*, 5, 1–2.

Thomson, H.C. (1925) The work of Sir Charles Bell in relation to modern neurology. *Brain*, 48, 349–457.

Todes, D.P. (2002) *Pavlov's Physiology Factory*. John Hopkins Press: Baltimore, MD.

9

MAPPING THE CEREBRAL CORTEX

Our observation confirms thus the opinion of Mr Bouillaud, who places in these (the frontal) lobes the seat of the faculty of articulation of speech.

Paul Broca

To locate the damage which destroys speech and to locate speech are two different things.

John Hughlings Jackson

Summary

By the mid-nineteenth century, few researchers interested in brain function would have accepted mental faculties could be localised to distinct regions of the cerebral cortex. For one thing, phrenology had been thoroughly discredited by the scientific community. And, for another, experimental research on animals had shown the brain to have a 'common action' causing it to act as an integrated whole. This later position was associated with a number of great physiologists, including Pierre Flourens, who while accepting the cerebral cortex was specialised for sensation, perception and intelligence, nonetheless denied these functions could be broken down into a single faculty. Rather, he regarded the cerebral cortex as a single system whose diffused parts 'concur, consent and are in accord.' This view was also one supported by the great majority of philosophers and theologians who believed the mind, or soul, was unitary and undividable. Bolstered by this great weight of authority, and with any opposition likely to be ridiculed, few would have argued otherwise. Yet, this theory of the cerebral cortex would be overturned in a surprisingly short period of time during the second part of the nineteenth century. In fact, the antecedents of this change were in place as early as the 1830s when doctors began to examine the behaviour of brain damaged individuals. The work of Frenchman Jean-Baptiste Bouillaud, in particular, was to show people with language disturbances, or aphasia, frequently had damage to their frontal lobes – leading him to propose this brain area served as the 'principle lawgiver of speech'. Although many remained sceptical of these claims, the situation was dramatically reversed in 1861, when the respected French doctor Paul Broca proved a localised area existed in the posterior region of the left frontal lobe for producing speech. It was a pivotal moment when new theories of the cerebral cortex had to be formulated, not least when the Englishman John Hughlings Jackson provided clinical evidence to show the right and left cerebral hemispheres had different mental functions. His work was supported when a second language centre was discovered by Carl Wernicke in 1874. But language was not the only behavioural function to be discovered with a distinct anatomical locus. In 1870, the Germans Eduard Hitzig and Gustave Fritsch used electrical stimulation to identify areas for eliciting movement – a finding confirmed by the Scotsman David Ferrier who extended their technique to other areas of the brain. In doing this, Ferrier would discover, albeit controversially, localised cortical areas for hearing, vision and even intelligence. Thus, by the end of the century, the concept of cortical localisation had been firmly re-established, although in a very different form to that proposed by Franz Gall and the phrenologists.

Language and the frontal lobes: Jean-Baptiste Bouillaud

Although phrenology had become discredited by the 1830s, not all investigators accepted the prevailing anti-localisation views of the times. One such person was the Frenchman Jean-Baptiste Bouillaud. A student of Francois Magendie in Paris, Bouillaud received his doctorate in 1823 and distinguished himself for research on rheumatism and heart disease. He was also one of the first practitioners to use the drug digitalis (derived from the foxglove plant) to treat high blood pressure. For this work, Bouillaud rose to become professor of the Hôpital de la Charité in 1831. He was also an admirer of Gall's work, and by the early 1820s as a young doctor, had began to notice some individuals with speech loss had damage to the frontal part of their cerebral cortex. This was not entirely a new finding as Gall had described two aphasic men with fencing foil wounds to their brain just above their eyes, leading him to place the faculty for language at that site. Bouillaud was intrigued enough, however, to examine the matter more methodically by looking at individuals with other types of brain damage – resulting in him building up a sample of some 29 cases by 1825. He discovered a striking pattern: in the 13 cases where there was speech disturbance, all had frontal lobe damage. Bouillaud reported his results to the Royal Academy of Medicine in Paris, and published a short monograph (1825), which argued frontal lobe damage produced a deficit in fluent speech, without affecting intelligence or language comprehension. Bouillaud also noted that the inability to speak was not related to difficulties in moving the tongue as damaged individuals could swallow and eat normally. Thus, the problem was specific to the poor execution of words. Concluding that the nervous force responsible for language production was in the frontal lobes, Bouillaud described the region as the 'principle lawgiver of speech'.

While these findings generated considerable debate among the members of the Royal Academy, Bouillaud's conclusions were not widely accepted. One reason was probably because of Bouillaud's known association with Gall. A more reasonable objection was due to the fact many of the doctors listening to the evidence, knew of patients who had frontal lobe damage without speech deficits. Indeed, some of these were well-known patients in the hospitals of Paris, including a soldier who had been shot through the head destroying much of his frontal lobe. Another was a syphilitic patient whose left hemisphere had been reduced to pulp. Undaunted, Bouillaud looked for other aphasic patients to support his case, 'affirming with an incredible vigour and steadfastness' the seat of 'articulate language' lay in the frontal areas of the brain. Yet, he found few supporters, no doubt concerned about the implications with its overtones of phrenology. Indeed, their scepticism seemed to be well-founded when one of the luminaries of French clinical pathology, Gabriel Andral, examined the records of 37 individuals with frontal lobe damage attending the Hôpital de la Charité between 1820 and 1831, and found speech impediments in only 21 of them. To complicate matters further, Andral also found 14 cases of speech loss in individuals whose brain damage lay outside the frontal lobes.

Some ten years later, Bouillaud presented a number of new cases, along with detailed autopsy reports, to further support his view that the anterior cerebral cortex was involved in the articulation of language. Again, his theories largely fell on deaf ears. Frustrated and increasingly strident in his views, Bouillaud announced a prize of 500 francs in 1848 to anyone who could prove a deep lesion of the frontal lobes did not cause speech deficits. Bouillaud's confidence was high since he had accumulated evidence from hundreds of aphasic cases. However, he was soon challenged by the French surgeon Alfred Velpeau who knew of a patient with a large bilateral tumour of the frontal lobes without any speech impediments. Velpeau even went as far to say, 'a greater chatter never existed'. Bouillaud strongly contested the claim, leading to an increasingly bitter argument between the two men. Indeed, for many neutral observers the issue was unresolved because the tumour damage in Velpeau's patient had spared other parts of the frontal lobes. Nonetheless, Bouillaud was reluctantly forced to pay the money in 1865.

Paul Broca and his speech centre

Despite his financial loss, Bouillaud received further support from his son-in-law Ernest Aubertin. Also working in Paris, Aubertin was searching for new aphasic patients and came across an interesting case in 1861. The patient had blown away the best part of his left frontal cranium during a suicide attempt after shooting himself in the head. While the injury had apparently left both speech and intellect intact, Aubertin discovered he could abruptly stop the patient in mid-sentence by applying upward pressure to the exposed frontal lobes with a spatula. However, speech returned immediately after the compression ceased. Aubertin described the patient to the recently formed Société d'Anthropologie in Paris, using it to support the theory of language localisation in the frontal lobes. However, for many, the findings again remained inconclusive – not least when it was known that some patients with left-sided frontal lobe injuries exhibited normal speech. However, one person listening to the debate was Paul Broca, the founder of the Société, and one of the most respected doctors in France. Although he had been relatively indifferent about the long running debate on the localisation of language, Broca was soon to come across his first aphasic patient. It was an encounter that would radically change people's attitudes to cortical localisation.

Broca is now recognised as one of the most eminent brain investigators of the nineteenth century. Born in the small town of Sainte-Foy-la-Grand, close to Bordeaux, Broca was the son of a doctor who served with distinction at the Battle of Waterloo. He graduated in medicine from the University of Paris when just 20 years old, and specialised in surgery, working in several Paris hospitals, before rising to the Chair of Surgical Pathology in 1868. Although

FIGURE 9.1 Paul Broca (1824–1880) the father of localisation theory who was the first scientist to discover a brain area for the articulation of language.

Source: Wellcome Library, London.

often remembered for his anthropological research (he was the first to describe Cro-Magnon man), Broca also made several important medical discoveries. These included recognising the venous spread of cancer, the nutritional causes of rickets, and the nature of degeneration in muscular dystrophy. He also wrote a 900 page book on aneurysms and pioneered the use of the microscope to detect early tumour formation. Much later in his career (1788) Broca identified a large gray mass of different structures, which he called the *le grand lobe limbique* (the limbic lobe) nestled under the cerebral cortex.[1] Although Broca believed this region served as the 'old' emotional part of the brain, we know today it is involved in a much wider range of behaviours including memory formation, motivation, and visceral functions.

Broca would have probably listened to Aubertin's presentation with little more than general interest, but a few days later came across his first aphasic patient at the Bicetre Hospital. Called Monsieur Leborgne, although more popularly known in the wards as 'Tan' since this was the only word he could utter, the patient had first been admitted to the hospital some twenty years earlier as a young man after suffering from an abrupt loss of speech. Despite this, his intellect and language comprehension were unimpaired at the time. Ten years after his admission, Leborgne lost the ability to move his right arm and leg following a stroke, which confined him to bed. It was coincidence Leborgne was transferred to Broca's ward in the wake of Aubertin's talk, as he had fallen seriously ill from a gangrenous infection from a bed sore. Six days later Leborgne died with Broca able to perform an autopsy on his brain. The results were so remarkable, that Broca quickly wrote a brief report and presented it to the Société d'Anthrologie just a few days later. He had found a large fluid-filled hole the size of a chicken's egg in the second or third convolution of the left frontal cortex – a site that Bouillaud and Aubertin had long been arguing was involved in the articulation of language. It has since become known as Broca's area.

Four months later, Broca gave a more comprehensive account of his findings to the Société d'Anatomie. In this talk, also published as a paper in 1861, Broca referred to the loss of speech suffered by Leborgne as '*aphémie*', and stressed it had been caused by the stroke to the posterior part of the frontal cortex. Importantly, this was not the site Gall had identified as the faculty of language (which was located just above the eyes) and Broca was keen to emphasise his discovery did not support phrenology. Nonetheless, Broca was ebullient in his praise of Bouillaud and Aubertin in their endeavours to bring many new instances of aphémie to the attention of the Société. Although Broca accepted a single case of this kind did not prove a speech faculty existed in the frontal cortex, he thought it 'extremely probable' that an area existed. Indeed, a few months later, Broca discovered another patient who was only able to utter a few simple words after collapsing from a stroke. When the autopsy of the brain was undertaken, it again revealed damage to the same area of the left frontal lobe. This was only the beginning for Broca who identified eight more aphasics with left sided damage of the frontal lobes by 1863. Describing the consistency of his results as 'remarkable', Broca was nevertheless cautious about drawing any definite conclusion from his findings. However, there was little doubt that the assertions of Flourens and other anti-localisationists would have to be revised.

Right-handedness and the dominant hemisphere

Broca had shown that aphémie was associated with damage to the left-side of the frontal lobe. Curiously, neither Bouillaud or Aubertin had made this anatomical left-sided connection with any conviction, perhaps because the phrenologists had always emphasised the importance of bilaterally in the brain. However, the unilateral location of the speech centre was a puzzle. How did it come about? In 1865, Broca attempted to answer this question by referring to the work of two French colleagues, Pierre Gratiolet and Francois Leuret. These researchers had examined the foetal development of the human brain, and noticed the left cerebral hemisphere grew slightly bigger than the right – a difference that first arose in the first months of pregnancy. It had also led the English neurologist John Hughlings Jackson to propose the left hemisphere was more likely to be 'educated' before its right-sided counterpart. Broca also realised this idea could account for why the capacity for language developed in the left hemisphere. If the left cortex developed in advance of the right during the first years of life, then it would be more likely to take on the complex and demanding function of speech.

The better developed left hemisphere also allowed Broca to make another important claim: it was responsible for causing the majority of people to be right-handed. This bold assertion was made partly on the basis that it was known the left-hand side of the brain exercised control over the right-hand sided movements of the body, and vice versa. Although the anatomy of this system was not fully understood at the time,[2] it had been established that the situation most likely occurred because fibres from higher areas of the brain, crossed in an area of the brainstem known as the pyramids, before passing on to the spinal cord. Having knowledge of this anatomy, Broca reasoned if the left hemisphere was dominant over the right hemisphere, then it would explain the preponderance of right-handedness. Similarly, left-handedness could be explained by a person having a more dominant right-sided brain. Broca stated these views formally in 1865 by declaring:

> The majority of men are naturally left-brained, and . . . exceptionally some among them, these people we call left-handers, are on the contrary right-brained.

The human propensity for being right-handed, therefore according to Broca, was due to a better developed left hemisphere. However, this raised other implications, not least the possibility of the left cerebrum being superior for intelligence and thinking. Indeed, it was not long before Broca was disparagingly referring to the right hemisphere as '*la gaucheie cérébrale*' or 'awkward brain'. This was a radical departure from traditional accounts of intelligence since it opposed the commonly held idea that symmetrical functioning of the brain was a necessary requirement for its mental unity.

Despite this, Broca was able to find individual cases where the seat of language was not located in the left hemisphere. For example, Broca had known of an epileptic woman at the Salpetriere Hospital with fluent speech, yet whose autopsy revealed damage to the same site of damage found in Leborgne. To explain this discrepancy, Broca proposed the right hemisphere could take over the responsibility of speech under certain circumstances – especially if the left hemisphere was injured or compromised in early development. Thus, Broca provided an explanation for instances where there was no damage to the left hemisphere in certain patients with aphasia. And, of course, such a person with right hemispheric language function would be most likely left-handed. Although Broca admitted he had constructed his theory in advance of all the facts, he was confident subsequent evidence would vindicate him. In this respect he was proven largely correct.[3]

For obvious reasons, the speech centre in the left posterior region of the frontal cortex is now known as Broca's area. However, Broca's priority for discovering this region would be contested by a relatively unknown doctor called Gustave Dax who claimed his father (Marc) had found it first. A country doctor from Sommiéres in the south of France, Marc Dax reported the cases of three aphasic patients with left hemisphere damage, to a medical conference in Montpellier in 1836. It also led to the publication of two short papers. Unfortunately, Dax died a year later, and his work was soon forgotten, despite being carried on by his son Gustave. Some 27 years later, in 1863, Gustave had collected information from a sufficient number of aphasic patients to submit his findings to the Academy of Medicine in Paris. This was two years after Broca had first reported the case of Leborgne. However, the paper by Dax was delayed in publication and did not appear until 1865. Somewhat embittered, Gustave Dax pointed out his father had made the same discovery as Broca a quarter of a century earlier – a claim causing considerable antagonism within the French academic authorities. Today, most observers favour the case of Broca. The original Dax papers had little impact (it seems they were 'lost' for many years) and Broca was only aware of them after his own had been published. Although Dax had associated the left hemisphere with speech, Broca's exact identification of the third convolution of the frontal cortex with language articulation is generally recognised as the more compelling evidence.

The minor hemisphere: John Hughlings Jackson

Around the same time as Broca was investigating aphasia and handedness, the Yorkshire born John Hughlings Jackson, at the newly founded National Hospital in London, was also becoming interested in the effects of brain damage on language. Sometimes described as the founder of English neurology, and widely renowned for instigating the modern study of epilepsy,

FIGURE 9.2 John Hughlings Jackson (1835–1911) who confirmed the importance of Broca's area for spoken language. He also showed the left and right cerebral hemispheres to have different functions and anticipated the discovery of the motor cortex by Fritsch and Hitzig.

Source: Wellcome Library, London.

Jackson had become drawn to aphasia after examining the behavioural symptoms of strokes to the middle cerebral artery,[4] which often caused a loss of speech. After learning of Broca's work, Jackson offered support from his own observations, reporting the symptoms of 31 patients with left-sided brain damage to the *British Medical Journal* in 1864. Like Broca, Jackson had come to recognise that left hemisphere damage was most likely to produce speech disturbances accompanied by right-sided hemiplegia or muscle weakness. However, Jackson modified his views two years later in 1866 when he argued that language was not as precisely located in the left hemisphere as Broca maintained. Instead, the right hemisphere also had some linguistic capability. This later claim was made on the basis that many aphasiac patients with damage to the left hemisphere, often blurted out involuntary emotional exclamations or swear words. Thus, Jackson proposed the left hemisphere was mainly involved in propositional (intellectual) speech, while the right was responsible for involuntary and emotional utterances. In other words, both hemispheres were capable of language, but produced it in different contexts – an idea different from the one proposed by Broca who viewed the left as being 'better educated' than the right.

Recognising the left and right cerebral cortices of the brain have different mental specialisations, Jackson referred to them as the major and minor hemispheres respectively. This did not mean, however, the right hemisphere served an unimportant role. For example, Jackson showed most aphasic patients with left-sided damage retained a capacity to recognise objects, even though they could not name them – a finding leading him to reason the right hemisphere has a superior perceptual function. He confirmed this idea in 1872 when Jackson identified a male patient with right-sided cerebral damage, who could not recognise other people, including his own wife. Despite retaining normal vision, the man also had difficulty when it came to identifying places and things. A few years later, Jackson found another patient, a 59 year-old lady called Eliza T., who suddenly lost her sense of direction. Although living in the same house for over 30 years, she was now unable to find her way home, or recognise her surroundings. An autopsy of her brain revealed a large malignant tumour in the posterior part of her right hemisphere. While this part of the brain lacked the verbal abilities of the left, it was clearly essential for recognising objects and people.

Although Jackson came to associate the two cerebral hemispheres with different abilities, he did not believe their functions were as strictly localised as they first appeared. He tried to explain his position in the late 1860s when he warned: 'While we may localise the *damage* that makes a man speechless, we do not localise language. It will reside in the whole brain.' Later, in 1874, Jackson expressed the idea again. 'To locate the damage which destroys speech and to locate speech are two different things.' In order to understand Jackson, it is important to realise he had been strongly influenced by the evolutionary ideas of the time. Believing simple reflexes were the basic unit of all behaviour, which became increasingly more complex from spinal cord to cerebral cortex, Jackson argued the left hemisphere had evolved with a greater role in movement, and the right in sensation. However, these functions were not exclusive to either hemisphere for they depended on the whole orchestration of the nervous system. This view of brain function was not as extreme as the anti-localisation position of Flourens, for Jackson accepted certain areas had evolved to become more important than others. He also conceded that brain damage at certain sites yielded predictable clusters of symptoms, with Broca's area being essential for language articulation. Despite this, Jackson compared Broca's area with the retina of the eye – pointing out that while the whole retina can see, its central point has the greatest visual acuity. In other words, Broca's area was not the only area of the brain to be involved in the expression of language.

Jackson's most outstanding contributions to neurology, however, were in the field of epilepsy. Most notably, he provided a new classification of epilepsy, based on the anatomical location of the seizure. It was, in effect, the first neuronal theory, which continued to guide doctors up until the 1970s. Even now, terms used by Jackson such as *grand mal seizures* remain in common usage. Jackson also described several new forms of seizure, including ones that give rise to unilateral symptoms – now known as *Jacksonian epilepsy*. Most impressive of all, Jackson's observations of patients with epilepsy enabled him to recognise that the muscle spasms accompanying seizures often followed predictable patterns. In other words, certain body parts were affected in a sequential order. From this, he guessed that the motor areas of the cerebral cortex must be organised in a way that mimics the topography of the body. This, of course, implied a localisation of function in the brain for movement. He would soon be shown to be correct when the German researchers Fritsch and Hitzig in 1870 discovered the motor cortex in the posterior part of the frontal lobes (see later).

A second language centre: Carl Wernicke

A central question following the discovery of Broca's area concerned the nature of the deficit resulting from its damage. Although he had described it as one of articulated speech, Broca was also aware this impairment could manifest itself in many different ways. Because of this, by 1866, Broca had developed a new classification dividing the aphasic syndrome into four types[5]. However, the neuroanatomy of language was to become further complicated in 1874, when a young German doctor called Carl Wernicke identified a second brain area associated with an alternative form of aphasia. Born in the German town of Tarnowitz[6] and a medical student at the nearby University of Breslau, Wernicke had also spent six months working in Vienna under Theodor Meynert. Although primarily regarded as the greatest neuroanatomist of his time, Meynert was also a clinician who diagnosed patients with neurological illnesses. In 1866, prior to Wernicke's arrival, Meynert had come across a female patient with bizarre unintelligible speech patterns, accompanied by a difficulty in comprehending language. When he performed an autopsy, a large lesion in the upper left temporal lobe was found.[7] Meynert was struck by the significance of this site for he had also shown the auditory nerve from the ears projected to the same area. From this he concluded that the temporal lobes must contain a 'sound-field' responsible for the recognition of speech.

Wernicke must have been greatly influenced by his brief stay in Vienna, for a year or so after his return to Germany in 1874, he published a short monograph[8] that described several other patients with this new type of aphasia. The symptoms were strikingly different from those reported by Broca, whose patients exhibited impoverished speech accompanied by pronunciation difficulties. Instead, those with Wernicke's aphasia showed fluent speech, except it was unintelligible, and composed of many nonsense words or inappropriate utterances. In addition, these aphasic patients could barely understand what was said to them, unlike those with Broca's aphasia who showed normal intelligence and language comprehension. Although Broca had come across some of Wernicke's symptoms before, he had not combined them into a pattern or syndrome. Wernicke realised, however, his deficits represented a new form of aphasic disorder, and one associated with damage to the left-sided temporal lobe – a region laying some distance from the frontal lobes.

By itself, this would have been a great achievement for a 26-year-old doctor with barely three years training in neurology, but Wernicke went on to propose a new theoretical model of language processing by the brain. Wernicke's great insight was to recognise that Broca's aphasia is basically an impairment in the vocal production of speech, whereas the temporal lobe deficit is one of comprehension. More specifically, Wernicke explained the comprehension process by arguing it began when auditory information passed from the ears to the 'sound field' located in the temporal lobes which contained a memory store for recognising words. From the temporal lobes, an internal verbal representation was then passed to the frontal cortex that contained the templates for controlling the production of verbal sounds by the tongue and larynx. Wernicke called the temporal lobe deficit 'sensory aphasia' and the frontal lobe one 'motor aphasia'. It was a brilliant attempt to tie together anatomical and functional findings in order to produce a general neuroanatomical theory of language.

Wernicke's new model also made some interesting predictions, especially as it implied the existence of a direct pathway passing from Wernicke's region to Broca's area. Although Wernicke was not aware of any such system of fibres,[9] he was able to predict the consequences of their damage. Using little more than his imagination, Wernicke reasoned if a patient had damage to this connection, they would still be able to comprehend language since the 'sound area' of the temporal lobe was intact. In addition, the same person would have fluent speech as the frontal motor area still functioned normally. The deficit, therefore, must involve the passing of linguistic information, and Wernicke deduced this would manifest itself by the person being unable to fluently or accurately repeat words spoken to them. We now know Wernicke was remarkably prescient. Patients with this type of brain damage (i.e. to the arcuate

 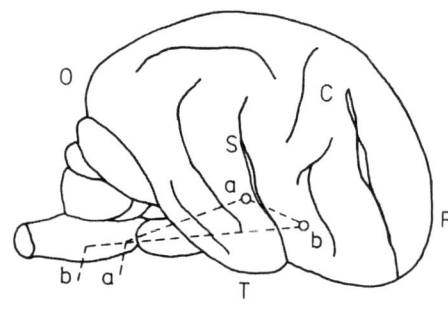

FIGURE 9.3 (A) Carl Wernicke (1848–1904). German doctor who discovered a second language centre in the temporal lobes in 1874 involved in verbal comprehension.

Source: National Library of Medicine.

FIGURE 9.3 (B) The processing of language envisaged by Wernicke, which shows the acoustic nerve projecting to a centre for language comprehension (labelled a) and a centre for motor imagery and production (labelled b) extending to brainstem areas.

Source: From Wernicke 1874.

fasciculus) do indeed have problems repeating verbal material, especially abstract words, despite being fully aware of their mistakes. Today, this type of neurological condition is called conduction aphasia.

Wernicke's theory had shown how language and cognition could be modelled by integrating neuroanatomy with clinical observation, and it also made a number of other predictions that were testable through observation and experimentation. One of these concerned the possibility of a brain region specifically responsible for the encoding of written words (i.e. reading). Wernicke suspected this site lay close to the speech comprehension area in the temporal lobes. He was proved correct in 1892 when Frenchman Joseph Dejerine described a patient who suddenly lost the ability to read, despite having no obvious visual deficit. The autopsy revealed a stroke to the angular gyrus – a site lying between the visual cortex and the temporal 'sound area'. Wernicke's model of language processing was not only accurate, but it has proven remarkably enduring, with its basic tenets still accepted today.[10] In addition to his work on aphasia, Wernicke described other medical conditions, including one resulting from thiamine deficiency commonly found among alcoholics. Today, this is known as *Wernicke's encephalopathy*. He also found time to write a textbook entitled *Foundations of Psychiatry* that he regarded as his most important work. Sadly, Wernicke died at the relatively young age of 56 from injuries sustained in a cycling accident.

The discovery of a motor area in the frontal cortex

Language was not the only function to become associated with a specific area of the cerebral cortex. In 1870, some four years before Wernicke described sensory aphasia, two German doctors, Eduard Hitzig and Gustave Fritsch, identified a region in the posterior part of the frontal cortex linked with producing movement. Or, more specifically, they found an area in the dog where electrical stimulation evoked predictable motor responses. At first, this discovery evoked general disbelief, as it was widely believed the cerebral cortex was insensitive to electrical stimulation. After all, this had been the view of the two greatest brain investigators of their respective centuries: Albrecht von Haller (in the eighteenth) and Pierre Flourens (in the nineteenth). Indeed, both had stimulated the outer grey surfaces of the cortex using a variety of methods without effect. Consequently, during the eighteenth century, the cerebral cortex was frequently dismissed as an insignificant 'rind', which is what the Latin word for *cortex* means. Flourens had a different view since he regarded the cerebral hemispheres as the site responsible for the 'will and sensations', with 'muscle movements' relegated to the striatum and cerebellum. Although most researchers agreed, there was some contrary evidence. For example, Giovanni Aldini had produced a variety of muscular responses, including those of the face and arms, from stimulating the cerebral cortex in dead human bodies (see Chapter 5). His contemporary Luigi Roland also noted limb movements when he inserted a wire from a voltaic pile into the cortex of a pig. The issue was further highlighted when Hughlings Jackson deduced from the way certain epileptic seizures 'marched' across the body, that a distinct brain site must be responsible for producing movement. Since he was confident a loss of consciousness arising from a seizure involved the cortex, this implied the existence of a motor region within the cerebral hemispheres.

The two Germans who were to discover the motor cortex had very different characters. Hitzig was educated at the University of Berlin and despite a Jewish background, was said to be a proud Prussian with an arrogant and stern demeanour. He also established a respected medical practice in an age when the therapeutic benefits of electrical shocks were in vogue. It was during one of these treatment sessions, Hitzig first discovered he could induce involuntary eye movements if applying a strong current to the back of the head – an effect causing him to wonder if the site for producing this reaction lay in the cerebral cortex. Fritsch had also contemplated much the same thing. Although a doctor by training, Fritsch was an adventurer who among his travels had lived in South Africa for several years where he undertook a detailed study of the social culture of the Bushman. He also made expeditions to Egypt, Syria and Persia, resulting in the publication of several books on physical anthropology.[11] At some point during his travels, perhaps while serving as a surgeon in the Prusso-Danish War in 1864, Fritsch was dressing an open head wound when his patient began to twitch violently – a result indicating the cortical surface was being excited in some way. In the late 1860s Fritsch moved to Berlin where he met Hitzig who was just beginning to experiment with electric stimulation on rabbits. The two men had mutual interests in the excitability of the cerebral cortex and agreed to co-operate.

Their experimental collaboration began in 1870 when Hitzig and Fritsch began electrically stimulating the surface of a dog's cerebral cortex. Although both men were members of the Berlin Physiological Institute, it had no facilities for this type of work, forcing them to work at Hitzig's house, where they sequested his wife's dressing table for the study.[12] Here a dog was tied down without anaesthesia, and a part of the cranium removed to expose the brain.[13]

FIGURE 9.4 A group of German physiologists and anatomists (circa 1880) that includes Gustave Fritsch (labelled 6 in the photograph) and Eduard Hitzig (labelled 7). These two men induced motor movements in dogs by direct electrical stimulation of the cerebral cortex, thereby showing it was not electrically unexcitable as had been believed.

Source: Wellcome Library, London.

Then by using a fine platinum electrode attached to a battery to produce low levels of electric current (just enough to be felt by the tip of the tongue) they probed the cortical surface. It was not long before their procedure elicited limb spasms on the opposite side of the body from where the stimulation was being applied to the frontal cortex. More precisely, weak currents produced twitching in small groups of muscles, whereas a stronger current caused more pronounced muscular contractions to spread over both sides of the body. As their probing continued, they localised the site of this movement to regions in the medial-posterior part of the frontal cortex.[14] Examining this area more closely, Fritsch and Hitzig realised there were four distinct areas corresponding to the neck, forearm, hind-limb and face.

To confirm this cortical region was responsible for producing bodily movement, Fritsch and Hitzig lesioned the sites that initially induced the motor responses. The logic was simple: if the site produced movement, then lesioning of this area should produce a motor deficit

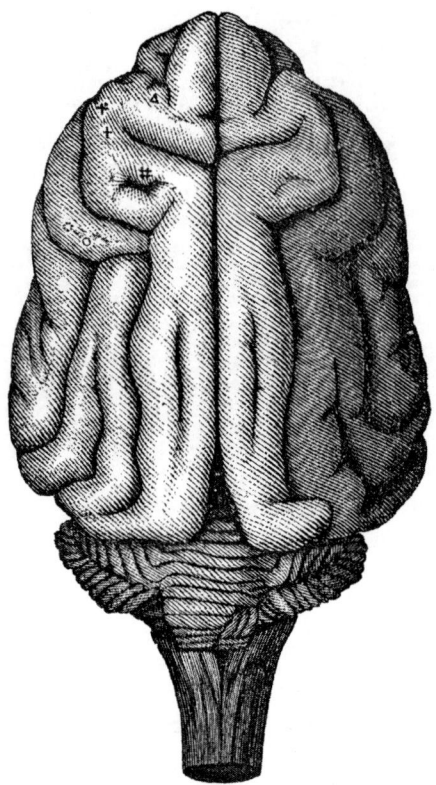

FIGURE 9.5 Dorsal view of the dog's cerebral cortex showing the areas that Fritsch and Hitzig stimulated to induce movement in 1870. Stimulation of the area marked by the triangle elicited neck movements; those of the forearm were obtained from stimulation of the site designated by the two crosses; hind-limb movements are marked by the cross-hatched square, and facial movements by the circle.

Source: Wellcome Library, London.

corresponding in some way to what they had observed through stimulation. The results confirmed their predictions. For example, when Fritsch and Hitzig excised the area for moving the right forepaw, the dog was unable to use this limb properly, causing it to limp when attempting to walk. Although the lesion did not produce total paralysis of the limb, it nonetheless impaired the strength and precision of its movement.

Fritsch and Hitzig published their findings in 1870. It not only provided further evidence for the theory of cortical localisation, but supported the observations of Hughlings Jackson made earlier with his epileptic patients. However, since Fritsch and Hitzig were unaware of the Englishman's work, they did not acknowledge him in their paper.[15] It was to be an omission that infuriated David Ferrier (see below). Nonetheless, by showing the cerebral cortex was electrically excitable, Fritsch and Hitzig had managed to open up a new avenue of brain research. Their work also raised several new questions. For example, the lack of complete paralysis following lesioning indicated other areas for movement must exist. In fact, Fritsch would claim in 1884 that the residual movement of a limb after removal of a centre was similar to the continuation of the bureaucracy's work when the governmental minister was on vacation.

This was akin to saying that the brain did not have one specific or executive command function, but was comprised of many individual parts that could also make important decisions if necessary. Although this was an idea supporting cortical localisation, it also implied the components of the brain functioned more like a democracy than an autocracy.

Evidence for localisation builds: David Ferrier

A new way of exploring the brain through electrical stimulation of the cortex was now possible, and one person to take full advantage of this technique was Scotsman David Ferrier. A highly prolific and at times controversial researcher, Ferrier studied philosophy at Aberdeen and psychology in Heidelberg, before graduating in medicine from Edinburgh in 1868. After a short period as a physician in Bury St Edmunds to complete his MD thesis, Ferrier moved to London in 1870, taking up a post as neuropathologist at King's College Hospital. Around the same time he also worked at the National Hospital for Paralysis and Epilepsy,[16] coming into personal contact with Hughlings Jackson – an experience that greatly inspired Ferrier to pursue research. In fact, he got the chance to do so when invited to work at the West Riding Lunatic Asylum in Wakefield, Yorkshire, by its superintendent James Crichton-Browne in 1873. This was one of the largest mental institutions of the Victorian era, housing over a thousand patients, while also providing its scientists with some of the best laboratory and animal facilities in the country. More importantly for Ferrier, he was given the remit of establishing the 'experimental proof of the views entertained by Hughlings Jackson . . . and to follow up the pathology of Fritsch and Hitzig.' It was a unique opportunity for an ambitious young researcher, and over the next three years Ferrier would extend the research findings of Hitzig and Fritsch in a number of important ways.

One of Ferrier's innovations was to use faradic stimulation – a type of alternating current, which allowed a more constant and less damaging current to be applied to the brain. This was a considerable advance over the galvanic stimulation used by Hitzig and Fritsch that had been administered in brief bursts and produced little more than muscle twitches. Using his new method, Ferrier soon confirmed, as Hughlings Jackson had implied, seizures could be induced by several seconds of electrical stimulation applied to the cerebral cortex. However, Ferrier also wanted to map the motor areas of the brain to confirm the findings of Hitzig and Fritsch. In doing this, Ferrier came to realise that if he anaesthetised his animals beforehand, stimulation could be applied for even longer periods. This also made the use of faradic stimulation much more versatile. At low intensities, it induced very fine movements such as closure of an eyelid, the flick of an ear, or a grasping of a paw. However, if its intensity was increased, behaviours became more complex including co-ordinated movement of the limbs and changes in head orientation. Ferrier also applied his stimulation to a large number of animals, including cats, rabbits, dogs, guinea pigs, rats, pigeons, and even jackals. Although his results supported Hitzig and Fritsch with movements being produced from stimulating discrete areas of the frontal cortex, the relative size and topography of the regions differed between species. This finding showed each type of animal had evolved its own unique type of motor specialisation.

The results from these experiments were initially presented in the *Journal of the West Riding Lunatic Asylum* in 1873, although they gained much wider recognition when Ferrier was invited to give the Croonian Lecture before the Royal Society the following year. This also allowed the work to be published in the Society's *Philosophical Transactions* – then one of the most

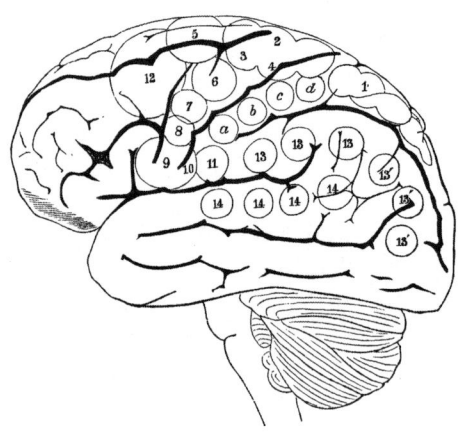

FIGURE 9.6 (A) Sir David Ferrier (1843–1928) who used electrical stimulation of the brain to examine cortical localisation.

Source: Wellcome Library, London.

FIGURE 9.6 (B) Map of the human brain, on which sites found from stimulation of the monkey cortex have been superimposed. These types of map were of great help to the early pioneers of neurosurgery.

Source: Ferrier 1876, page 304.

prestigious scientific journals in Britain. However, this was not going to be without its difficulties. Ferrier had not forgiven Hitzig and Fritsch for overlooking the work of his mentor Hughlings Jackson in their seminal paper of 1870. Consequently, Ferrier made scant reference to the Germans in return – a situation that caused the paper to be rejected by the editors of the journal. Nor was Ferrier inclined to change his mind on the matter. When asked to rewrite the paper, he concentrated solely on his work with monkeys. In this way, Ferrier avoided mentioning the work of Hitzig and Fritsch, which had been performed solely on dogs.[17]

Ferrier had extended his research to monkeys in 1874 after being sponsored by the Royal Society – a move attracting the wrath of many animal activists.[18] Despite this, Ferrier hoped his work would help him locate more accurately the motor cortex in humans. This had practical benefits for the surgeons of the time who were beginning to pioneer brain operations for the removal of tumours and blood clots (see Chapter 13). Identifying the motor areas in humans was important: if their location was known, they could be avoided by surgeons, thereby reducing the likelihood of complications such as paralysis. In fact, Ferrier discovered nineteen different areas in the monkey's cerebral cortex that produced movement, including those for arm retraction, walking, flexion and extension of the wrist, mouth opening, sneering and blinking. Further, these areas corresponded with the frontal sites found in other mammals and birds. Ferrier referred to this area as the *chief centre of voluntary movement*. However, it was also apparent the monkey's brain was more complex than in other animals, for Ferrier also discovered that certain integrated actions could be evoked from areas of the parietal and temporal lobes.

Stimulating the human brain

In the same year as Ferrier began his work on monkeys, a sensational and disturbing report appeared in *The American Journal of Medical Sciences*, describing the effects of electrical brain stimulation on a conscious human being. The author of this paper was a 42-year-old hospital doctor from Cincinnati called Robert Bartholow, previously best known for his work on male sexual dysfunction, and a guidebook for administering drugs by the newly invented syringe. In early 1874, Bartholow received into his care a 30-year-old Irish female domestic servant called Mary Rafferty. She was dying from a malignant ulcer of the scalp, caused by the friction of a piece of whalebone in her wig, which she wore to hide a disfiguring burn mark from a childhood accident. The ulcer had caused a large part of her posterior skull to waste away, exposing the cortical surface of the brain underneath. When she entered hospital, Rafferty was described as being in a cheerful mood, although undernourished and feeble-minded. Unfortunately, her condition was also incurable and rapidly deteriorating. Recognising a unique opportunity to address some of the questions raised from previous stimulation studies in animals, not least whether the cerebral cortex was electrically excitable, an issue still remaining contentious at the time, Bartholow received permission from Rafferty (despite her 'feeble-mindedness') to experiment on her exposed cerebral cortex with direct electrical current.[19]

Although Rafferty's skull was intact in its anterior regions, which meant Bartholow could not access the *chief centre of voluntary movement* as identified by Ferrier, he was able to reach large parts of the superior parietal lobe. It was enough to interest Bartholow who arranged his experiment to take place over the course of six experimental sessions, spread across four days. After establishing the electrodes produced no 'mechanical irritation' in the first session, Bartholow stimulated the dura mater overlying the left cortex with a mild faradic current. It was a procedure which Ferrier had previously shown was likely to produce violent convulsive movements, and Rafferty's reaction confirmed the Scotsman's findings. More specifically, the stimulation caused her right arm and fingers to be rigidly flung out, legs extended forward, and the head strongly deflected to the right. Stimulation to the right dura mater produced much the same effect on the left side of the body. While this procedure was likely to have caused Rafferty considerable pain for the dura mater contains pain receptors, unlike brain tissue, Bartholow does not comment on the matter. Nor did he state the size of the electrode by which he administered the current.

If this was not enough for one day, Bartholow now turned his attention to stimulating the exposed brain tissue. Starting with faradic current applied to the left hemisphere, the patient was forced to suddenly thrust out her right arm and leg – a response that initially caused her some amusement. However, it was soon followed by Rafferty complaining of unpleasant tingling sensations in her right arm, which she rubbed vigorously. Bartholow now proceeded to stimulate her right hemisphere with an increased strength of current. At this point, Rafferty cried out in distress and out-stretched her left hand 'as if in the act of taking hold some object in front of her.' Undeterred, Bartholow continued the stimulation. Rafferty's lips turned blue, her pupils dilated, and she suffered a five-minute convulsion, resulting in a period of unconsciousness that lasted some 20 minutes. Upon waking, Rafferty complained of feeling unwell, and still exhibiting intermittent seizures was taken back to bed to recuperate. Two days later, Bartholow tried to reinstate his experiment with galvanic stimulation. However, Rafferty's condition had worsened with a right-sided paralysis accompanied by painful tingling sensations. Although Bartholow had to abandon his plans, he would make one last effort the following day. However, Rafferty had become decidedly worse in health with frequent lapses into unconsciousness. Nor would she recover. Later that evening, Rafferty suffered a massive seizure and fell into a coma from which she would not wake – mercifully dying the following day.

The autopsy, which was necessary to confirm the site of the electrode placements, took place after Rafferty's brain had been hardened in a solution of chromic acid for 24 hours. It was reported that while the needle tracks could be traced on both sides, they were also contaminated by liquefied matter, showing pus and liquid had seeped in from the diseased surface of the brain. Bartholow's paper outlining his behavioural findings was published a few months later in April 1874. It attracted much criticism and disapproval, although Ferrier's reaction was more favourable, with him replying in *The London Medical Record* (May 1874) that the experiment was of 'great interest in relation to the physiology of the brain.' However, two years later, he condemned the work by stating the procedure 'is not to be commended or likely to be repeated.' Even Bartholow was forced to admit it would be 'in the highest degree criminal' to undertake such a procedure again. Yet, it appears to have done Bartholow no professional harm. Five years later, he took the Chair at Jefferson Medical College in Philadelphia, where he was to spend the rest of his career.

Cortical areas for hearing and sight

Among Ferrier's many discoveries was the existence, in the monkey at least, of areas mediating movement that lay outside the boundaries of the frontal cortex. One of these sites was the upper part of the temporal lobes which lay adjacent to the Sylvian fissure[20] (i.e. a region lying roughly behind the ears). When Ferrier stimulated this area with an electric current, it produced a sudden turning of the head that resembled a response of an animal startled by an unexpected sound. It was a reaction implying a part of the temporal lobes might be involved in hearing, and to examine this possibility further, Ferrier bilaterally lesioned the region in monkeys, and tested their reaction to a variety of stimuli. He found the monkeys were completely unresponsive to sound – a finding confirmed when two animals with unilateral lesions were shown to be deaf in only one ear. Thus, Ferrier had apparently managed to localise the function of hearing to the upper temporal cortex.

These findings, however, did not go uncontested. Other researchers set about testing Ferrier's findings and obtained different results. For example, the Italian Luigi Laciani working with dogs, found the superior temporal lesion only produced a short-lasting deafness. Part of the problem was the poor survival times of Ferrier's monkeys who tended to die soon after surgery from infections. To overcome this problem, Ferrier set about improving his use of antiseptics, and was able to repeat his experiment on a monkey that lived for well over a year. The hearing loss was so complete that Ferrier took the monkey to the 1881 International Medical Congress in London, where he astonished the audience by firing a gun next to its head – an act that did not cause the animal to flinch. Although Ferrier's findings continued to be disputed, we now know he had discovered the general location of the primary auditory cortex (the main area of the cerebral cortex to receive input from the ears). This area was also close to the 'sound field' which had been identified by Meynert and Wernicke as important in the recognition of speech.

However, there was to be a greater furore when Ferrier claimed he had located the brain area responsible for vision. This assertion was made in 1875 when Ferrier identified a region in the monkey parietal lobes called the angular gyrus,[21] lying close to the occipital lobes,[22] where electrical stimulation produced eye movements and blinking. It was as if the monkey was experiencing visual sensations. More controversially, Ferrier claimed he had discovered the long sought after visual cortex. The question of vision was one of the oldest in brain research and can be said to have begun in 500 BC when Alcmaeon of Crotona followed the optic nerves into the brain. Others throughout the centuries did much the same to locate the visual centre of the brain, with the Italian Giovanni Santorini in 1724, placing it in a small region of the thalamus which later became known as the lateral geniculate nucleus. However, after Flourens had experimentally established decortication produced blindness, the primary seat of vision was located in the cerebral cortex. Flourens, of course, denied localisation of function, insisting that vision was mediated by the entire cortex. However, this idea was rejected in 1855 by the Italian Bartholomeo Panizza who noticed that strokes to the occipital region produced blindness and visual deficits. Panizza also unilaterally blinded a dog and traced the resulting degeneration of the optic nerve back into the occipital cortex. More tellingly, lesions of this region produced complete blindness, whereas the removal of a single hemisphere produced blindness in the opposite eye. Although this was strong evidence for a localised visual area, Panizza's research was largely ignored at the time – perhaps because the views of Flourens were still in the ascendant.

Ferrier was aware of Panizza's research, but his own findings did not support the idea of a visual area in the most posterior regions of the cerebral cortex (i.e. the occipital cortex). In fact, Ferrier was unable to produce eye movements when he stimulated the areas of the brain identified by Panizza, nor did his occipital lobe lesions produce any significant visual impairment. Consequently, Ferrier continued to insist it was the angular gyrus, lying on the border with the parietal lobes, and not the hind-most areas of the cortex, which was involved with vision. His theory was also supported when Ferrier lesioned the angular gyrus in monkeys, with its unilateral removal apparently causing blindness in the opposite eye. Ferrier even reported the case of a bilaterally lesioned monkey who could not see a cup of tea (of which it was 'exceedingly fond') when it was placed in front of the animal's eyes. Ferrier presented these findings to the Royal Society in 1876, and published them the same year in his widely acclaimed book *The Functions of the Brain*. He was confident enough to proclaim the angular gyrus was the site in the brain responsible for the sense of vision.

Ferrier's mistake

Ferrier's pronouncement, however, would be fiercely attacked two years later by the Berlin professor Hermann Munk. Like Panizza, Munk had demonstrated blindness after bilateral lesions to the occipital cortex, but not from the angular gyrus. However, Munk also made another important claim: a unilateral lesion did not cause blindness in the opposite eye as Ferrier stated – but in the opposite visual field of the retina. In other words, if a lesion was made to the left side of the cortex, the animal would be unable to see out of the right-sided part of both eyes. This implied the optic nerves to the brain were more complex than previously suspected. On top of this, Munk found that small lesions to certain regions of the occipital lobes could produce interesting visual deficits not reported by Ferrier. This included seeing objects, but not recognising their meaning. For example, lesioned animals often paid no attention to water or food, despite being hungry. They also appeared indifferent to danger as evidenced when a lighted match was bought up to their eyes without them showing fear or backing away. Yet, such dogs could walk around normally and avoid bumping into obstacles. Munk explained these results by proposing the dog had lost its store of visual memory images – or what he called 'psychic blindness'. Put another way, they had been reduced to the status of a puppy with no recollection of visual experience. However, these deficits were not permanent. Munk found that a dog could re-learn the significance of its visual world within four to six weeks of re-training. And, even total blindness could be reversed if the animal was given sufficient time to recover, providing parts of its occipital cortex had been spared.

The differences in opinion between Ferrier and Munk were to result in an acrimonious and long-running dispute. Munk would be the one ultimately proven correct. There are several reasons why Ferrier was misled by the angular gyrus. One lay in his poorer surgical techniques. Most of his monkeys only survived for a few weeks after surgery, providing Ferrier a restricted period to examine their behaviour. In contrast, Munk was able to study his animals for months or even years. This gave Munk a great advantage for he soon realised the effects of angular gyrus lesions on blindness were not permanent. Another problem lay with Ferrier's removal of the occipital lobe. He had used a landmark on the surface of the cortex called the lunate sulcus to define its boundary with the parietal lobe. However, this was an imprecise marker, leading to several important parts of the occipital cortex being spared. This included the calcarine fissure. In fact, we now know the calcarine fissure is crucial for vision, for it is the part of the occipital lobes receiving direct visual input from the eyes.[23] Even a small amount of residual cortex left here provided an animal with some degree of vision. These were understandable errors, although this would not restrain Munk in his criticism of Ferrier. Reporting to the Berlin Physiological Society in 1881, Munk lambasted the Scotsman's work by declaring: 'there was nothing good to be said about it,' and that 'Mr Ferrier has not made one correct guess (and) all his statements have turned out to be wrong.' These criticisms were so harsh that the American William James would feel obliged in his *Principles of Psychology* (1890) to reply: 'Munk's absolute tone about his observations and his theoretical arrogance have led to his ruin as an authority.'

A test of cortical localisation

Despite the increasing evidence supporting cortical localisation, which by the late nineteenth century extended to language, movement, audition and vision, there were still opponents to the idea. One critic was the Strasburg professor Friedrich Goltz. He had first become interested in the issue of localisation during 1869 when discovering a decerebrate frog was able to swim, hop and jump out of hot water. A headless frog was also able to croak if the appropriate part of its spinal cord was stimulated. It seemed as if the brain wasn't needed for many behaviours, and during the 1870s, Goltz extended his interest to examining the effects of cerebral cortex removal in dogs. Although his operative procedure clearly affected the intelligence of the animal (Goltz wrote his dogs acted like 'idiots'), they were not robbed of their senses. For example, the blare of a trumpet would rouse a decorticated dog from its sleep, and it was able to turn away from a bright light. The animal could also walk, albeit somewhat awkwardly, and while it did not seek out food, it was able to lap milk and eat without difficulty. Goltz decorticated three dogs, exhibiting them at scientific meetings throughout Europe. The behaviour of his animals appeared to provide strong evidence against the idea of cortical localisation. Although Goltz did not deny the possibility of deficits occurring after damage to specific areas of the cerebral cortex, he believed no one single area was crucial for a particular function. In fact, he tended to favour the theory that the results of any lesion could be explained as a general defect in attention – which he held was a general responsibility of the whole cortex.

In 1881, Goltz presented two of his dogs at the International Medical Congress in London, which provided a forum for the world's most eminent medical researchers to publicise their work. This was a huge spectacle attended by some 120,000 people, including more than 3000 delegates from over 70 countries. One of the sessions concerned the cortical localisation of function, with Goltz opening the proceedings by exhibiting a dog with the greater part of its cerebral cortex removed. It was clear to all the animal was capable of running around, and could see, hear, and feel pain. Thus, the cerebral cortex was seemingly unimportant for these behaviours. The next presentation, however, was provided by Ferrier. This was a small group of monkeys, each with a small localised lesion to the cerebral cortex – resulting in a specific impairment such as movement, audition or vision. The behavioural evidence from these animals clearly contradicted Goltz. In the ensuing debate, Ferrier accused his adversary of not fully destroying the cortex in his dogs, resulting in some degree of residual function. It was an accusation leading the chairman of the proceedings to ask Goltz and Ferrier whether they would be prepared to sacrifice one of their animals, so their brains could be examined by an independent panel of judges. Both men agreed, and the work was carried out by scientists at Cambridge, including John Langley and a young Charles Sherrington. Their findings, published later in the year, showed Ferrier had been correct in the estimation of his lesions, although Goltz had seriously underestimated the amount of cortex removed from his dog. For most observers, this vindicated Ferrier's concept of cortical localisation.

There was yet to be a final twist to the saga. The exhibits shown by Ferrier had not gone unnoticed by members of the anti-vivisection league – an influential body who had achieved significant popular support by persuading Parliament to bring about the formation of the Cruelty to Animals Act in 1876. This was the first of its kind in the world, and it stipulated researchers could be prosecuted for cruelty unless they were licensed and provided justification for their work. The Act also required an experiment involving the infliction of pain upon animals could only be performed when '*the proposed experiments are absolutely necessary for the due instruction of the persons to save or prolong human life.*' Three months after the congress, Ferrier received a summons to appear at Bow Street police station, charged with performing surgical experiments on monkeys without a license. Interestingly, for reasons that have since become obscure,[24] Ferrier had applied for a license, but been refused despite his previous experience working with animals. Nonetheless, Ferrier was proven innocent and acquitted, for the operations had been undertaken by colleague and licensee Gerald Yeo at King's College. It was a highly publicised case, however, bringing into focus the dilemmas concerning the benefits and costs of medical research. This is an issue as relevant today as it was in the nineteenth century.

The story of Phineas Gage

Another brain area to be examined by Ferrier were the frontal lobes. This region, of course, had given rise to localisation theory with Broca's discovery of a language centre in 1861, and the motor areas by Fritsch and Hitzig in 1870. However, both these regions were relatively small and lay in the more posterior borders of the frontal lobes. This left a much larger expanse of frontal cortex more anteriorly. It was also an area that fascinated many researchers – not least because this part of the human brain is better developed than in other species. To give an example: the frontal lobes in human beings occupy about 30 per cent of the cortical mass compared to about 17 per cent in chimpanzees and only 3.5 per cent in rats. Consequently, the frontal lobes are generally regarded as the most recent part of the brain to have evolved and likely to make us uniquely human. This situation led many investigators during the second part of the nineteenth century, including the eminent English philosopher and evolutionary biologist Herbert Spencer (who had coined the expression 'survival of the fittest') to speculate that the anterior parts of the frontal lobes must be responsible for mankind's higher functions, such as abstract thinking, moral responsibility and personality.

When Ferrier set about examining the frontal lobes in the 1870s, he found their most anterior regions were unresponsive to electrical stimulation. Their ablation, however, caused 'a decided alteration in the animal's character and behaviour'. In particular, lesioned monkeys changed from being attentive and naturally curious, to emotionally apathetic, indifferent and weak-willed. Although these changes were not easy to define precisely, it appeared as if these animals had lost the capacity of intelligence and attention – deficits which reminded Ferrier of dementia. It led him to conclude that the frontal cortex was the 'substrata of those psychical processes that lie at the foundation of higher intellectual operations.' In other words, they underpinned intelligence. In this respect, Ferrier was even supported by Goltz, who admitted dogs with frontal lesions could be recognised by their stupid facial expressions and lack of normal fear. Yet, this view was far from unanimous with many researchers unable to find any obvious impairment after frontal lobe damage. This was perhaps best summed up by the German Jacques Loeb who stated in 1900 that, 'There is perhaps no operation which is so harmless for a dog as the removal of the frontal lobes.' At the very least, the behavioural deficits were not as severe as one might reasonably expect from such a highly evolved and conspicuous brain structure.

Ferrier presented his experimental work on the frontal lobes in his Goulstonian Lectures given to the Royal College of Physicians in 1878.[25] However, these lectures would be better known for a human case study which continues to intrigue many today. This is the story of Phineas Gage who had parts of his frontal lobes blown away in a horrific accident. Gage was a foreman of a railway construction gang working in New England, whose job was to flatten the ground for the laying of railway tracks. This involved the 'tampering down' of sand over explosive with a large rod. However, on one fateful September morning in 1848, Gage accidentally dropped his tampering iron, igniting the blasting powder, and sending the rod through his left cheek and out the top of his head. The force of the blast was so powerful, it caused the rod measuring over a yard in length and weighing some 13 pounds,[26] to land some 30 metres away. Gage was thrown unconscious to the ground, although he regained consciousness a few minutes later without any obvious impairment. Carried to a nearby hotel on an ox-cart, Gage was then examined by the physician John Harlow[27] who found his patient sitting upright in a chair. The injury was clear to see with Harlow observing 'the pulsation's of the brain' through a wound in his skull, resembling an 'inverted funnel'. Harlow removed several bone fragments, placed the larger ones back into the skull, and dressed the injury. Despite his traumatic ordeal, Gage's mind seemed to be unaffected, and he walked away from his treatment unaided.

Gage was convinced he would be back at work within a few days, but he soon became delirious and fell into a coma. Another danger was an infection of the brain – a risk requiring his wound to be cleaned and drained on a regular basis. However, gradually Gage recovered his full physical fitness and within two months was declared cured. The only visible sign of his accident was a pronounced hollow in the crown of his skull, and blindness in the left eye. Despite his apparent well-being, however, Gage was never re-employed by the railroad company. While physically capable of work, Gage's personality had undergone a dramatic change. Prior to the accident, Gage had been widely regarded as responsible, hard-working and considerate. He had even been complimented as 'the most efficient and capable man' in the employment of the company. Following the accident Gage's behaviour became increasingly capricious, disinhibited and impulsive. Gage soon lost his friends by having 'little deference for his fellows' and caring nothing for social graces, made worse with an inclination to use obscene language. Now considered erratic, unpredictable and unable to form cohesive plans, Gage had become according to Harlow, 'a child intellectually with the animal passions of a strong man.'

Little is known about Gage's life in the years after his accident, although he kept his tapering rod, which apparently never left his sight, and is believed for a short while to have been an attraction in Barnum's American Museum in New York. According to one source, for an extra 10 cents, visitors were allowed to lift up a lock of Gage's hair to reveal the pulsating brain beneath a layer of skin.[28] Gage also appears to have found work as a stagecoach driver in New Hampshire and Chile. But according to his mother, he was never able to hold down a job for very long, and always found something that did not suit him. Some 12 years after his accident Gage began suffering from seizures, probably contributing to his death in 1860, at the age of 38. Although no autopsy was performed, Gage's body was later exhumed in 1867, and his skull sent to Harlow, who estimated the tampering iron had destroyed the best

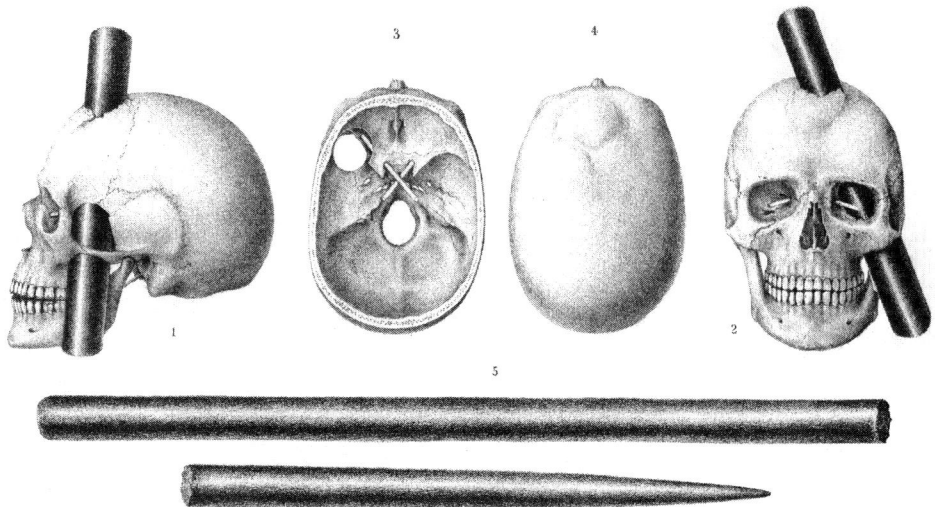

FIGURE 9.7 The trajectory of the tampering iron that passed through the skull and brain of Phineas Gage in 1848.

Source: Taken from Bigelow 1850 (the image was kindly supplied by the Warren Medical Museum).

part of Gage's left anterior frontal lobe.[29] Harlow's reporting of the case was of special interest for Ferrier who believed the changes in Gage's behaviour had some correspondence with those of his lesioned monkeys. Both were considered to be weak-willed, apathetic and unable to pay attention. Ferrier also cited Gage as an example of how a frontal lobe injury can produce a marked personality change without any alteration in sensory or motor function. Thanks to Ferrier, the case of Phineas Gage is now one of the most famous in clinical neuropsychology. His story has also entered popular culture, becoming the subject of plays, films, TV programmes and even YouTube sketches. Gage's skull and tampering iron are now on display in the Museum of the Harvard Medical School.

While Gage lives on in the thoughts of psychologists, Ferrier's place in the pantheon of great neuroscientists is a mixed one. A character who often courted controversy, Ferrier made a number of well-publicised mistakes in his career, not least by localising the function of vision in the angular gyrus situated in the parietal lobes. Despite this, Ferrier more than any other nineteenth century figure, substantiated the idea of cortical localisation, whose basic tenets are widely supported today. Charles Sherrington, for one, would laud his achievements in 1892 by writing that Ferrier had placed cerebral localisation at the centre of neurological interest, and provided the basis for a 'scientific phrenology'. Much of Ferrier's work was also covered in two widely acclaimed books *The Functions of the Brain* in 1876 which describes his experimental results (this was expanded and revised in 1886), and *The Localization of Brain Disease* in 1878 where he used the concepts of cortical localisation to pave the way for the development of neurosurgery (see Chapter 13). For this work, Ferrier would be awarded with many honours including a knighthood in 1911. Ferrier was also one of the founders of the journal *Brain* (first published in 1878) along with John Hughlings Jackson, Sir John Bucknill and Sir James Critchton-Browne.

Notes

1. This area is now recognised to include the *hippocampus* and *amygdala* in the medial temporal lobes, the *cingulate cortex* on the underside of the cortical mantle, and the *hypothalamus* in the basal forebrain.
2. A few years later, in 1870, Fritsch and Hitzig would discover the motor cortex. This is now known to be the source of the cortico-spinal tract, which is the motor pathway crossing in the brainstem.
3. The situation is more complex than Broca envisioned. Research by Rasmussen and Milner (1977) has shown that 96 per cent of right-handers have language lateralised to the left hemisphere. For left-handers, this figure drops to about 70 per cent. The remaining left handers either have language lateralised to the right (as Broca believed) or have language localised to both hemispheres.
4. One of the three major paired arteries that supply blood to the cerebral cortex, especially its outer or lateral aspects.
5. These were (1) alogia characterised by a general reduction in intelligence; (2) verbal amnesia characterised by unintelligible speech that often has no bearing to what is trying to be expressed; (3) aphemia characterised by poor pronunciation; and (4) alalia characterised by loss of neural control over the speech apparatus including tongue and vocal cords.
6. Now Tarnowskie Gory, in Western Poland.
7. In layman's terms, the temporal lobe is the part of the cerebral cortex that lies laterally (at the side) of the hemisphere (medial to the ears).
8. Der Aphasische Symptomencomplex (the symptom-complex of aphasia).
9. Actually such a link was already known. In 1809, Johann Christian Reil had identified a group of fibres running close to the lateral (Sylvian) fissure which coursed through the temporal, parietal and frontal regions. This pathway was later described in more detail and called the arcuate fasciculus by the German anatomist Karl Friedrich Burdach in his three volumed work (*Vom Bau und Leben des Gehirns und Rückenmarks*) published between 1819 and 1826.

10 It is more commonly known as The Wernicke–Geschwind model of language.
11 In 1908, at the age of 70 years, Fritsch also published a remarkable work on the comparative racial morphology of the human retina in which he introduced the term *fovea*.
12 Although most accounts state the operations were carried out on Frau Hitzig's dressing table, one source (Taylor and Gross 2003) has said they were performed on Frau Fritsch's table.
13 Hitzig and Fritsch would later use surgical anaesthesia when undertaking such experiments. They presumably did not use it in their first experiments for fear that it would make the cortex 'unexcitable'.
14 In humans, this region lies in the precentral gyrus, just forward of the central sulcus, which is the main point of division between the frontal and parietal lobes. A motor function for this region would be confirmed in humans through the work of Canadian Wilder Penfield (see Chapter 13).
15 This is perhaps not surprising. Jackson published much of his work in obscure journals. For example, his speculations on the motor cortex had been made in the *Transactions of St Andrew Medical Graduates' Association*, which could not have been known in Germany. Even had this paper been seen outside England, Jackson's numerous writings were frequently accused of being difficult and ambiguous.
16 Now the National Hospital for Neurology and Neurosurgery.
17 In later works, including his books, Ferrier would be more complementary in recognising the achievements of Hitzig and Fritsch.
18 The first animal welfare organisations came into being during the nineteenth century. The first such organisation in the world, the Society for the Prevention of Cruelty to Animals (SPCA) was formed in 1824. It later became the Royal SPCA (RSPCA) in 1840 as a result of the patronage of Queen Victoria.
19 It is not exactly clear what methods Bartholow used to receive her permission. In a letter justifying his actions to the *American Journal of Medical Sciences* in 1874, Bartholow simply states that Rafferty 'consented' to the experiments.
20 This is the fissure that separates the temporal lobes from the parietal lobes.
21 We have already seen, that in 1892, the Frenchman Joseph Dejerine found that damage to the angular gyrus also produced alexia.
22 The occipital lobes (relating to the occipital or skull occiput bone) are the smallest of four paired lobes in the human cerebral cortex and are located in its most posterior portion (i.e. at the back of the cortical mantle).
23 In fact, this region of the brain (i.e. the primary visual cortex) is recognisable because it contains a fine white line known as Gennari's stripe. It was first discovered by the Italian Francesco Gennari in 1776.
24 Ferrier's first experiments in London had actually been carried out in sheds at the back of his house, but following protests from the antivivisectionists, he had to transfer this work to King's College where facilities were provided for him.
25 These lectures had been initiated in 1639 at the bequest of Thomas Goulston.
26 The tampering rod was actually 3 feet 7 inches long, and 1 1/4 inches in diameter, tapering down to about a 1/4 inch at the other end.
27 It was Harlow's second report on Gage published in 1868 that Ferrier used as the basis of his talk.
28 See Douwe Draaisma's book *Disturbances of the Mind*.
29 A study in 1994 using computer technology by Hanna and Antonio Damasio used Gage's skull to digitally reconstruct the rod's path through his brain. It was shown to enter the skull below the left eye and exit through the top of the head, causing damage to both frontal lobes.

Bibliography

Benson, D.F. (1993) The history of behavioral neurology. *Neurologic Clinics*, 11, 1–9.
Benton, A.L. (1964) Contributions to aphasia before Broca. *Cortex*, 1, 314–327.
Benton, A.L. (1984) Hemispheric dominance before Broca. *Neuropsychologia*, 22, 807–811.
Benton, A.L. (1991) The prefrontal region in early history. In Levin, H., Eisenberg, H. and Benton, A.L. (eds) *Frontal Lobe Function and Dysfunction*. Oxford University Press: New York.
Berker, E.A., Berker, A.H. and Smith, A. (1986) Translation of Broca's 1865 report *Archives of Neurology*, 43, 1065–1072.

Bigelow, H.J. (1850) Dr Harlow's case of recovery from the passage of an iron bar through the head. *American Journal of Medical Science*, 39, 12–22.
Carlson, C. and Devinsky, O. (2009) The excitable cerebral cortex: Fritsch G. Hitzig. Uber die elektrrische erregbarkeit des grosshirns. Arch Anat Physiol Wissen 1870: 37, 300–332. *Epilepsy and Behaviour*, 15, 131–132.
Catani, M. and Mesulam, M. (2008) The arcuate fasciculus and the disconnection theme in language and aphasia: history and current state. *Cortex*, 44, 953–961.
Colombo, M., Colombo, A. and Gross, C.G. (2002) Bartolomeo Panizza's observations on the optic nerve. *Brain Research Bulletin*, 58, 529–539.
Critchley, M. and Critchley, E.A. (1998) *John Hughlings Jackson: Father of English Neurology*. Oxford University Press: Oxford.
Damasio, A. (1994) *Descartes' Error*. Vintage: London.
Damasio, H., Grabowski, T., Frank, R., Galaburda, A.M. and Damasio, A.R. (1994) The return of Phineas Gage: clues about the brain from the skull of a famous patient. *Science*, 264, 1102–1105.
Draaisma, D. (2006) *Disturbances of the Mind*. Cambridge University Press: Cambridge.
Eggert, G.H. (1977) *Wernicke's works on Aphasia: A Source Book and Review*. Moulton: The Hague.
Eling, P. (2006) Meynert on Wernicke's aphasia. *Cortex*, 42, 811–816.
Ferrier, D. (1876) The *Functions of the Brain*. Putnam: New York.
Ferrier, D. (1878) The Goulstonian lectures on the localization of cerebral diseases. *British Medical Journal*, 1, 443–447.
Ferrier, D. (1878) *The Localization of Cerebral Disease*. Smith & Elder: London.
Ferrier, D. (1886) *The Functions of the Brain*. 2nd edn. Smith & Elder: London.
Geschwind, N. (1967) Wernicke's contribution to the study of aphasia. *Cortex*, 3, 448–463.
Giannitrapani, D. (1967) Developing concepts of lateralization of cerebral functions. *Cortex*, 3, 353–370.
Gibson, W.C. (1962) Pioneers in localization of function in the brain. *Journal of the American Medical Association*, 180, 944–951.
Glickstein, M. (1985) Ferrier's mistake. *Trends in Neurosciences*, 8, 341–344.
Greenblatt, S.H. (1984) The multiple roles of Broca's discovery in the development of the modern neurosciences. *Brain and Cognition*, 3, 249–258.
Gross, C.G. (1998) *Brain Vision Memory: Tales in the History of Neuroscience*. MIT Press: Cambridge, MA.
Gross, C.G. (2007) The discovery of the motor cortex and its background. *Journal of the History of the Neurosciences*, 16, 320–331.
Harlow, J.M. (1848) Passage of an iron rod through the head. *Boston Medical and Surgical Journal*, 39, 389–393.
Harlow, J.M. (1868) Recovery from the passage of an iron bar through the head. *Publications of the Massachusetts Medical Society*, 2, 327–347.
Harrington, A. (1985) Nineteenth-century ideas on hemisphere differences and 'duality of mind'. *Behavioral and Brain Sciences*, 8, 617–634.
Harrington, A. (1987) *Medicine, Mind and the Double Brain*. Princeton University Press: Princeton, NJ.
Harris, L.J. and Almergi, J.B. (2009) Probing the human brain with stimulating electrodes: The story of Robert Bartholow's (1874) experiment with Mary Rafferty. *Brain and Cognition*, 70, 92–115.
Heffner, H.E. (1987) Ferrier and the study of auditory cortex. *Archives of Neurology*, 44, 218–221.
Iniesta, I. (2011) John Hughlings Jackson and our understanding of the epilepsies 100 years on. *Practical Neurology*, 11, 37–41.
Jefferson, G. (1953) The prodromes to cortical localisation. *Journal of Neurology, Neurosurgery and Psychiatry*, 16, 59–72.
Joynt, R.J. and Benton, A.L. (1964) The memoir of Marc Dax. *Neurology*, 14, 851–854.
Macmillan, M. (2008) Phineas Gage: unravelling the myth. *The Psychologist*, 21, 828–831.
Marshall, J.C. and Fink, G.R. (2003) Cerebral localization, then and now. *Neuroimage*, 20, S2–S7.
Marx, O. (1966) Aphasia studies and language theory in the nineteenth century. *Bulletin of the History of Medicine*, 40, 328–349.

Morgan, J.P. (1982) The first reported case of electrical stimulation of the human brain. *Journal of the History of Medicine and Allied Sciences*, 37, 51–64.

Neylan, T.C. (1999) Frontal lobe function: Mr Phineas Gage's famous injury. *Journal of Neuropsychiatry and Clinical Neuroscience*, 11, 280–283.

Pearce, J.M.S. (2003) Sir David Ferrier MD, FRS. *Journal of Neurosurgery and Psychiatry*, 74, 787.

Pearce, J.M.S. (2003) The nucleus of Theodor Meynert (1833–1892). *Journal of Neurosurgery and Psychiatry*, 74, 1358.

Philips, C.G., Zeki, S. and Barlow, H.B. (1984) Localization of function in the cerebral cortex: Past, present and future. *Brain*, 107, 327–361.

Schiller, F. (1992) *Paul Broca: Explorer of the Brain*. Oxford University Press: Oxford.

Steinberg, D.A. (2009) Cerebral localization in the nineteenth century: the birth of a science and its modern consequences. *Journal of the History of Neurosciences*, 18, 254–261.

Stookey, B. (1954) A note on the early history of cerebral localisation. *Bulletin of the New York Academy of Medicine*, 30, 559–578.

Stookey, B. (1963) Jean-Baptiste Bouillaud and Ernest Aubertin. *Journal of the American Medical Association*, 184, 1024–1029.

Taylor. C.R. and Gross, C. G. (2003) Twitches versus movements: a story of motor cortex. *History of Neuroscience*, 9, 332–342.

Tesak, J. and Code, C. (2008) *Milestones in the history of Aphasia*. Taylor & Francis: Hove.

Tizard, B. (1959) Theories of brain localization from Flourens to Lashley. *Medical History*, 3, 132–145.

Walker, A. E. (1957) The development of the concept of cerebral localization in the nineteenth century. *Bulletin of the History of Medicine*, 31, 99–121.

Walker, A.E. (1957) Stimulation and ablation: their role in the history of cerebral physiology. *Journal of Neurophysiology*, 20, 453–449.

Wernicke, C. (1874) *Der aphasische Symptomencomplex: Eine Psychologiische Studie auf anatomisher Basis*. Cohn & Weigert: Braslau.

Whitiker, H.A. and Ethlinger, S.C. (1993) Theodor Meynert's contribution to classical nineteenth century aphasia studies. *Brain and Language*, 45, 560–571.

Wilgus, J. and Wilgus, B. (2009) Face to face with Phineas Gage. *Journal of the History of Neurosciences*, 18, 340–345.

Young, R.M. (1968) The functions of the brain: Gall to Ferrier (1808–1886). *Isis*, 59, 251–268.

Young, R.M. (1990) *Mind, Brain and Adaptation in the Nineteenth Century*. Oxford University Press: Oxford.

10
THE RISE OF PSYCHIATRY AND NEUROLOGY

The goal of pathological anatomy cannot be attained without an incessant association between the lesion under study in all its minutest details of development and the pathological events unfolding at the bedside of the patient.

Jean Marie Charcot

I call dementia praecox schizophrenia because I hope to show that the split of the several psychic functions is one of its most important characteristics.

Eugene Bleuler

Summary

The study and treatment of mental illness, otherwise known as psychiatry, has a much longer history than neurology where disorders are linked to some specific dysfunction of the nervous system. Indeed, certain conditions now classified as psychiatric illnesses such as mania, melancholia and hysteria have been recognised since the time of Hippocrates. In contrast, an understanding of the importance of the nervous system in health and disease did not occur until the eighteenth century. The origins of this development can be said to have begun with the coining of the term 'neurologie' by Thomas Willis, which entered the English lexicon in 1681. Although the term meant the 'doctrine of the nerves' it gained a much broader definition in the next century when the disharmony of the nervous system was implicated in a wide range of somatic and mental disorders. Another pivotal point was the publication of Giovanni Morgagni's great work *De sedibus* in 1761, which decisively broke with old humoral theories of health. In its place would be the recognition that disease arose from 'the cry of suffering organs'. Morgagni's method of establishing the anatomical location of a disease by correlating the life histories of his patients with post-mortem findings was to provide a powerful means of detecting new illnesses and their organic basis. This anatomo-clinical method, however, would be most famously used by Jean Marie Charcot who took over the Parisian asylum Le Salpêtrière in 1862. At the time patients in these types of institution suffered from all kinds of nervous and mental diseases, including insanity, epilepsy and general paralysis of the insane. However, in less than 20 years, Charcot would bring order to this chaos by defining a number of neurological conditions, establishing their anatomical basis, and constructing an accurate system of classification. His work resulted in the world's first professorship of nervous diseases being created for him at the University of Paris in 1881. It was the moment when clinical neurology became an area of medical specialisation. However, the development of psychiatry required a different approach since many mental illnesses do not have an organic lesion, and their wider variety of symptoms make classification far more difficult. Nonetheless, progress was made through the endeavours of the German Emil Kraepelin who first identified dementia praecox (schizophrenia) in 1887. Regarded by many as the founder of medical psychiatry, Kraepelin's classification of psychiatric disorders forms the basis of those used today by such authorities as the American Psychiatric Association and The World Health Organization.

A new concept of disease: Giovanni Battista Morgagni

Our modern concept of disease is largely a result of changes that took place in the eighteenth century. This is a period sometimes known as the 'Enlightenment' when a new belief in humanitarianism, reason and scientific progress swept away the long antiquated teachings of the past. These changes manifested themselves in many ways, not least in medicine, where old concepts involving the Hippocratic humours gave way to a more rational understanding of illness and its treatment. Arguably, the most important person to bring about this change was the Italian Giovanni Battista Morgagni. Despite his humble background, Morgagni began studying medicine in Bologna in 1698 at the age of 16, where he came under the guidance of the great anatomist Antonio Valsalva.[1] Yet, it appears even at this early age, Morgagni was questioning what he was being taught. This scepticism would lead Morgagni to be elected president of the Academia degli Inquieti (Academy of Restless Persons) in 1705, which was a reputable academic institution whose members questioned traditional teachings and were keen to advance science through their own endeavours. A year later, in 1706, Morgagni published his first major treatise *Adveraria anatomica* (*Anatomical Writings*). Renowned for its precision and anatomical detail, *Anatomica* was later extended into five volumes, making Morgagni one of the most respected anatomists in Italy. After a spell in Venice and serving time as a physician in his home town of Forli, Morgagni was appointed Professor of Anatomy at Padua in 1715. This was the most prestigious Chair in Italy whose previous occupants included Vesalius and Fallopio. Morgagni's great reputation as an anatomist would attract students from all over Europe who gave him the epithet of 'his anatomic majesty'.

From all accounts, Morgagni was a tireless physician, teacher and researcher, although he did not produce any more major works over the next 45 years. But this did not mean he was being unproductive. On the contrary, Morgagni was carefully studying the symptoms of disease and learning more about pathological anatomy through autopsies. The fruits of this labour were to result in his monumental five-volumed work entitled *De sedibus*[2] in 1761, better known in English as *The Seats and Causes of Disease Investigated by Means of Anatomy*. Published when Morgagni was 79 years old, the work represented a lifetime of observation, dissection and careful reflection. It was also a major step forward in medical thinking for the book's main argument rejected traditional humoral beliefs by proposing the cause of all illness lay with the breakdown of the bodily organs – a new and radical concept for the mid-eighteenth century. Interestingly, the book may never have been written had Morgagni not gone on a holiday in 1740 with a young 'dilettante' friend, who took much interest in 'scientific matters'. Encouraged by his curiosity, Morgagni began a correspondence that would lead to an exchange of some 70 letters, over a period of 20 years, describing the clinical histories of his patients followed by post-mortem findings. Helped by his friend's comments and queries, his collection of letters was then revised to form the basis of Morgagni's great treatise.

De sedibus is basically an extensive medical compendium that makes explicit reference to some 700 patients whom Morgagni had personally examined and treated. It also combines his observations with the work of Swiss anatomist Théophile Bonet whose *Sepulchretum sive anatomica practica* (1679) provided an additional survey of over 3000 autopsies.[3] Emphasising the morbid anatomy of disease, *De sedibus* is comprised of five books, each dealing with a different organ system. For those with neuroscience interests, the most relevant is the first,

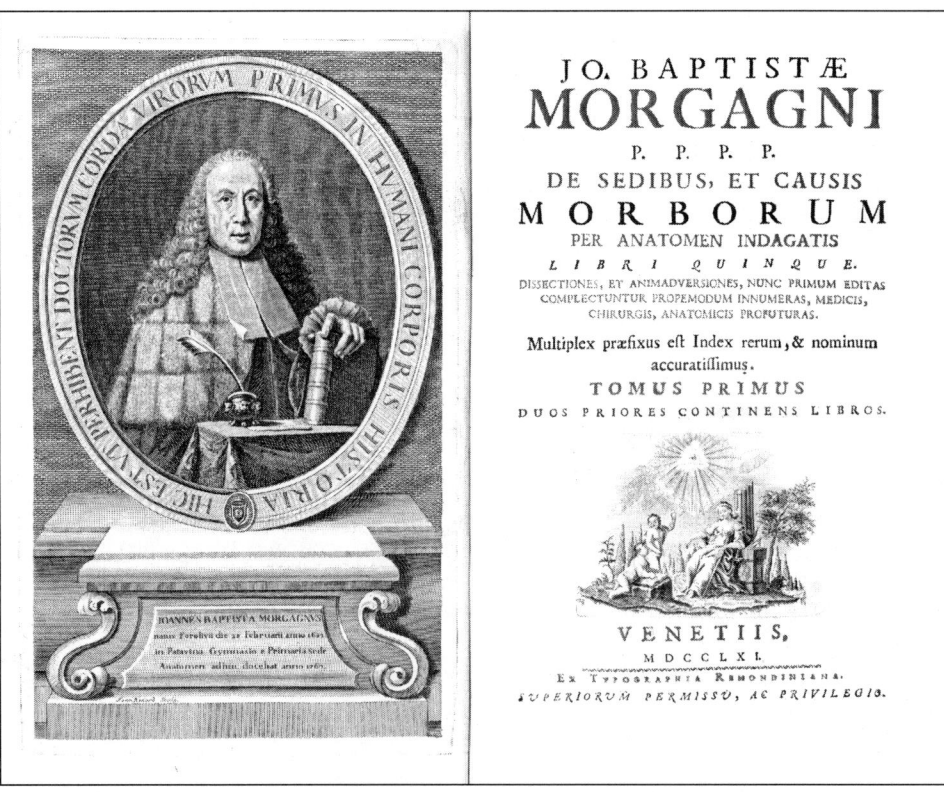

FIGURE 10.1 Giovanni Battista Morgagni (1682–1771) who wrote *De sedibus, et causis morborum per anatomen indagatis* (1761) that is based on the clinical and post-mortem records of seven hundred patients. *De sedibus* replaced the old humoral theory of health with one associating disease with damage to the organs of the body.

Source: Wellcome Library, London.

which deals with maladies of the head. This includes a detailed analysis of several brain disorders including infantile cerebral palsy, cerebral tumours, inflammation and fluid on the brain (hydroencephalus). Another disorder described by Morgagni is apoplexy (stroke), which he describes as resulting from a softening of the brain, or a cerebrovascular accident (i.e. haemorrhage) bought on by a large number of possible causes including 'anxieties of the mind', 'straining at stool' and sneezing. Morgagni is also aware that injury to the cerebral hemispheres, including stroke, can produce a hemiplegia (paralysis) on the opposite side of the body.

While it can be argued that Morgagni had not discovered anything fundamentally new about the brain, *De sedibus* was nevertheless a turning point, encouraging physicians to view disease as an abnormality arising from a localised region or organ, rather than an imbalance of some humour disturbing the whole body. In other words, it persuaded medical practitioners of something we now take for granted: the symptoms of disease are caused by afflictions affecting some part of the body's anatomy. Or as Morgagni memorably put it: disease was 'the cry of suffering organs'. However, perhaps just as important, were the implications for understanding how disease arose. From Morgagni's perspective, the body was a finely tuned apparatus, with each organ contributing to the overall function of the whole. Consequently, an organ could break down through disease, causing the entire body to dysfunction or stop working. The task of the anatomist, therefore, was to identify how various types of disease arose in localised organs of the body, and use this knowledge to establish more rational and effective forms of treatment. These were ideas that quickly gained acceptance, leading to the first chairs of pathologic anatomy in several European universities by the start of the nineteenth century. As we shall see, this would also be an important step in the later development of both neurology and psychiatry.

The rise of nervous disorders in the eighteenth century

The search for new rational systems of medicine in the eighteenth century also led to a growing commercialisation of health, with the rise of the medical practitioner or 'quack'. This was especially true in Britain where sick people began consulting doctors more frequently to demand new remedies and drug treatments. One common class of illness were the nervous complaints, although they were very different from the ones that are recognised today. The three most common 'nervous' disorders of the age were hysteria, hypochondria and dyspepsia – all linked with a bewildering range of physical and mental ailments. Hysteria, for example, was believed to be an affliction of women, associated with wandering movements of the womb (an idea dating back to Hippocrates) and responsible for excessive emotion, irrational speech, paralysis and convulsions.[4] Hypochondria was no less serious, being a chronic condition causing depression, lethargy, worry, and morbid fears of going insane. From our modern perspective, dyspepsia was perhaps the most difficult to comprehend as it was characterised by chronic indigestion and flatulence. However, it frequently led to palpitations, swellings, vertigo, fainting and extreme nervousness. These disorders were also surprisingly common. For example, in the seventeenth century, Thomas Sydenham estimated hysterical disorders, including hypochondria, were responsible for affecting up to one-sixth of the diseases to which mankind was prone. Less than a century later this type of complaint was said to afflict one-third of the English by George Cheyne.

These illnesses also placed the physician in an awkward position as patients rarely showed any physical signs of disease. Consequently, there was no obvious or rational form of treatment. Traditionally, such conditions had been explained through an imbalance of humours. More precisely, their origin arose from the uterus in hysteria, the spleen in hypochondria, and stomach in dyspepsia. As the humoral theory fell from favour, other interpretations had to be sought, with the nervous system becoming increasingly implicated in health and disease. A key factor in this development was the work of Thomas Willis who outlined a number of nervous diseases in his great trilogy of books.[5] His influence was made greater when Samuel Pordage introduced the word 'neurologie' into English in 1681 after translating *Cerebral anatome* from Latin. Although the term had simply meant the study of the nerves, this came to have a broader meaning in the eighteenth century as the nervous system was implicated in co-ordinating the body's physiology. Another factor was the changing conception of the nerve fibre. Up until the seventeenth century, nerves were widely believed to be hollow and provide conduits for the passage of animal spirits. However, in the eighteenth century, they were regarded as hard and solid, making them prone to physical damage. This also encouraged the idea that the nervous system could be put under strain which forced it to 'break down' in some way.

As nervous disorders became increasingly associated with physical and mental aberrations, it led to the commonly used expression 'the nerves' meaning something akin to emotional excitability or anxiety. One of the first writers to popularise this way of thinking was the Scottish physician George Cheyne whose *An Essay on Health and Long Life* published in 1724 had made him a well-known figure. This small treatise was partly confessional since Cheyne admitted he had ruined his own health and become obese with hard-drinking and high-living. Nonetheless, the book's main message was to persuade its readers to take responsibility for their own health. Another of Cheyne's famous works was *The English Malady* which appeared in 1733. It was again a personal account of his battle with diet and depression, although the 'English malady' referred to nervous agitation bought on by the stresses of urban living and an immoderate and luxurious lifestyle. This, according to Cheyne could lead to serious nervous ailments such as melancholy, hysteria, hypochondria, and insanity. *The English Malady* also defined nervous illnesses as those where the 'system of the nerves and their fibres are evidently relax'd and broken'. Although Cheyne may not have invented such terms, he helped increase their popularity and a proliferation of expressions including 'frayed nerves' or 'shattered nerves' became widely used to explain a wide range of ailments relating to mind and body.

Cheyne had shown, albeit in a popularist way, that an important connection existed between temperament and the nervous system. This association, however, would be given a new perspective by Edinburgh professor William Cullen. One of the most prominent figures of the Scottish Enlightenment, Cullen realised the nervous system controlled most physiological actions of the body. In fact, he regarded 'nervous energy' and not animal spirits as the force responsible for governing the very processes of life itself. Extending this idea to health, Cullen recognised that excesses or deficiencies in nervous energy could cause many illnesses. Or, as Cullen put it: 'life is a function of nervous energy . . . and disease mainly nervous disorder.' Since the brain provided the main source of nervous power in the body, it inevitably followed that it must be the most important organ determining the health of the individual.

Cullen had provided a new way of understanding nervous disease, and he introduced his ideas in a four-volumed work entitled *Synopsis nosolgia methodicae*. Appearing in 1769, just eight years after *De sedibus*, this was Cullen's attempt to offer a new way of classifying disease. However, the book also introduced a term, which was to have a major impact on the development of psychiatry. The term was neurosis, which according to Cullen described a group of diseases characterised by a general 'disharmony' of the nervous system, where no pathological lesion could be observed.[6] In fact, Cullen defined four categories of neurosis: Comata (apoplexy and stroke), Adnamiae (involuntary reflexive actions), Spasmi (convulsions) and Vesaniae (an old Latin term for insanity including melancholia and mania). This last category is of particular interest for Cullen was proposing some forms of madness might have their origin in the 'nervous energy' of the brain. It is a premise on which much of modern psychiatry is founded.

The discovery of Parkinson's disease

In 1817, a London doctor called James Parkinson published a small pamphlet entitled *Essay on the Shaking Palsy*, which described a new progressive debilitating condition characterised by resting tremor, paralysis and festinating gait. Known today as Parkinson's disease, this disorder is recognised as one of the most common degenerative diseases of the brain. It also continues to attract the attention of modern day specialists who seek to more fully understand its causes and improve treatment. Yet, Parkinson was probably not the first to observe the disease since there is a possible reference to the illness in an Egyptian medical papyrus (1350–1200 BC) where a king is depicted with a slackened mouth and propensity to dribble. An account closely resembling Parkinson's disease is also described in Shakespeare's *Henry IV*, and perhaps even a visual portrayal in Rembrandt's sketch *The Good Samaritan*. However, it was Parkinson who first grouped the symptoms of tremor and paralysis into a specific neurological disease. He did this by correctly deducing resting tremor was not a secondary consequence of some other affliction, such as alcoholism or old age, as had been previously thought. Nor was his 'species of palsy' the same as that occurring from apoplexy or brain compression. Instead Parkinson saw it was associated with a specific disturbance of gait, which therefore made it likely to be a manifestation of an underlying brain disease in its own right.

Parkinson was born in the Shoreditch area of London, the son of an apothecary and surgeon. Although qualifying as a surgeon in 1784, Parkinson would spend the rest of his working life as a parish physician, taking over the family practice after the death of his father.[7] He also served as an attending doctor at the local asylum, and was warden of his local church. Like many educated gentleman of his times, Parkinson had an eclectic range of interests, being a member of several scientific associations including the British Geological Society (he was a founder member) and the Medical Society of London – to whom he presented a vivid account of two men hit by lightening in 1789. Parkinson was also a prolific author with books on palaeontology, geology and popular medicine. Another intriguing thing about Parkinson was his involvement in English politics. The early eighteenth century was a time of great social change with the reverberations of the French revolution still taking place, and the American colonies breaking away from Britain. Parkinson was sympathetic to these developments – so much so, that he published a number of political pamphlets critical of King and government under the pseudonym 'Old Hubert'.[8] His activities even led to Parkinson being subpoenaed and examined under oath by the Privy Council concerning his knowledge about a conspiracy to assassinate King George III in 1794. Although there is no evidence he was implicated in the plot, Parkinson only testified after receiving assurances he would not be forced to incriminate himself. It is also interesting to note that despite being a distinguished member of his community, no picture or portrait of Parkinson is known to exist.[9]

However, it is for his brief 1817 monograph of just 66 pages, written in the last years of his life, Parkinson will always be remembered. Parkinson had first noticed this 'tedious and most distressing malady' in one of his patients. He was particularly curious because the patient had led a life of remarkable temperance and sobriety. Nor did they ever suffer from rheumatism, pains in the head, or seizures – all known causes of abnormal movement. After being alerted to its symptoms, Parkinson began recognising them in others. In fact, he was to observe the disorder in two other of his patients, along with three individuals met by chance in the streets – one of whom he could not examine personally but 'only seen at a distance'. The naming of the disorder as *shaking palsy* (or its Latin equivalent *paralysis agitans*) was apt, as the essential features of the disease is a rigid paralysis of the limbs accompanied by marked tremors when attempting to move. This tremor was so violent at times that Parkinson vividly wrote it could not only 'shake the bed-hangings, but even the floor and sashes of the room.' Parkinson also outlined how the disease progressed, with its onset usually heralded by a tremor in one arm, leading to a stooped posture, slow impoverished movement, flexed posture and muscle weakness. These symptoms eventually caused the victim to become bed-ridden and helpless.

Parkinson was not in a position to study the brains of his patients, so could only speculate on the lesion's site. Nonetheless, he realised the disorder spared the 'senses and intellect' of its victims which led him to reason the injury producing the disorder did not occur in the cerebral cortex as this would have affected mental reasoning. Instead, Parkinson placed the suspected lesion in the medulla oblongata of the brainstem – an area that he thought would affect the mobility of the upper spinal cord. In fact, Parkinson was so convinced of his prognosis, he called for post-mortem examinations to be undertaken in order to identify the brain pathology.[10] Parkinson even used his medulla theory to invent a cure for the disease by recommending blood-letting from the neck followed by blistering and inflammation of the skin. He also suggested inserting small pieces of cork into the blisters to cause them to produce a 'purulent discharge'. These efforts, he believed, would divert blood and pressure away from the medulla.

Although Parkinson's monograph appears to have received favourable reviews from the British medical establishment when first published, it soon became forgotten. One reason may lie with the fact that Parkinson's essay never went beyond a first edition, and few copies were ever published. In fact, only five copies are known to exist today, making it one of the most valuable medical texts known to collectors (it is worth in excess of £10,000). Moreover, Parkinson's death in 1824, coming just seven years after his essay, was an insignificant event, with the medical world leaving no obituary in the journals of the time. However, it should not be forgotten that Parkinson was only a provincial doctor after all, and perhaps his radical socialist political views did not help his cause. Parkinson's work, however, would become better known in France where his account of the disorder was to be acknowledged by the great neurologist Jean Charcot, who named the disorder '*La Maladie de Parkinson*' in his honour.[11] Somewhat curiously, the disease would still be referred to as *paralysis agitans* by the English surgeon William Gowers in his best selling neurology textbook of 1888 – long after Charcot had invented the eponym in 1869.

The Napoleon of the Salpêtrière: Jean Martin Charcot

If there is one person responsible for creating modern neurology as a distinct area of medical specialisation, it is surely Jean Martin Charcot. The son of a working-class Parisian carriage builder born in 1825, Charcot was the second of four boys whose father, it is said, could only afford to pay for the education of one. Thus he decided the oldest would take over the family workshop, the two youngest were to join the army, and the most studious (Jean Martin) was to pursue a career in the medical profession. It is not entirely clear if this was the occupation best suiting Charcot, for he was also a talented artist who drew constantly – a skill he later used as a clinician when recording behavioural abnormalities in his patients. Nonetheless, Charcot would rise from his humble beginnings in the egalitarian times of post-revolutionary France to found the world's first neurological school at the Salpêtrière in Paris, and arguably become its greatest doctor. Yet, his rise was not without its setbacks. Beginning his medical studies in Paris in 1844, Charcot failed his first competitive examination for an internship (i.e. hospital residency) in 1847. Rectifying the matter a year later, Charcot finished his training in 1852 after serving four one-year rotations at different hospitals in Paris and completing a MD thesis on joint disease and rheumatoid arthritis. Even so, Charcot was to spend the next ten years working routinely as a doctor in several Parisian hospitals and treating patients in various out-patient facilities.

One of the hospitals where Charcot served his internship was the Salpêtrière.[12] First built in 1603, it had originally been used as a store in central Paris for ammunitions deriving its name from 'saltpetre' (a chemical in gunpowder), which it manufactured. Re-located to the left bank of the Seine in 1634 for safety reasons, it was converted into a shelter for the poor after a royal edict in 1656. Although the shelter had no obligation to care for the ill, it attracted the disabled and insane, necessitating it to become an asylum. At its peak in the early nineteenth century, Salpêtrière was the largest asylum in Europe with over 8000 inmates housed in 45 buildings. It was also effectively a town in its own right, covering 100 acres with streets, squares, gardens and an old church. The Salpêtrière also achieved international recognition in 1793 for being the place where Philippe Pinel unshackled many of its patients from their chains – an event widely seen as ushering in a new humane form of 'moral treatment' for the mentally ill. This practice was soon adopted elsewhere including Germany, where Johann Christian Reil in 1808 coined the term 'psychiatry'.

By the mid nineteenth century, however, the Salpêtrière had fallen into ruins, with its main asylum of 4500 female patients in a chaotic and badly administered state. Nonetheless, Salpêtrière seems to have left a deep impression on the young and aspiring Charcot, who saw in it a unique opportunity to examine a large population of patients, most of whom were committed for life, suffering from a 'pandemonium of human infirmities'. Sensing there was much to gain, Charcot would return there in 1862 as chief of its medical services, accompanied by friend and colleague Alfred Vulpian.[13] They immediately set to work examining the patients' disabilities and organising them into wards, earning them the epithet 'the Castor and Pollox of experimental physiology and pathology' from the *British Medical Journal* in 1879. Unlike Vulpian who only stayed five years, Charcot remained at the Salpêtrière for the rest of his career, taking advantage of the great abundance of clinical material, to establish himself as the world's greatest expert on diseases of the nervous system. A small stout man with a big head and bull neck, Charcot's authoritative command of the hospital would sometimes lead him to be referred to as the Napoleon of Salpêtrière.

The anatomo-clinical method

When Charcot returned to Salpêtrière, it housed two types of female patient. The first were the insane or mentally deranged (the *aliénées*) who made up about one third of the asylum's

FIGURE 10.2 Jean Marie Charcot (1825–1893). Salpêtrière's 'Napoleon' who took advantage of its 'pandemonium of human infirmities' to discover new neurological conditions and understand better those which already existed.

Source: Wellcome Library, London.

population. The remainder were those who lived in the hospital because their illnesses, which included 'paralyses, spasms and convulsions', precluded them from working. It was this second group that Charcot was most interested in. Despite being housed in ramshackled dormitories of around 75 to 125 women, Charcot realised his patients gave him the possibility of establishing a large database of new neurological diseases. In effect, he was 'in possession of a kind of museum of living pathology whose holdings were virtually inexhaustible'. But first, Charcot and Vulpian had to bring order to their hospital. They did this by undertaking detailed clinical examinations of their patients, which they recorded using notes and other supporting documentation. This was updated until the patient's death. In addition, Charcot and Vulpian set up a pathology laboratory in an old disused kitchen where they could perform autopsies. At first this included little more than a microscope and several neural stains used to analyse the diseased sections of brain tissue and spinal cord. With time, it would become a splendid pathological museum. From the very beginning of their leadership, Charcot and Vulpian's aim was clear: they wanted to correlate the various behavioural symptoms they were observing in their patients, with the physiological abnormalities at death. It became known as anatomo-clinical method, and was to prove a highly effective means by which to classify patients of the Salpêtrière into more relevant diagnostic categories on the basis of their nervous system pathology.

The anatomo-clinical method was not new. It had been pioneered by Morgagni and used by many others. But in the hands of Charcot it provided a very effective means of classifying different categories of illness. Although helped by the large number of 'neurological' patients at his disposal, its success was more due to Charcot's reliance on clinical observation, which depended on collecting information from a number of different sources. Interestingly, unlike other house doctors of the time, Charcot rarely attended patients in the wards. Instead, they were bought to his consulting room where the patient was undressed and their medical history read out. It is said Charcot would often drum his fingers on his desk before examining the patient himself, or requesting they perform some action. Charcot also made meticulous notes from these meetings, sometimes sketching or photographing what he observed, or using data from gait and tremor recordings to build up a dossier about the condition. For Charcot, careful clinical observation was crucial because he was aware different diseases often shared similar symptoms. Nonetheless, perhaps because of his artistic inclination, Charcot appears to have been blessed with an astute ability for observing the distinguishing characteristics of a disease. This would be instrumental in him identifying previously unknown conditions, and greatly add to the understanding of those already known.

The discovery of multiple sclerosis

One of Charcot's most important contributions to the classification of neurological disease was his description of *la sclérose en plaques* (multiple sclerosis) in 1868. Charcot was not the first to observe this illness. Indeed, the earliest reference to this disease may well go back to the fourteenth century when Ludwina of Sciedam of the Netherlands, who would be canonised as a saint in 1890, began suffering from increasing muscle weakness following an ice-skating accident when aged just 15. First manifesting itself by walking difficulties, her illness deteriorated over the next 37 years causing a loss of balance, visual deficits and paralysis. Viewing her periods of remission as a divine message from God, Ludwina accepted her illness as a mission to treat others and acquired fame as a healer and holy woman. Other accounts of this illness include Augustus d'Este, an illegitimate grandson of King George III of England, who recorded his symptoms in a diary over a 26-year period. In his case, the illness began with visual disturbances, progressing to severe recurrent episodes of motor disability and weakness. While the disease was familiar to some nineteenth century physicians, it was poorly understood and frequently confused with many other conditions.

Charcot first became interested in this disorder during the mid 1860s when he encountered three patients at Salpêtrière who had suffered from motor and sensory impairments of the lower extremities for many years. Around the same time, he also came across a cleaning lady with an awkward gait and peculiar tremor which arose during purposeful movement. Charcot was sufficiently curious to employ the women as his personal house-maid, despite her tendency to break his expensive porcelain! Charcot initially diagnosed the woman with *paralysis agitans* (i.e. Parkinson's disease). However, when the woman died in 1866, Charcot found the autopsy did not support his initial diagnosis. In fact, there were hardened and discoloured plaques scattered throughout the white matter of her brain and spinal cord. Charcot was not the first to observe this type of pathology. Similar plaques had been found in patients with movement difficulties by a young Scottish physician called Robert Carswell in 1838, and the Parisian professor Jean Cruveilhier around the same time. Nonetheless, Charcot realised he was not dealing with *paralysis agitans* and set about finding more patients with plaques to better delineate the clinical symptoms associated with this type of illness.

Over the next few years, Charcot was assisted in his attempts to understand *la sclérose en plaques* by a number of his students – one of whom was a young German undertaking his

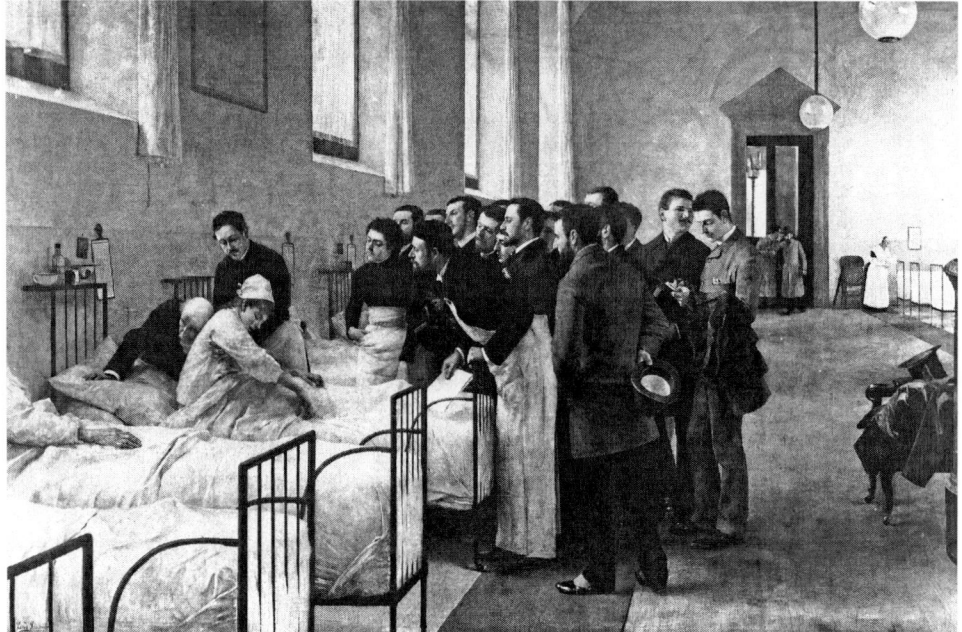

FIGURE 10.3 Painting by Luis Jimenez y Aranda entitled 'Charcot at the Salpêtrière' (1889) exhibited in the Provincial Museum of Art in Seville.

Source: Wellcome Library, London.

doctorate called Leopold Ordenstein. He was to make a vital discovery when realising paralytic patients could exhibit two different types of tremor: those occurring when the patient was at rest (resting tremor), and those arising when the person attempted to move (voluntary tremor). More importantly, the autopsy findings revealed patients with intentional tremor always had sclerotic plaques whereas those with resting tremor did not. Thus, Ordenstein had found a definitive way of distinguishing between *la sclérose en plaques* and *paralysis agitans*. This was a landmark discovery as both diseases were easily confused with each other. Although it was Ordenstein's discovery (he would publish the findings in his doctoral thesis of 1868), Charcot received most of the credit, for he had presented three cases of *la sclérose en plaques* to colleagues in 1865. The disease would become more widely known through his lectures, published in English in 1877. By then, Charcot was able to show the disorder was due to degeneration of myelin (the fatty white substance surrounding nerve fibres) within the spinal cord and brain. His clinical and pathological description was so thorough, the German Julius Althaus called for *la sclérose en plaques* to be called Charcot's disease in 1877. However, the illness would be more widely known as disseminated sclerosis in England, at least until the 1950s when the term 'multiple sclerosis' became more common.

Other neurological conditions discovered at the Salpêtrière

The reliable diagnosis of *la sclérose en plaques* also enabled Charcot to formulate a better clinical description of *paralysis agitans*. Although Parkinson had described resting tremor as a cardinal sign of his disorder, Charcot observed this symptom fluctuated greatly and sometimes disappeared completely. Thus, he came to believe the term *paralysis agitans* was inappropriate as patients with the disorder did not always exhibit shaking. Perhaps because of this, Charcot began referring to the condition as *La maladie de Parkinson* in 1869. Charcot also noticed the tremor was typically accompanied by three other major symptoms: slowness of movement, balance impairment and muscle rigidity. While the latter had seemingly been missed by Parkinson, Charcot realised that muscular rigidity contributed greatly to the patient's stooped posture and immobilisation. Other features identified by Charcot included the thumb moving over the fingers during tremor, a blank facial expression, and small handwriting. All of these symptoms were easy to observe, which allowed Charcot to confidently present patients with Parkinson's disease in his lectures, thereby helping to bring the condition to wider medical attention.

However, when it came to undertaking an autopsy, Charcot could find no sign of any damage in the nervous system of those with Parkinson's disease. Despite this, he was confident there was a lesion in the brain that one day would be discovered. Charcot also provided a more effective form of treatment than Parkinson who had advocated blood-letting. Instead, Charcot used a number of drug treatments including hyoscine (scopolamine) derived from the deadly nightshade plant. This has an effective depressant effect on the nervous system, and is still used today in treating the early stages of the disease. Later, in 1892, after observing long car or train journeys were often beneficial for individuals with Parkinson's disease, Charcot invented an apparatus that hoisted the patient into a suspended harness that mimicked the rocking motion of a train.

The 'divide and classify' methods used by Charcot to correlate clinical symptoms with pathological anatomy also led to the discovery of motor neuron disease, otherwise known as amyotrophic lateral sclerosis. Charcot first became aware of this disorder in the 1860s when he examined the symptoms of two groups of patients: those with spasticity (i.e. characterised by involuntary muscle spasms), and those with atrophied or wasting muscles. After performing a number of autopsies, Charcot realised these two groups had different types of spinal cord damage. The patients with spasticity showed degeneration of the lateral columns of the spinal cord,[14] whereas those with muscular dystrophy had damage to the anterior horns of the spinal cord.[15] From these observations, Charcot suspected the cause of the muscular wasting in the later group lay with the lack of neural stimulation that occurred after anterior horn degeneration. However, as Charcot undertook more autopsies he found a small number of patients who exhibited both spasticity and dystrophy, and they showed damage to both the lateral columns and anterior horns. Charcot realised he had identified a new disorder, characterised by a rapidly progressive wasting disease beginning in mid-life and usually fatal within a few years. We now know a number of famous people have suffered from motor neuron disease, including baseball player Lou Gehrig, composer Shostakovich and actor David Niven. The Nobel-prize winning physicist Stephen Hawking is another victim, although his illness is a rare form with an unusually slow progression.

One of the most fascinating disorders to be discovered at the Salpêtrière was Tourette's syndrome, named after Georges Gilles de la Tourette,[16] the son of a merchant, who travelled to Paris to begin his internship under Charcot in 1881. A doctor with a penchant for inventing new treatments (he was to design a vibrating helmet apparently successful in the treatment of facial neuralgia and vertigo), Gilles de la Tourette was first put to work on a project examining hypnotism in the management of hysteria. He soon established himself as one of Charcot's most trusted assistants, and in 1884 was given a special task after Charcot came

FIGURE 10.4 Georges Gilles de la Tourette (1857–1904) who discovered the disorder that bears his name in 1885.

Source: Wellcome Library, London.

across a paper by the American George Beard, which described the behaviour of men with French ancestry who suddenly barked and jumped in response to a sudden noise. Reasoning such men were also likely to exist in France, Charcot asked Gilles de la Tourette to find new cases. Within a year, Tourette had discovered nine subjects that exhibited strange tics and abrupt vocal utterances, often accompanied by facial grimaces or urges to imitate other people's actions. Tourette also realised the disorder explained the behaviour of a famous French noblewoman, the Marquise de Dampierre, who became a recluse after developing an impulsive tendency to shout out obscenities. Charcot was so impressed with this new condition he called it *La maladie des tics de Gilles de la Tourette*. Unfortunately, Tourette did not achieve the fame he perhaps deserved. Two years later, a young women and ex Salpêtrière patient, believing she was under his hypnotic influence, shot him three times as he sat in his office. Although Gilles de la Tourette recovered from his injuries, he became increasingly capricious and was fired from his hospital post in 1901. Diagnosed with syphilis soon after, he was admitted to a psychiatric hospital in Switzerland where he died in 1904.

Charcot's investigations into hysteria

In the later part of his career, Charcot turned his attention to the subject of hysteria – a term then widely used to describe extreme emotional behaviour in women. Originally linked with movements of a wandering womb by Hippocrates, it was a belief that remained remarkably persistent up until the nineteenth century, despite the protestations of some who recognised hysteria could also occur in males. Although hysteria remains a difficult disorder to define, it is generally recognised as a condition where the patient suffers from physical abnormalities arising from a psychological or mental cause. However, perhaps the most puzzling aspect of hysteria, especially for nineteenth century physicians, was that it took so many bizarre forms. Its symptoms, for example, frequently included excessive laughing or crying, wild bodily movements, paralysis, numbness, or temporary deafness and blindness. In addition, many patients presenting with hysteria show a heightened sensitivity to touch, fainting and a predilection for drama and deception – all of which were highly intermittent and exacerbated by stress. As a young doctor, Charcot would have been aware of hysteria, and his professional interest in the illness arose when put in charge of a ward of women suffering from convulsive disorders. Although most were genuine epileptics, some were recognised as hysterics who had learned to imitate their epileptic attacks. The cause of this behaviour was a mystery. While some doctors felt such symptoms had an organic basis in the brain, others felt the hysteric was simply malingering or acting. Indeed, many hysterics exhibited symptoms with no conceivable neurological explanation. One example was 'glove anaesthesia' where patients reported no feeling in their hand, yet felt sensation in the wrist and arm. Because of this, hysterics at Salpêtrière tended to be treated by the physicians who specialised in mental illness away from the wards governed by Charcot.

In 1870, a bureaucratic decision was made to restructure the Salpêtrière that resulted in hysteric patients being housed in Charcot's wards. It encouraged Charcot to confront the challenge of hysteria 'head-on' with methods he had used to understand neurological illness. In fact, within a few years, Charcot was seemingly obsessed with understanding this puzzling condition. He was also interesting others in hysteria too, partly because the disorder readily lent itself to theatrical demonstrations that he exploited in his Tuesday lectures. These 'shows' had been initiated in the early 1870s, and proved so successful a purposefully built amphitheatre seating 600 students was constructed to present them. It was also the first theatre in Europe to project photographic slides onto a large screen. Both educational and entertaining, the lectures typically used subjects from outpatient clinics whom Charcot had not met before. With the aim of diagnosing the disorder, Charcot would interview the patient and then perform some type of test to demonstrate the symptoms. For example, in one instance, Charcot had his patient stand behind a lighted screen, wearing flumes of feathers, in order to illustrate different types of tremor. Largely improvised, and not without humour, Charcot sometimes used mimicry to point out the characteristics of different illnesses, or he used his artistic skills to sketch the patients' behaviour on a blackboard. The Tuesday lectures also provided a forum for Charcot to speculate on new ideas, or introduce rare diseases that were poorly characterised.[17] These were to attract many eminent researchers to Paris, including Alfred Binet, William James and Sigmund Freud, along with many curious members of the public who travelled from far and wide to view them.

Although the lectures presented a vast array of different neurological disorders, the ones on hysteria always attracted most attention. Part of this was due to the fact Charcot had discovered hysterical patients were particularly susceptible to hypnosis – a discovery not lost on Sigmund Freud who would later use this technique as a means of uncovering unconscious desires, thereby founding the discipline of psychoanalysis. The incorporation of hypnosis into the talks also contributed to the visual spectacle, especially when Charcot found many hysteric

FIGURE 10.5 Painting entitled *Une Lecon Clinique a la Salpetrie* by Andre Brouillet (1887), which depicts Charcot explaining how to diagnose hysteria while two nuns wait to catch the patient (Blanche Wittman) as she collapses.

Source: Wellcome Library, London.

patients had areas on their body which could be pressed to trigger, or alleviate, the hysteric symptoms. One such zone lay over the lower abdomen at the point where the ovaries were situated, and Charcot frequently used it as a means to demonstrate the sudden lability of hysterical reactions. Some of his patients also became star performers in their own right, the most famous being a woman called Blanche Marie Wittman who was first diagnosed with hysteria in 1878. Known as the 'Queen of the hysterics', Wittman's astonishing emotional, cataleptic, and somnambulistic feats were reported in both medical journals and the wider press, even leading to rumours she was Charcot's mistress.

By the late 1870s, Charcot had come to regard hysteria as an inherited condition, which developed in individuals with a 'susceptible' neurological system, and whose symptoms were triggered by shock or highly traumatic personal event. It was therefore similar to a physical disease and warranted a search for its biological cause. However, most doctors dismissed these ideas. Furthermore, the attention given to hysteric patients was to have a number of unforeseen consequences. This was most readily seen when hysteria became the most common complaint at Salpêtrière, rising from an incidence of 1 per cent in the 1840s to over 20 per cent by the early 1880s. Even more disturbing, hysteria became increasingly prevalent throughout France, where it became known as 'the illness of the age'. For many, it was implausible that the rise of hysteria reflected a neurological disorder, and a more convincing explanation was given by Charcot's pupil Joseph Babinski who pointed out that hysterics were more likely to be highly suggestible. In other words, if the patient had come to the

Salpêtrière with vague complaints of distress and feeling unwell, the doctors there, including Charcot himself, may well have persuaded them they were hysterical. This was an idea shown to be entirely feasible by the work of Freud, who confirmed verbal suggestion was a powerful means of affecting the mental health and physical behaviour of certain patients.

Charcot's legacy

Appointed chief physician at the Salpêtrière in 1862, within 20 years or so, Charcot had transformed it from a 'grand asylum of human misery' to one of the most famous and internationally acclaimed neurological institutes in the world. Organised into specialised wards with excellent research laboratories and teaching facilities, the Salpêtrière attracted students from all over the world. Charcot had also risen from a highly respected doctor to be an internationally acclaimed expert on diseases of the nervous system. All of this was not overlooked by the French government who awarded him 200,000 francs in 1881 to create the world's first Chair devoted to neurological disease at the University of Paris. It was the moment when clinical neurology became an established area of medical specialisation. Charcot's influence on the new discipline was manifold in many ways. In particular, his anatomo-clinical method had shown how it was possible to discover new neurological diseases, that hitherto were confused or even unsuspected. In fact, prior to Charcot, relatively few illnesses were recognised as having a neurological basis – the few exceptions being apoplexy epilepsy, neurosyphilis, paraplegia, congenital idiocy and 'brain fever'. After Charcot, physicians were better able to separate brain and spinal cord dysfunction; nerve and muscle disease; and psychiatric abnormalities. Charcot had almost single handedly laid the foundations for modern neurological classification.

Charcot also achieved fame for his teaching. The Tuesday lectures have already been mentioned, but he would later supplement these with a more formal presentation on a Friday. This lecture was prepared meticulously in advance, with the notes given to an assistant who edited them for publication. These notes would eventually form a large corpus of work called the *Oeuvres Complètes*, which were published in nine volumes between 1877 and 1890. Some of these notes were also translated into English as *Lectures on the Localisation of Cerebral and Spinal Diseases* (1883) and *Lectures on the Diseases of the Nervous System* (1877–1889). This forms a large body of neurological thought that helped to establish the foundation of neurology. Charcot's influence was perhaps best summed up by Joseph Babinski who remarked: 'To take away from neurology all the discoveries made by Charcot would be to render it unrecognisable.'

History has treated Charcot less kindly regarding hysteria. The exotic allure of the disorder and the theatricality of his clinical demonstrations was to make Charcot many enemies – a situation not helped by his alleged arrogance and dislike of being contradicted. Criticism of Charcot also came from writers and social commentators such as Tolstoy and Maupassant who lampooned the idea of hysteria as a physical brain disease. Their views were justified as shown by the fate of Blanche Marie Wittman. Following Charcot's sudden and unexpected death in 1893, at the age of 67 after spending a day walking in the countryside of Burgundy, her hysterical bouts of convulsions, paralysis and delirium abruptly came to an end. The same was true for many other patients under Charcot's care. The reason for Wittman's transformation was simple: Salpêtrière no longer had any use for a Queen of Hysterics. Despite this, Wittman did not immediately leave the hospital where she had lived for 16 years. Instead she was employed in the photography laboratory, and promoted to technician in the radiology department. Her expertise in this field would lead her to work for Marie Curie who discovered radioactivity and the only scientist to have won two Nobel prizes. The story of their relationship was also dramatised in a novel by Per Olov Enquist published in 2007. This describes Wittman as a tragic figure who loses both her legs and an arm through the effects of radiation damage, dying as a limbless but beautiful torso.[18] The real truth is less sensational: she lost her fingers, hand, and parts of her forearm. Nonetheless, Wittman died in 1913 from cancer believed to have been bought on by radiation exposure.

A new way of classifying mental illness: Emil Kraepelin

Charcot had achieved much of his success through his innovative way of classifying neurological symptoms, and others would adopt a similar strategy with psychiatric disorders. The most influential person to do this was Emil Kraepelin, widely recognised today as the founder of modern psychiatry. Born in the North German town of Neustrelitz, Kraepelin according to one biography was the son of an alcoholic actor, who left home leaving his opera singing wife to bring up two sons and a daughter. Despite these early setbacks, Kraepelin developed an interest in biology from his elder brother Karl[19] which inspired him to study medicine. Beginning his studies in 1874 at the University of Wurzburg, Kraepelin turned to psychiatry after attending a course run by William Wundt, creator of the world's first psychological laboratory in Leipzig. Four years later, after writing his medical thesis, Kraepelin moved to Munich to work with anatomist Bernhard von Gudden.[20] Unfortunately, these were difficult times for Kraepelin, who was put in charge of a men's psychiatric ward with 150 dangerous and violent patients. His research did not provide any compensation either since an area of blindness in Kraepelin's left eye impaired his use of a microscope making him unsuitable for neuroanatomical work. After four unhappy years, Kraepelin took up a post in Leipzig working in a medical clinic under Paul Flechsig. However, the two men clashed and Kraepelin was fired after just a few months.[21] Now more than ever determined to pursue psychology, despite having to use his own savings, Kraepelin joined Wundt's laboratory where, among other things, he examined the effects of various drugs on the brain.

Kraepelin now found himself in a position where he had no money, satisfactory employment, and little chance of further training. It was also a situation exacerbated by a recent engagement to his childhood sweetheart. Brooding on the hopelessness of his situation, Kraepelin accepted an offer to write a short introduction on psychiatric disorders. He appears not to have been enthralled by the task, nor was the book extensive, for Kraepelin apparently wrote it during the Easter vacation of 1883. Kraepelin could not have known it, but his small *Compendium der psychiatrie* would evolve into the *Textbook of Psychiatry*, which now provides the foundation for all the major classification systems in psychiatric use today.[22] It also ensures that Kraepelin's medically oriented views continue to dominate modern-day psychiatry. However, in its first incarnation, the *Compendium* was little more than a reinstatement of contemporary German thinking about mental disorders. Nonetheless, it made Kraepelin much more aware of the deficiencies existing in psychiatric knowledge at the time, and he was particularly appalled at the lack of agreement within psychiatric diagnosis. Although a number of classification systems existed in the mid-nineteenth century, they were often based on subjective criteria, which led to some highly questionable conditions such as 'masturbatory

FIGURE 10.6 Emil Kraepelin (1856–1926) who developed a classification of psychiatric disorders, which forms the basis of many used today. He was also the first investigator to recognise what we today know as schizophrenia.

Source: Wellcome Library, London.

insanity' or 'wedding-night psychosis'. Kraepelin could do little to change this situation except to emphasise the need for psychiatry to be more objective. However, Kraepelin's interest in psychiatric classification had been roused – a situation that would tempt him back into clinical work from experimental psychology.

In 1886, after working as head of an asylum in Dresden, Kraepelin was unexpectedly offered a professorial position at the University of Dorpat in Estonia. He was just 30 years old, and despite being unable to speak Estonian or Russian, his dream of undertaking psychiatric research could now be fulfilled. Placed in charge of 80 psychiatric beds, Kraepelin set up a laboratory for psychological research. He also began writing the second and third editions of his textbook, encouraging him to more seriously consider alternative approaches to psychiatric classification. At the time, mental illnesses were generally diagnosed by a predominant symptom such as melancholia or insanity, and attributed to a psychological trauma such as a loss of a friend, or weak nerves. However, Kraepelin was beginning to realise that any one symptom by itself did not constitute a defining feature, since most mental disorders shared a number of similar abnormalities. For him, a new approach was clearly needed, and Kraepelin set about experimenting with alternative ideas. One of his first innovations was to compare more closely the course and progression of different unrelated disorders, and to group them into 'curable' or 'incurable'.

After serving five years of what he referred to as 'exile' in Estonia, Kraepelin accepted a professorship at the University of Heidelberg in 1891. This university was much more at the heart of German academic life, and Kraepelin took over a large psychiatric clinic with hundreds of patients supported by a well-equipped laboratory for pathological investigations. Still determined to follow the progression of different illnesses, Kraepelin began compiling specially designed index cards (called *Zählkarte*), which recorded the 'essential characteristics of the clinical picture' on patients entering his clinic. These patients were followed up a year or two later, even if they had left the hospital, to assess whether the initial diagnosis had correctly predicted their outcome. Within a few years Kraepelin had compiled a large data bank from over a thousand patients. While they primarily provided a means by which Kraepelin could judge the success of his clinical methods, the cards were also an invaluable resource allowing him to group and regroup the symptoms found in his patients. In fact, this was apparently a favourite pastime of his when on vacation! Kraepelin was not only attempting to plot the course of his patient's illnesses, but also trying to establish new categories of mental disorder by classifying patterns of behaviour into syndromes. That is, establishing whether certain constellations of symptoms frequently occurred together.

The discovery of schizophrenia

Today, schizophrenia is regarded as the archetypal mental illness. Indeed, if one is asked to imagine 'madness' then it is likely something resembling schizophrenia will come to mind. That is, a psychotic mental disorder (i.e. an impairment in 'reality testing') characterised by strange, disorganised and delusional thinking, accompanied by hallucinations and a reduced ability to feel normal emotion. Although these symptoms were known to nineteenth century psychiatrists, they were considered to be part of other disorders. However, as Kraepelin attempted to make sense of the information collected from his diagnostic cards, he began to see they were related. The first step in this discovery occurred in the early 1890s when Kraepelin noted a large proportion of his patients, whatever their earlier clinical presentation, ended up in a state of presenile dementia. Closer inspection of these patients revealed they suffered predominantly from one of three disorders: These were: *dementia praecox* (where deterioration of mental function occurred in teenagers); *catatonia* (a state characterised by stupor and bouts of excitement); and *dementia paranoides* (identified by paranoid delusions). Kraepelin introduced this idea in the fourth edition of his *Textbook* (1893), and proposed all led to degeneration of the brain. Kraepelin made an even bolder move in the fifth edition of his *Textbook*, published three years later, when he combined these three disorders into one single entity – dementia praecox (meaning early dementia). This reflected his belief that they all arose in young people and caused irreversible mental deterioration. He also divided dementia praecox into four subtypes: (1) simple, marked by apathy and slow social decline; (2) paranoid, characterised by delusions; (3) catatonic, as shown by poverty of movement; and (4) hebephrenic, characterised by 'silly' or 'child-like' thinking. Although he did not fully realise it, Kraepelin had invented what turned out to be schizophrenia.

Kraepelin made another far reaching change in the sixth edition of his *Textbook* (1899) when he divided all psychiatric illnesses into thirteen groups. Although most of his categories were already well-established, including entities such as neurosis and mental retardation, one grouping was new. This was a common cluster of mental disorders known as the *psychoses* – essentially a large collection of illnesses where no organic cause could be found. Kraepelin now split these into manic depression and dementia praecox. This was a significant development for manic depression was an amalgamation of two illnesses that had been regarded as separate for over 2000 years – namely *melancholia* and *mania*. Thus, Kraepelin was the first to realise they were related in some fundamental way. Despite this, he thought manic depression was an intermittent and non-deteriorating illness, in contrast to dementia praecox which caused irreversible degeneration of the mental faculties. Importantly, this made the diagnosis and treatment of mental illness much more straightforward: if patients were sad or euphoric with marked changes in emotion, they were manic-depressive and likely to get better; if they lacked emotion and exhibited thought disturbances, they had dementia praecox and were unlikely to improve. Today, no psychiatrist would accept these prognostic outcomes as both types of illness, in most cases, can be successfully treated. Nonetheless, the division between manic depression and dementia praecox is one still maintained in the classification of psychiatric disorders.

Kraepelin's reformulation of dementia praecox significantly improved the classification of mental illness, but the term was rejected by the Swiss-born psychiatrist Eugen Bleuler in 1911. A doctor strongly influenced by the theories of Freud, Bleuler worked with psychotic patients, often interacting with them on a daily basis in his Zurich clinic. This intimacy led Bleuler to realise the irreversible process of dementia did not always occur as Kraepelin had insisted. Rather, Bleuler observed a marked mental improvement in many of his patients – a finding that also contradicted the idea of the illness being caused by an organic brain dysfunction. Instead, Bleuler came to recognise the defining feature of dementia praecox as disorganised thinking resulting from the splitting of the emotional and intellectual functions of the personality. He called this phenomenon schizophrenia from the Greek *schizo* ('split') and *phrene* ('mind'). Bleuler also defined the main symptoms of schizophrenia in terms of four 'A's: (1) a blunted 'affect' resulting in diminished emotional responding; (2) a loosening of 'associations' causing disorganised thought; (3) 'ambivalence' or an inability to make decisions; and (4) 'autism' referring to a preoccupation with one's own self or thoughts. Bleuler further pointed out the symptoms of schizophrenia could be *positive*, characterised by overactive thinking or behaviour, or *negative* with impoverished thought processes and action. A hundred years later, our modern-day conception of schizophrenia is still one closely tied to the formulation provided by Bleuler.

A new type of dementia: Alois Alzheimer

Kraepelin also played a key role in naming and establishing a form of dementia that is now recognised as one of the greatest public health threats facing our modern world. This is Alzheimer's disease, a progressive, incurable and degenerative illness that causes severe loss of memory, mood changes and decline of bodily functions – named after the German Alois Alzheimer who first described its pathological features in 1905. Alzheimer was born in the Bavarian town of Markbreit, and studied medicine at Berlin and Würzburg, before submitting his dissertation on the glands producing ear wax in 1888. A man of striking appearance with a scar running from his left eye to his chin from a sabre duel incurred as a young man (this is never shown in his pictures), Alzheimer was also a heavy cigar smoker who apparently always left a trail of cigar stubs on his rounds. After qualifying as a doctor, Alzheimer moved to Frankfurt where he worked in the city's municipal mental asylum. Here, he examined the neuropathology of syphilis,[23] which was then a major cause of mental dysfunction (known as general paralysis of the insane). Indeed, this illness accounted for half of all admissions to psychiatric institutions and at least 10 per cent of all dementia, up until the end of the nineteenth century. It had also struck down a number of famous individuals including Beethoven, Nietzsche, Maupassant, and Randolph Churchill. After examining his patients during the day, Alzheimer spent his evenings in the laboratory. Here he became great friends with the renowned histologist Franz Nissl. By all accounts they were kindred spirits spending much time hunched over their respective microscopes, examining slides of tissue taken from autopsies. It was a productive time for Alzheimer who demonstrated that general paresis is associated with a softening atrophy of cortical tissue due to vascular damage. In turn this gradually strangled the blood supply to the brain. Indeed, this was widely accepted as the prevailing cause of senile dementia until the late 1960s.

In 1901, a 52-year-old woman called Auguste Deter was brought into Alzheimer's clinic by her husband. No longer able to look after herself, she had become bewildered, restless and paranoid. The initial clinical examination showed the woman to have profound memory problems, unable to remember her surname, and giving confused answers to simple questions. Her condition also rapidly declined over the next five years – so much so, that Deter lay in bed dazed, incontinent, and in a state of 'total feeble-mindedness'. She died in April 1906, although by now Alzheimer had moved to a laboratory in Munich where Kraepelin was director. However, Alzheimer was sufficiently interested in the case for Deter's brain and medical dossier be sent to him. He also decided to examine her nervous tissue with a relatively untried silver impregnation technique that strained nerve fibres, developed a few years previously by the Pole Max Bielschowsky. It was a fortuitous choice for when Alzheimer examined the brain under his microscope, he observed marked atrophy of the cerebral cortex. More importantly, Alzheimer noticed large numbers of tangled nerve fibres, accompanied by a widespread deposition 'of a peculiar matter' that resembled tiny starch particles.[24] Despite his wealth of experience, these pathological signs were unfamiliar to Alzheimer, raising the possibility that a new degenerative disorder had been discovered. Alzheimer presented his findings in November 1906 to a medical conference in Tübingen, and published them a year later in a three-paged article.

Following Alzheimer's report, a number of similar cases were soon published in the medical literature. Some of these came from a colleague in Munich, the Italian Gaetano Perusini, who searched the institution's records to find patients who had shown a similar mental decline to Deter. He found three (aged 45, 63 and 65 years) whose brains had also been kept for neuropathological investigation. Importantly, all showed the tangled nerve fibres and starch-like plaques spread through the cerebral cortex as described by Alzheimer. Interestingly, none of these patients appeared to exhibit vascular damage that was more typical of 'normal' or senile dementia. Another relevant study was published in 1907 by Oskar Fischer at the University of Prague, who identified plaque material in 12 out of 16 individuals – although in his sample, all patients had suffered from a much slowed mental decline and memory loss. This led Fischer to believe the plaques were a symptomatic feature of senile dementia in general.

Kraepelin was suitably impressed by this new condition, believing it had been delineated both clinically and pathologically, to include it in his eighth edition of his *Textbook of Psychiatry* (1910). He also named it Alzheimer's disease. Curiously, Kraepelin emphasised it as a pre-

 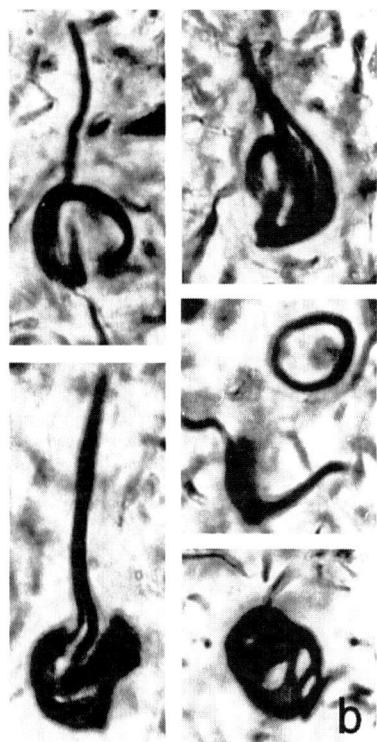

FIGURE 10.7 (A) Alois Alzheimer (1864–1915). German psychiatrist who discovered neurofibrillary tangles and plaques in a severely demented patient in 1905, later named Alzheimer's disease by Emil Kraepelin in 1910. The original of this portrait is kept in the historical library of the Max Planck Institute of Neurobiology. (B) Drawings from Alzheimer's laboratory of various forms of neurofibrillary tangles seen in the nerve cells of those with dementia.

Source: Both images supplied by Professor Manuel Graeber, University of Sydney. Permission obtained for reproduction of the portrait by the Max Planck Institute.

senile form of dementia whose symptoms began around the age of 50 years. This was despite the fact that two of the patients described by Perusini were in their 60s, and Fischer had shown cortical tangles and plaques in those with much milder cognitive decline. Whether it was appropriate for Kraepelin to define Alzheimer's disease as a pre-senile dementia (i.e. distinct from senile dementia beginning at around 65 years) has been questioned. In fact, it appears Kraepelin only knew of three pre-senile cases fitting the clinical profile provided by Alzheimer. It has also led to the accusation that Kraepelin made the decision for less than genuine reasons – perhaps to enhance the prestige of his laboratory. Whatever the truth of the matter, Kraepelin's characterisation of Alzheimer's disease led it to be regarded as a rapidly progressive type of dementia that began in middle age. Unfortunately, Kraepelin's mistake would result in the illness being largely overlooked as a health concern in favour of vascular dementia until the early 1970s. Only then would it be finally recognised as the most common form of dementing illness in the Western world. Today, it is known to afflict some 5 per cent of people over the age of 65 years and 20 per cent over 80, making it the fourth leading cause of death behind cancer, heart disease and stroke.

Notes

1 Valsalva is famous for his anatomical work on the ear. For example, he coined the term Eustachian tubes (a tube linking the middle ear with the nasal passage of the nose) and correctly established its function (e.g. to maintain the correct pressure in the ear).
2 *De sedibus, et causis morborum per anatomen indagatis.*
3 This was a monumental work, effectively providing a data base of 3000 autopsies with clinical reports, including among others the work of Johann Wepfer and Thomas Willis.
4 The word hysteria is derived from the Greek *hystera* meaning uterus.
5 *Cerebri anatome* (1667), *Pathologiae cerebri* (1667) and *De anine brutorum* (1672).
6 By the end of the nineteenth century in the hands of Freud, a *neurosis* would become a purely psychological disorder with its cause arising in the unconscious.
7 It is interesting to note that Parkinson was not a fully qualified physician.
8 The most famous of these was entitled *Revolutions Without Bloodshed; or Reformation Preferable to Revolt* and known to have sold about 2000 copies.
9 Parkinson's anonymity is further enhanced by the fact his grave is no longer identifiable, although a memorial to him exists in the church (St Leonard's in Shoreditch) where he was baptised, married and buried. This was erected in 1955 to mark the 200th anniversary of his birth.
10 In this respect Parkinson was to be eventually proven wrong although the lesion would not be fully identified until the 1960s when degeneration of a small darkly pigmented nucleus called the substantia nigra, which sits in the upper most portion of the brainstem (or midbrain) was found to cause the disease.
11 Even Charcot found it difficult to obtain a copy of Parkinson's essay, and it seems he eventually obtained a copy after a frustrating search from a Dr Windsor who was librarian at the University of Manchester.
12 The Salpêtrière still exists, and was the hospital that Princess Diana was taken to after her fatal car accident in 1997.
13 Vulpian is best known for his work on the adrenal medulla that led to the discovery of adrenaline, and for introducing the use of the microscope in post-mortem pathology.
14 This is the part of the cord that contains descending pathways by which the brain controls voluntary movement.
15 This is the part of the cord from which the nerves controlling the body's skeletal muscles emerge.
16 Tourette once described himself as 'ugly as a louse but very intelligent'.
17 Some of the first examples of Huntington's chorea and Friedreich's ataxia were briefly shown in the Tuesday presentation.
18 *The Story of Blanche and Marie* by Per Olov Enquist (Vintage books).
19 Later to become director of the Natural History Museum in Hamburg.
20 Von Gudden was also personal physician to King Ludwig II of Bavaria who suffered from a persecution complex and was believed to be insane. In June, 1886, both Ludwig and Gudden were to be found drowned, possibly murdered, in Lake Starnberg close to Berg Castle. Their deaths remain a mystery.
21 The reasons for this are shrouded in mystery. Kraepelin reports he had been fired after making 'derogatory remarks about (Flechsig's) official oath', although another source holds that Flechsig accused Kraepelin of neglecting his patients.
22 The final eighth edition in 1923 would comprise four volumes and over 3000 pages.
23 It is believed syphilis was brought back from the Americas by the sailors of Columbus, with the first major outbreak occurring in 1493–1494 during a conflict between Spain and France in Naples. Often manifesting itself with genital sores, leading to ulceration that could eat into the bones and face, the long-term consequences of the disease can lay dormant for many years before producing dementia and insanity.
24 These were later labelled as senile plaques.

Bibliography

Alexander, F.G. and Selesnick, S.T. (1966) *The History of Psychiatry*. Harper & Row: New York.
Berrios, G.E. and Hauser, R. (1988) The early development of Kraepelin's ideas on classification: a conceptual history. *Psychological Medicine*, 18, 813–821.

Berrios, G.E. and Porter, R. (1995) *A History of Clinical Psychiatry: The Origin and History of Psychiatric Disorders*. The Athlone Press: London.

Bick, K.L. (1994) The early history of Alzheimer's disease. In Terry, D., Katzman, R. and Bick, K. (eds) *Alzheimer's Disease*. Raven Press: New York.

Boller, F. and Forbes, M. (1998) History of dementia in history: an overview. *Journal of the Neurological Sciences*, 158, 125–133.

Braceland, F. (1957) Kraepelin, his system and his influence. *American Journal of Psychiatry*, 113, 871–876.

Bynum, W.F. (1964) Rationales for therapy in British psychiatry: 1780–1835. *Medical History*, 18, 317–333.

Compston, A. (2005) Jean-martin Charcot on 'sclérose en plaques (multiple sclerosis). *Advances in Clinical Neuroscience and Rehabilitation*, 5, 28–29.

Decker, H.S. (2004) The psychiatric works of Emil Kraepelin: a many-faceted story of modern medicine. *Journal of the history of the Neurosciences*, 13, 248–276.

DeJong, R. (1937) George Huntington. *Annals of Medical History*, 9, 201–211.

Draaisma, D. (2006) *Disturbances of the Mind*. Cambridge University Press: Cambridge.

Duvosin, R. (1987) History of Parkinsonism. *Pharmacology and Therapeutics*, 32, 1–17.

Engstrom, E.J. (2003) *Clinical Psychiatry in Imperial Germany*. Cornell University Press: London.

Garrison, F.H. (1925) Charcot. *International Clinics*, 4, 244–272.

Gelfand, T. (1999) Charcot's brains. *Brain and Language*, 69, 31–55.

Goetz, C.G. (2000) Amyotrophic lateral sclerosis: early contribution of Jean-Martin Charcot. *Muscle and Nerve*, 23, 336–343.

Goetz, C.G. (2011) The history of Parkinson's disease: early clinical descriptions and neurological therapies. *Cold Spring Harbour Perspectives in Medicine*; 1: a008862.

Goetz, C.G., Bonduelle, M. and Gelfand, T. (1995) *Charcot: Constructing Neurology*. Oxford University Press: New York.

Hakosaio, H. (1991) The Salpetriere hysteric – a Foucauldain view. *Science Studies*, 4, 19–33.

Hare, E. (1991) The history of 'nervous disorders' from 1600 to 1840, and a comparison with modern views. *British Journal of Psychiatry*, 159, 37–45.

Havens, L.L. (1965) Emil Kraepelin. *The Journal of Nervous and Mental Disease*, 141, 16–28.

Hustvedt, A. (2011) *Medical Muses: Hysteria in Nineteenth-Century Paris*. Bloomsbury: London.

Johnstone, R.W. (1959) William Cullen. *Medical History*, 3, 33–46.

Kent, D. (2003) *Snake Pits, Talking Cures, and Magic Bullets: A History of Mental Illness*. Twenty First Century Books: Minneapolis, MN.

Knoff, W.F. (1970) A history of the concept of neurosis with a memoir of William Cullen. *American Journal of Psychiatry*, 127, 120–124.

Kyziridis, T.C. (2005) Notes on the history of schizophrenia. *German Journal of Psychiatry*, 8, 42–48.

Lehmann, H.C., Hartung, H.-P. and Kieseir, B.C. (2007) Leopold Ordenstein: on *paralysis agitans* and multiple sclerosis. *Multiple Sclerosis*, 13, 1195–1199.

McHenry, L.C. (1969) *Garrison's History of Neurology*. Charles C. Thomas: Springfield, IL.

Maurer, K. and Maurer, U. (2003) *Alzheimer: The Life of a Physician and the Career of a Disease*. Columbia University Press: New York.

Maurer, K., Volk, S. and Gerbaldo, H. (1997) Auguste D and Alzheimer's disease. *The Lancet*, 349, 1546–1549.

Morris, A.D. (1989) *James Parkinson: His Life and Times*. Birkhauseer: Berlin.

Müller, U., Fletcher, P.C. and Steinberg, H. (2006) The origin of psychopharmacology: Emil Kraepelin's experiments in Leipzig, Dorpat and Heidelberg (1882–1892). *Psychopharmacology*, 184, 31–138.

Murray, T.J. (2009) The history of multiple sclerosis: the changing frame of the disease over the centuries. *Journal of the Neurological Sciences*, 277, 53–58.

Palha, A.P. and Esteves, M.F. (1997) The origin of dementia praecox. *Schizophrenia Research*, 28, 99–103.

Parkinson, J. (1817) *An Essay on the Shaking Palsy*. Sherwood, Neely and Jones: London.

Parry Jones, W. (1987) 'Caesar of the Salpétrière': J-M Charcot's impact on psychological medicine in the 1880s. *Bulletin of the Royal College of Psychiatrists*, 11, 150–153.

Pearce, J.M.S. (2005) Historical descriptions of multiple Sclerosis. *European Neurology*, 54, 49–53.

Riese, W. (1945) History and principles of classification of nervous diseases. *Bulletin of the History of Medicine*, 18, 465–512.

Schutta, H.S. (2009) Morgagni on apoplexy in De Sedibus: A historical perspective. *Journal of the History of the Neurosciences*, 18, 1–24.

Shepherd, M. (1995) Two faces of Emil Kraepelin. *British Journal of Psychiatry*, 167, 174–183.

Shorter, E. (1992) *From Paralysis to Fatigue: A History of Psychosomatic Illness in the Modern Era*. The Free Press: New York.

Shorter, E. (1997) *A History of Psychiatry: From the Era of the Asylum to the Age of Prozac*. John Wiley: New York.

Stern, G. (1989) Did Parkinsonism occur before 1817? *Journal of Neurology, Neurosurgery and Psychiatry. Special Supplement*. 11–12.

Stone, M.H. (1997) *Healing the Mind: A History of Psychiatry from Antiquity to the Present*. Pimlico: London.

Stott, R. (1987) Health and virtue: or how to keep out of harm's way. Lectures on pathology and therapeutics by William Cullen *c*. 1770. *Medical History*, 31, 123–142.

Tally, C. (2004) The history of the (re)naming of multiple sclerosis. *Journal of the History of the Neurosciences*, 13, 351.

Tan, S.Y. and Shigaki, D. (2007) Jean-Martin Charcot (1825–1893): pathologist who shaped modern neurology. *Singapore Medical Journal*, 48, 383–384.

Tedeschi, C.G. (1974) The pathology of Bonet and Morgagni. *Human Pathology*, 5, 601–603.

Tubbs, R.S., Steck, D.T., Mortazavi, M.M., Shoja, M.M., Loukas, M. and Cohen-Gadol, A.A. (2012) Giovanni Battista Morgagni (1682–1771): his anatomic majesty's contributions to the neurosciences. *Childs Nervous System*, 28, 1099–1102.

Tyler, K.L. and Tyler, H.R. (1986) The secret life of James Parkinson (1755–1824): The writings of Old Hubert. *Neurology*, 36, 222–225.

Weber, M.H. and Engstrom, E.J. (1997) Kraepelin's 'diagnostic cards': the confluence of clinical research and preconceived categories. *History of Psychiatry*, 8, 375–385.

Wender, P.H. (1963) Dementia praecox: the development of a concept. *American Journal of Psychiatry*, 119, 1143–1151.

Yahr, M.D. (1978) A physician for all seasons: James Parkinson 1755–1824. *Archives of Neurology*, 35, 185–189.

Yonace, A.H. (1980) Morgagni's letters. *Journal of the Royal Society of Medicine*, 73, 145–149.

Young, A.W. (1935) Franz Nissl 1860–1918, Alois Alzheimer 1864–1915. *Archives of Neurology and Psychiatry*, 33, 847–852.

11
SOLVING THE MYSTERY OF THE NERVE IMPULSE

All of science is either physics or stamp collecting.

Earnest Rutherford

In 1933, while investigating the nervous system of the squid, I noticed certain transparent tubular structures of about a millimetre in diameter. At first I took them to be blood-vessels . . .

John Zachary Young

Summary

The nerve impulse, also known as the action potential, is crucial to our understanding of how the brain works. Your ability to consciously read these words; to reason and think; to laugh and cry; to engage in voluntary action, and much more, depends on millions upon million of tiny electrical impulses coursing through the axon fibres of your nervous system. Thus, if neuroscientists are ever to understand the brain, and how it produces the miracle of human behaviour, they will have to decipher this incredible cacophony of electrical information. However, as the twentieth century dawned, little was known about nerve impulses. At best, investigators knew the voltage inside the resting nerve cell was negative compared to its outside, and this difference momentarily reversed when an impulse passed down the axon. In fact, Emil Du Bois-Reymond referred to this as a 'wave of relative negativity', and its velocity estimated at around 27 metres per second by Hermann von Helmholtz in 1850. It was a speed implying the involvement of physio-chemical events, although it would be another 50 years before a plausible theory emerged to account for it. This came from Julius Bernstein who explained the resting negativity of the neuron by attributing it to the unequal distribution of charged particles, called ions, existing between the cell's interior and exterior. In turn, he hypothesised, the nerve impulse was created when the cell's membrane temporarily 'broke down' allowing a transient flow of positively charged ions into the axon. It was an ingenious theory, although at the time there was no way of verifying the electrochemical events Bernstein had predicted. Further progress would have to wait until after the First World War, when new technology, including amplifiers and oscilloscopes were bought to bear on the problem. It was to result in what some have called the heroic age of neurophysiology, with vital contributions to our understanding of the action potential from a number of scientists who would become Nobel Prize laureates. The most important breakthrough came when the Oxford biologist John Zachary discovered the giant squid axon in 1936. This was sufficiently large enough for simple wires, or electrodes, to be implanted inside the nerve cell, allowing the fine currents of action potentials and their ionic correlates to be recorded. This endeavour would become most associated with the work of British physiologists Alan Hodgkin and Andrew Huxley, who published a series of precise mathematical equations describing the formation of the action potential in chemical and physical terms in 1952. Their explanation of the nerve impulse has never seriously been challenged since, and is widely recognised as the culmination of one of the greatest research endeavours in science, which began with Galvani's discovery of animal electricity in 1791.

Action potentials and ions: Julius Bernstein

The beginning of electrophysiology can be said to have begun with the invention of the galvanometer by Christian Oersted in the 1820s, which measured the presence of an electrical current by a simple deflection of a magnetic compass needle (see Chapter 5). In the hands of the Italian Carlo Matteucci this apparatus showed biological tissue was capable of producing an electrical current that flowed through nerve fibres. Although many questions remained, it was in effect, confirmation of animal electricity first proposed by Luigi Galvani in 1791. In 1842, the German Emil Du Bois-Reymond noticed something else: a nerve fibre 'at rest' had a positively charged outer surface compared to its interior. In other words, there was a voltage difference between the negative inside and positive outside of an unexcited nerve cell. Today, this is known as a membrane potential.[1] This difference, however, became briefly reversed when a nerve impulse passed down the fibre. So, for example, if an electrode recorded from the surface of the fibre, then one would see the galvanometer's needle swing fleetingly from positive to negative. Du Bois-Reymond believed this 'wave of relative negativity' represented the nerve impulse. Later in 1850, Hermann von Helmholtz, using reaction time methodology with frog nerve and muscle, measured the speed of this impulse at about 27 metres per second. Although fast, it was much slower than the flow of electricity down a wire. This implied that an active biological process was working, which Helmholtz called an 'action potential'. It also bought the question of the wave of relative negativity back into focus: for if its speed was similar to the nerve impulse, this would be strong evidence that the two things were different manifestations of the same phenomenon.

The person to show that the wave of negative relativity followed exactly the same time parameters as the nerve impulse was the German Julius Bernstein, then at the University of Heidelberg. A previous student under Du Bois-Reymond, and assistant to Helmholtz, Bernstein managed this feat by inventing an apparatus in 1868 called the 'differential rheotome' (literally meaning 'flow cutter'). This device was essentially a specialised galvanometer that recorded electrical activity from a nerve fibre, after a variable delay and for a tiny fraction of a millisecond.[2] Since the sampling interval could be made to be extremely brief, Bernstein was able to examine the voltage characteristics of any portion of the negative variation. Importantly, this also allowed him to build up a picture of its time course. His results indicated the negative variation had a rise time of about 0.3 milliseconds and lasted for about 0.9 milliseconds. In addition, it moved at a speed of around 28 metres per second, which was very similar to the velocity of the nerve impulse measured by Helmholtz. In effect, Bernstein had provided what is now generally regarded as the first quantitative depiction of the action potential.

Another problem that concerned Bernstein was trying to understand the difference in voltage between the inside and outside of the nerve cell when it was at rest. By the late nineteenth century many scientists had come to believe the answer to this problem lay with differing concentrations of chemicals with electrical charges known as ions.[3] An ion is an atom that has lost, or gained, an electron. To understand why this gives the atom an electrical charge, it is necessary to realise atoms are always neutral – because its positively charged nucleus is counterbalanced by negatively charged electrons surrounding it. Consequently, a loss of an electron will make the atom positively charged. Indeed, this is a frequent occurrence for certain atoms such as sodium (Na), and potassium (K), which have a single electron in their outer orbit. Thus, they are easily turned into ions (Na+ or K+). Another important biological ion is chloride (Cl). However, its atomic structure means it is likely to gain an electron making it more negative (Cl–). As physiologists began to understand this type of pharmacology, they realised that different concentrations of ions inside and outside the nerve cell could be the key to explaining its electrical characteristics. Unfortunately, this was not easy to examine. Although some cells in the body, such as muscle, were known to have different levels of potassium and sodium ions in their interior and extracellular compartments, resulting in voltage differences, nerve cells were far too tiny for their ionic chemistry to be measured in this direct way.

Despite these difficulties, in 1902, Bernstein attempted to explain how the movement of ions could explain the formation of an action potential. To begin, he postulated that the negative voltage inside the nerve cell was due to a greater amount of negative ions compared to its outside. Consequently, the nerve was in a state of disequilibrium. However, under normal circumstances, the membrane of the nerve cell acted as a barrier to stop ion flow, thereby maintaining this difference. Viewed in this way, the nerve cell was like a battery, storing both positive and negative charges, before its terminals were connected. Bernstein's great insight, however, was to realise the nerve cell's membrane might become permeable to ions. That is, if there was a temporary breakdown in the membrane's resistance, then positive ions would move into the cell, causing its negatively charged interior to become neutral (zero).[4] If this happened, the sudden change in voltage would appear as a wave of relative negativity on its surface. This theory also neatly explained why the nervous impulse was much slower than an electrical current flowing through a cable. If Bernstein was correct, then the action potential moved in small jumps as the membrane temporarily became more permeable to ions along its axon. While some of the finer detail of Bernstein's theory was later shown to be wrong,[5] his basic idea concerning ion flow turned out to be correct. Bernstein's 'membrane theory' was the first modern plausible physico-chemical explanation of the nerve impulse.

The all-or-nothing law

Around the same time as Bernstein was formulating his ionic theory of the action potential, Liverpool and Oxford neurophysiologist Francis Gotch was trying to establish a better depiction of a nerve impulse. Gotch had been one of the first investigators to work with the capillary electrometer, invented in 1870, which measured the electrical response of a nerve fibre to a given stimulus. This apparatus was essentially a thin U-tube of half-filled mercury, with a small amount of diluted sulphuric acid on top. By passing an electrical current from a nerve through this mixture, the mercury could be made to move a short distance up one of the sides of the U-tube. And, by shining a light through the mixture and positioning a small strip of film behind it, small movements could be recorded onto a moving trace of paper. This was a significant improvement over the galvanometer's needle deflection. Although the response was little more than a featureless blip (typically 1 millimetre high and 2 millimetres long), it provided a more accurate visual record of the nervous impulse. By using this new instrument, Gotch confirmed what Bernstein had shown: the action potential only lasted a few thousandths of a second (milliseconds). Moreover, it had a tiny voltage of about 130 thousandths of a volt.

But Gotch also noticed something else: the shape of this 'blip' (i.e. its voltage intensity and duration) was always consistent, whether it had been produced by a strong or a weak stimulus. In other words, the nerve impulse appeared to be all-or-nothing – it either occurred in full, or not at all. There was no 'in-between' or graded response. This phenomenon was not entirely new, for it had been described in the heart muscle by American physiologist Henry Pickering Bowditch in 1871. Thus, a heart muscle fibre will always contract to its fullest extent, providing the stimulus that caused it, is above a certain threshold. However, Gotch was the first to realise it might also apply to the nervous system. In addition, Gotch also observed if a second stimulus was applied to a nerve fibre before the first 'blip' had dissipated, then no second impulse was produced. In other words, there was a brief moment when the nerve fibre was insensitive to further stimulation – an effect Gotch called the refractory period. These were important developments for it provided clues about the method of coding used by the nervous system. The implication was this: each nerve impulse was of the same duration and intensity, but separated by tiny gaps of time much like a very simple version of Morse code.

The all-or-nothing law was an important advance in changing opinion about how nerve cells transmit information. Prior to Gotch's discovery, it was still believed by some investigators that each nerve cell contained its own unique form of energy – an idea first formulated by the German physiologist Johannes Müller in the early nineteenth century. As evidence, Müller pointed out that while electrical stimulation can affect all sensory organs (e.g. eye, ear, tongue etc), every sensory nerve reacts to it differently. For example, one nerve transmits it as light, another as sound, and another one as taste. Thus, for Müller, there is something unique about each type of sensory nerve, regardless of how it is stimulated. This theory also refuted the idea physical and nerve energy are identical, thereby tacitly supporting the possibility of vitalism or some spiritual force. However, the discovery of the all-or-nothing law offered strong evidence against Müller's specificity hypothesis, by showing it was the frequency of the nervous signal that encoded information. That is, all nerve cells use the same energy, but differ in their patterning of impulses.

The all-or-nothing principle was extended to skeletal muscle fibres by Cambridge professor Keith Lucas during 1904 and 1905. He sought to answer a simple question: how does a muscle

Solving the mystery of the nerve impulse 271

FRANCIS GOTCH, D.SC. OXON. & LIVERP., LL.D. ST. AND., F.R.S.,

FIGURE 11.1 Francis Gotch (1853–1913) who first showed the all-or-nothing nature of the nerve impulse. He also discovered the refractory period that occurs between successive impulses.

Source: Wellcome Library, London.

produce a graded change in contraction? The problem can be illustrated when we flex our arm. If we pick up a heavy object, our muscles have to contract more strongly than when we hold a light one. How do our muscles make this adjustment? There were two possibilities: (1) all muscle fibres contract together at the same rate; or (2) only some fibres contract – but when they do so, they contract in an all-or-none fashion. Lucas examined this question by taking a frog's cutaneous dorsi muscle, and dissecting it into small strips that contained between twelve to thirty fibres. Lucas chose this muscle because he knew an individual nerve fibre projecting to it innervated about twenty muscle fibres. He then stimulated the nerve pathway with an electric current. The strips of muscle responded by showing clear step-like jumps in the intensity of the contraction as the stimulation was increased. It was a result proving some muscle fibres were contracting maximally in an all-or-nothing manner, while the remainder did not respond at all. This finding also suggested that nerve impulses were firing in an all-or-none manner. However, Lucas was reluctant to make this claim, preferring instead to wait until more definitive proof was available. Sadly, this did not happen as Lucas's life was tragically cut short in 1916, when he was killed at the age of 37 in a mid-air collision over Salisbury Plain while testing equipment for the Royal Air Force.

Thermionic valves and amplifiers

In 1912, a young Cambridge student with an exemplary first class degree in Natural Sciences called Edgar Douglas Adrian, joined Lucas's laboratory to investigate whether nerve impulses followed the all-or-none law. Much inconsistent evidence had emerged concerning the possibility. One objection had come from the discovery that the electrical strength of an action potential weakened if passed through a cooled or anaesthetised section of nerve fibre. This implied a graded or partial response was possible – at least in some instances. To determine whether this was indeed the case, Adrian exposed a small segment of frog nerve to alcohol vapour, and then followed the passage of a nerve impulse along the fibre. The results showed that while the alcohol 'block' produced a decrement in the force of the signal, it was only short-lasting. In other words, the nervous impulse was like a burning train of gunpowder: as long as its embers were not totally extinguished, it would be recharged back to its full 'strength'. This supported the all-or-none law. It also showed the flow of energy along a nerve fibre was not a passive process. Rather, there appeared to be a self-regenerative mechanism at work allowing the full force of the impulse to be reinstated.

Although his research was put on hold by the First World War, in which Adrian served as a doctor at St Bartholomew's Hospital, he returned to Cambridge in 1919 to take over Lucas's laboratory. This was a much more exciting time to be involved in neurophysiological work. Prior to the war, many researchers had been frustrated by the limited technology of their times. Galvanometers were slow and inaccurate, and the capillary electrometer provided little detail. Both were also unsuitable for measuring the tiny currents existing in nerve fibres. However, this situation improved greatly in the post-war years when the thermionic valve (or vacuum tube) became available, able to amplify a tiny nervous impulse some fifty times without distortion.[6] The first person to do this was Harvard professor Alexander Forbes. An expert in electronics and radio communication, Forbes used his war-time knowledge to build the first vacuum-tube amplifier for neurophysiological work in 1918. This apparatus was to revolutionise research over the next thirty years until replaced by the transistor. Adrian quickly realised the importance of the new device, and invited Forbes to Cambridge in 1921 to help him construct his own. In return, Adrian taught Forbes to fly an aeroplane.

Around the same time, others were beginning to realise that the vacuum tube's magnification of the nerve impulse could be significantly increased if relayed through a series of amplifiers. This procedure was first undertaken in 1921 by Herbert Gasser and Joseph Erlanger working at Washington University in St Louis who constructed an apparatus capable of increasing a nerve signal by 5000 times. This was now an enhancement finally enabling researchers to make reasonably accurate recordings of an action potential. However, Gasser and Erlanger improved the technology even further by using a cathode ray oscilloscope that allowed the time and voltage parameters of the action potential to be viewed on a screen. And, by placing electrodes onto an exposed nerve fibre and connecting them to an amplifier, the extracellular currents associated with the action potential could be recorded with a high degree of accuracy.

It was not long before Gasser and Erlanger made some important discoveries. For example, it became apparent that peripheral nerve fibres (such as those extending from skin and joints to the spinal cord) are actually composites of many hundreds of nerve axons. Furthermore, Gasser and Erlanger identified three types of fibre within these bundles which transmitted nerve impulses at different speeds. The fastest to conduct information were called A-fibres with a speed of around 100 meters per second (mps). These were followed by B-fibres with a velocity of around 10 mps. And, the slowest were the unmyelinated C-fibres which conveyed impulses at about 2 mps. Each type of fibre was also found to be associated with a different function, with the sensory transmission of muscle sense and touch being mediated by A and B-fibres, and low intensity pain messages by C-fibres. Thus, it was clear the peripheral nervous system prioritised information concerning motor reflexes and movement over pain signals. Gasser and Erlanger would win a Nobel Prize for their work in 1944.

Recording from single nerve fibres: Edgar Adrian

Back in Cambridge, Adrian was using the new amplifying powers of the thermionic valve to analyse the nature of nerve conduction. In one study, he recorded activity from a nerve that was still attached to the gastrocnemius muscle of the frog's leg. By hanging a 50-gram weight from the muscle, Adrian found it increased the number of impulses in the nerve fibre. These went from ten impulses per second at rest, to 50 impulses per second when fully stretched. In other words, the firing rate encoded the amount of force being placed on the muscle. It was further confirmation of the ideas first proposed by Francis Gotch that emphasised the neural message as a patterned code. However, Adrian also found this effect was short-lasting, with the number of impulses diminishing in frequency after only two or three minutes. It suggested the coding of the nervous system was preferentially designed to detect new information. Indeed, the significance of this soon became clear when Adrian examined fibres conveying touch from the skin. The results were similar: when pressure was applied to the skin, it caused the frequency of impulses to increase in the nerves passing to the spinal cord – although this burst of activity quickly subsided if the touch remained constant. This phenomenon is now known as 'habituation' and recognised as an important means of allowing information to be ignored that is no longer of any significance.

Up to this point, all research on the electrophysiological properties of nerve cells had been performed on peripheral fibres, which as Gasser and Erlanger had shown, were actually bundles of much finer individual axons. This meant every recorded nervous event was a composite response – often involving hundreds of nerve cells firing in unison. Although this had given rise to such theories as frequency coding, and the all-or-none principle, these laws still had to be confirmed at the level of the single nerve fibre. However, this was seemingly impossible. Researchers had no way of being able to dissect out a tiny single axon, embedded and entwined among hundreds of others, within a given fibre. However, in 1925, Adrian was working with a young Swedish researcher called Yngve Zotterman, who wanted to isolate the detector within muscle tissue that encodes stretch – otherwise known as the muscle spindle.[7] In order to do this, Adrian and Zotterman began progressively cutting away strips of muscle, and were delighted when they found it was possible to shave away the tissue until they were down to a single spindle. It was an important moment for the two researchers also knew the spindle was attached to a single nerve axon. Furthermore, it was possible to record from this fibre even though it lay in a bundle containing a mass of inactive axons. With this new preparation, Adrian became the founder of single cellular recording by being the first to take recordings of action potentials from individual nerve fibres.

The response obtained by Adrian appeared as small rapidly occurring spikes on the screen of his oscilloscope. Furthermore, they confirmed what many had suspected: the nerve impulse was always the same height and travelled at the same speed, regardless of the stimulus

FIGURE 11.2 Edgar Douglas Adrian (1889–1977). The first person to record from single axons and prove all neurons encode information by using a form of frequency coding.

Source: Wellcome Library, London.

that had caused it. Thus, Adrian confirmed the all-or-none principle for nervous transmission. His results also showed a single axon could transmit up to 400 nerve impulses a second. This was a high figure, although not totally unexpected, as it was known the action potential only lasted a few milliseconds. Clearly, the axon could convey a lot of information in a short period of time. However, it was the patterning of these responses that was important. Put simply, Adrian demonstrated larger amounts of stimulation increased the number of spikes per second. That is, the number of impulses in a single axon was proportional to the strength of the sensory stimulus (or its change). It was almost as if the nerve cells were encoding information by Morse code – except they were limited to using patterns of dots. In fact, a memorable analogy would be used by Edgar Douglas Adrian in 1927 who likened the coded nervous message to a stream of bullets fired from a machine gun. Indeed, if the nervous impulses are fed through an amplifier and played through a loudspeaker this is exactly what they sound like.

The code used by the nervous system was fundamentally simple. In the immediate years after confirming the all-or-none law, Adrian extended his single cell recordings to sensory neurons transmitting touch information from the skin's cutaneous receptors, including those involved in pain. He also examined certain motor neurons including the phrenic nerve controlling the diaphragm. No matter what type of axon he examined, they always encoded information in the same way, that is, by the patterning of small and brief 'all-or-none' electrical impulses. In 1932, Adrian was confident enough to state that it was a 'universal' coding mechanism used by all types of nervous system. The old concept of specific nerve energy, proposed by Müller, was redundant.

In 1932, at the age of 43, Adrian was awarded the Nobel Prize along with Sir Charles Sherrington 'for their discoveries regarding the functions of neurons.' Continuing his work at Cambridge, Adrian turned his attention to a number of other sensory systems including olfaction and the ear's vestibular apparatus. He also wrote several influential books.[8] Perhaps his most influential achievement during this period however was to recognise the importance of the electroencephalograph (EEG). This apparatus had been invented by the German Johannes (Hans) Berger in 1924, who initially used two silver wires under the patient's scalp, one at the front of the head and one at the back, to record the brain's electrical activity (later he would use two sheets of tin-foil as recording electrodes). Berger was then able to record the tiny electrical voltages, as small as one ten thousandth of a volt, with a galvanometer, whose activity he managed to photograph in bursts of three second duration.

Remarkably, Berger found the cerebral cortex did not exhibit a random mass of neural 'noise' as might be expected. Instead, it showed a regular rhythm, which in a resting conscious person with eyes closed, was around ten cycles ('beats') per second. Berger called this the alpha rhythm. This rhythm became faster (12 cycles per second or more) when the person was aroused (the beta rhythm). The apparatus had significant clinical potential, but for reasons that are unclear, the German medical authorities reacted to Berger's findings with indifference and hostility. According to some biographers, this was because Berger was opposed to Nazi rule, which led to his removal from the Swiss University of Jena in the late 1930s – an event likely to have contributed to his suicide by hanging in 1941. However, some commentators have recently given alternative reasons.[9] Whatever the truth of the matter, one thing is clear: the EEG only became widely known when Adrian, after reading about Berger's research in 1932, demonstrated it to the Physiological Society in Cambridge in 1934. Following this, the EGG quickly became used as an important tool for diagnosing epilepsy, and an experimental means for measuring brain activity.

FIGURE 11.3 Johannes 'Hans' Berger (1873–1841) who was the first to discover alpha waves by recording electrical potentials from under the scalp of human subjects in 1924.

Source: Picture originally taken by Dr Frederic A. Gibbs, Chicago.

The giant squid comes to the rescue

By the late 1930s, much had been learned about the nature of nervous conduction. The all-or-none principle had been established, and the nervous system shown to use a form of frequency coding to encrypt information. These were big steps forward. However, the biggest question of all remained unsolved: namely the electro-chemical events underlying the action potential. This ignorance was hardly surprising. The microscopic size of neurons and their axons, which in mammals are no bigger than a few microns (thousandth of a millimetre) in diameter, meant it was impossible to record voltage changes taking place inside the cell. Nor were researchers able to measure the tiny amounts of electrical current passing through the membrane during the nervous impulse. Both required probes small enough to be placed both inside and outside the cell. Bernstein's theory, which proposed a difference in ion or chemical composition between the interior and exterior of the nerve cell as the source of the electric current, also remained unanswered because of the same problem. Thus, even with the development of modern vacuum-tube amplifiers and oscilloscopes, the small size of neurons meant many fundamental questions concerning the action potential were impossible to address.

Somewhat unexpectedly, the solution to these problems would be reported to the scientific world in 1936 at the Cold Spring Harbour Symposium[10] by Oxford biologist John Zachary

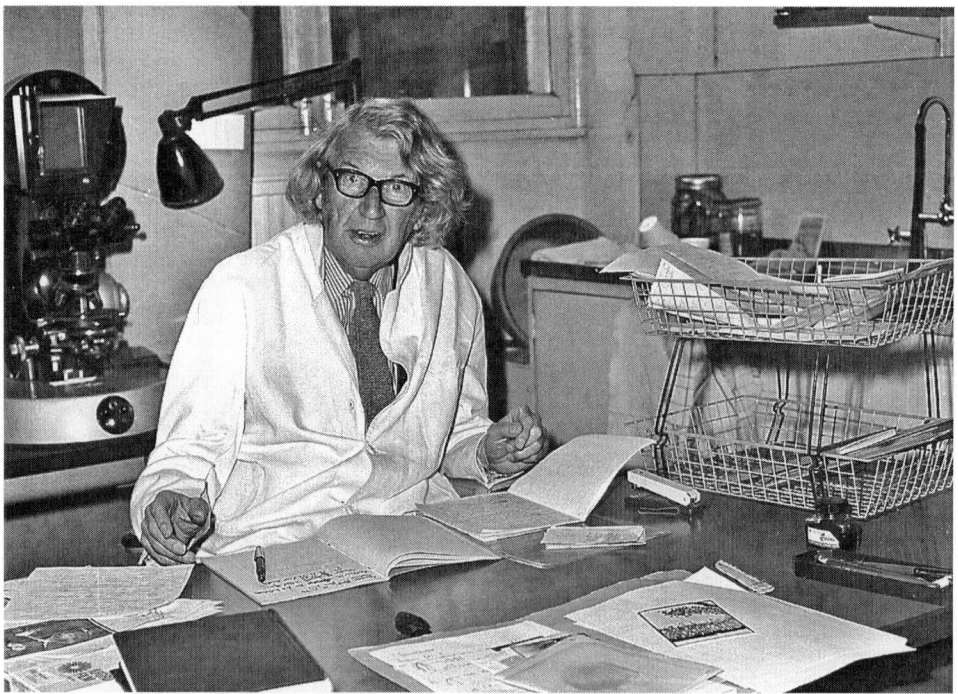

FIGURE 11.4 John Zachary Young (1907–1997) who reported the existence of the squid's giant axon in 1936. This was to be the key for some of the century's most important discoveries in nerve physiology.

Source: Wellcome Library, London.

Young. A man of imposing stature, friendly presence and infectious enthusiasm, who often wore a red tie to reflect his socialist sympathies, Young was an expert on the nervous system of cephalopod molluscs – a group of animals including the octopus, squid and cuttlefish. In 1933, Young was working at the Marine Biological Association in Plymouth with the North Atlantic long-finned squid (*Loligo pealii*), when he noticed their mantle cavities (the part of the body that protects its gills) contained large transparent tubular structures. At first, Young assumed these were blood vessels, but on closer inspection began to suspect they could be neural ganglia. In 1936, a Rockefeller Foundation Fellowship allowed Young to work in the United States where he confirmed these were indeed ganglia capable of conducting action potentials. In fact, they formed part of the system controlling the mantle muscle and siphon enabling the squid to move quickly away from threatening situations. Of greater significance was the fact these axons were about a millimetre in diameter. In other words, they were over a hundred times larger than those found in any mammal.

The 'giant axon' as it became known, was the vital breakthrough electrophysiologists were hoping for. Its diameter was not only large enough to allow simple wires, or electrodes, to be implanted inside, allowing voltage changes during the action potential to be recorded, but it was possible to 'squeeze' the cytoplasm out of the axon to measure its composition. Thus, it offered an effective means of testing Bernstein's ionic hypothesis. It was also a relatively simple procedure for investigators to free the axon from the squid's mantle with forceps and a dissecting microscope, and keep it alive in a bath of salt water. In fact, the squid proved so robust at being manipulated this way, experimenters could often test the isolated axon for up to 12 hours or more. Investigators now had the means by which they could begin to reveal the secrets of the nervous impulse. Later, in 1973, the Nobel Prize winning physiologist Alan Hodgkin, who would take full advantage of the giant squid axon in his work, helped put its discovery into perspective:

> It is arguable that the introduction of the squid giant nerve fibre by J.Z. Young in 1936 did more for axonology than any other single advance during the last forty years. Indeed, a distinguished neurophysiologist remarked recently at congress dinner (not, I thought, with the utmost tact), 'it's the squid that really ought to be given the Nobel Prize.

Testing Bernstein's membrane hypothesis

In the summer of 1939, Kenneth Cole and Howard Curtis working at Woods Hole, Massachusetts, and Alan Hodgkin and Andrew Huxley[11] in Plymouth, England, began examining the electrical properties of the giant squid axon. Although both parties worked independently, Hodgkin had worked for a short period in Cole's laboratory and a friendly rivalry existed between them. The basic experimental technique adopted by the two groups involved inserting an extremely fine electrode (platinum or silver wires) into the axon, while placing a second electrode on the surface, so its tip was just across the membrane from the other. This allowed them to measure the voltage differences at the two sites during an action potential. One of the first findings was reported by Cole and Curtis who used a device called a Wheatstone bridge to measure the resistance of the membrane to electrical flow. They found the axon's membrane suddenly reduced its electrical resistance some 40-fold in response to an action potential – which is another way of saying the membrane became 40 times more permeable to the passage of electrically charged ions at that point in time. This finding lent considerable support to Bernstein's chemical hypothesis that had proposed the action potential was caused by the 'breakdown' of the membrane to the passage of ions. Furthermore, as Bernstein had shown, the breakdown appeared to be very brief – taking place over a period of just 3–4 milliseconds.

A more critical examination of Bernstein's theory, however, took place when the two groups began to record the voltage differences between the inside and outside of the axon. The first step was to test the axon in its resting state (i.e. before the action potential), and this soon produced some unexpected findings. For example, when Hodgkin and Huxley measured the potential difference between the electrode tips, they found the axon's interior was approximately −45 millivolts compared to the outside. Although it has been known since the nineteenth century that the inside of the resting nerve cell was negative, this disequilibrium was far greater than expected. In fact, it was a difference in voltage that placed the axon under considerable electrical tension. Nonetheless, this resting potential was confirmed by Cole and Curtis who estimated the figure at −51 millivolts. Cole and Curtis also showed this difference was partly due to the potassium concentrations inside the axon, which were higher

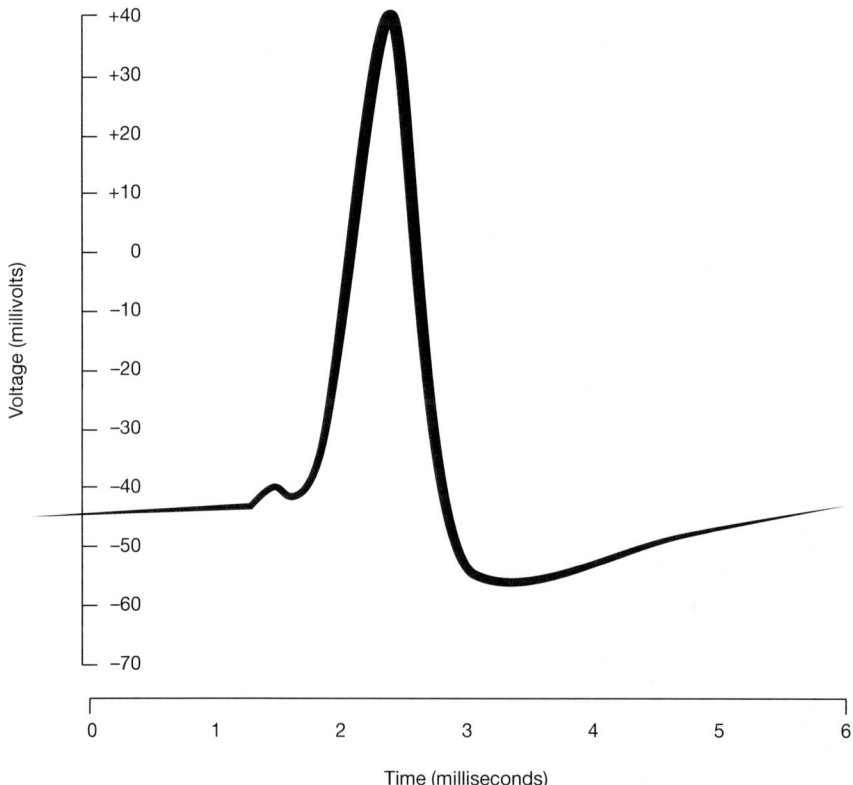

FIGURE 11.5 An early depiction of the action potential taken from an oscilloscope trace and recorded by a microelectrode implanted in the giant squid axon by Hodgkin and Huxley in 1939. For the first time the nerve cell's resting potential was shown to be about −45 millivolts, which shoots to +40 millivolts and back in a process lasting approximately 3 milliseconds during the action potential.

Source: Image redrawn from the original by Charlotte Caswell.

than those on the outside. Again, Bernstein's membrane theory had predicted much the same thing.

However, Bernstein's theory was not supported when Hodgkin and Huxley examined the voltage changes taking place inside the squid axon *during* an action potential. Bernstein's theory had predicted the voltage inside the axon should become zero as the membrane broke down, but Hodgkin and Huxley found the current flow into the axon was much greater than predicted – resulting in an increased voltage, or over-shoot, of about +45 millivolts. In fact, when the entire action potential was plotted over time, it could be seen to be an event lasting no more than 2–3 milliseconds, with the voltage inside the axon jumping sharply from about −45 millivolts at rest, to +40 millivolts at its peak, before dropping back to its resting value. This overshoot indicated there was some sort of 'extra' process at work, other than a passive breakdown of the membrane. Unfortunately, the solution to the problem had to be put on hold for several more years because of the Second World War. In Britain, Hodgkin turned his attention to the development of radar, while Huxley worked on naval gunnery. Even when the war was finished, the resumption of scientific activities was slow. There were fuel and food shortages in Britain, and the Plymouth laboratory had been damaged by German bombs. Consequently, Hodgkin and Huxley did not commence their partnership again until 1947.

The role of sodium in the action potential

Hodgkin returned to work in 1945 (Huxley was to return a year later) and began to consider the reasons why such a marked reversal of membrane potential occurred during an action potential. Bernstein had argued that potassium (K+) ions were responsible for creating the current flowing across the axon's membrane. However, Hodgkin also began to suspect the involvement of sodium (Na+) ions. For one thing, sodium ions were known to occur in high concentrations outside the cell where they served an important role in maintaining the electrolytic balance of the body. Furthermore, by the 1940s, researchers knew the excitation of muscle depended on the presence of sodium ions in the extracellular fluid, and if muscle was excited by sodium ions then why not the nerve axon? In 1947, Hodgkin tested this idea with the help of pharmacologist Bernard Katz. Using a simple procedure, they examined the effects of sodium ions on the formation of an action potential by altering their concentration in the fluid bathing the axon. They obtained significant results: lowering the sodium levels reduced the strength of the action potential, while eliminating sodium from the fluid abolished the impulse completely. In contrast, increasing the sodium levels made the squid axon more excitable. This was strong evidence the flow of sodium ions into the nerve cell were involved in creating the positive 'spike' (i.e. depolarisation) so characteristic of the action potential.

Hodgkin and Katz also discovered another feature of the axon's depolarisation that would later turn out to be crucial for understanding how an action potential was initiated. Put simply, they showed the action potential was only set into motion if the axon's initial resting value (i.e. −45 millivolts) was raised by another 15 millivolts (i.e. to about −30 millivolts). Thus, −30 millivolts was the critical *threshold value* for the action potential to be generated. As Hodgkin and Katz had already shown, this was the point where the membrane would lower its resistance to sodium, thereby setting into motion the train of events reversing the voltage inside the cell from negative to positive. In fact, Hodgkin and Katz were to demonstrate two types of sodium flow into the cell. A fairly gentle one producing the threshold (or trigger) for the action potential to occur; and a second more intense one responsible for causing the rapid depolarisation of the axon to about +40 millivolts.

The action potential explained

Another important objective was to measure the concentration of ions which accompanied the electrical currents passing in and out of an axon during the action potential. Even with the giant squid axon, this was initially impossible to do since the action potential was too fast and currents too small. However, the problem was solved by the invention of the 'voltage clamp' technique first used by Kenneth Cole in 1949, and later adopted by Hodgkin and Huxley[12]. The procedure is one that involves an electronic monitoring system where two fine wires (electrodes) are implanted inside the axon – one measuring the voltage inside the cell, and the other producing a current that can alter this to a level determined by the experimenter. Thus, the technique enables the researcher to set the potential inside the cell, and then to maintain it (i.e. 'clamp it') without necessarily triggering the action potential. Although this is obviously an artificial situation, it nevertheless allows the current flow across the membrane to be accurately recorded at the clamped point. Alternatively, the method allows the nerve fibre's action potential to be halted instantly at any point during its existence – thereby allowing a millisecond by millisecond visual examination of its characteristics to

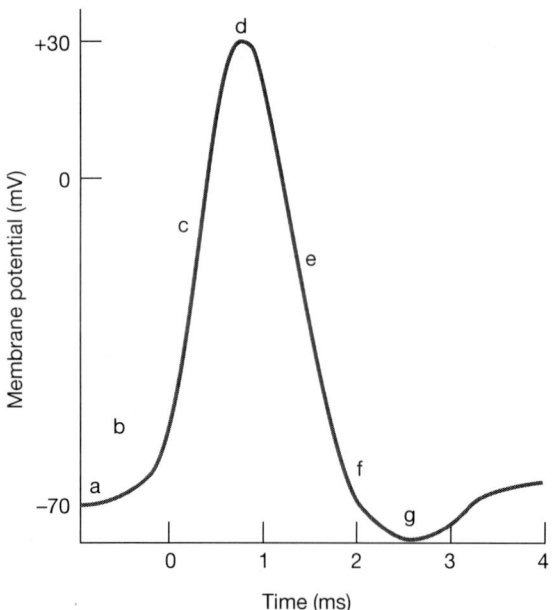

FIGURE 11.6 A typical account of the action potential that was first established by the work of Hodgkin and Huxley. If the resting potential (presented here as −70 millivolts) is increased to about −55 millivolts, sodium ions will enter the cell causing the membrane potential to peak at about +50 millivolts. As this point is reached, the sodium channels close and potassium ions move out of the axon, bringing the membrane potential back to its resting state.

Source: From Wickens 2009.

FIGURE 11.7 (A) Alan Lloyd Hodgkin (1914–1998) and (B) Andrew Fielding Huxley (1917–2012) whose collaboration resulted in a series of five papers published in 1952, which provided a series of mathematical equations that fully described the ionic basis of the action potential.

Source: The Royal Society.

be observed. This is important, for it allows the experimenter to examine what is happening to the membrane's permeability at any moment, and judge the flow of ionic current necessary to produce the voltage changes being observed.

From 1948 to 1951, Hodgkin and Huxley working in Plymouth during the summer and Cambridge in winter, used the voltage clamp technique to compile a large amount of data describing what happens during an action potential. In particular, they used the method in conjunction with other procedures, such as varying the ion concentrations surrounding the axon, or by applying drugs known to interfere with the movements of specific types of ion across the membrane. Perhaps their most crucial finding was to show that the upward spike of the action potential (i.e. its positive depolarisation) was caused by an inward current of sodium ions, whereas the downward part of the spike returning the voltage back to negative was caused by the outward flow of potassium ions. Thus, the nerve cell used two currents separated in time and not one as envisaged by Bernstein. But Hodgkin and Huxley also realised something else: the membrane did not simply breakdown to allow the passage of these ions. Rather, there were special channels, or pores, which 'actively' transported sodium and potassium ions across the membrane to allow for these two different but synchronised currents to occur. This was something Bernstein could never have imagined, and it made the axon's membrane significantly more complex to understand.

These were truly groundbreaking experiments. However, Hodgkin and Huxley did not stop there for they used their data to create a series of mathematical equations to fully describe the formation of the action potential. How these calculations were produced has become part of physiological folklore. The initial analysis took place in the bitter cold winter of 1947 when Huxley, unable to use the computer at Cambridge (the only one then owned by the university), was forced into 'cranking a Brunsviga calculating machine with mitten-covered hands'.[13] The task would be an enormous assignment taking Huxley about six months to complete all the calculations.[14] Huxley's endeavours, however, were to result in what has been called the most important theoretical achievement in modern neurobiology. These equations were published in a series of five landmark papers in the *Journal of Physiology* in 1952, and they outlined all the key biophysical properties of the action potential, including its form, duration, amplitude and velocity, along with the strength of the ionic movements necessary for the response. The papers even went further by modelling the current and voltage changes in terms of hypothetical pores and channels that were either open or closed, while using a statistical approach to estimate the numbers of ions entering and leaving the cell.

Hodgkin and Huxley's description and explanation of the action potential has proven so accurate, it has never been seriously challenged. Their collaboration is also recognised as one providing an exemplary balance between experiment and theory, perhaps unmatched anywhere else in the history of neuroscience. For their brilliant work, Hodgkin and Huxley were awarded the Nobel Prize for Physiology or Medicine in 1963 (along with John Eccles). Both would also be knighted (Hodgkin in 1972 and Huxley in 1974) although surprisingly, despite remaining at Cambridge, they would not collaborate again after their classic work. Hodgkin continued his investigations into the giant squid axon, while Huxley turned his attention to skeletal muscle. Nonetheless, their remarkable achievements represent a culmination of a research endeavour that began with Galvani in 1791, and many other great minds of their day.

The role of calcium

Among the many things Hodgkin and Huxley had shown regarding the action potential was the importance of positively charged sodium ions crossing the membrane into the axon, and the outward movement of potassium ions to bring the potential back to its resting state. In the same year as Hodgkin and Huxley's series of papers were published, evidence for another type of ion with a different type of function began to emerge – this time for calcium. The importance of calcium in human physiology had long been known. In the nineteenth century, the British physiologist Sydney Ringer had developed a solvent containing sodium, potassium and calcium salts that could be used to keep a perfused frog's heart alive (i.e. Ringer's solution). In a slightly modified form, this solution was also found to be isotonic with blood and effective as an intravenous drip. However, following Hodgkin and Huxley's work, calcium would also be implicated in neurotransmitter release, by Bernard Katz and his collaborators working at University College in London.

Katz was a German by birth, whose father was a Russian fur trader and mother a Polish Jew. However, recognising the dangers of Nazi Germany, he left his home country in 1935 after graduating in medicine from the University of Leipzig. Possessing no more than a League of Nations stateless certificate and four English pounds, Katz arrived in London and was taken under the wing of eminent physiologist Archibald Hill. Three years later, Katz received his Ph.D. from University College, and won a Carnegie Fellowship soon after (1939), which he spent in Sydney working with John Eccles.[15] After the war, Katz returned to University College where he succeeded Archibald Hill as professor in 1952. In London, Katz would also begin working with British neurophysiologist Paul Fatt, examining the frog's neuromuscular junction.[16] Using the newly invented microelectrode, which they implanted into the end plate of the muscle, Fatt and Katz found a transient small rise in voltage lasting a few milliseconds after they stimulated its motor neuron. They called this the end plate potential (EPP). Although the amplitude of the EPP was small (e.g. around 5 or 10 millivolts) it normally caused the muscle to contract. Fatt and Katz also found the EPP was abolished by the drug curare, which was known to block receptors for acetylcholine. In contrast, muscle excitability was increased by eserine, which increased levels of acetylcholine.[17] At the time the idea of chemical neurotransmission at the synapse was still highly controversial (see next chapter). Nonetheless, these results provided good support for its existence – at least at the neuromuscular junction.

It was during this research, Fatt and Katz noticed a new phenomenon. While studying end plate potentials in the muscle cells, the oscilloscope sometimes showed 'blips' of extremely small voltages (e.g. around 0.5 millivolts) at random intervals in the absence of any motor neuron stimulation. Because these blips were so small, Fatt and Katz called them 'miniature end plate potentials' (mEPPs). They also initially believed they arose as an artefact from the electrical 'noise' of their instrumentation. This idea was shown to be wrong, however, when Fatt and Katz applied curare to the end plate – an action eliminating the mEPPs. In other words, the 'noise' was being produced by chemical activity (most likely acetylcholine) at the synapse. This was confirmed when they cut the motor nerves innervating the motor end plate and found this abolished all mEPP activity in the muscle cells. Thus, the mEPPs were a biological phenomenon.

In order to examine these tiny potentials further, Fatt and Katz removed the sodium (Na+) from the solution bathing the neuromuscular junction – which abolished both the EPPs and 'blips' (mEPPs). This finding was not surprising for sodium was well-established as an ion with electrical properties, and its influx into the end plate known to cause the contraction of muscle. However, when Fatt and Katz reduced the levels of calcium (Ca+) they obtained an unexpected result: this significantly reduced the amplitude of the EPPs, but had no effect on the mEPPs. Although it was initially unclear what was happening, Fatt and Katz were to discover that calcium ions have a crucial physiological role: they entered the axon terminals of the motor nerve upon arrival of a nerve impulse where they trigger the release of acetylcholine. Thus, when the action potential arrives at the nerve endings of the motor nerve, after being propagated down the length of the axon, the influx of sodium is replaced by calcium ions. And, it is this specific ion which causes acetylcholine to be secreted into the synapse. This process is called exocytosis and is now recognised to be an essential feature of all neurotransmitter release.

The quantal release of neurotransmitter

The existence of the mEPPs at the muscle end plate, however, remained a mystery. What was causing these tiny electrical events? Further research by Fatt and Katz was to show that these were the result of a tiny package of acetylcholine being intermittently released from the presynaptic ending. In other words, the mEPPs were random events, occurring spontaneously and independently of any calcium influx at the motor nerve's axon terminals. Although they did not appear to serve any obvious function, Fatt and Katz suggested mEPPs might help in some way to inform the muscle's end plate that the synaptic connections at

FIGURE 11.8 Bernard Katz (1911–2003) who won the Nobel in 1970 for his work on the 'quantal' release of neurotransmitters. His work was also important for establishing the role of calcium in neurotransmitter release.

Source: The Royal Society.

the neuromuscular junction were still 'working'. Then, Fatt and Katz made another breakthrough. On closer inspection of the EPPs, they realised each one was an integral multiple (0, 1, 2, 3 or more) of the mEPP. This was an important and highly informative discovery, since Fatt and Katz could now deduce that the EPPs were the sum result of very small packages (or what they called 'quanta') of acetylcholine secreted from the axon endings. Each was the size of a mEPP. Or put another way: acetylcholine was not being released in a continuous secretary flow, but in the form of small units, or packages, of fixed size.

One question arising from this discovery was how much acetylcholine is stored in a quantal package? Since the amounts were far too small to be measured directly, Katz addressed this problem with colleague Jose del Castillo using a specialised form of statistical analysis. Their calculations indicated a large number of acetylcholine molecules were released in one quantum – perhaps as many as 10,000. They also estimated about 200 quanta were needed to create a typical EPP. Shortly after this computation, the first detailed pictures of the neuromuscular junction were produced by the electron microscope in 1954. It lent further support to the theory of quantal release, for the micrographs revealed the axon terminals to contain many small 'vesicles' that provided the repositories for storing acetylcholine molecules. It also became apparent that neurotransmitter release occurred when these vesicles fused with the membrane causing their contents to be released into the synapse – a process directly triggered by the influx of calcium ions into the axon ending.

The discovery of ion channels

Bernstein's original chemical theory of the nerve impulse had envisaged the total breakdown of the membrane's resistance to ion flow. However, this idea had been shown to be wrong by Hodgkin and Huxley. In fact, at the heart of their theory lay an alternative theory – namely the possibility of voltage-dependent ion channels for sodium and potassium ions. That is, when the inside of the nerve cell reached a certain threshold potential, specialised channels for sodium and potassium ions would open in the membrane, allowing ionic movement in and out of the axon. However, Hodgkin and Huxley had only been able to infer the existence of these ion channels from their mathematical analysis. This was far from satisfactory, but they had no alternative, for ion channels were too small to observe directly, even with an electron microscope. Indeed, the width of the neural membrane is extremely thin. It has been estimated that if a typical mammalian cell was magnified to the size of a watermelon, the membrane would be about as thick as a piece of office paper. Consequently, at the time of Hodgkin and Huxley, the actuality of ion channels remained hypothetical.

Although their existence was far from proven, investigators speculated on the mechanism by which the ion channel opened. Presumably, the ion channel was some type of specialised device in the membrane, which only allowed the appropriate ion through at the right moment. How it managed this feat was unclear. One possibility was the channel contained some type of electrical 'gate'. Under resting conditions the 'gate' would be closed, but when the voltage threshold value was reached inside the cell, it was then opened. At this point the ions would flow in (or out) until the gate was closed again, but there were alternative possibilities including one where the ions were carried through the membrane on some type of transporter system. In fact this was the option that Hodgkin and Huxley had preferred. Unfortunately, there seemed to be no way the ion channel question could ever be answered or resolved.

However, in the early 1970s, two researchers working at the Max Planck Institute in Germany, Erwin Neher and Bert Sakmann, attempted to isolate a tiny part of muscle membrane that was small enough to contain an individual ion channel. To do this, they took advantage of a newly developed micropipette with an extremely small opening of between 0.5 to 1 micrometers.[18] By using its suction force, Neher and Sakmann were able to tear away a tiny piece of membrane. They called this the patch-clamp method. Neher and Sakmann realised any ions flowing through this membrane must take place through the channel. Moreover, the current, no matter how small, could be measured by an electrode placed in the pipette, attached to a very sensitive amplifier. Unfortunately, when Neher and Sakmann attempted to do this, they were unable to establish a complete electric seal between pipette and membrane. Thus, the current flowing through the pipette was not identical to the current flowing through the membrane. Despite this, Neher and Sakmann managed to produce square-like 'blips' on their oscilloscope, which appeared to provide a measure of the current passing through a single ion channel. The traces also showed the channels opened in an all-or-none manner. It was an encouraging start considering the limitations of their technique.

Frustratingly, Nehar and Sakmann's attempts to perfect the seal were continually thwarted until one day in 1980, when they noticed its effectiveness was dramatically increased if a gentle suction was applied to the inside of the pipette. By doing this, they increased the electrical resistance between the wall of the micropipette and membrane over 1000 times. It was now sufficiently tight to characterise individual ion channels with a precise amplitude and time resolution, and the results they obtained with this method exceeded all expectations. For example, Neher and Sakmann detected currents of approximately 1–2 picoamperes (pA), which is equivalent to some 0.0000000000001 amps. They also observed a single ion channel activated by acetylcholine opening and shutting repeatedly – with its switching time being less than ten picoseconds (a picosecond is one millionth of one millionth (one trillionth) of a second). In fact, by careful analysis, Neher and Sakmann were even able to determine that the main sodium current in cultured muscle cells had a mean amplitude of 1.6 pA, which only lasted for 1 millisecond. From this they could calculate that approximately 10,000 sodium ions flowed through a channel activated by acetylcholine in any one millisecond.

The patch clamp method has proven remarkably versatile and shown two basic types of ion channel exist: those opened by a neurotransmitter attaching itself to a receptor (ligand-gated); and those opened by a change in the internal voltage of the cell (voltage-gated). Each group also has many different channels, even for a single type of ion. Consequently, channels are now described in terms of their ionic selectivity (e.g. sodium, potassium, calcium and chloride) and their mechanism (voltage-gated, ligand-gated). This has also thrown new light on old research findings. For example, as Hodgkin and Huxley hypothesised, we now know the squid axon contains two types of channel, which are permeable to sodium and potassium ions. Both are voltage-gated, thereby containing a sensor capable of detecting changes in the cells resting potential. Despite this, the two channels operate differently: the sodium channel opens before the potassium channel, but the latter remain open for longer. However, the patch clamp method has done more than answer old questions; it has revolutionised our understanding of electrophysiology. Before its invention, high resolution recording could only be undertaken with relatively large cells, which needed to be penetrated with two microelectrodes. Now recordings can be made from almost any type of cell, including those taken from slices of the brain. Moreover, by applying different neurotransmitters, or drugs, to the pipette, the electrical characteristics of these cells to chemical stimulation can be established. Given the ubiquity and importance of ion channels, it is not surprising the invention of the patch clamp method has led to a plethora of new knowledge about the brain. Erwin Neher and Bert Sakmann were awarded the Nobel Prize in Physiology or Medicine in 1991.

Notes

1 Today, we know that virtually all plant and animal cells, or at least those surrounded by a membrane, show a negative voltage in their interior compared to the exterior. In most instances this allows parts of the cell to function as a battery providing 'power' for certain membrane function. However, in nerve cells and muscle cells the negative interior serves as a means of transmitting electrical signals.
2 A millisecond is one thousandth of a second.
3 The term 'ion' had first been used by Michael Faraday in the 1830s to refer to a charged particle produced by the process of electrolysis. However, it soon became clear that ions can be formed without an electrical current. This occurs, for example, when sodium chloride (salt) dissociates into positive sodium ions and negative chloride ions after being placed in water.
4 It is a fundamental scientific law that a positive charge is always attracted to a negative charge (or vice versa). Similarly, an area of high concentration will always be attracted to an area of low concentration (or vice versa). Both forces explain why positive ions outside the cell are attracted to the negative interior.
5 Bernstein believed that at rest the membrane was permeable to positive potassium ions, which leaked out of the nerve cell to make it negative. When an action potential was created, Bernstein hypothesised that the membrane broke down to *all* ions – which then caused a temporary flow of positive ions into the cell. We now know that sodium ions, which are predominantly found in the extracellular fluid, have a much more important role than Bernstein envisaged.

6 The thermionic valve was invented in 1904, and developed by the Royal Air Force to enhance radio signals.
7 This is a long thin sensory detector embedded in muscle tissue, whose significance had been first recognised by Charles Sherrington in the 1890s (see Chapter 8).
8 These include *The Basis of Sensation* (1928); *The Mechanism of Nervous Action* (1932); and *The Physical Background of Perception* (1947).
9 There have been accusations that Berger was actually a member of the SS, and served on a Court for Genetic Health which imposed sterilisations on the Jews. See Ernst Klee, *Das Personenlexikon zum Dritten Reich: Wer war vor und nach*, (1945) 41.
10 The Cold Spring Harbour Laboratory is located in the village of Laurel Hollow, New York. Originally formed in 1890, it is a major centre of biological research which employs over 400 scientists. It has also given rise to eight Nobel Laureates.
11 Andrew Huxley was half-brother of the writer Aldous Huxley and grandson of the famous nineteenth century biologist Thomas Huxley – otherwise known as Darwin's bulldog, who championed the theory of evolution.
12 It is unclear who first invented the voltage clamp method that allowed the voltage to be controlled across the neural membrane. While it was first being tried by Kenneth Cole at Woods Hole in 1947, it is known that Hodgkin and Huxley were aware of the procedure during the war.
13 The Brunsviga machine inputs numbers by means of hand adjusted levers, with output then shown as digits on a series of wheels. These values then have to be transcribed onto paper.
14 Huxley later recalled that it took him around eight hours of solid computation just to plot a five millisecond description of a given variable.
15 During the war, Katz would join the Australian Air Force as a radar officer.
16 The region where the motor nerve makes synaptic contact with a specialised part of the muscle known as the end plate.
17 Eserine inhibits an enzyme called acetylcholinesterase (ACh), which acts to break down acetylcholine. By inhibiting this enzyme, acetylcholine levels are increased.
18 A micrometer is one thousandth of a millimetre.

Bibliography

Adrian, E. (1965) The activity of the nerve fibres: Nobel lecture, December 12, 1932. In: *Nobel Lectures: Physiology or Medicine 1922–1941*. Elsevier: Amsterdam.
Bennett, M.R. (2001) *History of the Synapse*. Harwood Academic: Amsterdam.
Bradley, J.K. and Tansey, E.M. (1996) The coming of the electronic age to the Cambridge Physiological Laboratory. E.D. Adrian's valve amplifier in 1921. *Notes and Records of the Royal Society of London*, 50, 217–228.
Cole, K.S. and Curtis, H.J. (1939) Electrical impedance of the squid axon during activity. *Journal of General Physiology*, 22, 649–670.
Cowan, W.M., Südhof, T.C. and Stevens, C.F. (2001) *Synapses*. John Hopkins University Press: London.
Darwin, H. and Bayliss, W.M. (1917) Keith Lucas 1789–1916. *Proceedings of the Royal Society of London*, Series B, 90, 31–42.
Debru, C. (2006) Time, from psychology to neurophysiology. A historical view. *C.R. Biologies*, 329, 330–339.
Erlanger, J. and Gasser, H.S. (1968) *Electrical Signs of Nervous Activity*. University of Pennsylvania Press: Philadelphia, PA.
Frank, R.G. (1994) Instruments, nerve action, and the all-or-none principle. *Osiris*, 9, 208–235.
Goldensohn, E.S. (1998) Animal electricity from Bologna to Boston. *Electroencephalography and Clinical Neuropsychology*, 106, 94–100.
Häusser, M. (2000) The Hodgkin–Huxley theory of the action potential. *Nature Neuroscience*, 3, 1165.
Hodgkin, A.L. (1972) The ionic basis of nervous conduction: Nobel lecture, December 11, 1963. In: *Nobel Lectures: Physiology or Medicine 1963–1970*. Elsevier: Amsterdam.
Hodgkin, A.L. (1976) Chance and design in electrophysiology: an informal account of certain experiments on nerve carried out between 1934 and 1952. *Journal of Physiology*, 263, 1–21.

Hodgkin, A.L. (1979) Edgar Douglas Adrian, Baron Adrian of Cambridge. *Biographical Memoirs of the Royal Society*, 25, 1–73.

Hodgkin, A.L. and Huxley, A.F. (1939) Action potentials recorded from inside a nerve fibre. *Nature*, 144, 710–711.

Horn, J.P. (1992) The heroic age of neurophysiology. *Hospital Practice*, July, 65–74.

Huxley, A.F. (1972) The quantitative analysis of excitation and conduction in nerve: Nobel Lecture, December 11, 1963. In: *Nobel Lectures: Physiology or Medicine 1963–1970*. Elsevier: Amsterdam.

Huxley, A. (2002) From overshoot to voltage clamp. *Trends in Neurosciences*, 25, 553–558.

Jasper, H.H. and Sourkes, T.L. (1983) Nobel laureates in neuroscience: 1904–1981. *Annual Review of Neuroscience*, 6, 1–42.

Jeng, J.-M. (2002) Ricardo Miledi and the calcium hypothesis of neurotransmitter release. *Nature Reviews Neuroscience*, 3, 71–75.

Katz, B. (1952) The nerve impulse. *Scientific American*, Nov, 55–64.

Katz, B. (1962) The Croonian Lecture: the transmission of impulses from nerve to muscle, and the subcellular unit of synaptic action. *Proceedings of the Royal Society of London. Series B, Biological Aspects*. 155, 455–477.

Katz, B. (1996) Neural transmitter release: from quantal secretion to exocytosis and beyond: The Fenn Lecture. *Journal of Neurocytology*, 25, 677–686.

Keynes, R.D. (1958) The nerve impulse and the squid. *Scientific American*, Dec, 83–90.

Keynes, R.D. (2005) J.Z. and the discovery of squid giant nerve fibres. *The Journal of Experimental Biology*, 208, 179–180.

Lenoir, T. (1986) Models and instruments in the development of electrophysiology, 1845–1912. *Historical Studies in the Physical and Biological Sciences*, 17, 1–54.

Lohff, B. (2001) Facts and philosophy in neurophysiology. The 200th anniversary of Johannes Müller (1801–1858). *Journal of the History of the Neurosciences*, 10, 277–292.

Lucas, K. (1912) Croonian Lecture: The process of excitation in nerve and muscle. *Proceedings of the Royal Society of London, Series B*, 85, 495–524.

McComas, A.J. (2011) *Galvani's Spark: The Story of the Nerve Impulse*. Oxford University Press: Oxford.

Marasco, D.D. (2011) The great era of English electrophysiology: from Francis Gotch to Hodgkin and Huxley. *Archives Italiennes de Biologie*, 149 (suppl), 77–85.

Messenger, J. (1997) John Zachary Young (1907–1997). *Journal of Marine Biological Association*, 77, 1261–1262.

Nicholls, J. and Hill, O. (2003) Bernard Katz: his search for truth and beauty. *Journal of Neurocytology*, 32, 425–430.

Nilius, B. (2003) Pflugers Archiv and the advent of modern electrophysiology. *Pflugers Archives: European Journal of Physiology*, 447, 267–271.

Piccolino, M. (2002) Fifty years of the Hodgkin–Huxley era. *Trends in Neurosciences*, 25, 552–553.

Robinson, J.D. (2001) *Mechanisms of Synaptic Transmission*. Oxford University Press: Oxford.

Schuetze, S.M. (1983) The discovery of the action potential. *Trends in Neurosciences*, May, 164–168.

Seyfarth, E.-A. and Peichl, L. (2002) A hundred years ago: Julius Bernstein (1839–1917) formulates his 'membrane theory'. *Neuoforum*, 4, 274–276.

Seyfarth, E.-A. (2006) Julius Bernstein (1839–1917): pioneer neurobiologist and biophysicist. *Biological Cybernetics*, 94, 2–8.

Shepherd, G.M. (2010) *Creating Modern Neuroscience: The Revolutionary 1950s*. Oxford University Press: New York.

Tansey, E.M. (1997) John Zachary Young. *Endeavour*, 21, 141–142.

Verkhratsky, A., Krishtal, O.A. and Petersen, O.H. (2006) From Galvani to patch clamp: the development of electrophysiology. *Pflugers Archives: European Journal of Physiology*, 453, 233–247.

Wickens, A.P. (2009) *Introduction to Biopsychology*. Prentice Hall: Harlow.

Young, J.Z. (1938) The functioning of the giant nerve fibres of the squid. *Journal of Experimental Biology*, 15, 170–185.

Young, J.Z. (1940) Giant nerve-fibres. *Endeavour*, July, 108–113.

12
THE DISCOVERY OF CHEMICAL NEUROTRANSMISSION

Adrenaline might then be the chemical stimulant liberated on each occasion when the impulse arrives at the periphery.

Thomas Renton Elliott

A remarkable conversion indeed. One is reminded almost inevitably of Saul on his way to Damascus when the sudden light shone and the scales fell from his eyes.

Henry Hallett Dale

Summary

Arguably, the greatest advance in understanding how the nervous system works during the twentieth century was the establishment of chemical neurotransmission. This discovery not only led to the realisation the brain is awash with a great variety of neurotransmitters, but also resulted in the introduction of the first antipsychotic and antidepressant drugs during the 1950s. It was a moment when progress in neuropharmacology can be said to have changed our world for ever. This situation, however, could have hardly been foreseen at the turn of the century. Although the synapse had been identified by Cajal, and the possibility of chemical communication speculated upon as early as 1878, the electrical option remained more plausible. After all, the electrical signal only had the smallest of distances to travel between nerve cells, and even if some power was dissipated in transit, it presumably could be regenerated again upon reaching the next cell. In contrast, chemical transmission appeared to be too slow and complex. Yet, over the coming years it would be the chemical theory that gained support, especially when investigators began to discover drugs that produced actions on the body closely mimicking nervous stimulation. This led to an extension of the neurotransmitter concept in 1905, when Cambridge professor John Langley proposed the existence of specialised receptors for adrenaline in the sympathetic nervous system. A decade later, the existence of receptors was established beyond reasonable doubt when Henry Dale identified separate nicotinic and muscarinic sites for acetylcholine throughout the body. However, the most vivid demonstration of neurotransmission was provided in 1920 when the Austrian Otto Loewi performed a classic experiment on the frog heart, showing the vagus nerve to secrete a chemical that slowed its beat (this later turned out to be acetylcholine). Despite this, many researchers, including the outspoken and forthright Australian John Eccles, were unwilling to broaden chemical neurotransmission to the brain and spinal cord. It was a situation, now known as the soup and sparks controversy, that caused much heated debate and consternation among the delegates of academic meetings and conferences, especially in the years following the Second World War. However, the issue was decisively settled in 1952 when Eccles was forced to admit, after performing the crucial experiment himself, that chemical transmission did indeed take place in the central nervous system. From this moment onwards, investigators would begin to look for neurotransmitters afresh in the brain and map out their pathways, opening up many new avenues of research in pharmacology and neuroscience.

The curious effects of curare

By the mid nineteenth century, it was becoming clear that certain drugs could exert their effects by acting directly on the nervous system. Perhaps the strongest evidence for this idea came from the renowned French physiologist Claude Bernard.[1] During the early 1840s, Bernard was a doctoral student and unpaid assistant to Francois Magendie at the Collège de France, where both men were trying to understand the effects of different drugs on parts of the body. In 1844, Bernard received an unexpected gift, when a friend gave him two arrows bought back from South America smeared with the poison curare.[2] This is a drug still used today by certain tribes of Amazonian Indians, who use it to coat darts normally shot from a blow-pipe, to bring down small prey. Although the poison causes muscle paralysis and respiratory failure within minutes, the prey can be eaten since the curare is quickly broken down in the stomach and alimentary tract. This drug had long fascinated Europeans since Sir Walter Raleigh came across it in his expedition up the Orinoco in 1595. Despite this, a concentrated extract of 'black pith' containing curare, was only bought back to Europe in 1745 by the Frenchman Charles Marie de la Condamine who returned after a ten-year expedition to the Amazon. He also demonstrated some of its lethal effects on chickens at the University of Leyden.

Curious to examine the effects of his new gift, it is said Bernard thrust one of the curare arrows into the thigh of a rabbit and watched as the animal became paralysed and died. Determined to understand how it produced these effects, Bernard then embarked on a series of experiments that was to take him over six years to complete. Interestingly, Bernard found the drug had to enter the bloodstream for it to exert its lethal action. He also discovered curare caused death by respiratory failure without producing convulsions or pain. Most puzzling of all, Bernard showed the heart continued to beat long after respiration had stopped. In fact, Bernard was able to show a curare treated animal could survive the poison, providing it was put on an artificial respirator, and enough time given for the drug to leave the body.

Bernard also attempted to understand how curare acted to produce muscle paralysis. To do this, he used a preparation where he pre-treated a frog with curare and then extracted its sciatic nerve still attached to the leg muscle. Bernard found this nerve was now insensitive to electrical stimulation as shown by its inability to cause muscle contraction. However, a stimulus applied directly to the leg muscle caused it to contract normally. This was an important observation for if a muscle from a curare treated animal could still twitch, then the drug's action must presumably be on the nerve. Bernard followed this experiment up with many more. In one experiment, he set up his frog nerve–muscle preparation again, but this time immersed the sciatic nerve in a bath of curare solution, but leaving its point of attachment with the muscle outside the bathing medium. When Bernard stimulated the nerve, he found the curare did not affect the nerve conduction with the leg vigorously twitching in response. However, if the neuromuscular junction was immersed in curare, then the stimulation had no effect. From this, Bernard concluded that the curare must act somewhere in the junction between the nerve and muscle.

Another important finding was that curare had no effect on sensory nerves. For example, Bernard prepared a frog with a ligature that severely interrupted the blood flow to the lower body. He did this so that curare administered into the upper part of the body, would be unable to reach the leg muscles. Indeed, as expected, this procedure produced a marked paralysis of the torso. Yet, when Bernard pinched the skin in the paralysed upper body, it also caused reflexive movements of the hind limbs. This could only mean one thing: while curare was preventing the neural message reaching the muscles (i.e. to cause paralysis), it was not blocking the sensory nerves conveying touch information to the spinal cord. Thus, the essential action

FIGURE 12.1 Claude Bernard (1838–1878) whose experiments using the South American poison curare showed it acted on a very specific part of the body – namely the junction between nerve and muscle.

Source: Wellcome Library, London.

of the poison was very specific; it only acted on the motor nerves, or their point of contact, controlling the musculature of the body. It was an exciting finding that significantly narrowed down the location of its pharmacological effect in the body.

The first statement regarding chemical neurotransmission

The site where the motor nerve's end plate comes to innervate muscle is known as the neuromuscular junction, and a good description of it had been provided by the German Wilhelm (Willy) Kühne[3] in 1869. However, at the time, Kühne was unable to establish whether the motor nerves made contact with the muscles, or fused with them in some way. Nonetheless, Bernard's research had clearly revealed an understanding of this site had important implications for explaining certain drug actions. In fact, Bernard favoured the idea that curare somehow blocked transmission along the nerve fibre – perhaps by having an anaesthetic action on its ending. However, in 1875, his pupil Alfred Vulpian, proposed curare acted by blocking the nervous impulse on the other side of the motor nerve fibre – that is, on the muscle itself.[4] This theory was shown to be all the more feasible in 1886 when Kühne was able to undertake a more detailed examination of the neuromuscular junction and clearly observe a cleft between the motor endplate and muscle fibre. In effect, Kühne had discovered the neuromuscular synapse[5] – a finding he was to report in his Croonian lecture to the Royal Society in 1888.

The discovery of the neuromuscular synapse by Kühne raised many questions concerning the nature of neurotransmission. Kühne, for example, assumed the close contact of the nerve fibre and muscle was sufficient 'to allow transfer of the excitation from the latter to the former'. In other words, he assumed the signal crossing the synapse was electrical in nature. However, a chemical message could not be ruled out. In fact, almost a decade before Kühne identified the synapse at the neuromuscular junction, Emil du Bois-Reymond in his two-volume work on the physiology of nerve and muscle in 1877, had outlined two possibilities. He wrote, what is generally regarded as the first outline of chemical neurotransmission:

> Of known natural processes that might pass on excitation, only two in my mind are worth talking about: either there exists at the boundary of the contractile substance a stimulatory secretion ... or some other powerful stimulatory substance, or the phenomenon is electrical in nature.

Bois-Reymond did not know what chemicals could act as synaptic messengers, although he suggested lactic acid and ammonia as two possible candidates. He had nonetheless given some degree of respectability to the possibility of chemical neurotransmission.

Mapping the sympathetic ('involuntary') nervous system

The *somatic nervous system* is the peripheral system of nerves responsible for carrying motor input to, and sensory information from, the musculature. By focusing their attention on the nerves innervating the skeletal muscles, both Bernard and Kühne had been examining part of this system. They were also helped by the fact that the somatic nervous system was reasonably well understood at the end of the nineteenth century. Indeed, Charles Bell and Francois Magendie had shown that the anterior roots of the spinal cord give rise to its outgoing motor neurons, while the posterior roots received the incoming sensory fibres. In addition, Sherrington had meticulously mapped the areas of the body innervated by each motor nerve. However, the body also contains a second system of nerves called the *autonomic nervous system*, which regulates activities normally beyond our conscious control such as heart rate, breathing, gastrointestinal activity, blood pressure and energy mobilisation (to name a few). Consequently, this system consists of a different set of motor neurons, which project to the smooth muscle of the body (this includes the blood vessels and bronchioles), as well as cardiac muscle and various glands. In addition, the autonomic nervous system also monitors visceral organs and blood vessels with sensory neurons that provide input information to the lower areas of the brain including the pons and medulla. Thus, the somatic and autonomic systems are completely independent from each other.

The earliest reference to the 'involuntary' system goes back to Galen who among other things identified the vagus nerve innervating the heart, and phrenic nerve controlling the diaphragm. Thus, Galen can be credited with being the first to associate nerve action with the motion of the heart and blood, and the process of respiration. However, it was Thomas Willis in the seventeenth century who first realised the body had a system of nerves, separate from those passing to skeletal muscles, which governed the passions and instincts of the individual. While Willis had some misconceptions about these nerves (he believed they arose from the cerebellum, which he associated with involuntary action) he nevertheless correctly identified them with a chain of ganglia running down each side of the spinal cord, extending from the base of the skull to the coccyx.[6] Willis also realised that the nerves leaving this ganglion spread out through the body to innervate a vast array of organs. Although he believed these bought all the various organs of the body into 'sympathy' with each other, he nevertheless referred to them as the intercostal nerves (meaning located or occurring between the ribs). However, in 1732, a Danish-born professor working in Paris called Jacobus Winslow introduced the more popular term 'sympathetic nervous system'. It would also lead to the trunk of nerves running down both sides of the spinal cord becoming known as the sympathetic ganglia.[7]

Progress into understanding the physiological role of the sympathetic system was made in the mid-nineteenth century when researchers began to realise some of its nerve fibres controlled blood circulation. For example, Albert Kölliker at Würzburg in Germany followed certain sympathetic nerves to the muscle layers embedded in the walls of arteries, while Charles Brown-Séquard in Paris showed stimulation of these nerves constricted blood vessels with increases in blood flow and pressure. Another interesting discovery was reported to a congress of Italian scientists in Naples, in 1845, by two brothers called Eduard and Ernst Weber. They had come across an unusual effect: electrical stimulation of the vagus nerve slowed the heart and could even stop it beating. This was the first time most physiologists had been alerted to the paradoxical fact that nerve stimulation was sometimes able to inhibit an autonomic activity rather than exciting it.[8] Soon after, others would find that heart rate could be increased by stimulating a different set of sympathetic nerve fibres. Thus, it was becoming clear certain organs of the body such as the heart, were innervated by both inhibitory and excitatory nerve pathways extending from the sympathetic ganglia.

At the time of their discovery, these excitatory and inhibitory fibres could not be anatomically distinguished from each other. This problem was to be solved by the Cambridge physiologist Walter Holbrook Gaskell. During the 1880s, Gaskell attempted to follow the pathways of what he called the 'involuntary nervous system' by cutting serial sections of the spinal cord and staining them with osmic acid. He chose this stain because it could distinguish the fine, small diameter, nerve fibres of the sympathetic system, from the much larger fibres, which innervated the skeletal muscles. Interestingly, Gaskell found the smaller involuntary nerves left the spinal cord in three major 'outflows'. These arose from the *cranial* or upper part of the spinal cord; the *thoracic/lumbar* part (i.e. the chest and lower back region); and the bottom or *sacral* part of the spinal cord. Further, these three systems had different anatomical characteristics. Although they all passed from the spinal cord covered in a white sheath of myelin, only the thoracic/lumbar fibres entered the nearby sympathetic ganglion. Curiously, these left the ganglion as unmyelinated fibres (this allowed sympathetic fibres to be classified as preganglionic and postganglionic). In contrast, the cranial and sacral fibres remained myelinated as they passed into the body. Gaskell had therefore succeeded in finding a simple anatomical distinction between the thoracic/lumbar fibres, and those emerging from the upper (cranial) or lower (sacral) parts of the spinal cord.

This anatomical difference, however, was also found to have an important functional significance when Gaskell realised the unmyelinated and myelinated fibres had opposite or antagonistic effects. For example, when he stimulated the thoracic/lumbar outflow nerves with an electric current, it excited the organs they innervated. Thus, stimulation mimicked the effects of bodily arousal, which included the rapid beating of the heart, increased respiration and vasodilation. In contrast, stimulating the cranial and sacral outflows inhibited the same organs, and bought the bodily state of arousal back to normal. It was clear, therefore, the involuntary nervous system was composed of two separate, but complimentary parts. However, Gaskell also made another remarkable discovery: he found that almost every organ in the body was innervated with both excitatory and inhibitory fibres. Although much of Gaskell's work was done on reptiles, including crocodiles and turtles, he knew that it would likely apply to mammals. His predictions were soon to be confirmed by Gaskell's colleague John Newport Langley.

Defining the sympathetic and parasympathetic systems: John Newport Langley

The son of a Newbury school teacher, John Langley attended Cambridge in 1871 to study mathematics and history, but changed to natural sciences after coming under the influence of Sir Michael Foster (who later became the university's first Professor of Physiology in 1883). Even before his graduation, Langley was working in Foster's laboratory, where he had been given the task of examining the physiological action of a substance called jaborandi juice (derived from the Brazilian shrub *Pilocarpus jaborandi*). This was known to contain an active alkaloid[9] called pilocarpine, which was one of a number of different alkaloids attracting the attention of pharmacologists at the time for their striking physiological effects. It did not take Langley long before he found intravenous injections of pilocarpine in cats and dogs slowed down their heart beat, and caused marked salivation. Langley knew the heart was innervated by the 'inhibitory' vagus nerve, and therefore assumed jaborandi was somehow acting on it. To test his theory, Langley applied curare to the vagus nerve, a procedure he believed would paralyse the nerve endings innervating the heart, thereby inhibiting the effects of jaborandi. However, this blocking effect did not occur. Instead, the physiological action of jaborandi was only inhibited by the prior administration of another alkaloid called atropine (derived from the *Atropa belladonna* plant). It was a puzzling effect and Langley could only explain it

FIGURE 12.2 John Newport Langley (1852–1925). Cambridge physiologist who more than any other person was responsible for mapping out the autonomic nervous system (a term he coined in 1898). He also formulated the concept of 'receptive substances' in 1905.

Source: Wellcome Library, London.

by assuming jaborandi and atropine did not act on the vagus nerve (which was unaffected by both drugs), but directly on the tissue of the heart itself.

In 1889, Langley and collaborator William Lee Dickinson, turned their attention to another alkaloid, nicotine, the main ingredient in tobacco which was associated with a wide range of physiological actions. Although this drug is known for causing many symptoms of arousal, Langley and Dickinson discovered an unexpected effect: if a solution of nicotine was applied directly to the sympathetic ganglion with a fine artist's brush (the nicotine easily permeates into the ganglion), it blocked the transmission of nerve impulses from the myelinated fibres of the thoracic/lumbar outflow to the unmyelinated ones. For example, one of the pathways extending from the thoracic/lumbar outflow was the superior cervical ganglion which innervates the face and head. When Langley and Dickinson stimulated the myelinated fibres entering this pathway in the cat, they found it caused the nictitating membrane of the eye to retract, the pupils dilate, and the hairs on the face and neck to rise. However, these effects were not produced if the superior cervical ganglion was treated with nicotine beforehand.

While its pharmacological action was as yet unknown, the blocking effects of nicotine proved to be an invaluable tool for mapping out the fibres of what Gaskell had called the involuntary nervous system. By combining a number of various techniques including electrical

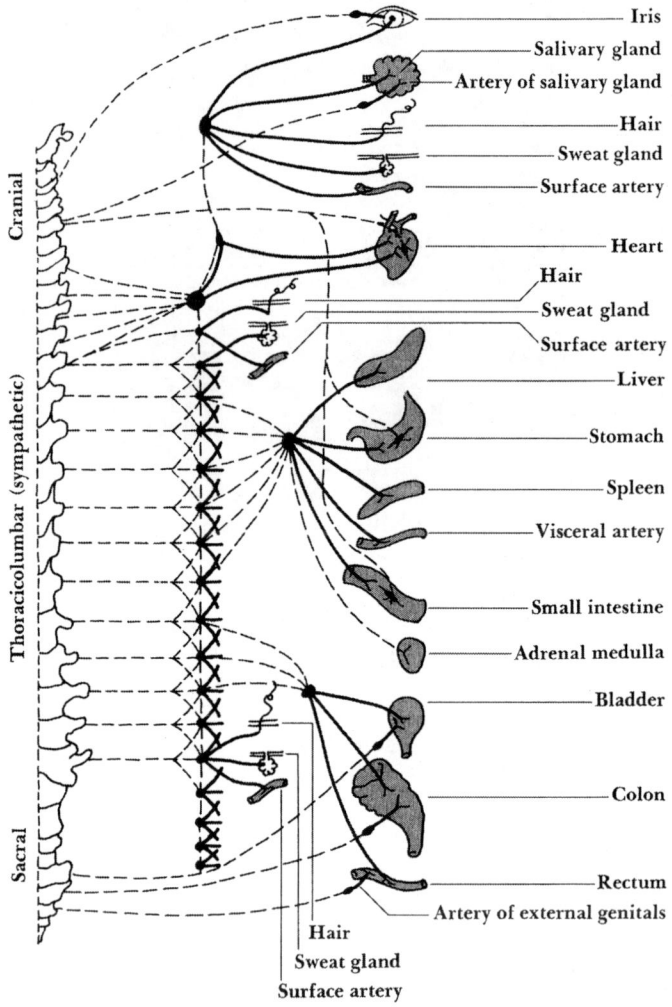

FIGURE 12.3 Diagram of the autonomic nervous system. Our understanding of this system is largely based on the work of John Langley who distinguished between the excitatory sympathetic (thoracic/lumbar) and inhibitory parasympathetic (cranial and sacral) systems. He also showed both systems contain pre-ganglionic and post-ganglionic fibres, although the site of the synapse is different for the sympathetic and parasympathetic systems.

Source: From *Gray's Anatomy of the Human Body*, 1918, Plate 839, 20th Edition.

stimulation, lesioning and nicotine blockade, Langley was able to follow all of its three main outflows (the cranial, thoracic/lumbar and sacral) to their bodily destinations. In doing this, he was then able to assess their specific physiological functions. Perhaps of greater significance, Langley would also start to use a completely new nomenclature to describe these systems, replacing the term 'involuntary' with 'autonomic' at the turn of the century. Later, in 1905, he would call the excitatory thoracic/lumbar system 'sympathetic', and the inhibitory cranial/sacral system 'parasympathetic'. These terms have continued to be used up to the present day.

Langley's mapping studies of the autonomic nervous system also revealed something else of fundamental importance: both the sympathetic and parasympathetic systems contained two sets of fibres, which he referred to as 'preganglionic' and 'postganglionic'. The pre-ganglionic fibres arose from the spinal cord, although those entering the sympathetic ganglion from the thoracic/lumbar system were short in length (as Gaskell had shown), whereas those leaving the spinal cord from the cranial and sacral outflows (i.e. the parasympathetic system) were much longer. In fact, these latter nerves extended out into the periphery where they projected to smaller ganglia, close to, or within, specific visceral organs, such as the colon and bladder, etc. In contrast, the postganglionic fibres, which left the sympathetic ganglion were invariably long, whereas the parasympathetic ones were always shorter as they were already close to their final destination (see Figure 12.3). Intriguingly, the existence of preganglionic and postganglionic fibres showed there were two synapses in both the sympathetic and parasympathetic systems: one in the ganglion, and one at the terminal of the postganglionic fibre, which innervated the target organ.[10]

A claim for neurotransmitters and receptors

In 1901, Langley turned his attention to adrenaline, which was known to be a substance produced by the adrenal glands. Evidence linking these glands with a chemical that produced stimulatory physiological effects had first emerged in the early 1890s when Yorkshire physician, George Oliver, began experimenting with adrenal extracts prepared by his local butcher. A dedicated amateur researcher with his own laboratory at home, Oliver's main interest was in blood pressure, a phenomenon he examined by inventing his own recording devices. During this work, Oliver had become aware that patients with adrenal gland damage, as occurs in Addison's disease, suffered from low blood pressure. Keen to understand why, Oliver persuaded his son to consume a small amount of adrenal tissue so he could measure the effects. To his surprise, it quickly produced a sudden and dangerously high increase in blood pressure. Realising the adrenal extract must contain an unknown chemical with significant biological actions, Oliver travelled to London to demonstrate its effects to University College professor Edward Scäfer who injected it into one of his laboratory dogs. Within minutes the dog's blood pressure had become so high it exceeded the capacity of the recording device to measure it. The active ingredient in the extract was later shown to be adrenaline by Japanese chemist Jokichi Takamine in 1901.

Langley's research on adrenaline led him to realise it produced effects on the body very similar to those following electrical stimulation of the sympathetic nervous system. For example, adrenaline raised blood pressure by constricting the smooth muscle found in the walls of arteries and veins. It also increased the heart beat, dilated the bronchial passages, and stimulated salivary secretion. How was adrenaline producing these effects? Initially, the evidence favoured the idea adrenaline was working independently of the nervous system. Indeed, Oliver and Scäfer provided support for this idea when they found an adrenal extract, perfused directly into the arterial system of a frog devoid of its central nervous system, caused a massive contraction of the smooth muscle. Langley was to demonstrate much the same effect in dogs, cats and rabbits, when he found adrenaline still exerted physiological effects after all the sympathetic nerve pathways had been severed. Thus, adrenaline appeared to be acting on the target tissue muscle directly.

FIGURE 12.4 Thomas Renton Elliott (1877–1961). A graduate student of Langley's who first proposed the concept of chemical transmission involving adrenaline in 1905.
Source: Wellcome Library, London.

However, this situation raised another important question: where was this adrenaline coming from? Although adrenaline was known to be produced by the adrenal gland, its actions on the targets of the sympathetic nervous system was too rapid for this glandular source to be the origin. An alternative explanation had to be sought, and one possibility was that adrenaline acted as a neurotransmitter which was released from the sympathetic nerve endings. By now the existence of synapses in the nervous system was not in serious doubt. In fact, Charles Sherrington had demonstrated their functional importance in mediating spinal reflex activity in the 1890s, and Langley had shown synapses to exist in the autonomic nervous system with his nicotine experiments. However, chemical transmission at the synapse was still a highly controversial theory, and while it is certain Langley must have considered such a possibility, he did not state it publicly. However, a medical postgraduate called Thomas Renton Elliott working in Langley's laboratory at the time, was not so reticent. In 1904, when addressing the Physiological Society in London, Elliott proposed the existence of neurotransmission involving adrenaline. There was also evidence to back his claims. Elliott pointed out, for example, that adrenaline only acted on tissue innervated by the sympathetic nerves, and had no effect on the parasympathetic system or skeletal muscle. And, in concluding his talk, Elliott made what is now regarded as the first clear modern statement about the feasibility of chemical neurotransmission:

'Adrenaline might then be the chemical stimulant liberated on each occasion when the impulse arrives at the periphery'.

A year later, Elliott published a more detailed account of his ideas in which he speculated that all synapses, whether in the autonomic nervous system, or those situated between the motor nerves and skeletal muscles, might use neurotransmitter substances. Although he thought it likely the chemical messenger in each instance was different, Elliott was convinced the terminals of the sympathetic branch of the autonomic nervous system liberated adrenaline.

Most physiologists, however, remained highly sceptical of chemical neurotransmission since many questions remained unanswered. For example, it was not clear how a nerve cell could manufacture a neurotransmitter. Nor could Elliott explain how a chemical might produce a biological effect on its target tissue or organ. However, in 1905, John Langley proposed an answer to the latter problem in a paper that attempted to explain the pharmacological actions of nicotine and curare. By now, Langley had extended his work with nicotine beyond the autonomic nervous system, to the somatic nervous system. More specifically, Langley had found nicotine also caused contraction of the skeletal muscles. For example, injections of nicotine into a chicken caused its leg muscles to stiffen and extend. This effect also occurred after the nerves to the muscles had been severed, indicating nicotine was not acting on the nerves themselves, but directly on the muscle. In fact, the excitatory action of nicotine on skeletal muscle was not dissimilar to the stimulatory effects of jaborandi on the heart (discussed above). And, the similarity was further highlighted when Langley found he could block the actions of nicotine by injecting the chicken first with curare. Thus, its action was not unlike the way atropine blocked the effects of jaborandi.

To explain these effects, Langley proposed that nicotine must act on some specialised part or 'receptive substance' of the muscle. In turn, the nicotine 'receptor' then used the chemical message in some way to initiate the biological effects inside the muscle cell. Langley also reasoned there must be two types of receptor: one capable of producing muscle contractions sensitive to nicotine; and the other sensitive to the inhibitory effects of curare that produced muscle paralysis. Although Langley did not literally use the word 'receptor', it was the first time this type of concept had been used to explain drug action. The receptor idea, however, was not entirely foreign as the term had been used successfully by Paul Ehrlich in 1900 in an immunological context. Consequently, there were many biologists who recognised the feasibility of Langley's idea. In fact, the receptor theory is one of the key breakthroughs in the modern history of pharmacology, allowing a better understanding of drug action, and supporting the theory of chemical neurotransmission.

Henry Dale investigates the properties of ergot

The concept of chemical neurotransmission was further supported by one of England's greatest pharmacologists Henry Hallett Dale. The son of a pottery manufacturer from Staffordshire, Dale began studying natural sciences at Cambridge in 1894 where he worked for a short period under John Langley. Completing his medical training at St Bartholomew's Hospital in 1904, and after being employed for a short time at University College, Dale took up a research position at the Wellcome Laboratories in London. This was an unusual choice for the Wellcome Institute was a commercial concern, established in the 1890s, to produce an antitoxin for diphtheria. The appointment was also against the advice of his university friends who feared it would compromise Dale's independence as an academic researcher. However, the Wellcome offered a secure income and career advancement, which Dale took full advantage of, by becoming its Director in 1906. He would hold this position until 1914 when joining the newly created National Institute for Medical Research.[11] A man who was affable, genial, and greatly revered by his colleagues, Dale's long and distinguished career that lasted over 50 years, was to be associated with two neurochemicals: histamine and acetylcholine. He was to identify both as natural constituents of animal tissues and neurotransmitter substances.

When Dale started work in his new job, he was encouraged by its owner Sir Henry Wellcome to examine the chemical constituents of a parasitic fungus which infected rye and other cereal grasses called ergot. This fungus had famously been known in medieval times to produce a form of poisoning called St Anthony's fire, that caused outbreaks of convulsions, hallucinations and intense burning sensations in its victims. These epidemics occurred most commonly in parts of central Europe where grains had to be stored over the winter period. Later, in the eighteenth century, ergot was shown to cause severe constriction of blood vessels and the uterus, leading to its adoption by midwives who used it to induce abortions and stop bleeding after childbirth. In the wrong hands, however, it could also result in gangrene with loss of limbs and even life. Despite this, at first, Dale was not particularly enthralled at the idea of making his first excursion into pharmacology by examining ergot. Yet, Sir Henry Wellcome's suggestion turned out to be a fortunate one. As Dale began to examine ergot, he realised it was a treasure trove of active pharmacological substances, many of which were isolated by his colleague and brilliant young chemist George Barger.

Dale's first breakthrough with ergot came by chance. One day he was asked to test an extract from the adrenal glands to see if it contained adrenaline or noradrenaline.[12] He tested the extract on a cat pretreated with ergot, and found to his surprise the animal's adrenal-induced blood pressure did not increase as expected. Clearly, there was a substance in the ergot inhibiting the effects of the adrenal extract. This turned out to be ergotoxine – a chemical that blocked the constricting effects of adrenaline on the smooth muscle of blood vessels. In fact, Dale had discovered the first adrenergic blocking agent, which are now a class of drug used to treat high blood pressure. Continuing his work with noradrenaline, Dale also realised this substance mimicked the effects of sympathetic nervous stimulation more faithfully than adrenaline itself. This led Dale to propose in 1910 it was noradrenaline and not adrenaline that was the chemical conveying synaptic information in the sympathetic nervous system – a prediction eventually proven to be correct in 1946 by Ulf von Euler in Sweden.

In 1910, Dale and Barger identified another chemical in ergot with potent physiological effects, which caused the smooth muscle of the uterus and bronchioles to constrict. It also produced a sudden fall in blood pressure, and many features of anaphylactic shock – a reaction associated with life-threatening hypotension and respiratory difficulties. In fact, they had discovered histamine, a naturally occurring chemical in the body, which initially was found in the walls of the intestine. Over the next decade Dale would show that histamine is distributed throughout the body, and released by biological tissues in response to injury where it causes blood vessels to dilate and capillaries to leak. Thus, it greatly aids the immune response. Histamine is also secreted as part of an allergic reaction where it can cause such symptoms as sneezing, itching and swelling. Later, this chemical would be shown to exist in the nerve cells of the brain (in the 1940s) and proven as a neurotransmitter (in the 1960s).

The discovery of acetylcholine

The neurochemical to be most associated with Henry Dale, however, was acetylcholine. The first clue this chemical had a role to play in the nervous system came when a related substance called choline was found to be present in the adrenal extract used by Oliver and Scäfer. Choline was not a new substance. It had been first discovered in ox bile by Andreas Strecker in 1864 and synthesised two years later. Nonetheless, its discovery in an extract of the adrenal gland suggested it exerted a biological function, and this was confirmed in 1906 when American pharmacologist Reid Hunt found choline caused blood pressure to fall. It was also an effect opposite to adrenaline. Interestingly, choline's action was blocked by atropine (the same chemical Langley had shown reversed the effects of jaborandi). These findings encouraged Hunt to synthesise a number of choline derivatives – one of which was acetylcholine. Remarkably, it proved to be 100,000 times more active than choline at reducing blood pressure. The extreme potency of this substance suggested that it might serve a pharmacological role

FIGURE 12.5 Henry Hallett Dale (1875–1968). Renowned for his work with the parasitic fungus ergot, which would lead him to discover histamine and acetylcholine. His work would also be important in establishing the chemical nature of neurotransmission.

Source: Wellcome Library, London.

in the body (i.e. only a small amount was needed to produce a biological effect). However, at the time, there was no evidence that acetylcholine, unlike choline, existed as a natural substance in the body.

Dale's interest in acetylcholine began in 1913 when he suspected ergot contained a compound called muscarine. This chemical, first extracted from the iconic red-capped and white spotted poisonous fly agaric mushroom (*Amanita muscaria*) in 1869, was of great interest to pharmacologists since it produced a wide range of physiological effects. These included a slowing of heart rate, reduced blood pressure, and pupil constriction. Intriguingly, these effects appeared to mimic the inhibitory action of the parasympathetic nervous system. In other words, they were opposite to the effects produced by the sympathetic nerves. However, after working with colleague Arthur Ewins, Dale realised the depressor substance in ergot was not muscarine, but acetylcholine. The implication, although Dale did not state it at the time, was that acetylcholine acted as a neurotransmitter in the parasympathetic system. It was not only a radical idea, but one beset with experimental difficulties since acetylcholine was highly unstable and quickly broken down by the body into inert substances. This made its detection, even if it was naturally present in the nervous system, extremely difficult.

Over the next year, Dale compared the actions of muscarine and acetylcholine at many sites throughout the body, leading him to realise they produced very similar physiological effects. This was largely because both substances acted on the same organs – although Dale found some notable exceptions. One of these was the neuromuscular junction where only acetylcholine, and not muscarine, caused muscle contraction. The neuromuscular junction also showed another interesting characteristic: the effects of acetylcholine were mimicked by nicotine. However, both acetylcholine and nicotine were blocked at this site by prior administration of curare. In contrast, the action of muscarine on the parasympathetic nervous system was not blocked by curare – but by the alkaloid atropine. Dale attempted to provide an explanation for these results in 1914 when he argued that all these drugs worked at the synapse for acetylcholine. More importantly, he proposed there were two types of cholinergic receptor with different chemical affinities: *muscarinic* and *nicotinic*. It was a tacit acceptance of the neurotransmitter concept, although Dale was reluctant to speculate further. In fact, this would be his last research on acetylcholine for some years due to the outbreak of the First World War. Instead he would focus on histamine, which was implicated in producing states of shock after battle wounds.

The situation regarding the effects of chemicals on the peripheral nervous system had now reached a complicated stage, and it is perhaps helpful here to recap the main findings of the previous 20 years or so. Adrenaline (which was known to exist naturally in the body) had been shown to mimic the effects of the sympathetic system by Langley and Elliott. In contrast, acetylcholine (as yet unproven as a naturally occurring substance) had been shown to mimic the effects of the parasympathetic system by Dale and his colleagues. To make things more complex, nicotine had been shown to exert effects in the sympathetic ganglion (by Langley), and to work on the neuromuscular junction by Dale through a cholinergic mechanism. Although the concept of neurotransmitters and receptors had been formally proposed by Elliott, there were few, including Dale, who were prepared to openly embrace the idea. For many, it was more likely, despite the known existence of synapses at certain sites within the nervous system, that neural communication was electrical rather than chemical. Further progress would have to wait until after the First World War.

Demonstrating chemical neurotransmission: Otto Loewi

The feasibility of chemical neurotransmission in the autonomic nervous system, was strengthened, if not proven, by a seminal experiment performed by Otto Loewi in 1920. A German of Jewish extraction from Frankfurt, who admitted as a student having been more interested in his philosophy lectures than those for medicine, Loewi was to become Professor of Pharmacology at the University of Graz in 1909. He was well acquainted with the controversies surrounding chemical neurotransmitters for he had met John Langley, Thomas Elliott and Henry Dale during a visit to England in 1902 and 1903. However, Loewi's main research interests at the time were concerned with general metabolism, and later renal and pancreatic function. In fact, there is little evidence to show that he had any practical interest in the workings of the autonomic nervous system, until the night of Easter Saturday, 1920, when Loewi woke from a dream with an idea for an experiment in his head. After scribbling down his thoughts, Loewi returned to sleep, but was unable to decipher his notes the next morning. Fortunately, the following night, Loewi experienced the same dream. This time Loewi cycled to his laboratory where he performed a simple experiment that took less than two hours. It was to be one of the most famous in pharmacology. Indeed, upon seeing the results the next morning, Loewi's research assistant is said to have remarked it would win him the Nobel Prize. Although some were sceptical of the exact story line concerning Loewi's dramatic account, including his great friend Henry Dale, the predictions of his assistant proved true.

The experiment Loewi performed was relatively simple. A heart was removed from a frog still attached to its parasympathetic (vagus) and sympathetic nerves, and placed into a small chamber containing a solution of saline. Loewi then stimulated the vagus nerve, which caused the beat to slow down – a standard procedure that had been undertaken many times before. Then, Loewi did something new: he extracted some of the solution from the chamber, and applied it to a second heart whose vagus nerve had been removed. This too immediately caused the heart to slow down its beat. In other words, the second heart acted as if it was in receipt of stimulation from the vagus nerve. Following this, Loewi returned to the first heart, and now stimulated the sympathetic branch – a procedure causing it to beat faster. Again, he extracted the bathing fluid and applied it to a second deafferented heart. This time the heat beat showed an acceleration.[13] There was only one feasible explanation: chemicals were being liberated from the stimulated nerves of the first heart, which, when applied to the second heart, caused its beat to change. Loewi called the inhibiting agent from the vagus nerve *vagusstoff*, and the accelerating agent *acceleransstoff*.

Loewi published his experiment in a four page article in 1921, boldly claiming he had demonstrated the existence of chemical neurotransmission. He also argued that the two chemicals liberated from the nerve endings exerted their effects on the 'end-organs' of the heart itself – a tacit acceptance of Langley's receptor concept. However, not all pharmacologists were impressed. Many still remained sceptical of neurotransmission, and their doubts were increased when Loewi's frog heart experiment proved hard to replicate by others. We now know Loewi was fortunate in having undertaken his study at a time of the year when the temperature was cold, as this increases the stability of neurotransmitters (which are rapidly broken down by deactivating enzymes). Indeed, had Loewi performed his experiment in a warmer month, he may not have obtained a positive result. Furthermore, the frog heart is now recognised as proving the best preparation for this type of experiment – again, another

FIGURE 12.6 Otto Loewi (1873–1961). A German pharmacologist whose famous experiment, allegedly performed on the night of Easter Sunday in 1920, convinced many of the feasibility of chemical neurotransmission.

Source: Wellcome Library, London.

fortuitous choice by Loewi. Nonetheless, on account of the difficulties in replicating the experiment, Loewi was asked to demonstrate his experiment at an International Congress in Stockholm in 1926. He is said to have successfully done this on 18 separate occasions. This went a long way to helping dispel the doubt surrounding the experiment's authenticity.[14] It also encouraged a much wider belief in the plausibility of chemical neurotransmission.

In the years following his experiment, Loewi set about trying to establish the identity of what he had called *vagusstoff* and *acceleransstoff*, which inhibited and stimulated the heart respectively. Although Loewi suspected the former to be acetylcholine, and the latter an adrenaline-like substance, it would take a large number of experiments before this was clarified. However, gradually over a five year period, support would emerge for the acetylcholine theory. For example, Loewi showed that physostigmine, which allows acetylcholine to exist in the synapse for longer periods before it is broken down, enhanced the inhibitory effects of *vagusstoff* on the heart. In addition, Loewi found atropine (the drug that Langley had shown blocked the action of muscarine on the slowing of the heart) acted to inhibit the action of *vagusstoff*, which again supported the involvement of acetylcholine. However, neither physostigmine or atropine had any effect on the accelerator substance secreted by the sympathetic nerve. In fact, this substance was blocked by the drug ergotoxine that Dale had shown reversed the effects of adrenaline. However, Loewi was reluctant to commit himself to the idea *acceleransstoff* was adrenaline. He would be proven correct for it was later shown to be noradrenaline by investigators in the 1940s – the same neurochemical Dale had shown mimicked the effects of sympathetic stimulation.

Confirming acetylcholine as a neurotransmitter

By the late 1920s evidence had accumulated to show acetylcholine was a neurotransmitter – at least in the peripheral nervous systems. It had, for example, been shown to be the same chemical as *vagusstoff* liberated from the vagal pathway to the heart. Acetylcholine was also inactivated by a substance similar to physostigmine called cholinesterase. Importantly, Loewi was to find small traces of cholinesterase in heart tissue in 1930. However, before acetylcholine could be fully established as a neurotransmitter, it would have to be identified as a naturally occurring chemical in the tissues of the body. This was no easy task. Only tiny quantities were produced from the nerve endings, and this was believed to be broken down almost immediately into choline and acetic acid. The problem seemed insurmountable. However, around the same time as Loewi was establishing acetylcholine as a neurotransmitter, Dale was working at the Medical Research Institute in London examining the effects of histamine. This substance was also suspected of having a neurotransmitter role, although like acetylcholine, it was not certain at the time if histamine was a natural constituent of the body. However, this doubt was dispelled in 1927 when Dale managed to isolate histamine from the liver and lungs. The success of this endeavour led Dale to more seriously consider the problem of detecting acetylcholine, and he set about this task in 1929 with chemist Harold Dudley. By visiting a local slaughter house and collecting spleens taken from freshly killed horses and oxen, Dale and Dudley managed to obtain seventy-one pounds of tissue, from which they extracted one-third of a gram of acetylcholine. It was a tiny amount, but enough to prove acetylcholine was a natural constituent of biological tissue.

By the early 1930s most pharmacologists accepted the parasympathetic branch of the vagus nerve secreted acetylcholine for chemical neurotransmission at the synapse. However, there was little convincing evidence to show that acetylcholine acted in a similar fashion elsewhere, despite being implicated in many other physiological actions. This situation changed in 1933 when Dale secured the appointment of Wilhelm Feldberg to work in his laboratory. A Jew who had worked at the University of Berlin, Feldberg was forced to leave his post as the Nazi party came to power. This alerted Dale for Feldberg had developed a technique using a contracting muscle taken from a leech that was extremely sensitive to tiny amounts of acetylcholine. Even more important, Feldberg had bought with him a drug called eserine (physostigmine) which, when added to the muscle preparation, increased its sensitivity to acetylcholine more than a million fold. It was a pharmacological technique that was sensitive enough to detect the presence of acetylcholine extracted from the fluids of animals such as cats and dogs.

Feldberg set about doing this in Dale's laboratory by stimulating a given nerve, and then collecting the blood drawn from a nearby vein – hoping it would reveal the incredibly small amounts of acetylcholine (i.e. when it was applied to the leech muscle). In fact, this method proved so successful over the next three years, Feldberg and Dale used it to determine the sites of acetylcholine release throughout the peripheral nervous system, resulting in the publication of twenty-four papers in the *Journal of Physiology*. It was ground breaking research showing acetylcholine was secreted at many more places in the nervous system than previously envisaged. As expected, acetylcholine was found to be secreted by the nerve endings of the parasympathetic nervous system, where it inhibited the functioning of the visceral organs. In addition, acetylcholine was found to be the chemical released at the synapses of the sympathetic system located between the preganglionic and postganglionic fibres (i.e. in the sympathetic trunk). And, adding to this, acetylcholine was confirmed as the neurotransmitter at the neuromuscular junction. Thus, acetylcholine was a neurotransmitter in both the autonomic and somatic nervous systems.

In 1936, Dale and Loewi, who had been good friends for over 30 years, shared the Nobel Prize 'for their discoveries relating to chemical transmission of nerve impulses'. However, life would soon change for Loewi who worked at the University of Graz. Two years after his award, the Germans marched into Austria and imprisoned Loewi, along with his two youngest sons. Incarcerated for two months, Loewi was only released after agreeing to give up all his securities, including his Nobel Prize money. Loewi then managed to flee to London where he stayed with Dale for several months. Although now in his 60's, Loewi received an offer of a lectureship at the New York University College of Medicine, which he was happy to accept. It enabled him to buy an apartment on the west side of New York City and obtain U.S. citizenship in 1946. He would spend much of this time giving lectures, writing articles, and inspiring a new generation of American pharmacologists by recounting the story of how he had discovered chemical neurotransmission. Loewi lived in New York until his death in 1961.[15] Dale outlived his great friend by some seven years. In the intervening years he had received many accolades, becoming the president of the Royal Society (1940–1945) and receiving a knighthood in 1948. In addition, the eponym 'Dale's principle' was named after him by John Eccles in 1954, which stated neurons only release one type of neurotransmitter (a law now know to be incorrect). Dale would die in a Cambridge nursing home at the age of 93 years in 1968.[16]

The soup versus sparks debate

In his 1936 Nobel Prize acceptance speech, Dale speculated on the possibility of identifying chemical neurotransmitters in the brain and spinal cord. However, proving the existence of chemical neurotransmission in the periphery, even with relatively accessible procedures involving the neuromuscular junction or vagal stimulation of the heart, had been difficult enough. Establishing the same principle in the central nervous system would be a much more formidable challenge. And, even if chemicals were found as candidates, Dale knew it would prove almost impossible to establish if they served as neurotransmitters or some other physiological role. To confound the issue, some believed chemical neurotransmission in the central nervous system was not feasible. The most outspoken critic of this idea was the Australian John Carew Eccles, who favoured the idea of electrical transmission at the synapse. His forthright manner would lead to much heated debate at scientific conferences during the late 1930s and 1940s, which became known as the soup and sparks controversy. Although these meetings often appeared highly charged and adversarial, with the strident Eccles not afraid to clash with the more diplomatic Dale – behind the scenes the two men were good friends who exchanged letters and shared experimental findings. Although Eccles had been forced to accept chemical transmission between the vagus nerve and heart during the 1930s, he believed this was an exception to the rule. The main reason lay in his belief that chemical

FIGURE 12.7 John Carew Eccles (1903–1997) who was an outspoken critic of chemical neurotransmission in the CNS. That is, until 1952, when one of his own experiments provided convincing evidence for its existence.
Source: Wellcome Library, London.

transmission was too slow for the contraction of muscle, or flow of information in the brain. Despite this, he did concede in 1939, that acetylcholine may have some secondary role at the neuromuscular junction – perhaps by raising the excitability of the muscle to electrical stimulation.

Eccles was a native and medical graduate from Melbourne, who had won a prestigious Rhodes scholarship allowing him to become a doctoral student under Charles Sherrington at Oxford in 1928. Here Eccles turned his attention to studying neurophysiology – apparently because of his deeper interest in the issue of how the mind interacts with the brain. After receiving his Ph.D. in 1929, Eccles took up several academic appointments in Oxford, before returning to Australia in 1937 to become director of a research institute in Sydney. Although his chief responsibility was providing a clinical pathology service, Eccles used his new position to examine the electrophysiological properties of the neuromuscular junction, while also attracting a number of talented researchers to his laboratory. Seven years later, Eccles moved to the University of Otago in Dunedin, New Zealand, where he undertook the first ever intracellular recordings of nerve cells in the central nervous system in 1950. This achievement was possible through the development of extremely fine glass micropipette electrodes, which had a tip diameter of about 0.5 microns (or about fifty-thousandths of an inch). These were so tiny Eccles could not even use a microscope to guide them into the cell as they were smaller than the wavelength of light. Instead, they had to be connected to a oscilloscope, and implanted into the cell by following the changes in electrical potential. Fortunately, the cell membrane was found to seal itself around the micropipette, allowing the neuron to function normally for many hours after its insertion.

Using his new technique, Eccles examined the responses of nerve cells in the spinal cord of anaesthetised cats. More specifically, he focused on the motor neurons passing out to the quadriceps muscle of the thigh. He chose this preparation because, as Sherrington had shown, the cell body of its motor neuron (located in the spinal cord) received sensory feedback from fibres that conveyed information about the muscle's stretch (i.e. from its muscle spindle). When Eccles stimulated this feedback pathway with a small electrical current, he found it failed to trigger an action potential in the motor neuron. However, it did cause the resting potential inside the cell to increase its voltage slightly. Eccles called this temporary effect an *excitatory post-synaptic potential* (EPSP). As he increased the stimulation, thereby activating larger numbers of sensory nerve fibres to the motor neuron, Eccles found the EPSP became more positive. In fact, if the sum of the EPSPs increased the resting potential inside the motor neuron by about +15 millivolts, then it triggered an action potential. Put simply, Eccles had shown activity at a single synapse between sensory fibre and motor neuron was insufficient by itself to create a nerve impulse. Rather, there had to be activity at a large number of synapses before the motor neuron was 'excited' enough to generate an action potential. This was much the same as saying that one EPSP did not produce an action potential. However, if many EPSPs occurred at the same moment, then one could be created.

A year later, in 1952, Eccles used the same basic procedure to perform one of the most important experiments in the history of neurophysiology. Again Eccles implanted the micropipette into a cell body of a motor neuron projecting from the spinal cord to the quadriceps muscle. This time, instead of stimulating the sensory nerve from its muscle spindle, he stimulated one coming from the hamstrings of the posterior thigh – that is, the opposing muscle to the quadriceps. It had been known since the late nineteenth century, mainly through the work of Sherrington, these two muscles worked in a reciprocal fashion. Consequently, when one contracted, the other relaxed. It was also clear that some synapses in the system must be acting in an inhibitory way to allow this to happen. The concept of inhibition posed a problem to Eccles who could only account for it by assuming an electrical excitatory input was being blocked from reaching the cell. However, when he stimulated the sensory nerve from the hamstrings, he observed an *inhibitory post-synaptic potential* (IPSP) in the quadriceps motor neuron. In other words, the negative resting potential in the cell was made even more negative for a brief period – making the cell even more resistant to firing. This was contrary to what Eccles had expected. More crucially, there was no way his theory (i.e. blocking the flow of electrical energy across the synapse) could explain the IPSP.

There was only one reasonable explanation: both excitatory and inhibitory synapses had to exist between the axon terminals of the sensory nerves and quadriceps motor neuron, which used chemical neurotransmission. It was a pivotal moment when Eccles was forced to admit he had been wrong regarding the existence of electrical chemical neurotransmission.[17] Considering his previous blunt and out-spoken views on the subject, there has perhaps never been a greater volte-face in the history of science. Dale, for one, could not resist commenting 'a remarkable conversion' had occurred that was similar to 'Saul on his way to Damascus when sudden light shone and the scales fell from his eyes.' Yet, while history will tell us Eccles was wrong for rejecting chemical neurotransmission, his crucial experiment had proven its existence. Moreover, from now on, he would devote himself to understanding how inhibition operating presynaptically is able to regulate neurotransmitter release in the central nervous system. This work was to prove instrumental in Eccles being awarded the Nobel Prize in 1963 along with Alan Hodgkin and Andrew Huxley. It is also interesting to note Eccles was a devout Roman Catholic and one of the few Nobel laureates with a strong belief in God. He even believed in divine intervention, writing 'that there is a Divine providence operating over and above the materialistic happenings of biological evolution.'

Neurotransmitters in the brain

The discovery of chemical neurotransmission in the spinal cord also led to increasing interest in the possibility of neurochemicals in the brain, and it was not long before evidence of norepinephrine and acetylcholine was found there. They also appeared to have important functional roles. For example, in the 1950s, norepinephrine levels were shown to be decreased in the hypothalamus of experimental animals after subjecting them to high stress by Marthe Vogt working in Edinburgh. In contrast, acetylcholine was found by J.F. Mitchell in Cambridge to be secreted in the cat's somatosensory cortex after the animal's paws had been stimulated. More intriguingly, pharmacologists also began to find evidence of new potential neurotransmitters in the brain. One of these was 5-hydroxtryptamine (serotonin). This chemical had first been discovered in blood serum in 1948, where it acted as a clotting factor. Five years later, in 1953, large quantities of serotonin were identified in the brain. Its neurotransmitter status was given impetus when lysergic acid diethylamide (LSD) was found to block the effect of serotonin on vascular smooth muscle. It led many to suspect that LSD was also acting on serotonergic-containing neurons in the brain. Another substance to attract attention was dopamine. Until the late 1950s, dopamine was only thought to be a chemical involved in the synthesis of noradrenaline. However, in 1958, the Swedish pharmacologist Arvid Carlsson discovered dopamine in the brain, including the striatum where most of it was concentrated. The importance of striatal dopamine would further be highlighted when its near total depletion was implicated as a cause of Parkinson's disease in the 1960s.

Today, some 50 years later, it is known that about 100 substances act as neurotransmitters, or have some neuromodulatory role, in the brain and spinal cord. These substances come in many different forms, including monoamines, amino acids, peptides and even gases. The story of their discovery and our modern chemical understanding of the brain, however, would require another book to be written. However, progress at proving the neurotransmitter status of even norepinephrine or acetylcholine in the 1950s and 1960s would be slow and difficult – not least because to be established as a neurotransmitter requires a number of criteria to be fulfilled. For the record these criteria are: (1) there have to be precursors and/or synthesis enzymes for the neurotransmitter; (2) the neurotransmitter must be shown to exist in the presynaptic endings of the axon; (3) it must exist in sufficient quantities to affect the postsynaptic cell; (4) there must be specialized receptors for the neurotransmitter; an (5) there must be specialized receptors for the neurotransmitter's deactivation. This list gives at least some indication of the difficulties facing the neuropharmacologist in proving chemical transmission.

The first step in establishing a neurotransmitter, however, is to identify it as a naturally occurring substance in neural tissue. This is not as straightforward as it first appears since neurotransmitters are unstable and broken down quickly after they are released. Consequently, because they can not normally be measured directly, pharmacologists have to find alternative ways of identifying them. An early approach to detect acetylcholine was developed in 1949 by George Koelle at John Hopkins University who used a stain for acetylcholinesterase (AChE). This is the enzyme that breaks down acetylcholine after it has been released into the synapse. Based on the reasonable assumption that AChE will be found wherever acetylcholine occurs, Koelle used this staining method to identify the location of acetylcholine in the brain. However, this technique was used to greater effect in the 1960s by Charles Shute and Peter Lewis in Cambridge who used the AChE stain in conjunction with a lesioning method that destroyed selective parts of the brain. By examining how certain lesions disrupted AChE staining, Shute and Lewis found that acetylcholine was localised in distinct pathways. For example, their work showed much of the cholinergic innervation of the forebrain (cerebral cortex, limbic system and striatum) came from fibres originating in the brainstem, or basal parts of the forebrain close to the hypothalamus.

A different technique was developed to map out noradrenaline. In the mid 1950s, Olavi Eränkö working at the University of Helsinki, discovered cells of the adrenal gland fluoresced in the colour green if treated with formaldehyde vapour and exposed to ultraviolet light. This also enabled the adrenaline within the tissue to be visualised under a microscope. It was a discovery prompting others to consider whether it could be extended to nerve cells in the brain. Two investigators who took up the challenge in the early 1960s were the Swedes Bengt Falck and Nils-Ake Hilarp. Within a few years they had not only successfully identified noradrenergic neurons, but showed that the highest concentrations of the chemical occurred in the axon terminals – a pattern clearly supporting its function as a neurotransmitter. This work was extended by Annica Dahlstrom and Kjell Fuxe who found a way of distinguishing between noradrenaline, dopamine and serotonin on the basis of their fluorescent colours. It was an important breakthrough, for their method could be combined with lesioning, to map out the pathways containing these substances in the brain. Their work revealed a number of new neurochemical pathways, most of which arose from small and clearly defined areas of the brainstem or midbrain. For example, the origin of most noradrenaline fibres in the forebrain arose from a tiny nucleus in the pons called the *locus coeruleus*. In contrast, most of the serotonin derived from a nearby nucleus called the *raphe*. Dahlstrom and Fuxe also showed these noradrenaline and serotonin containing fibres have a widespread distribution throughout in the forebrain, which overlap in most regions. However, the dopamine containing pathways, which arise from the substantia nigra and ventral tegmental area located in the midbrain, projected to more localised regions of the brain, including the striatum, frontal cortex and ventral striatum (nucleus accumbens). These discoveries have helped revolutionise our understanding, especially in regards to how drugs affect behaviour and psychological processes.

The first drug treatments for mental illness

During the 1950s, a revolution would take place in the fields of psychopharmacology and psychiatry with the introduction of psychoactive drugs to treat mental illness. A notable day in this respect would be the 26 March 1954, when the Food and Drug Administration in the US approved the use of chlorpromazine (Thorazine) as a medication for the treatment of schizophrenia. It was the first time a drug had been specifically marketed to treat a mental illness, and within a decade it was to be followed by many others, including antidepressants and anxiolytics. Today, it is difficult to imagine a world without drugs for mental illness, but prior to the 1950s the situation was different. Anyone unfortunate enough to be suffering from a serious mental illness had few options. They could perhaps be given a barbiturate to make them sleep, or an amphetamine-like stimulant to revitalise the brain. For severe depression, electroconvulsive shock (ECT) or even a frontal lobotomy were not uncommon attempts at a cure. Such treatments were the desperate attempts of well-meaning psychiatrists to do something to help their patients. Nonetheless, many mentally ill patients were placed in institutions, often for long periods of time, with little hope of recovery. Indeed, by the mid 1950s the number of institutionalised people had reached record levels. This situation changed dramatically, however, with the introduction of chlorpromazine that calmed agitated mental patients and enabled them to carry on relatively normal lives outside the confines of mental institutions. The success of this drug is shown by the fact that it led to a significant decline in the number of hospitalised people with mental health problems: the number of patients in US asylums dropping from a high of 559,000 inpatients in 1955, to 452,000 just a decade later.

Although the discovery of chlorpromazine was largely serendipitous, it also depended on astute clinical observation. Chlorpromazine belongs to a class of drugs known as phenothiazines, and these had first come to the attention of chemists working in the textile industry as dye fixatives in the late nineteenth century. They had even been used as worming agents in veterinary practice. More importantly, in the 1930s, some phenothiazines were found to have anti-histamine properties. This was the era when Dale was undertaking his classic work on histamine, with pharmaceutical companies beginning to recognise its commercial opportunities. One company to begin synthesising antihistamines, including different types of phenothiazines, was the French owned Rhone-Poulenc (now Sanofi-Aventis). It would lead to the development of promethazine, synthesised by Paul Charpentier in 1947. Although initially used as an antihistamine to stabilise blood pressure in operations requiring anaesthesia, promethazine was also found to calm anxious patients before surgery. This was an interesting effect, not least because there was a need at the time for new sedatives to be introduced in psychiatry. Encouraged by this new effect, in 1950, Charpentier introduced a chlorine atom into one of the rings of promethazine, thereby forming chlorpromazine. Given to the French neurosurgeon Henri Laborit for testing, chlorpromazine produced such a remarkable state of calmness and indifference in his patients, he recommended it to two psychiatrists, Jean Delay and Pierre Deniker. They soon established it was highly effective at treating schizophrenia. Chlorpromazine not only rendered difficult patients more manageable, but reduced core psychotic symptoms including hallucinations, delusions and grandiose thoughts. It also differed from other sedatives as it did not induce coma at high doses. Delay and Deniker published their findings in 1952, leading to the use of chlorpromazine in Europe (where it was marketed as Largactil)[18] and the US in 1954. Within ten years, it is estimated chlorpromazine had been administered to fifty million patients.

In 1958, a second type of antipsychotic drug was developed by Paul Janssen, founder of Janssen Pharmaceutica in Belgium, called haloperidol. Believing that if he could find a drug that reversed amphetamine intoxication, then he would have a potential substance for overcoming psychotic behaviour, Janssen came across a butyrophenone compound (not a phenothiazine), which had a similar but more powerful action than chlorpromazine. Testing the drug on his first patient, a 16-year-old boy with paranoid schizophrenia, its effect was reported by Janssen to be 'amazing'. The patient would be maintained on no more than 1 mg of the drug per day and go on to graduate in architecture, marry, and father two children. Furthermore, haloperidol was shown to produce fewer side effects than chlorpromazine, which would increase its popularity, especially in Europe. Today, it remains the first line treatment for schizophrenia.

Chlorpromazine and haloperidol not only gave many patients a new lease of life, but provided opportunities for neuroscientists to better understand the biological causes of their illness. In other words, if the pharmacological effects of these drugs could be explained, then they would most likely provide vital clues about the underlying neurochemistry of schizophrenia. The first important clue, however, came from reserpine that had been isolated from the *Rauwolfia serpentina* plant by the Swiss owned Ciba Drug Company during the early 1950s. This drug, like chlorpromazine and haloperidol, calmed down psychotic patients, although it produced a number of troublesome side effects which severely limited its use. These included severe reduction in blood pressure and a Parkinson-like rigidity. Nonetheless, reserpine appeared to have a simple pharmacological action, for by the late 1950s, it had been shown to deplete the brain of noradrenaline, dopamine and serotonin. Of these, dopamine depletion appeared to be most important for its antipsychotic effect – largely because reserpine could antagonise the effects of amphetamine (a drug known to cause the release of dopamine from nerve endings, and also produce psychotic symptoms). However, early studies with chlorpromazine or haloperidol showed that neither drug lowered brain levels of dopamine. It was a peculiarity that initially puzzled researchers until a novel explanation was proposed by Avrid Carlsson and Margit Lindqvist in 1963. They suggested chlorpromazine and haloperidol worked on the brain by 'blocking' dopamine receptors. In other words, both drugs reduced dopamine activity in the synapse without depleting it. This idea could not be experimentally verified in 1963. However, Carlsson and Lindqvist would be proven correct in the early 1970s when new discoveries allowed receptors to be examined in greater detail. This work would also lead to the dopaminergic theory of schizophrenia (still the best theory we have) and Carlsson's Nobel Prize in 2000 (along with Eric Kandel and Paul Greengard).

The first antidepressants

Some three years after the introduction of chlorpromazine in 1954, the first antidepressant became commercially available. This was iproniazid (marketed under the trade name of Marsilid) and like chlorpromazine its discovery owed more to good fortune than planned intent. The beginnings of antidepressant pharmacology can be said to begin with a substance called hydrazine that was used towards the end of the Second World War in Germany as a propellant for the V2 rocket. After the war, large stocks of hydrazine were taken over by pharmaceutical companies interested in seeing if any of its derivatives had clinical uses. One such derivative of hydrazine was called isoniazid, synthesised by the Hoffmann-La Roche drug company, which turned out to be an effective agent in the treatment of tuberculosis. This was a significant development for tuberculosis was a significant cause of death at the time, and it led to other derivatives being tested, one of which was called iproniazid. Although this too had antitubercular properties, it also became apparent that iproniazid was helping many seriously ill and emaciated patients with lung infections to increase weight and feel more optimistic. It was even reported by the *Associated Press* in 1953, that at one New York hospital, iproniazid caused tuberculosis patients to dance and be in a party mood. This situation did not go unnoticed by a number of American psychiatrists, including Nathan Kline, who began clinical trials with iproniazid on patients with various types of metal illness. The results were impressive with about 70 per cent of patients showing a significantly enhanced mood. Since the drug was being used as a treatment for tuberculosis, it was straightforward to market iproniazid for depression, and it was introduced for this purpose as Marsilid in 1957. Within one year, more than 400,000 would be treated with this drug in the US.

Moreover, unlike the antipsychotic drugs, researchers had a good idea of how iproniazid worked in the brain. In 1952, the biochemist Albert Zellar, working in Chicago, had found iproniazid inhibited the enzyme monoamine oxidase (MAO). This is a substance responsible for breaking down monoamine neurotransmitters (noradrenaline, serotonin and dopamine) after their release into the synapse. Thus, by inhibiting MAO, iproniazid appeared to be exerting its antidepressant effect by increasing the levels of these neurotransmitters at their receptors. It was the first clue that antidepressive drugs might be working by correcting a 'chemical imbalance' in the brain, which also provided the underlying cause of the illness. In other words, the success of MAO inhibition in treating depression, implied this illness occurred when levels of the aminergic neurotransmitters were abnormally low, or when their neurochemical systems in the brain became under active.

Marsilid was not the only antidepressant to be marketed in the 1950s. The success of the Rhone-Poulenc drug company in developing effective antipsychotic agents had led to other companies, including the Swiss based Geigy, to also synthesise antihistamine compounds. One of these was simply known as G-22355, later to be known as imipramine. Although developed as an antipsychotic agent (it closely resembled chlorpromazine) when tested by the Swiss psychiatrist Roland Kuhn, it was found to be much more effective in treating withdrawn and depressed patients, than those suffering from hallucinations and delusions. This was a totally unexpected finding and at first treated with some scepticism by the medical community. Nonetheless, its mood enhancing effects were repeated in many other studies, leading to imipramine being launched in Switzerland in 1957, and marketed as an antidepressant across Europe under the name Tofranil a year later. Since then, imipramine has maintained its status as one of the most effective antidepressants up to the present day.

Imipramine was a completely different type of drug than Marsilid since its molecular structure contained three rings – leading to this class of agent becoming known as the tricyclic antidepressants. Furthermore its pharmacological action was different since it did not inhibit MAO. Yet, imipramine was found to increase the levels of noradrenaline and serotonin in the brain. The important question was how? The answer was provided by Jules Axelrod in 1961 working at the National Institute of Health in Bethesda. Axelrod was to show the endings of sympathetic noradrenergic neurons have a specialised uptake pump that removed excess neurotransmitter from the synapse – a discovery leading him to win a Nobel Prize in 1970. Axelrod also demonstrated imipramine blocked this process (it was later shown to block serotonin and dopamine uptake pumps as well). By acting in this way, imipramine slowed down the removal of monoamines in the synapse, thereby increasing their concentration. These discoveries would also lend support to the monoamine theory of depression, which held that depressed mood was due to low levels of noradrenaline and serotonin occurring in the brain. This remains, even today, the best theory we have to explain the cause of depression.

Notes

1. Bernard is best remembered today for introducing the concept of the *milieu intérieur* (sometimes known as homeostasis) as a basic requirement of life. The term *milieu intérieur* basically refers to the fact that while the external environment constantly changes, the interior milieu of the body must be kept under strict limits if life is to be maintained.
2. Derived from two types of woody vine (*Strychnos toxifera* and *Chondrodendron tomentosum*).
3. An ex student of Bernard's in Paris, Kühne had replaced von Helmholtz as professor at Heidelberg in 1871.
4. In this respect, Vulpian would be proven correct, although it would be another 50 years before this was demonstrated to be true.
5. The synapse was named by Sherrington in 1897 (see Chapter 7).
6. Both Galen and Willis believed this ganglion derived from the brain, although this was shown to be false in 1727 by Francois-Pourfour du Petit who showed that it originated from below the cranium.
7. Although these two ganglia are separate from the spinal cord, they are nevertheless attached to it by a mass of white nerve bundles derived from the spinal nerves.
8. Weber and Weber were not the first to show inhibition in the nervous system. Charles Bell had shown that opposing nervous forces acted on the muscles of the eye, and Volkmann (1838) had shown vagal inhibition before Weber and Weber. However, it was the work by Weber and Weber that attracted most interest.
9. Alkaloids are a group of chemical compounds, which contain mainly nitrogen atoms. They also occur naturally in plants, many of which have been used since ancient times for their recreational or therapeutic purposes.
10. In addition to the sympathetic and parasympathetic components, Langley would add a third system to the autonomic nervous system. This was the enteric nervous system, which is intrinsic to the walls of the gastrointestinal tract. Because it is largely independent of the functioning of the sympathetic and parasympathetic system, it is sometimes referred to as the body's second brain.
11. Dale would be the director of National Institute for Medical Research from 1928 to 1942.
12. Confusingly, the substance called adrenaline by most of the world is known as epinephrine to Americans. To add to the confusion, noradrenaline is correspondingly known as norepinephrine. The general reader, however, only has to know that they are slightly different structurally.
13. Curiously, Loewi never provided a diagram of his experimental set-up, and it is known that in some instances he only used one heart in his experiments – applying the solutions to its different ventricles. It is also reported in some accounts (although it is probably false) that Loewi used an apparatus where the two hearts in separate chambers shared the same solution through a interconnecting tube.
14. Valenstein in his book *The War of the Soup and Sparks* also reports that in one instance Loewi was asked to stand at one end of the room and give instructions to a demonstrator standing at the other who was in charge of the heart preparation. This was to eliminate the possibility that some substance from Loewi's own fingers or fingernails were not contaminating the assay.
15. In 1973, at the centenary of his birth, Austria honoured Loewi by printing a special stamp.
16. One of the more unusual stories concerning Dale was that he was given the task of analysing several unidentified pills belonging to Hitler's deputy Rudolph Hess after his bizarre flight from Berlin to Scotland in 1941. Dale was to establish that the pills contained atropine, which was then commonly used to treat travel sickness.
17. Actually, electrical synapses do exist and were first demonstrated in the crayfish by Furshpan and Potter in 1957. They have since been demonstrated in vertebrates.
18. Apparently meaning 'large activity'.

Bibliography

Ackerknecht, E.H. (1974) The history of the discovery of the vegetative (autonomic) nervous system. *Medical History*, 18, 1–8.

Andersen, P. and Lundberg, A. (1997) Obituary: John C. Eccles (1903–1997) *Trends in Neurosciences*, 20, 324–325.

Bacq, Z.M. (1975) *Chemical Transmission of Nerve Impulses: A Historical Sketch*. Pergamon Press: New York.

Bennett, A.E. (1968) The history of curare into medicine. *Anaesthesia and Analgesia*, 47, 484–492.
Bennett, M.R. (2000) The concept of transmitter receptors: 100 years on. *Neuropharmacology*, 39, 525–546.
Burke, R.E. (2006) John Eccles' pioneering role in understanding central synaptic transmission. *Progress in Neurobiology*, 78, 173–188.
Cannon, W.B. (1929) *Bodily Changes in Pain, Hunger, Fear and Rage*. Appleton-Century-Crofts: New York.
Cannon, W.B. (1934) The story of the development of our ideas of chemical mediation of nerve impulses. *The American Journal of the Medical Sciences*, 188, 145–159.
Cowan, W.M., Südhof, T.C. and Stevens, C.F. (2001) *Synapses*. John Hopkins Press: Baltimore, MD.
Dale, H.H. (1953) *Adventures in Physiology: A Selection of Scientific Papers*. Pergamon: London.
Dale, H.H. (1962) Otto Loewi 1873–1961. *Biographical Memoirs of Fellows of the Royal Society*, 8, 67–89.
Davenport, H.W. (1991) Early history of the concept of chemical transmission of the nerve impulse. *The Physiologist*, 34, 129–190.
Dupont, J.-C. (2006) Some historical difficulties of the cholinergic transmission. *Comptes Rendus Biologies*, 329, 426–436.
Eccles, J.C. (1961) The Ferrier Lecture: The nature of central inhibition. *Proceedings of the Royal Society of London, Series B*, 153, 445–476.
Eccles, J.C. (1965) The synapse. *Scientific American*, 212, 56–66.
Feldberg, W. (1969) Henry Hallett Dale 1875–1968. *British Journal of Pharmacology*, 35, 1–9.
Fick, G.R. (1987) Henry Dale's involvement in the verification and acceptance of the theory of neurochemical transmission: a lady in hiding. *The Journal of the History of Medicine and Allied Sciences*, 42, 467–485.
Fishman, M.C. (1972) Sir Henry Dale and the acetylcholine story. *Yale Journal of Biology and Medicine*, 45, 104–118.
Fletcher, W.M. (1926) John Newport Langley in memoriam. *Journal of Physiology*, 61, 1–15.
Gray, R. (1918) *Gray's Anatomy of the Human Body*, 20th edn. Lea & Febiger: Philadelphia, PA (re-edited by Warren H. Lewis).
Langdon-Brown, W. (1939) W.H. Gaskell and the Cambridge Medical School. *Proceedings of the Royal Society of Medicine*, 33, 1–12.
Langley, J.N. (1903) Sketch of the progress of discovery in the eighteenth century as regards the autonomic nervous system. *Journal of Physiology*, 50, 225–258.
Leake, C.D. (1965) Historical aspects of the autonomic nervous system. *Anesthesiology*, July-August, 623–624.
Lee, M.R. (2005) Curare: the South American arrow poison. *Journal of the Royal College of Physicians of Edinburgh*. 35, 83–92.
López-Munoz, F. and Alamon, C. (2009) Historical evolution of the neurotransmitter concept. *Journal of Neural Transmission*, 116, 515–533.
López-Munoz, F. and Alamon, C. (2009) Monoaminergic neurotransmission: the history of the discovery of antidepressants from 1950s until today. *Current Pharmaceutical Design*, 15, 1563–1568.
Maehle, A.-H. (2004) 'Receptor substances': John Newport Langley (1852–1925) and his path to a receptor theory of drug action. *Medical History*, 48, 153–174.
Maehle, A.-H. (2009) A binding question: the evolution of the receptor concept. *Endeavour*, 33, 134–139.
Minz, B. (1955) *The Role of Humoral Agents in Nervous Activity*. Charles C. Thomas: Springfield, IL.
Morgan, C.T. (1965) *Physiological Psychology*. McGraw-Hill: New York.
Nozdrachev, A.D. (2002) John Newport Langley and his construction of the autonomic (vegetative) nervous system. *Journal of Evolutionary Biochemistry and Physiology*, 38, 537–546.
Parascandola, J. (1980) Origins of the receptor theory. *Trends in Pharmacological Sciences*, 4, 189–192.
Parnham, M.J. and Bruinvels, J. (eds) (1983) *Discoveries in Pharmacology, vol. 1: Psycho and Neuropharmacology*. Elsevier: Amsterdam.
Pearce, J.M.S. (2009) Links between nerves and glands: The story of adrenaline. *Advances in Clinical Neuroscience and Rehabilitation*, 9, 22–28.

Prüll, C.-R., Maehle, A.-H. and Halliwell, R.F. (2009) *A Short History of the Drug Receptor Concept*. Palgrave Macmillan: Basingstoke.

Rubin, R.P. (2007) A brief history of great discoveries on pharmacology: in celebration of the centennial anniversary of the founding of the American Society of Pharmacology and Experimental Therapeutics. *Pharmacological Reviews*, 59, 289–359.

Sheehan, D. (1936) Discovery of the autonomic nervous system. *Archives of Neurology and Psychiatry*, 35, 1081–1115.

Simmons, J.G. (2002) *Doctors and Discoveries: Lives that Created Today's Medicine*. Houghton Mifflin: Boston, MA.

Skandalakis, L.J., Gray, S.W. and Skandalakis, J.E. (1986) The history and surgical anatomy of the vagus nerve. *Surgery*, 162, 75–85.

Snyder, S.H. (1986) *Drugs and the Brain*. Scientific American library: New York.

Tansey, E.M. (2003) Henry Dale, histamine and anaphylaxis: reflections on the role of chance in the history of allergy. *Studies in History and Philosophy of Biological and Biomedical Sciences*, 34, 455–472.

Tansey. E.M. (2006) Henry Dale and the discovery of acetylcholine. *Comptes Rendus Biologies*, 329, 419–425.

Todman, D. (2008) John Eccles (1903–97) and the experiment that proved chemical synaptic transmission in the central nervous system. *Journal of Clinical Neuroscience*, 15, 972–977.

Valenstein, E.S. (2002) The discovery of chemical neurotransmitters. *Brain and Cognition*, 49, 73–95.

Valenstein, E.S. (2005) *The War of the Soup and Sparks: The discovery of Neurotransmitters and the Dispute over How Nerves Communicate*. Columbia University Press: New York.

Weatherall, M. (1990) *In Search of a Cure*. Oxford University Press: Oxford.

Willis, W.D. (2006) John Eccles' studies of spinal cord presynaptic inhibition. *Progress is Neurobiology*, 78, 189–214.

Zigmond, M.J. (1999) Otto Loewi and the demonstration of chemical neurotransmission. *Brain Research Bulletin*, 50, 347–348.

13
NEUROSURGERY AND CLINICAL TALES

When one of these flashbacks was reported to me by a conscious patient, I was incredulous. For example, when a mother told me she was suddenly aware, as my electrode touched the cortex, of being in the kitchen listening to the voice of her little boy who was playing outside in the yard.

Wilder Penfield

What it comes down to is that modern society discriminates against the right hemisphere.

Roger Sperry

Summary

While the history of cranial surgery, or trepanation, goes back to prehistoric times, surgical intervention on the brain itself is a relatively recent phenomenon. This practice first arose in the late nineteenth century when a number of developments took place making it safer and more feasible. One was the invention of anaesthesia in 1846, and another was the introduction of carbolic acid to limit infection by British surgeon Joseph Lister in 1867. However, a less obvious prerequisite for the development of neurosurgery was the need to understand better the behavioural functions of the brain. Although it may have been obvious to a mid-nineteenth century surgeon that a patient had a tumour or piece of shrapnel embedded in their brain, the difficulties of knowing where such damage lay, hidden from view, prohibited an operation from being undertaken. This problem was overcome, however, by the work of investigators such as Paul Broca and David Ferrier, who began to identify localised regions in the cerebral cortex for functions such as language and movement. As these areas were established, surgeons became increasingly more confident at reading the behavioural signs of their patients prior to surgery to give them clues where lesions might reside. The first surgical procedure on the brain was undertaken by William Macewen at the Royal Glasgow Infirmary in 1879, who successfully removed a tumour from the surface of the cerebral cortex. This was soon followed by more celebrated cases in London by Richard Godlee and Victor Horsley. From this point onwards, the growth of neurosurgery would be surprisingly rapid, and by the early part of the twentieth century had firmly established itself as a specialised branch of medicine. This was not only of benefit to a wide variety of patients, but in some cases allowed the surgeon to gain remarkable new insights into the workings of the human brain. One investigator to do this was Wilder Penfield who founded the Montreal Neurological Institute in 1934 for the surgical treatment of epilepsy. His stimulation of the brain in awake, conscious patients prior to surgery, among other things, led to the discovery of the motor and somatosensory cortices with a map-like representation of the body. Another surgical development with implications for understanding brain function was the severing of the corpus callosum. In the hands of Roger Sperry, the testing of callosum lesioned patients showed that the right and left cerebral hemispheres have different mental and emotional characteristics. Despite many successes, there have also been failures, and the most famous is the case of Henry Molaison (HM) who had the best part of his medial temporal lobes removed for the relief of epilepsy in 1953. His profound anterograde amnesia, while of great interest to psychologists and neuroscientists, is a salutary warning that neurosurgery can go wrong and should not be taken lightly.

Recovery from brain injury: Francois Quesnay

Surgery is one of mankind's oldest preoccupations, with its origins going back to prehistoric times when humans first started to make and handle stone tools (see Chapter 1). Ever since then, surgical techniques have increasingly become more sophisticated as technological innovation has advanced. However, the foundations of our modern surgery today can be seen to lie more with developments in the eighteenth century, which also saw technological progress through the rise of the industrial revolution. Otherwise known as the era of Enlightenment, this was a time when rational thought helped transform surgery from an often precarious 'art' into a scientific discipline capable of treating many diseases and ailments. This can be seen, for example, in the case of amputation where the eighteenth century surgeon was often surprisingly successful in removing limbs without causing death. Despite having no effective anaesthetics, or antiseptics, the surgeons were nonetheless generally skilled enough to control bleeding through the use of ligatures, drain wounds to avoid inflammation, and use sutures to stitch tissues together for healing. All of this was a hundred years before the role of germs in disease and infection was understood.

Yet, even in the eighteenth century, there were few surgeons prepared to extend their expertise to the brain. Even the relatively simple operation of opening the cranium through trepanation was considered highly risky. The reason was simple: the risk of fatality through infection in the crowded and unclean hospitals at the time was too great, with mortality as high as 90 per cent in some places. This meant that the great majority of head injuries were left untreated. However, one of the few doctors to take an alternative view was French surgeon Francois Quesnay. Although born in a prosperous farming family close to Versailles in 1694, Quesnay received no formal schooling, and only learned to read and write after being taught by his gardener. Despite this, Quesnay managed to teach himself the intricacies of Greek and Latin, and became sufficiently educated in medicine by the age of 16 to be apprenticed as a barber surgeon. It would lead him to Paris in 1712 where Quesnay won a place to study medicine at the College de Saint Côme. Recognised for his surgical expertise, Quesnay would rise from a country doctor to be personal physician to Louis XV in 1749. This period proved to be a prosperous one for Quesnay who was to save the King's son and heir from smallpox in 1852 – a feat for which he received a large sum of money and a title of nobility.[1]

This was also a time when several European wars were raging with French involvement, and Quesnay treated many soldiers with wounds returning from battle. His observations would lead him to write several celebrated essays including one entitled 'Remarks on Wounds of the Brain' published in 1743 which described the effects of brain damage from bone fragments, bullets and sword injuries. Contrary to what was widely believed at the time, Quesnay realised many injuries to the brain were not fatal. This was particularly true when damage occurred to the frontal lobes. For example, Quesnay describes a brigadier who received a musket ball above his eyebrows which penetrated 'two fingers length' into the brain, yet did not produce any morbid symptoms. Similarly, another soldier survived a penetration of lead shot into the frontal regions for many years without any noticeable mental or behavioural deficits. In contrast, damage to the cerebellum was invariably deadly, and injuries to the corpus callosum always ruinous to the mind of the individual.

Quesnay also supported these observations with his own experimental findings on dogs. This included driving nails through their brains – a procedure apparently causing the animals no serious long-term impairment. From this work and many other observations, Quesnay realised the brain was insensitive to pain, and its wounds healed 'almost as rapidly as wounds of other viscera.' Consequently, Quesnay became convinced that brain surgery was feasible under certain circumstances. This included the removal of brain tissue to search for an abscess, or some other foreign body, especially if causing pain and paralysis. He also believed that brain tumours could be removed with little serious or lasting impairment, and should be attempted if it meant saving a person's life. Quesnay's one reservation, however, was that brain surgery should take place in the person's home, and not a hospital which was likely to have an 'unwholesome air'. Unfortunately, Quesnay was alone in his views. Few in the medical establishment were willing to agree with him, and he was criticised in Britain for his speculative approach to surgery. Quesnay's later rejection of medicine for economics also meant his ideas were soon forgotten and would be virtually unknown to subsequent neurosurgeons. Strong opposition to neurosurgery would remain until the late nineteenth century.

The beginnings of modern neurosurgery

It is a matter of some debate when the first modern operative procedure on the brain took place, but for many it was performed in 1879 by the head surgeon at the Royal Glasgow Infirmary, William Macewen. There were a number of reasons why the time and place was right for such a procedure. One was the invention of anaesthesia in the 1840s, which reduced the impact of pain and opened up the possibility of new and daring surgical operations on the body. Another was the pioneering work of surgeon Joseph Lister (also at Glasgow Infirmary) who developed the use of carbolic acid to sterilise surgical instruments and clean wounds during the 1860s. This greatly reduced the risk of infection and increased the chances of survival for patients. Indeed, Macewen was to become one of the first surgeons to follow in Lister's footsteps by wearing a sterilised gown, and spraying his operating theatre with carbolic acid prior to surgery. However, there was another reason why brain surgery was

FIGURE 13.1 William Macewen (1848–1924). Scottish surgeon at the Royal Glasgow Infirmary who performed the first modern operative procedure on the brain in 1879.
Source: Wellcome Library, London.

becoming an increasingly feasible proposition in the second part of the nineteenth century. The work of experimental researchers such as David Ferrier had provided support for the idea of cortical localisation with discrete regions for functions such as language, sensation and movement. Now armed with 'maps' of these sites, surgeons could establish with better accuracy where a tumour, abscess or injury, was likely to be located in the brain by carefully examining their patient's behaviour. In turn, they could be confident of finding the lesion in the brain if they took a chance and performed the operation.

Macewen had first contemplated brain surgery as early as 1876 when diagnosing a suspected abscess in a young boy, but was not given permission by the child's physician to operate. Sadly, the boy died and his autopsy revealed an abscess where Macewen had predicted. Three years later, however, Macewen would perform the first brain intervention by removing a tumour in a 14-year-old girl called Barbara Wilson. The girl was suffering seizures of the right arm and face, which Macewen correctly deduced was from a left frontal lobe lesion in the vicinity of the motor cortex. To confirm this, Macewen trepanned above the suspected site, and in doing so he detected a tumour of the dura mater.[2] It was half an inch in thickness at its centre and had spread over much of the surrounding tissue. Macewen successfully removed the growth, restoring the patient back to perfect health, with the girl living another eight years before dying from natural causes. For reasons that are not clear, Macewen did not immediately publish the results of this operation. Nonetheless, the success of the surgery encouraged Macewen to perform several other procedures, including ones to remove blood clots, abscesses and splinters of bone. When Macewen eventually published the results of his surgery in 1888, he had performed brain surgery on 21 patients. 18 patients had made a full recovery with three deaths.

Around this time, similar operations were starting to be performed in London. An impetus behind this development was the Seventh International Medical Congress in London, August 1881,which according to some authors should be recognised as the most important event leading to the modern founding of neurosurgery.[3] This was the conference that had witnessed the famous public dispute between David Ferrier and Friedrich Goltz concerning the possibility of cortical localisation – an argument settled decisively by an independent panel in favour of Ferrier (see Chapter 9). Also watching the proceedings of this event were a number of British doctors interested in the potential uses of cortical localisation for neurosurgery. These included Rickman Godlee and Victor Horsley. Ferrier's demonstration of cortical localisation convinced many that a reliable map of the cerebral cortex could now be established. It would not only assist surgeons in their quest to understand behavioural symptoms, but to locate brain damage. Moreover, Ferrier's use of surgical anaesthetic and antiseptic techniques had resulted in the long-term survival of his monkeys after invasive procedures on the cerebral cortex. It was an encouraging sign that full recovery following brain surgery was now a viable proposition in humans.

The first London operation took place in November 1884, when surgeon Rickman Godlee at University College Hospital, was called upon by physician Alexander Hughes Bennett to operate on one of his patents. This was a 25-year-old epileptic farmer suffering from progressive paralysis to the left side of his body that had destroyed the power of his arm. From his understanding of cortical localisation, Bennett suspected a tumour in the right motor cortex. Godlee agreed to undertake the surgery, and with John Hughlings Jackson and David Ferrier in attendance, he administered chloroform anaesthesia to the patient under a mist of carbolic acid. Godlee also calculated the site of the motor cortex by drawing lines on the scalp using anatomical landmarks. The bone was then removed by splitting three trephine holes with a mallet and chisel. Although no tumour was visible on the brain's surface, an exploratory cortical incision revealed a glioma deeper in the tissue. By stopping the region's blood flow with a hot cautery, the tumour was excised without too much difficulty. Although the patient survived the procedure and initially did well, he died a month later from an infection of the brain's wound.

The Bennett–Godlee operation created a medical and public sensation with *The Times* publishing a leading article on the case, proclaiming it was the first of its kind (it wasn't). It also bought into focus the great benefits to medicine derived from the use of animal experimentation. This was a highly contentious issue in the late nineteenth century for vivisection was highly unpopular, with animal protection groups enjoying political influence. In fact, the first such organisation in the world, the Society for the Prevention of Cruelty to Animals (SPCA) had been formed in 1824 in England, and became the RSPCA in 1840 as a result of the patronage of Queen Victoria. In the wake of the Bennett–Godlee procedure, *The Times* contributed to the public debate by publishing more than 60 letters on the issue. The case was also reported in the medical literature where it persuaded other surgeons to perform similar operations. However, apparently unknown to Bennett and Godlee, their operation was not the first of its kind, since Macewen had removed a superficial meningioma a few years earlier. Although Bennett would be less than charitable towards Macewen, accusing him of methods that 'rendered exact localisation impossible' and being 'fortunate in finding the growth so confined,' he nevertheless had to accept the Scotsman's priority.

The first official neurosurgeon: Victor Horsley

The person to do most in establishing neurosurgery as a medical specialisation was Victor Alexander Haden Horsley. The son of London artist John Callcot who is said to have invented the Christmas card, Horsley was born in 1857 into an aristocratic family with Royal connections (Queen Victoria was his Godmother). Despite his privileged background, Horsley espoused many radical socialist causes including the vote for women, national insurance and free health care. He was also a life-long and vociferous critic of drinking alcohol and smoking. Following his father's advice, Horsley took up medicine, graduating in 1881 from London's University College, with a gold medal for his surgical skills. After receiving his FRCS[4] two years later, Horsley moved to the Brown Institute where he undertook experimental research. His output was extensive. Although primarily concerned with localisation of function, Horsley also demonstrated that removal of the thyroid gland caused myxoedema (swelling of the skin), as well as performing the first successful experimental hypophysectomy (removal of the pituitary gland). This was all the more astonishing for during this period Horsley was also an assistant surgeon at University College Hospital. And it would be for his surgical skills that Horsley was appointed in 1886 as a surgeon at The National Hospital for Nervous and Mental Disease in London.[5] This institution, established in 1860,

FIGURE 13.2 Sir Victor Horsley (1857–1916) the first neurosurgeon at the National Hospital for Nervous and Mental Disease in London. Here he was to perform many pioneering operations including the first successful removal of a tumour from the spinal cord.

Source: Wellcome Library, London.

was the world's first dedicated to treating diseases of the nervous system, and following Godlee and Bennett's success, it had decided to employ a dedicated brain surgeon. The 29 year-old Victor Horsley fitted the bill – a position he combined with a professorship at University College.

Horsley's first operation was performed in May 1886, on a young man suffering from status epilepticus – a life-threatening illness characterised by repetitive, persistent seizures. As a young boy, the patient had been run over by a cab in Edinburgh leaving him with a depressed skull fracture and right-sided paralysis. Although surgeons had healed the wound by removing bone fragments from the brain, the patient began to have seizures in his teenage years. These became progressively worse, so that by the time of his admission to the National Hospital, he was experiencing several generalised fits every day. The need to operate was also imperative as the disorder was causing mental deterioration. Horsley began the operation by removing the skull bone and raising the dura mater to expose the brain underneath. It soon revealed a deep red-coloured scar, some two inches long and one and a half inches wide, which Horsley cut and lifted away. The surgery was a success, freeing the young man of his seizures and paralysis. It also encouraged Horsley to undertake the operation on two further patients. He would take all three to the Annual Meeting of the British Medical Association at Brighton in August 1886. It was to firmly establish brain surgery as a legitimate form of treatment for severe forms of epilepsy, leading to a marked shift of attitude in the medical community. By the end of his first year at Queen's Square, Horsley had performed eleven intra-cranial operations with only one death.

A year later Horsley performed what Sir William Osler, then the best-known physician in the English-speaking world, called 'the most brilliant operation in the whole history of surgery.' The patient was a 45 year-old army officer, called Captain Gilbey, who had been injured in a carriage accident in 1884, which also killed his wife. At first, Gilbey experienced few problems from the injury, although over the next two years his legs developed a paralysis and become spastic. These symptoms indicated the presence of a spinal tumour – a virtual death sentence in the nineteenth century for nobody had ever managed to remove one with success. To make matters worse, when Horsley performed the surgery in 1887, the lesion was not found at the expected level of the spinal cord, a situation requiring him to remove several higher laminae before the tumour could be exposed. Nonetheless, Horsley's excision of the growth was successful. A year later it was reported the patient was working 16 hours a day and spending much of his time on his feet and walking.

It was the beginning of a triumphant surgical career for Horsley. By 1900 he had performed a series 44 operations including ones for the treatment of seizures and spinal tumours. These also included the first procedures to remove pituitary gland tumours and new methods for reducing swelling and blood loss of the brain. However, there was more to come, for in 1908 Horsley helped invent the stereotaxic instrument with physiologist Robert H. Clarke. This was an apparatus allowing investigators to target any region inside the brain hidden from direct view. It was a simple but clever device: by first identifying a certain landmark on the skull, and then using it as a starting point for the positioning of a surgical tool (say one to undertake a biopsy, lesion or an electrode to produce stimulation), the instrument could be lowered into the brain, using three dimensional co-ordinates taken from an anatomical atlas. Although the original apparatus was only intended as an experimental device to lesion the monkey's cerebellum, Clarke had recognised the possibility of developing it for clinical use. Consequently, he submitted a patent for a human stereotaxic frame in 1912.

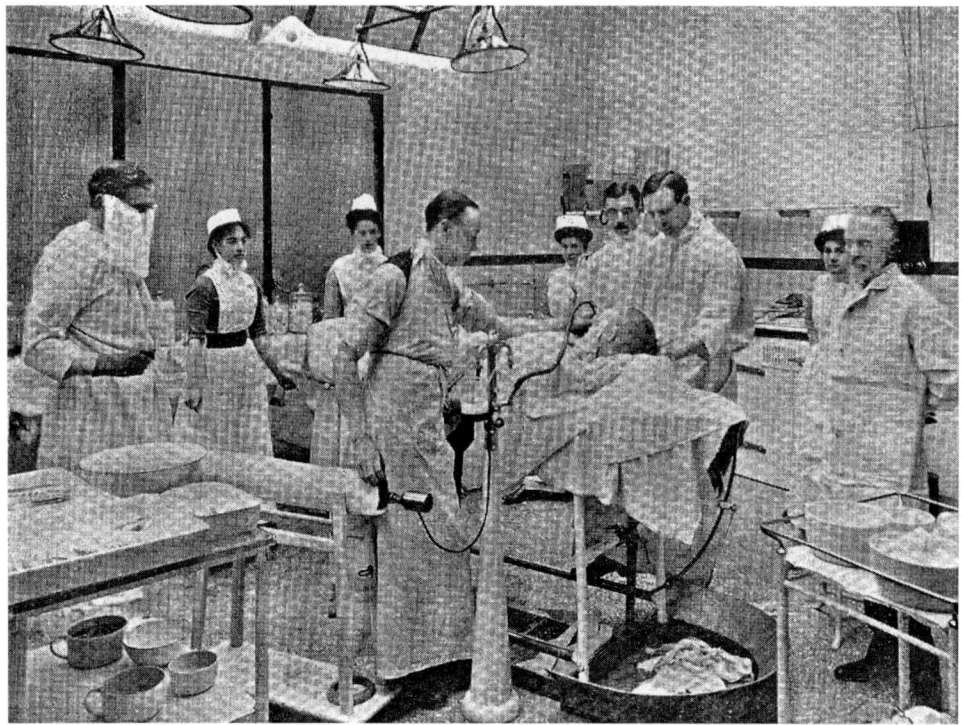

FIGURE 13.3 The operating theatre in the National Hospital for Nervous and Mental Disease, Queen's Square, London, 1906.

Source: Wellcome Library, London.

Horsley, it seems, was much more sceptical about the idea. Although it would take some 30 years to be developed, Clarke's optimism proved well-founded when the stereotaxic frame was first used for human operations in the 1940s. It revolutionised brain surgery, enabling procedures to be performed deep in the brain, including areas such as the thalamus and striatum, which had previously been out of reach. Any brain structure (in theory) could now be accessed for operative purposes. The stereotaxic technique also proved to be an important tool in the hands of experimental researchers who used it to assess the effects of lesions or electrical stimulation/recording in a variety of laboratory animals. Sadly, Horsley would not see these developments take place. Volunteering for service at the outbreak of the First World War, he died serving as an army officer in Iraq from heat stroke in 1916. Eccentric to the last, Horsley contributed to his own downfall by refusing to wear a hat.

Learning from epilepsy: Wilder Penfield

As neurosurgery progressed during the early part of the twentieth century it not only opened up a number of new therapeutic interventions, but provided a powerful new means of exploring the mysteries of the human brain. One doctor to take advantage of this possibility was Wilder Graves Penfield. Described in 1956 as the greatest living Canadian despite being American by birth,[6] Penfield's most notable life time achievement was to establish the internationally acclaimed Montreal Neurological Institute, which specialised in the surgical treatment of epilepsy. It was in this medical role that Penfield developed a remarkable procedure where he stimulated and explored the human brain with an electric current in fully conscious patients prior to their surgery. The procedure was necessary if the surgeon was to avoid removing areas of the brain that might lead to behavioural impairments potentially worse than the seizures themselves. Only requiring a local anaesthetic applied to the scalp, this preliminary exploration enabled surgeons to remove tissue with greater confidence. Penfield was also fascinated by how the physiological brain is able to give rise to the psychological mind, and he used his technique to identify regions of the cerebral cortex involved in a variety of mental and behavioural functions. A great believer in the idea that knowledge of the brain would ultimately allow man to understand himself better, Penfield once famously referred to epilepsy as 'his great teacher'.

Born into a medical family with several generations of physicians, Penfield's early years proved difficult when aged just eight, his father's practice failed, resulting in the separation of his parents. Despite the upheaval, Penfield is said to have exhibited a singular 'tenacity of purpose' throughout his school years, with his reward being a place at Princeton where he studied English literature. Although Princeton allowed Penfield to indulge in his love of sports (he excelled in football and wrestling), his long-held ambition was to win a Cecil Rhodes Scholarship, which would allow him entrance to the University of Oxford. After receiving his degree in 1913, Penfield won his scholarship a year later (after one failed attempt) when he decided to study medicine. Oxford was an inspirational experience for Penfield, especially after coming under the tutelage of Charles Sherrington who instilled in him a deep fascination with the mysteries of the 'unexplored' brain. With two years at Oxford, and a period of war service in America, Penfield completed his medical degree at John Hopkins in 1918. The following year he moved to Boston where he served as an apprentice to brain surgeon Harvey Cushing. Returning to work at the National Hospital in London to complete the final year of his Rhodes Scholarship, Penfield become fascinated with epilepsy. He was so intent in learning more about this condition that he turned down lucrative offers to become a full-time surgeon in America, preferring instead to take up a number of temporary and less well-paid appointments that combined surgery with research.

In 1928, at the age of 37, Penfield finally agreed to become a surgeon at the Royal Victoria Hospital in Montreal, accompanied by an academic post at nearby McGill University. His research had also led him to become interested in the formation of scar tissue following injury to the brain, and how this could instigate the development of epileptic seizures. Recognising that many types of specialist were needed if epilepsy was to be fully understood and successfully treated, Penfield also began to consider the possibility of creating an institute where doctors, neurosurgeons and pathologists could all work together. However, soon after his appointments in Montreal, Penfield learned of the work of Otfrid Foerster in Germany who had begun removing scar tissue from the brains of his patients. Keen to learn about his new surgical procedure, Penfield visited Germany with his young family for a six-month stay. His new skills would soon be put to the test upon arriving back in Montreal, for Penfield's sister Ruth had developed a tumour in her right frontal lobe. Sadly, Penfield found the tumour was malignant and extended into parts of the brain that he could not reach. Ruth passed away three years later. The experience, nonetheless, galvanised Penfield into founding the Montreal Neurological Institute in 1934, greatly helped by grants from the Rockefeller Foundation and the government of Quebec.

Stimulating the human brain

One of the problems facing Penfield in his attempt to treat epilepsy by undertaking surgical operations on the brain, was not so much identifying the scar tissue causing the seizures, but to remove it in such a way that it did not cause any behavioural or mental impairment for the patient. Surprisingly, much of the brain is behaviourally unimportant and can be removed without producing any overt deficit. However, there are exceptions that occur most frequently when lesions encroach into areas responsible for movement and language. In such instances removal of tissue can lead to paralysis and speech impediments. How to detect and avoid lesioning these areas? To do this, Penfield with colleague Herbert Jasper, developed a technique where the exposed brain is stimulated with a mild electrical current in a fully awake patient prior to surgery. The aim was simple: if the stimulation produced a sudden limb movement, or caused the patient to cry out, then this indicated the presence of a motor or language area to be avoided by the surgeon. While the procedure may appear grisly, it is relatively safe and does not cause discomfort, for the brain does not contain pain receptors. It is also straightforward to administer, requiring only a local anaesthetic to the scalp, which allows part of the cranium to be removed. Once the brain is exposed, the surgeon stands behind the seated patient and asks them to report their experiences while they stimulate the cortical surface with an electrode. Each time the current evoked a response, Penfield placed a numbered ticket onto the cortex to record the site, but in most instances the patient was unresponsive. Only occasionally would an impression such as a tingling sensation in the limb, a sudden movement, a vocalisation, or abrupt arresting of speech be induced. These were the sites Penfield avoided when later undertaking the actual surgery on the brain.

The Montreal procedure as it became known, was a major advance in the medical treatment of epilepsy, with removal of scar tissue becoming a standard and safe procedure in hospitals around the world. This is because Penfield's results were impressive, obtaining a perfect, or nearly perfect success, in around 45 per cent of patients; a good outcome in 20 per cent; and negligible improvement in 35 per cent of cases. The procedure was also undertaken on a large sample of patents, for over the course of his directorship, Penfield would perform over 2000 operations, with a mortality rate of less than 1 per cent. Today, some 80 years or so after the procedure was introduced, the success rate of the operation remains similar to that achieved by Penfield. For example, in a recent British study that examined the long-term outcome of 615 adults given surgery to treat epilepsy[7], it was found that 63 per cent of patients were free of seizures at two years, 52 per cent at five years, and 47 per cent at ten years.

However, the Montreal procedure had another benefit: it provided fascinating information about the workings of the human brain. The first report of this work appeared in 1937 when Penfield and Edwin Boldrey analysed the results from their first 163 operations. Their stimulation was mainly targeted at the deep fissure separating the frontal lobes from the parietal lobes known as the central sulcus – a region including parts of the motor cortex first identified in dogs by Fritsch and Hitzig in 1874. Penfield and Boldrey confirmed the existence of the motor cortex in humans by showing it was contained within a strip of tissue known as the precentral gyrus, which lay just anterior to the central sulcus. Stimulation of this area not only produced a variety of involuntary movements, but was organised in a map-like way with an 'upside-down' representation of the human body. For example, stimulating the most upper part of the precentral gyrus produced movements in the feet. At the stimulation electrodes were moved down, movements followed in the legs, abdomen and trunk, hands and head.

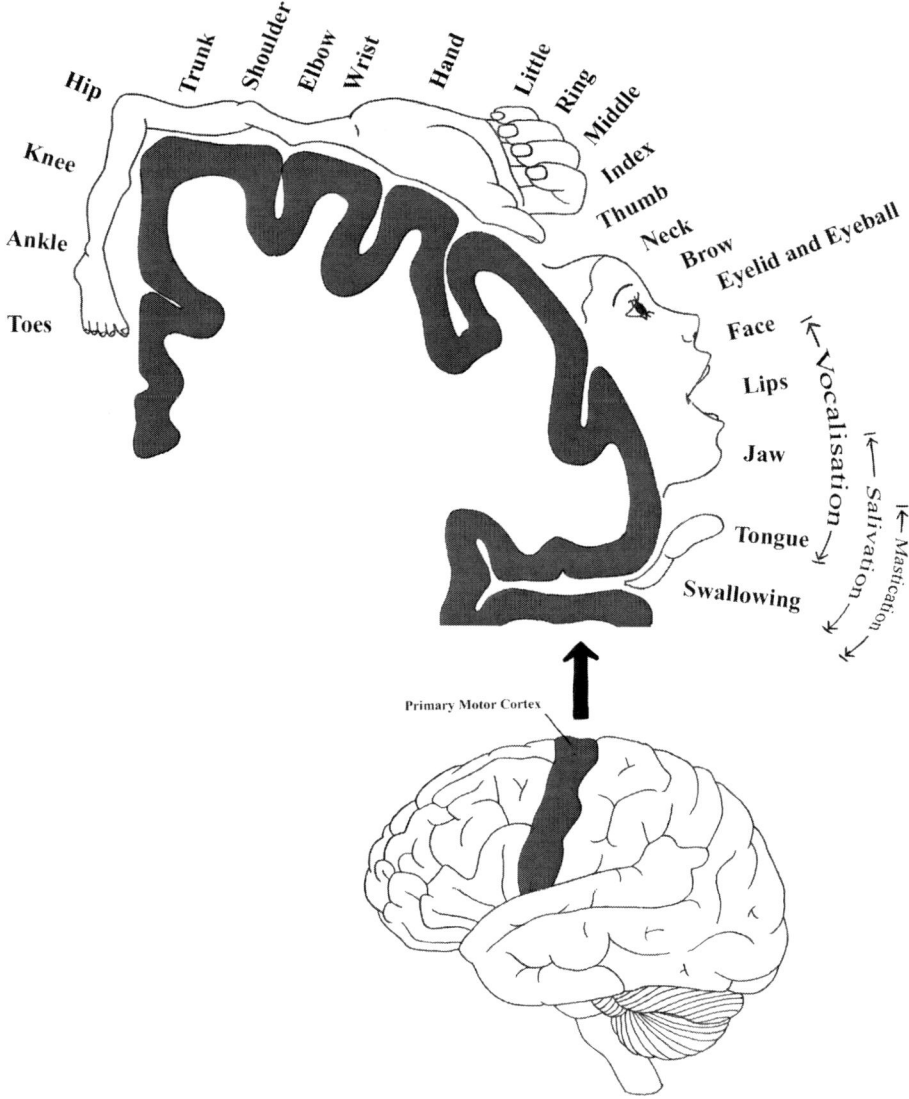

FIGURE 13.4 The homunculus of the human motor cortex, which was first mapped out in the 1930s by Penfield and Boldrey.

Source: Redrawn from Penfield and Rasmussen, 1950, by Charlotte Caswell.

It was also apparent some parts of the body were more fully represented than others, with the face and hands having the largest amounts of tissue dedicated to them. To help illustrate this, Penfield and Boldrey drew a homunculus, or 'little man', to caricature the relative extent of each region. Another region to be similarly examined was the somatosensory cortex. This area lies on the other side of the central sulcus in the parietal lobes, and receives sensory tactile and proprioceptive feedback from the skin and muscles of the body. It too was found to be organised with a map of the body resembling the one in the motor cortex.

Another concern of Penfield's was the exact location of the brain's language centres. He found where these lay by getting the patient to read aloud, and then seeing if any arrest of speech occurred from brain stimulation. He confirmed, as many had done before, speech production and comprehension was located predominantly in the left cerebral hemisphere. In addition to Broca's and Wernicke's language areas, Penfield identified a third speech-related region, called the supplementary motor cortex, which lay anterior and adjacent to the motor cortex. This showed that language was anatomically more complex than had previously been supposed. Penfield also realised that no two patients had an exact correspondence of language areas (unlike the ones for movement that always lie in the precentral gyrus). In fact, the brain areas with language functions were widely distributed throughout the cerebral cortex, which made it even more important for them to be carefully mapped out prior to surgery. In the 1950s, Penfield also began undertaking operative procedures on children – his first patient being a four-year-old with a malignant tumour in the central region of the brain. From this work, Penfield would find that complete recovery of language skills was often possible in children, even after extensive damage to the brain. He also noticed children learn language more easily than adults with this ability declining at about ten years. Going beyond his surgical findings, Penfield used this discovery to promote the educational benefits of teaching a second language in Canadian schools to children early in life.

Exploring the stream of consciousness

In the 1950s, Penfield turned his attention to operating on patients believed to have seizures arising from the temporal lobes. This type of epilepsy, first described by English neurologist John Hughlings Jackson in 1876, is very different from the dramatic generalised convulsions where the person turns rigid and falls to the ground with violent shaking. Instead, seizures emanating from the temporal lobes typically produces a 'dreamy state' (or absence) accompanied by a sense of 'reminiscence' that something has already occurred – a feeling known as *déjà vu*. Other symptoms may include an aura such as a strange taste in the mouth, peculiar odour, or a palpable sense of impending fear. The seizure also typically elicits automated or unconscious motor movements, as shown by chewing-like movements of the mouth which can develop into more complex forms of behaviour such as playing with the buttons on one's shirt. In some instances they even result in the person wandering away from the immediate situation to another place. Remarkably, Jackson had localised this form of epilepsy to the temporal lobe from the work of David Ferrier, who had shown that electrical stimulation of its most anterior regions in monkeys produced a number of oral automatisms similar to those shown in humans with seizures.

Although absence seizures were associated with the temporal lobes, in most cases there was no sign of brain damage to warrant a surgical procedure. Thus, Penfield only rarely turned his attention to examining this part of the brain. Nonetheless, in 1931, Penfield was examining a woman with focal epilepsy by stimulating the left temporal lobe when she suddenly exclaimed she was seeing herself re-enacting giving birth to a baby girl – an event that had happened many years previously. Penfield thought little of this response at first, seemingly believing it was a chance event. In fact, it would be another five years before Penfield came across a similar type of reaction during temporal lobe stimulation. This time it happened in a 14-year-old girl who reported experiencing herself walking through a meadow and being followed by a man holding a snake. It was not only a memory verified as genuine by her mother, but one that sometimes provided a warning of her daughter's impending seizure. This was important for Penfield since it suggested the memory's anatomical locus was close to the site of the brain producing the epilepsy. In other words, it was a potentially suitable place for excision. However, Penfield also realised his discovery had further implications for understanding the storage of memory in the brain, and the very nature of human consciousness itself.

By the early 1960s, Penfield had extended his observations to over 500 patients with temporal lobe stimulation. Although 'psychical responses' were not common, occurring in only about 17 per cent of instances, Penfield managed to distinguish two types. The first resembled the auras preceding the start of the seizure. Penfield called these responses *interpretive* because they were essentially an alteration (or illusion) in the patient's interpretation of their current situation. The second were vivid memories from the past. Penfield called these responses *experiential* as the person appeared to be re-experiencing the event in some way. For example, one patient reported: 'Oh a familiar memory – in an office somewhere. I could see desks. I was there and someone was calling to me – a man leaning on a desk with a pencil in his hand.' In another, a man reported himself standing on a 'street corner', and when asked where he had replied 'corner of Jacob and Washington.' Although patients were fully aware these memories had been elicited by Penfield's stimulation procedures, the experience was nevertheless perceived as being very real. This occurred when one patient heard orchestral music and assumed a record was being played in the operating theatre. Penfield also noted these episodes followed a real time sequence of events, which moved forward in time until the stimulating electrode was withdrawn.

These were remarkable observations and Penfield drew several conclusions from them. He believed, for example, that his stimulation was activating real memories, which were always preserved in amazing detail. Another belief was these memories flowed in what he called the stream of consciousness – analogous to a film in the inner mind where sensory images and thoughts are projected 'on the screen of man's awareness'. Because many of the evoked memories were also seemingly arbitrary or random, Penfield also proposed nothing is ever forgotten and long-term memory contains a record of every experience. Although he initially hypothesised the temporal lobes provided the storehouse of these memories, he later changed his mind believing they were located elsewhere in the brain. Nonetheless he would identify the temporal lobes as providing the vital mechanism for human memory and consciousness. As he put it in his own words in 1960:

> Hidden in the interpretive area of the temporal lobe there is a key to a mechanism that unlocks the past and seems to scan it for the purpose of autonomic interpretation of the present. It seems probable also this mechanism serves us as we make conscious comparison of present experience with similar past experiences.

Today, many cognitive psychologists disagree with Penfield's interpretation of his results. For one thing, they point out, many of his elicited memories were clearly hallucinatory. Others point out the implausibility of recording every memory in exact detail. In fact, most researchers would argue that human memories are a reconstruction rather than a permanent record. Nonetheless, Penfield's research remains some of the most intriguing in the history of neuroscience.

Penfield in later life

Penfield retired in 1960, at the age of 70, to begin a second career as a writer of historical novels and medical biography. By now he was internationally famous as a pioneering surgeon, intrepid researcher, and founder of the Montreal Neurological Institute, which had transformed the medical treatment of epilepsy. For this, Penfield was awarded with numerous honours, including the British Order of Merit in 1953 (which is only conferred on 24 living persons) and a companion of the Order of Canada in 1967. More interestingly, his retirement allowed Penfield the freedom to speculate on his philosophical views concerning the brain. He would do this, for example, in books such as *The Mystery of the Mind* (1974), which he wrote when in his 80's. He still had some surprises in store for his readers. Convinced the secret of the brain remained the most important and unexplored enigma in the whole of science, Penfield admitted as a young man he had been inspired by the belief that the neurological processes of the brain would one day explain the mind. However, by the end of his life, he had changed his opinion. Indeed, he pointed out that the conscious willing to do something, or believing in something, had never been invoked in a person by the means of electrical stimulation, and Penfield doubted it ever would. Or, as he vividly put it, 'There is a switchboard operator as well as a switchboard.' Thus, for Penfield, there was a second force, spiritual in nature, in control above the mechanical operation of the brain. It would lead him to postulate the brain and mind are entirely different things – a dualist

FIGURE 13.5 Wilder Penfield (1891–1976) who founded the Montreal Neurological Institute in 1934 and pioneered the surgical treatment of epilepsy.

Source: Courtesy of the Penfield Archive Neurological Institute, McGill University.

view most famously associated with Descartes, and one shared by his mentor Charles Sherrington. This is not a position, however, supported by most neuroscientists today. Penfield died at the age of 85 in 1976, just three weeks after completing his autobiography *No Man Alone* – a book espousing his belief in the importance of team work as a means to improving surgery, and ultimately in understanding the mysteries of the brain.

The Wada technique

In the aftermath of the Second World War, a young Japanese doctor called Juan Wada was called upon to treat a patient with a bullet wound to the brain. The wound was producing a life-threatening condition called status epilepticus, characterised by repetitive or continual seizure activity. In his desperation to halt the fit, Wada injected sodium amytal, a short-acting anaesthetic, into the carotid artery supplying blood to the left side of the brain where the injury was located. To Wada's relief, the procedure terminated the seizures, although it also caused an abrupt but temporary loss of speech for five to ten minutes. The aphasic effect of the anaesthesia should not have surprised Wada, for it had long been known that the language centres of the brain were more likely to be in the left hemisphere than the right. Nonetheless, Wada quickly recognised the potential usefulness of his new technique – not only for examining the localisation of language, but also helping surgeons assess the likely sequelae of administering electroconvulsive therapy (ECT), which was widely used in those days for treating depression. Consequently, Wada repeated the procedure on 15 more patients to assess its usefulness and safety. After publishing his results, Wada would be invited to the Montreal Neurological Institute to demonstrate his technique where it was recognised as an important new means of assessing the localisation of language and cognitive function, prior to the surgical treatment of seizures.

The use of the Wada test, however, was quickly used to answer other questions about brain function. In particular, researchers recognised it could assess the degree to which handedness was linked to the lateralisation of language. Ever since the work of Broca in the mid-nineteenth century, it had been known that the cerebral localisation of language has a tendency to be different in people who are right or left handed (see Chapter 9). In other words, there were some instances where people had dominant language function in the right hemisphere, rather than the left. This was also associated with being left-handed. The Wada test provided an effective means of examining this relationship. That is, by being able to selectively anaesthetise either one of the hemispheres, the surgeon could establish the dominant side for language, and then determine if it was related to the handedness of the patient. This question was pursued by Penfield's successor at the Montreal Neurological Institute, Theodor Rasmussen and psychologist Brenda Milner. They reported the results of their studies, based on the examination of over 200 subjects, in 1966. It confirmed the idea that the cerebral superiority of language has a tendency to be different in people who are either right or left-handed. For example, the Wada test revealed 92 per cent of the right handers and 70 per cent of the left handers exhibited language lateralised to the left hemisphere. However, in the left handers, 15 per cent had language strongly lateralised to the right, and 15 per cent had mixed dominance. Because 90 per cent of people are right-handed, a simple calculation shows over 90 per cent of the population will have a left hemispheric dominance for language. This leaves a minority (i.e. about 10 per cent) showing a right-sided specialisation, or bias, for speech-related functions.

Mental illness under the surgeon's knife

The emergence of modern neurosurgery had been a great success, resulting in the removal of tumours, abscesses and scar tissue from the brain and spinal cord. However, there were others interested in applying similar techniques to psychiatric disorders where the problem lay in the mind of the individual. In fact, this type of surgery would first take place in the late nineteenth century through the efforts of a Swiss doctor called Gottlieb Burckhardt, director of a medical asylum housing disturbed psychotic patients, located on the north-eastern tip of Lake Neuchâtel. Although aware of neurosurgical developments taking place in Britain, Burckhardt was also influenced by German compatriot Frederick Goltz, who had shown that decortication in dogs caused few behavioural effects, other than to produce a more placid and less excitable animal. Realising a similar calming effect may have benefits for his patients, Burckhardt undertook cortical excisions on six disturbed patients in 1888 whose natural chances of recovery 'could not be reasonably anticipated'. By using a knife to cut away tissue after removing the cranium, Burckhardt targeted Broca's area and the temporal lobe language area, in an attempt to reduce the hallucinatory voices affecting his patients. The benefits were somewhat mixed with Burckhardt reporting partial success in three individuals, no improvement in one, and failure in the remaining two cases (with one patient dying from convulsions and the other committing suicide). Although Burckhardt considered these results worthy of further exploration, his views were not supported by medical colleagues who strongly criticised his procedures.

It would be nearly another 50 years before others were to adopt similar procedures to treat mental illness – this time on the frontal lobes. This area of the human brain, which makes up about 40 per cent of its cortical mantle had, of course, given rise to localisation theory in the nineteenth century. However, many investigators had become further intrigued about the frontal lobes after learning of Phineas Gage, first described by David Ferrier in his Goulstonian Lectures of 1878 (see Chapter 9). Gage had lost a considerable part of his frontal lobes after an explosion had caused a tampering iron to pass through his head. Although surviving the injury and exhibiting no obvious behavioural impairment, Gage had nonetheless become increasingly child-like and frivolous in the years after. From this, and supported by his own work on animals, Ferrier had come to the conclusion that frontal lobe damage produced deficits in intelligence and attention. However, there was much contradictory evidence. The famous American surgeon Moss Starr, for one, had examined a number of patients with frontal lobe damage during the 1880s and 1890s, finding a 'mental instability' in only half of cases. In some instances there appeared to be no mental abnormality at all. Indeed, Karl Lashley had shown that lesions of the frontal cortex in monkeys only had a temporary effect on the solving of puzzles (e.g. pressing a lever to escape from a box) and any deficit was soon overcome with training.

It was against this backdrop, two Americans, Carlyle Jacobsen and John Fulton from Yale, started to examine the effects of frontal lobe lesions on problem solving in chimpanzees. This proved harder than first anticipated for one of their monkeys called 'Becky' was so aggressive and volatile, that the two investigators considered her untrainable. Despite this, they attempted to remove her frontal lobes in a two-stage operation. Although the first procedure produced little behavioural effect, when Jacobson ablated the remainder of the cortex, a profound change occurred: the monkey became tame, easy-to-handle and amenable to testing. Jacobsen reported these findings to the International Neurological Congress in London in 1935. One

FIGURE 13.6 Egas Moniz (1874–1953) who performed the first modern frontal lobotomy (with the help of surgeon Almeider Lima) in 1935 that led to a Nobel Prize in 1949. Portugal commemorated his birth philatelically in 1974 with a set of postage stamps.

Source: Shutterstock Images.

person listening in the audience was the Portuguese surgeon Egas Moniz[8] who realised a similar operation might be used to treat emotionally disturbed patients. Unable to undertake the operation himself because of his gout deformed hands, Moniz persuaded colleague Almeider Lima to do the surgery instead. This would take place soon after, for within three months of hearing Jacobsen's talk, the first frontal lobotomy was performed in a Lisbon hospital on a female patient suffering from severe depression and paranoia. Lasting no more than 30 minutes, Lima bore two holes into the skull, and injected an alcohol solution into the white matter of the frontal area, which destroyed the fibres linking the lobes with the rest of the brain. The operation proved to be a great success. There were no side effects and the patient was said to be significantly less anxious and paranoid as a result. Moniz even went so far to describe it as a 'clinical cure'.

Moniz soon refined his surgical technique by designing a special knife called a 'leucotome' with a movable and retractable steel wire loop. After drilling holes into the temples of the skull, the leucotome was inserted into the brain, the loop opened, and then twisted to cut out cores of tissue. Using this method there was no need to use an alcohol solution to destroy fibres. Within four months of undertaking his first leucotome operation, Moniz presented the results to the medical world, with data from 20 patients suffering from a variety of illnesses including schizophrenia, obsessional compulsive disorder and depression. He reported that the psychosurgery (a phrase he coined) had led to a full recovery in seven patients, produced a favourable outcome in seven others, and produced little improvement in the remainder. All patients survived the operations without serious impairment. A year later, Moniz gave a similar report on 18 more patients. The results astonished many psychiatrists. Mentally ill patients who had once been regarded as incurable were now, it was claimed, leading reasonably normal lives. Not surprisingly, many other surgeons around the world began to adopt the procedure. It also resulted in Moniz being the first Portuguese to be awarded the Nobel Prize in 1949 for his discovery of the therapeutic value of leucotomy in certain psychoses.

Ice pick surgery

The most influential practitioners of psychosurgery were the American physicians Walter Freeman and James Watts who performed their first neurosurgical procedure in 1936, operating on the frontal cortex of a woman suffering from 'agitated depression'. Encouraged by their success, Freeman and Watts operated on more patients and simplified their technique by drilling holes in the side of the skull, through which they could insert a thin knife to sever the fibre tracts linking the frontal cortex with lower brain areas. However, in 1948, Freeman went one step further by inventing a new technique called the transorbital lobotomy, which involved a sharp instrument resembling an ice-pick being positioned beneath the upper eyelid, and driven through the base of the cranium by a mallet. So positioned, the instrument could then be swept around to sever the connections between the frontal cortex and thalamus – an operation believed to dampen emotional reactivity. This procedure had a number of 'advantages' for ice pick surgery was quick (sometimes taking only ten minutes) and done with a local anaesthetic allowing it to be performed in mental hospitals without surgical facilities. There are even accounts of Freeman undertaking the procedure in his own doctor's office. Freeman was so enthusiastic about this new type of surgery, he travelled around the United States in his own personal van called a 'lobotomobile', which he drove to various hospitals.[9] The procedure also became commonplace in the United States, with estimates suggesting it was performed on some 50,000 patients during the 1940s and 1950s. More than 3000 of these would be supervised by Freeman alone before his retirement in 1970. His patients included 19 children, one of whom was four years old.

Unlike neurosurgery, which is accepted as a legitimate branch of medicine, the use of psychosurgery for the treatment of mental illness has always been controversial – both for ethical reasons (not least because it is undertaken in the absence of any neuropathology) and for its alleged rate of success. Indeed, fierce criticism and public debate during the 1950s and 1960s prompted the formation of several governing bodies to evaluate the viability and effectiveness of psychosurgical procedures with an eye to encouraging stricter legislation. Yet, the benefits of psychosurgery were often reported positively. For example, a survey in England and Wales showed over 10,000 patients had undergone a form of psychosurgery from 1942 to 1954, with over two-thirds exhibiting improvement. However, it also became clear in other studies that these operations often produced a variety of mental and behavioural impairments not immediately apparent after surgery. While the patient could often lead some type of independent existence, they were otherwise apathetic, lacked initiative, and acted without social restraint. It also emerged, as psychological testing became more sophisticated, that many lobotomised patients had difficulty following instructions, solving problems and making plans. Fortunately, by the late 1950s, the need for psychosurgery markedly declined as effective drug treatments became available. Consequently, by the 1970s, a number of countries had banned psychosurgery, including several US states, although this has not happened in the United Kingdom where the practice is rare but legal (perhaps one or two instances each year). For some it remains a barbaric practice, merely transforming an insane person into a more idiotic one, and therefore an unnecessary mutilation of the brain.

The split-brain technique

Around the same time as Freeman and Watt were performing their first lobotomies, a new surgical technique was being considered for the relief of epilepsy. This was the severing or splitting of the corpus callosum, otherwise known as commissurotomy, to halt the spread of seizure activity in the brain. The corpus callosum, which stands out as a white arched structure between the two cortical hemispheres, has a long history. First recognised by Galen who referred to it as a 'callous' since it resembled hard or thickened skin, the corpus callosum was also mentioned by Vesalius in 1543, who thought it did little more than support the brain mass above and maintain the shapes of the ventricles. By the seventeenth century, the corpus callosum had become implicated in mental function when Thomas Willis recognised it as a pathway allowing cerebral spirits to travel from one hemisphere to another. It was even regarded as 'the seat of the soul, which imagines, deliberates and judges' by Italian Giovanni Maria Lancisi in 1712. However, it was not until the later part of the eighteenth century the true function of the corpus callosum was identified when anatomists such as Giovanni Morgagni and Felix Vicq d'Azyr realised it was a bundle of nerve fibres (i.e. a commissure) connecting the grey regions of the two cerebral hemispheres. Today, we now know the corpus callosum contains in the region of 300 million 'white' myelinated nerve fibres, which allows the two cerebral hemispheres to communicate with each other.

The first surgical lesions of the corpus callosum in humans were carried out by the American neurosurgeon Walter Dandy at John Hopkins University during the mid 1930s. One of the great pioneers of neurosurgery, Dandy is renowned for introducing several important techniques into surgical practice. These include injecting air into the ventricles (air ventriculography) in 1918, allowing surgeons to accurately locate growths in the brain with X-rays, and the first clipping of an intracranial aneurysm in 1937, which for many marks the birth of cerebrovascular surgery.[10] Said to be a fast and dextrous surgeon who at his peak carried out over a thousand operations each year, Dandy partially severed the corpus callosum to gain access to third ventricle tumours located within the midbrain. By splitting the corpus callosum, he could reach deep down into the brain by parting the two cerebral hemispheres. Although this was a major ordeal, Dandy was surprised by the lack of any behavioural impairment following the surgery. In fact, all his patients appeared to be perfectly normal. At the time there was still a considerable amount of mystique surrounding the corpus callosum. Nonetheless, Dandy was confident enough to proclaim: 'no symptoms follow its division' and 'this simple experiment at once disposes of the extravagant claims to function of the corpus callosum.'

With these clinical findings in mind, William Van Wagenen in the late 1930s, working at Rochester, New York, began examining the effects of severing the corpus callosum in epileptic patients, for whom all other conventional therapy had failed. Assuming the operation would reduce the spread of seizures from one cerebral hemisphere to the other, a finding supported by animal experimentation, Van Wagenen performed a commissurotomy on ten patients. The results, reported in 1940, were variable with the operation eliminating or reducing the occurrence of generalised seizures in seven patients. Most encouragingly, there appeared to be no obvious behavioural drawback from the operation. This led Van Wagenen to repeat the procedure in many more patients over the subsequent years with similar success. Yet, curiously, no other surgeons took up the operation. The reasons for this are not clear, although the somewhat unpredictable benefit of the procedure, accompanied by the increasing realisation that many generalised seizures originate in the brainstem or thalamus, which bypass the corpus callosum, may have been key factors.

Nonetheless, Van Wagenen had amassed a sizeable group of patients with corpus callosum lesions, and decided to enlist the help of psychologist Andrew Akelaitis to assess the behavioural consequences of his operation more carefully. Although Akelaitis was to be thorough, testing the patients on a wide range of neurological and psychological tests between 1940 and 1945, he could find little evidence of any behavioural impairment. The patients experienced no sensory and motor abnormalities, or had problems with speech and language comprehension. They also performed normally on all tests of intelligence. One unusual effect, but only found in two of the patients, was a conflict between the right and left hands. For example, one patient reported trying to open a door with his right hand, while shutting it with his left. However, in both cases, this 'opposition' effect proved temporary. Considering the large size and prominence of the corpus callosum in the brain, and its known role in connecting the two cerebral hemispheres, these were surprising findings. It would lead Karl Lashley in 1951 to remark, perhaps not entirely in jest, that the only known function of the corpus callosum was to stop the two cerebral hemispheres from sagging!

Presenting information to individual hemispheres: Roger Sperry

One investigator to turn his attention to the intriguing problem of the corpus callosum in the 1950s was Roger Sperry. A native of Hartford, Connecticut, Sperry graduated in English from Oberlin College, Ohio, and followed this degree with a Masters in Psychology. It would, however, be his move to Chicago in 1941 to undertake a Ph.D. under the zoologist Paul Weiss where Sperry was to begin gaining a wider reputation. However, this was not concerned with the corpus callosum, but the growth of the nervous system – a subject he also pursued as a post-doctoral student at Harvard and the Yerkes laboratories in Florida. The question of how nerve cells grow to reach their correct destination in embryonic development had attracted investigators, including Ramon y Cajal during the nineteenth century. By the 1940s it was also generally believed that the wiring of the nervous system was strongly affected by learning experience which formed new connections and pathways. Sperry first began to explore the issue in rats by switching their flexor and extensor muscles, but later studied amphibians when discovering the newt's optic nerve regenerated back from the eye to the brain (or more specifically the tectum) after being cut. Moreover, the growth of this pathway restored full vision again.[11] In one set of acclaimed experiments, Sperry sectioned the optic nerves of amphibians and fish, and then rotated the eyes around by 180 degrees. Although the animals regained their vision they viewed the world as 'upside down' and 'right-left reversed'. More importantly, when Sperry traced the regenerated optic nerves, he found the original pattern from eye to tectum was always maintained. In other words, the retinal ganglion cells made exactly the same anatomical connections with the optic tectum as had occurred in earlier development – even when the axons had obstacles in their way, or were closer to other synapses. It was strong evidence for the idea that the neural circuits of the brain are 'hard-wired' with learning having no role in determining neural development. It also led Sperry to propose the chemoaffinity theory, which held the nerve cell was tagged with a specific chemical, as was the tectum, which led the axon to its correct, predetermined, destination.

After spending seven years in Chicago, where he would contract tuberculosis after dissecting a monkey, Sperry was appointed Professor of Psychobiology at the Californian Institute of Technology in Pasadena in 1954. Here, he began taking a greater interest in the corpus callosum with the assistance of doctoral student Ronald Myers. This was not an entirely new venture, for Sperry had been introduced to the mysteries of the corpus callosum after working with Karl Lashley. However, Sperry bought a greater sophistication to examining its function. Using cats as subjects, Sperry wanted to establish a method whereby he could present visual information solely to each hemisphere. This is not as straightforward as it first appears, however, since each of the cat's hemispheres (as in a human) receives input from both eyes. Consequently, to solve the problem, Sperry had to lesion both the optic chiasm (the point where the optic nerves cross to the other side of the brain) as well as the corpus callosum. Only by doing this could he arrange for each hemisphere to receive visual input from the same-sided eye. Now, by training a cat with a patch over an eye on a visual discrimination task, where the animal had to press a lever for a food reward whenever it saw a certain stimulus, Sperry was able to teach one half of the brain. This was shown to be successful when the patch was placed over the other eye, for the cat was then unable to perform the task. Interestingly, the initial results showed learning for each hemisphere to be identical – indicating two separate brains could exist side by side inside the cranium.

Around the same time, Philip Vogel and Joseph Bogen, two neurosurgeons working in a hospital close to the Californian Institute of Technology, were beginning to re-examine the benefits of severing the corpus callosum for seizures. They had also started to perform complete commissurotomies on their patients by cutting all the callosum's fibres. Since Vogel and Bogen were familiar with Sperry's experimental work on cats, they asked if he might be interested in evaluating and testing their subjects – a task Sperry gave to a young graduate student called Michael Gazzaniga. But how to go about testing the functions of each individual hemisphere? Sperry knew these patients still retained an intact optic chiasma. Thus, visual information, even if presented to one eye, would go to both sides of their brain. In other words, despite having a severed corpus callosum, both hemispheres would know what was happening. Fortunately, Sperry came up with an ingenious solution. He devised a procedure whereby the patient was required to stare at a fixation point in the middle of a screen, which was followed by a brief visual stimulus (a tenth of a second or less) to the right or left visual field of the eye. This exposure was long enough for the subject to start processing the stimulus, but not sufficient for them to turn their eyes to it. Because of the way the optic nerve is anatomically organised, this meant a stimulus presented to the left visual field passed exclusively to the right hemisphere, and vice versa. In addition, Sperry would design a second task where the patient was blindfolded and asked to identify objects placed in their hands. In humans, the pathways conveying the sensations of touch, unlike the visual system, completely cross to the opposite side of the brain. Thus, Sperry had a second method of selectively presenting information to a single cerebral hemisphere.

The two different personalities of the brain

The results from Sperry's innovative experimental procedures on split-brain patients have provided some of the most fascinating and astonishing findings in the history of biological psychology. Put simply, Sperry was to demonstrate that the right and left cerebral hemispheres have different functions, specialised abilities and even distinct personalities. Thus, when they are 'split' from each other, the two cortices act as independent and separate 'brains'. The first patient tested by Sperry, assisted by Gazzaniga in 1962, was a former Second World War paratrooper, known as WJ, who started having seizures after being hit over the head with a rifle butt. Over the years his fits had got progressively worse leading to a commissurotomy in 1961. Testing soon confirmed differences between the two hemispheres – especially the superiority of the left for language. For example, when Sperry and Gazzaniga presented a written word such as 'spoon' or 'key' to the left hemisphere, WJ was able to read it, vocalise it, and understand its meaning. However, no recognition occurred when the word was presented to the right hemisphere, with the patient reporting they had seen nothing other than a flash of light, or sometimes nothing at all. The left hemisphere also had the ability of writing by being able to scribble down answers to simple questions through its control over the right hand.[12] This was not a skill shared by the left hand.

However, WJ's right hemisphere had its own unique abilities. For instance, when the left hand was allowed to search by touch alone a tray of objects for at item (e.g. a spoon) that had been pictorially presented to the right hemisphere, WJ had no difficulty locating the correct object. This even occurred when he was unaware of any visual input being presented to him. WJ's right brain clearly possessed an awareness that was independent and separate from that of the separated 'left brain'.

Over the next decade or so, Sperry and Gazzaniga tested several more split-brain patients. Without exception, the left hemisphere proved dominant for language. In addition, the left proved superior at solving analytical problems for it was more rational and logical than the right. Yet, the right hemisphere was not entirely without language for some individuals could understand certain words and even read simple sentences. The right cortex was also more specialised when it came to spatial thinking and solving puzzles. It also constantly outperformed the left at recognising faces and pictures. Furthermore, the right hemisphere (as demonstrated by the performance of the left hand), showed superior ability at drawing pictures, learning mazes, completing jigsaws and solving geometric puzzles.

Intriguingly, the two hemispheres also showed different temperaments with the right being more emotional. This was demonstrated when a nude picture was presented to the right hemisphere – an event typically causing the person to blush and giggle, although they could not explain why. In some instances this even caused the individual to act in a contradictory manner, which has led some investigators to propose the two hemispheres have different types of consciousness. On one task, for example, where a split-brain patient was required to arrange a group of blocks to make a particular pattern with his left hemisphere, it was found that the left hand (right hemisphere) persistently tried to take over the task – so much, Gazzaniga had to wrestle with the right hand to stop it solving the problem. A similar situation sometimes arose when a split-brain patient attempted to read a book with their left hand. Although they may have found the book interesting, the subject often found themselves putting it down – presumably because the right hemisphere is poor at reading and sees little point in holding it. Gazzaniga has even described the case of one patient who would pull his trousers down with his right hand, only to pull them up with his other hand! The role of the corpus callosum, which had for so long puzzled investigators, was now becoming clearer: it linked the two different and specialised halves of the brain allowing them to function as one. This could be most dramatically seen in the case of patient Paul S, who unlike all other split-brain patients tested by Sperry, had language in both hemispheres. When asked what jobs they preferred, Paul's left hemisphere said that he wanted to be a draftsman, while his right hemisphere replied it wanted to be a racing driver. Paul was also tested at the time of the Watergate scandal, and while his left hemisphere liked Richard Nixon, his right hemisphere disliked him.

The man who instantly forgot: the case of HM

The development of neurosurgery in the twentieth century has been a medical success story benefiting vast numbers of people world-wide. However, there have been failures and also rare instances where mistakes have occurred through the actions of less than scrupulous practitioners. The most famous case is that of Henry Gustav Molaison, simply known as HM to others during his lifetime, so his anonymity could be maintained. HM was a mild mannered, 27-year-old American, who on a summer morning in 1953 underwent a surgical operation for the relief of increasingly severe epilepsy – only to become the single most studied and written about patient in the history of psychology. Little could anybody have realised his name (or rather initials) would be mentioned in almost 12,000 journal articles over the course of his life. Nor could HM ever fully understand his unique contribution to scientific knowledge as he forgot the events of his life almost as quickly as they occurred. As HM once confessed in an interview with Brenda Milner: 'Right now I am wondering. Have I done or said anything amiss? You see, at this moment everything looks clear to me, but what happened just before? That's what worries me. It's like waking from a dream. I just don't remember.' Unable to work or look after himself, HM was nonetheless a willing and regular participant in behavioural experiments at the Massachusetts Institute of Technology, resulting in an unprecedented amount of information concerning his amnesic condition. His remarkable story, which first came to the attention of the wider world in 1957, was to be instrumental in forming the discipline of Cognitive Neuropsychology (a branch of science concerned with how psychological processes are mediated by the brain) and inspired a new renaissance into the biological basis of learning and memory.

By all accounts HM was an average child growing up in a working class district of Hartford, Connecticut (the same birthplace as Roger Sperry) who enjoyed playing with friends and swimming in the local reservoir. This happy world was shattered at the age of nine when HM was knocked down in a cycle accident, an incident causing him to be unconscious for several minutes, and producing an injury requiring 17 stitches to the face and head. This is almost certainly the event that initiated HM's epileptic fits at the age of ten, which became progressively more severe as he got older. By the time he reached his early 20s, HM was having about ten 'blackouts' a week (a sign of temporal lobe epilepsy), which could develop into generalised seizures. Despite near toxic levels of medication, his dreams of becoming an electrician were over. In fact, his seizures were so debilitating that HM was even forced to give up working as an assembler in a car factory. Nor did he receive any help or affection from his alcoholic father who was horrified to have a 'mental' in the family. In 1953, a local doctor referred HM to a neurological hospital where the renowned neurosurgeon William

Scoville practised. An expert in performing lobotomies, Scoville had developed the orbital undercutting psychosurgical procedure in 1948, where holes are drilled into the skull above the eyes, allowing the frontal lobes to be lifted. This, in turn, allowed their fibres with the rest of the underlying brain to be observed and severed. Despite his fame, Scoville did not fit the typical picture of a doctor. Said to be ambitious and adventurous, he had also become known to the police for speeding in his red Jaguar and taking high-risk adrenaline-fuelled pranks such as climbing the George Washington Bridge at night.[13] Sadly, some of these reckless character traits were to become manifest in Scoville's approach to HM's surgery.

Scoville recommended a bilateral excision of his medial temporal lobes to limit HM's seizures. His parents duly agreed and the operation took place in August, 1953 with HM only receiving a local anaesthetic applied to the scalp, which meant that he was fully conscious throughout the procedure. To reach the medial temporal lobes, Scoville cut across the forehead and lifted the skin to expose the cranium. He then bore two holes into the bone using a hand drill. Peering through the holes, Scoville accessed the medial temporal region by pushing the frontal lobes up with a spatula.[14] This enabled the tissue to be cut away with a knife and removed using a straw-like suction pump. In fact, Scoville extracted a lump of tissue about the size of a 'tennis-ball' from both sides of the temporal lobe, including the greater part of the hippocampus, along with the amygdala and entorhinal cortex. Although Scoville claimed he had removed these areas before, this was the first time the operation had been performed bilaterally. During the surgery, Scoville also decided to place metal clips inside HM's brain to prevent the blood vessels from bleeding. This would also allow the boundaries of the lesion to be assessed by X-rays for future reference. Scoville did not know this act would preclude HM from being later scanned by high resolution MRI scanners for many years, since the magnetic waves might have caused the clips to move leading to injury or death.[15]

As an intervention to limit HM's seizures, Scoville's surgery was successful. Unfortunately, the improvement came at a terrible price, for it became apparent HM had developed a profound memory problem. This was first evidenced after surgery when he instantly forgot meeting doctors and friends whom he had spoken to just a few minutes earlier. Unfortunately, the problem did not get any better. Tragically, HM had lost the ability to create any new permanent memories – an impairment known as *anterograde amnesia*. The repercussions of this would be extremely debilitating in all sorts of ways. For example, HM could not make any new friends (or at least those he recognised) including researchers such as Suzanne Corkin whom he was to meet scores of times. He also forgot the daily events of his life as soon as they occurred, places he had visited, or any historical facts or information. Even simple questions such as the name of the current president, or his own age, were beyond HM's capabilities, although he would attempt a guess. At one point HM was employed for a while in a menial job where he was required to put cigarette lighters onto presentation cards. Despite the repetitive nature of the task that lasted some weeks, HM could never remember his daily work, or what he was expected to do the next day.

Although the operation also caused a period of retrograde amnesia for several years preceding HM's surgery, it had fortunately not destroyed his memories for events from his earlier life. Consequently, HM could remember his childhood and adolescence. Thus, he remembered his father came from Los Angeles, his mother from Ireland, and events of the Second World War. Nor had the operation affected his intelligence, since HM had an above average IQ

FIGURE 13.7 Henry Molaison (HM) (1926–2008). A young American who underwent an operation by William Scoville in 1953 to remove his medial temporal lobes in an attempt to treat his severe epilepsy. It resulted in a profound anterograde amnesia, which caused HM to forget the events of his life almost instantaneously.

Source: Courtesy of Suzanne Corkin.

scoring 118 on the Wechsler Adult Intelligent Scale. Furthermore, the operation had not significantly impaired his short-term memory. This allowed HM to 'keep things' in his mind – an ability enabling him to hold a conversation, read the newspaper, watch the television, and do crosswords. In fact, his IQ rose slightly in the years after his operation, perhaps as a result of his love for puzzles. However, the moment HM was distracted, the memory of what he was attending to disappeared from his thoughts. Conscious of his condition and apologetic for his memory loss, HM also had a sensitive nature and was likely to be embarrassed by a compliment. This made him very endearing to the people who met him. Indeed, one of his regular psychologists, Brenda Milner, recounts an incident when a fellow researcher commented how interesting they thought HM was. This was overheard by HM who blushed and turned away mumbling he didn't think he was deserving of such praise.

In the years following his operation, it became clear that HM's memory deficit was more complex than first believed, as evidenced by the fact that he showed learning on certain types of task. For example, HM showed improved performance on a mirror drawing task over successive days, which entailed drawing around a complex shape using visual guidance from a reflected source. HM could also learn to trace his finger through a diagrammatic maze, and he showed improved recognition of pictures when given a 'clue' to help trigger recall. On a more practical level HM also learned to use a walking frame when he sprained his ankle. However, all of this learning was at a subconscious level for HM had no conscious memory of coming across these tasks or objects before. Indeed, one way of better understanding this situation is to say HM had an intact *procedural memory* (i.e. a memory allowing him to do things automatically and without conscious intent), but did not have *declarative memory* (i.e. a memory of past events that can be 'recalled' to consciousness). HM's amnesia also led researchers to focus their attention on the involvement of the medial temporal lobes, especially the hippocampus, in memory processes. The fact that HM could remember events from his earlier life shows the hippocampus is not the site where memories are stored. However, his lack of post-surgical learning, clearly implicated the hippocampus in the storage, or 'consolidation' of new experiences. Today, the hippocampus is at the centre of biological research in memory, although at the time of HM's operation it was widely thought to be a brain structure involved in the sense of smell.

HM was a man who dressed neatly, spoke articulately, and enjoyed a joke. He was also said to be kind, charming and extremely mild-mannered. Yet, HM lived in the past, forced to exist in a world that did not move forward, where Truman was always president, and news of his parent's death, especially his mother, continued to evoke grief (HM was looked after by his mother until 1980, when aged 54 years, he was placed into a care home). He was also unable to judge his age, the current date, or estimate time beyond 20 seconds – possibly a blessing in disguise for this may have caused the events of HM's day to pass by more quickly. Yet, his amnesia for factual information was not as complete as is often reported. For example, after living in a new bungalow for eight years, HM was able to draw an accurate floor plan, and find his way home within a distance of two to three blocks. He also showed some learning of major people and events, as occurred when he confused a picture of Mohamed Ali with Joe Louis. He even recalled some limited but very confused information about President Kennedy's assassination and the space shuttle disaster. Interestingly, when HM was finally given an MRI scan in the late 1990s, it was found some of his hippocampus had been spared. Whether this explains his odd fragments of memory is not known. HM died in 2008 at the age of 82. It was only then his true identity was disclosed and a picture of him released to the world. After HM's death, his brain was dissected into 2000 slices in 2009, and digitised as a three-dimensional brain map, enabling it to be preserved. He is likely to be a legend who is never forgotten.

Notes

1 Around 1850, Quesnay would turn his attention to the subject of economics, and publish his 'tableau économique' in 1758, which is regarded as the first work that describes the workings of the economy in an analytical way. For this reason Quesnay is better known today as an economist than a doctor.
2 The outermost of the three layers of the meninges that surrounds the brain and spinal cord.
3 See Lyons, A.E. (1997). In Greenblatt, S.H., Forcht Dagi, T. and Epstein, M.H. (eds) *A History of Neurosurgery in its Scientific and Professional Contexts*.
4 The Fellowship of the Royal College of Surgeons (*FRCS*) is a professional qualification that allows doctors to practise as a surgeon in the United Kingdom and Republic of Ireland.
5 Now the National Hospital for Neurology and Neurosurgery.
6 This accolade was made by Writer Eric Hutton in the Canadian magazine *Maclean's*.
7 de Tisi, J.I., Bell, G.S., Peacock, J.L., McEvoy, A.W., Harkness, W.F., Sander, J.W. and Duncan, J.S. (2011) The long-term outcome of adult epilepsy surgery, patterns of seizure remission, and relapse: a cohort study. *Lancet*, 378, 1388–1395.
8 Moniz is also known for developing the technique of cerebral angiography. This is a technique that allows brain tumours to be observed through the displacement of blood vessels highlighted by an injection of a contrast agent such as sodium iodide.
9 Some of Freeman's first frontal lobotomies were performed in Spencer State Hospital, West Virginia. In an interview published in October 1980, the superintendent of the hospital, a Dr Thomas

C. Knapp, reported that Freeman just 'dropped in' one day to perform the operation. Knapp also recalled: 'He was a big name in neurology and he had all the proper papers and signatures – all I could do was watch. It was a real grisly thing.'
10 An aneurysm is a weakness in a wall of an artery, which forms a balloon-like structure. Before Dandy's technique was developed, a ruptured intracranial was nearly always a fatal condition.
11 This effect is generally only found in lower animals such as amphibians, e.g. frogs and newts. Mammals are unable to regenerate a severed nerve pathway.
12 The motor cortex of the left hemisphere controls the right hand, and vice versa.
13 See Rolls, G. (2005) *Classic Case Studies in Psychology*. Hodder Arnold: London.
14 Years later, brain scans would show that HM's frontal lobes remained slightly pushed up and squashed.
15 In fact, the clips turned out to be non-ferromagnetic, and in 1997, more than 40 years after his surgery, HM's surgical lesion was examined using modern scanning techniques.

Bibliography

Bakay, L. (1985) Francois Quesnay and the birth of brain surgery. *Neurosurgery*, 17, 518–521.
Benson, D.F. and Zaidel, E. (eds) *The Dual Brain*. The Guilford Press: New York.
Clarke, R.H. and Horsley, V. (1906) On a method of investigating the deep ganglia and tracts of the central nervous system. *British Medical Journal*, 2, 1799–1800.
Cooper, I.S. (1983) Sir Victor Horsley: father of modern neurological surgery. In Clifford Rose, F. and Brnum, W.F. (eds) *Historical Aspects of the Neurosciences: A Festscrift for Macdonald Crichley*. Raven Press: New York.
Corkin, S. (2002) What's new with the amnesic patient H.M.? *Nature Reviews Neuroscience*, 3, 153–160.
Corkin, S. (2013) *Permanent Present Tense*. Allen Land: London.
Corkin, S., Amaral, D.G., Gonzalez, Johnson, K.A. and Hyman, B.T. (1997) H.M.'s medial temporal lobe lesion: findings from magnetic resonance imaging. *The Journal of Neuroscience*, 17, 964–979.
de Tisi, J.I., Bell, G.S., Peacock, J.L., McEvoy, A.W., Harkness, W.F., Sander, J.W. and Duncan, J.S. (2011) The long-term outcome of adult epilepsy surgery, patterns of seizure remission, and relapse: a cohort study. *Lancet*, 378, 1388–1395.
Finger, S. and Stone, J.L. (2010) Landmarks of surgical neurology and the interplay of disciplines. In Finger, S., Boller, F. and Tyler, K. L. (eds) *Handbook of Clinical Neurology*, vol. 95. Elsevier: Amsterdam.
Feindel, W. (1982) The contribution of Wilder Penfield to the functional anatomy of the human brain. *Human Neurobiology*, 1, 231–234.
Gazzaniga, M.S. (1967) The split brain in man. *Scientific American*, 217, 24–29.
Gazzaniga, M.S. (2005) Forty-five years of split-brain research and still going strong. *Nature Reviews Neuroscience*, 6, 653–659.
Hankinson, J. and Amador, L.V. (1956) Stereotaxic Surgery. *Postgraduate Medicine*, 32, 28–39.
Harris, L.J. (1995) The corpus callosum and hemispheric communication: an historic survey of theory and research. In Kitterle, F.L. (ed.) *Hemispheric Communication: Mechanisms and Models*. Lawrence Erlbaum: New York.
Joynt, R.J. (1974) The corpus callosum: history of function regarding its function. In Kisbourne, M. and Smith, W.L. (eds) *Hemispheric Disconnection and Cerebral Function*. C. Thomas: London.
Lyons, A.E. (1997) The crucible years 1880–1900: Macewen and Cushing. In Greenblatt, S.H., Forcht Dagi, T. and Epstein, M.H. (eds) *A History of Neurosurgery in its Scientific and Professional Contexts*. The American Association of Neurological Surgeons: New York.
Lyons, J.B. (1966) *The Citizen Surgeon*. Dawnay: London.
Lyons, J.B. (1967) Sir Victor Horsley. *Medical History*, 11, 361–373.
Macmillan, M. (2004) Localisation and William Macewen's early brain surgery part 1: the controversy. *Journal of the History of the Neurosciences*, 13, 297–325.
Macmillan, M. (2005) Localisation and William Macewen's early brain surgery part II: the cases. *Journal of the History of the Neurosciences*, 14, 24–56.
Mashour, G.A., Walker, E.E. and Martuza, R.L. (2005) Psychosurgery: past, present and future. *Brain Research Reviews*, 48, 409–419.

Meador, K.J., Loring, D.W. and Flanigan, H. (1989) History of epilepsy surgery. *Journal of Epilepsy*, 1, 21–25.

Milner, B., Branch, C. and Rasmussen, T. (1966) Evidence for bilateral speech representation in some non-right handers. *Transactions of the American Neurological Association*, 91, 306–308.

Paget, S. (1919) *Sir Victor Horsley. A Study of his Life and his Works*. Constable: London.

Pearce, J.M.S. (1983) The first attempts at removal of brain tumours. In Clifford Rose, F. and Brnum, W.F. (eds) *Historical Aspects of the Neurosciences: A Festscrift for Macdonald Crichley*. Raven Press: New York.

Penfield, W. (1955) The twenty-ninth Maudsley lecture: The role of the temporal cortex in certain psychical phenomena. *The Journal of Mental Science*, 101, 451–465.

Penfield, W. (1958) Pitfalls and success in surgical treatment of focal epilepsy. *British Medical Journal*, March, 669–672.

Penfield, W. (1967) Epilepsy: the great teacher. *Archives of Neurology. Scandinavia*. 43, 1–10.

Penfield, W. (1975) *The Mystery of the Mind*. Princeton University Press: Princeton, NJ.

Penfield, W. (1977) *No Man Alone*. Little Brown: Boston, MA.

Penfield, W. and Rasmussen, T. (1950) *The Cerebral Cortex of Man*. Macmillan: New York.

Pereira, E., Green, A. and Cadoux-Hudson, T. (2010) Neurosurgery through the looking-glass. *Advances in Clinical Neuroscience and Rehabilitation*, 10, 32–36.

Powell, M. (2006) Sir Victor Horsley – an inspiration. *British Medical Journal*, Dec, 1317–1319.

Richmond, C. (2009) Henry Molaison: A unique case in the study of amnesia. *The Guardian*, 5 February.

Rolls, G. (2005) *Classic Case Studies in Psychology*. Hodder Arnold: Oxford.

Sachs, E. (1958) *Fifty Years of Neurosurgery*. Vantage Press: New York.

Sauerwein, H.C. and Lassonde, M. (1996) Akelaitis' investigations of the first split-brain patients. In Code, C. (ed.) *Classic Cases in Neuropsychology*. Psychology Press: East Sussex.

Sperry, R.W. (1964) The great cerebral commissure. *Scientific American*, 210, 42–52.

Springer, S. and Deutsch, G. (1989) *Left Brain, Right Brain*. W.H. Freeman: New York.

Stone, J.L. (2001) Dr Gottlieb Burckhardt – the pioneer of psychosurgery. *Journal of the History of Neurosciences*, 10, 79–92.

Temkin, O. (1971) *The Falling Sickness*. John Hopkins Press: Baltimore, MD.

Thomas, D.G.T. (1989) The first generation of British neurosurgeons. In Rose, F.C. (ed.) *Neuroscience Across the Centuries*. Smith-Gordon: London.

Valenstein, E.S. (1973) *Brain Control. Critical Examination of Brain Stimulation and Psychosurgery*. John Wiley: New York.

Valenstein, E.S. (1986) *Great and Desperate Cures: The Rise and Decline of Psychosurgery and other Radical Treatments for Mental Illness*. Basic Books: New York.

Valenstein, E.S. (1997) History of psychosurgery. In Greenblatt, S.H., Forcht Dagi, T. and Epstein, M.H. (eds) *A History of Neurosurgery in its Scientific and Professional Contexts*. The American Association of Neurological Surgeons: New York.

Whitaker, H.A,. Stemmer, B. and Joanette, Y. (1996) A psychosurgical chapter in the history of cerebral localisation: the six cases of Gottlieb Burckhardt (1891). In Code, C. (ed.) *Classic Cases in Neuropsychology*. Psychology Press: East Sussex.

Wieser, H.G. (1998) Epilepsy surgery: past, present and future. *Seizure*, 7, 173–184.

14

SURVEYING THE LAST 50 YEARS AND LOOKING AHEAD

One fact about the future of which we can be certain is that it will be utterly fantastic.

Arthur C. Clarke

We used to think our future was in the stars. Now we know it's in the genes.

James Watson

Summary

It is probably the case that more has been learnt about the brain over the last 50 years than in the rest of human history put together. Although estimates of the speed of progress vary, it is generally accepted that scientific and technological advancement is increasing at least exponentially – that is, doubling at a constant rate.[1] If we assume this expansion is taking place every ten years (a conservative estimate) then a simple calculation shows our knowledge of the brain increased a thousand-fold during the twentieth century.[2] As impressive as this figure appears, the exponential model shows that we will have doubled this within the first ten years of the twenty-first! Clearly, writing a history of neuroscience in a single chapter that tries to detail all the significant developments of the last 50 years or so, is an impossible task. Such an attempt can only highlight the very highest peaks of this highly varied mountain range. Despite this, two advances have taken place that stand out above all others. The first took place in 1953 when Watson and Crick described the molecular structure of DNA. It proved to be one of the most momentous discoveries in the history of science, heralding a new dawn in biology and medicine, which changed our world forever. If anything, its continuing impact, especially in regards to stem cell technology, is going to mean that brain research will overshadow all other biosciences in the coming years. The second big advance has been no less profound: it is the rise of the digital computer, which has led to the development of non-invasive scanning techniques, artificial intelligence and computer–brain interfaces. Such is the potential of these new developments, it has recently led the American Congress and European Union to invest huge sums of money in long-term and large-scale multidisciplinary brain projects with the promise of new neuroscientific discovery. The most pressing of these needs is in the understanding and treatment of neurological and degenerative disease – subjects which have been one of the great successes of neuroscience over the last 50 years, and which this chapter will primarily focus on. However, it should also not be forgotten that the human brain is incredibly complex with many mysteries, including its capacity for consciousness and free will, which in theory should not exist in a predetermined material world. Thus, it is not outside the realms of possibility that the development of neuroscience will one day solve these most perplexing puzzles and perhaps even change the nature of the universe as we know it.

Neuroscience comes of age: Parkinson's disease

Our understanding of neuroscience has led to many important achievements over the last 50 years and if one needs further proof they only need to consider the number of Nobel Prizes in this area during this period.[3] However, if there is a breakthrough moment when neuroscience came of age, fulfilling its potential to transform lives, then it surely took place in the 1960s when the brain site for Parkinson's disease was discovered – a breakthrough leading to the first successful drug treatments. As we have seen (Chapter 10), the disease was first described by James Parkinson in 1817, and named in his honour by Jean Marie Charcot in 1869. Although both suspected a brain lesion, evidence for this was not obtained until 1893 when Paul Blocq and Georges Marinesco in Paris undertook a post-mortem on a patient who had suffered from a one-sided Parkinson-like tremor. In doing this, they discovered a tumour 'slightly larger than a hazelnut' that had destroyed a small dark nucleus lying in the upper brainstem called the substantia nigra.[4] Unfortunately, the lesion had also compressed adjacent areas of the brain, including the peduncles of the cerebellum, which were also known to be involved in postural movement. A year later, another of Charcot's students Edouard Brissaud also highlighted the nigra in Parkinson's disease – a hypothesis seemingly confirmed in 1919 when the Russian Constantin Trétiakoff in his doctoral thesis examined this structure in 54 patients with a variety of neurological disorders. These included nine patients with Parkinson's disease, and all had nigral degeneration. However, the situation was now complicated by the discovery that other brain areas were also compromised in Parkinson's including parts of the striatum.

The uncertainty over the lesion site in Parkinson's disease would only become clearer in the 1950s when 3-hydroxytyramine (which became better known as dopamine) was discovered in the human brain by Kathleen Montagu working in Runwell laboratory in Wickford, near London. Although dopamine had been first synthesised in 1910 by G. Barger and J. Ewens, and shown to have weak effects on the sympathetic nervous system by Henry Dale, it was thought to be an intermediate compound formed during the synthesis of noradrenaline and adrenaline. However, this position changed when Avrid Carlsson and his colleagues at the University of Lund showed dopamine was depleted in the animal brain by a drug called reserpine – an effect that also produced a loss of movement similar to that found in Parkinson's disease. Carlsson also found L-dopa, a precursor of dopamine, reversed the behavioural effects of reserpine and restored movement in rats. This implicated dopamine as a neurotransmitter and further support came in 1959 when high levels were found in the striatum – a brain region with relatively little noradrenaline. These findings led Carlsson in the same year to propose that Parkinson's was due to a deficiency of dopamine.

Around the same time, Oleh Hornykiewicz in Vienna was examining how dopamine was distributed in the human post-mortem brain and he also discovered (in 1960) that this neurochemical was almost non-existent in the striatum of patients with Parkinson's disease. This was a ground breaking discovery for which Carlsson, but not Hornykiewicz, would win the Nobel Prize in 2000.[5] But why was this loss occurring in Parkinson's disease? Many suspected the substantia nigra was involved, not least because nigral lesions in experimental animals produced tremor and rigidity, which closely resembled the disorder. Indeed, decisive proof would come a year or two later when Bengt Falck and Nil-Ake Hilarp in Sweden developed histofluorescence techniques that allowed dopamine containing neurons to be visualised (see Chapter 12). Soon after, in 1964, their colleague Nils-Erik Andén showed nigral lesions led to the loss of fluorescence in the striatum. The result was clear: the substantia nigra was the source of the dopamine to the striatum, and nigral degeneration the pathological lesion producing Parkinson's disease.

L-dopa therapy

The discovery linking Parkinson's with dopamine depletion in the brain had exciting implications for it suggested a cure was possible if a way of replenishing this neurotransmitter could be found. Although peripheral administration of dopamine was not feasible since it did not cross the blood–brain barrier, Carlsson had shown L-dopa *did* reach the brain where it was taken up by striatal axon terminals. This led to the first trials examining L-dopa by Hornykiewicz in the early 1960s, who injected it into Parkinson patients – a procedure improving their symptoms for several hours. Benefits also occurred with oral administration. Unfortunately, L-dopa only produced a transient improvement with higher dosages causing a 'toxic delirium' with confusion, nausea and hypotension. These disappointing results, however, were overcome in 1967 when New York doctor George Cotzias gave patients small doses of L-dopa regularly every few hours, which he gradually increased over subsequent weeks until high doses could be tolerated. Cotzias also reduced L-dopa's toxicity by using a dopa decarboxylase inhibitor to inhibit its peripheral breakdown. The results were dramatic with some patients experiencing an almost miraculous relief of symptoms after years of immobility. Not surprisingly, it soon led to L-dopa becoming the first line treatment for Parkinson's disease – a position it still holds today.

The discovery of nigral degeneration in Parkinson's disease and the resulting treatment of the illness had raised the curtain to show some of the promise that brain research was capable of achieving. Unfortunately, as was too readily apparent to the victims, L-dopa was not a cure. Despite initially benefiting the majority of patients, the effectiveness of L-dopa typically declined after several years of use, with the return of debilitating immobility and tremor. Attempts to overcome this by increasing the dosage nearly always led to unpleasant side effects such as hypotension, nausea, fidgeting and disorientation. Perhaps more importantly, L-dopa does not stop the underlying degeneration of the substantia nigra. Thus, the real battle in the fight against Parkinson's disease had therefore barely begun.

The beginning of the genetic revolution

Parkinson's disease may have been the highlight of brain science during the late 1950s and 1960s, but there was a greater revolution taking place in the biosciences over much the same period. The beginning of this epoch can be pinpointed to a specific day, namely 25th April 1953, when James Watson and Francis Crick published a paper in *Nature* that revealed the molecular structure of deoxyribonucleic acid (DNA). It was a profound discovery, for Watson and Crick had solved the puzzle of inheritance – that is, how genetic material is passed from one generation to another. More specifically they demonstrated how DNA is made up of two chains (composed of phosphate and the sugar deoxyribose) that swivel around each other in the shape of the double helix. Holding the two helical strands together are pairs of simple molecules, like rungs of a ladder, called bases. Although DNA only contains four types of base (adenine, guanine, cytosine and thymine) there are huge numbers on each strand of DNA. However, Watson and Crick noticed something else: the bases bonded in a highly selective way. In fact, adenine paired with thymine, and cytosine with guanine. From this, Watson and Crick deduced if the two strands of DNA unwound, then each of the bases would only act as a 'magnet' for its own complimentary base. Consequently, if the bases attracted new partners the result would be a new strand identical to the old one. In other words, DNA was able to duplicate itself – a vital requirement for creating new cells and ultimately a new organism.

When the structure of DNA was first described, researchers knew genes were somehow responsible for making proteins – i.e. structurally complex three dimensional chemicals made up of long chains of smaller units called amino acids. These chemicals were also known to be vital for life with a wide range of functions in the body that included enzymes, receptors, antibodies and messenger substances. The elucidation of DNA indicated that our genes were simply composed of bases. But to confuse matters, while the alphabet was very simple (A, T, G and C), the message was potentially complex since there were astronomical numbers of bases on each strand of DNA.[6] An important step towards making sense of this chaos came in 1961 when Crick showed that only certain base triplets of DNA (known as codons) provided a code for making a single amino acid. Since proteins contain no more than 20 different types of amino acid, it also meant there were a limited number of possible triplets within our DNA that could specify these chemicals.[7] In other words, our genes are not simply long random base stretches: rather they are a series of 'three digit' sequences (such as CAG)[8] which each act as a blueprint for making amino acids.

An equally important question concerned how the cell went about turning the codons encrypted in DNA into fully functional proteins. By the late 1950s the suspicion had began to centre on another nucleic acid called ribonucleic acid (RNA),[9] which was single stranded and much shorter than DNA. Indeed, its involvement in protein synthesis was confirmed in 1961 when the South African Sydney Brenner (then working in Cambridge) identified messenger RNA (mRNA) as the transport molecule that carried genetic instructions from the DNA into the cell's cytoplasm. With this discovery the various steps of protein synthesis were quickly established. In short we now know it begins when part of the DNA is 'unzipped' to allow a 'replica' of mRNA to be formed. This carries a code from the nucleus to seek out a structure in the cell called a ribosome where the assembly of the protein will take place. At the ribosome, the codons of the messenger RNA are exposed one by one, allowing another type of RNA, called transfer RNA (tRNA) to attach complementary bases

FIGURE 14.1 James Watson and Francis Crick who discovered the molecular structure of DNA in 1953 – arguably the most important scientific advance of the twentieth century.

Source: Science Photo Library.

to them. Once a codon is filled up, an amino acid is formed, and these are joined together with a peptide bond. When all the transcripts of the messenger RNA have been filled and bonded, the protein is then transported away to fulfill its biological role.

The rise of molecular neuroscience

The chemical deciphering of DNA and subsequent understanding of protein synthesis, or what Crick referred to as the central dogma of molecular biology (i.e. 'DNA makes RNA makes protein') was to impact on neuroscience. Problems that had previously been understood at the cellular level could now be approached in terms of specific molecules. Although its impact was slow to affect brain science, by the late 1970s a highly specialised area of research known as molecular neurobiology had emerged. Strongly reductionist in its outlook, and influenced by developments in neuropharmacology, investigators were now interested in establishing how the different cells of the nervous system use molecules, or even parts of molecules, to accomplish a wide range of diverse ends. This work included how signal transduction occurs in neurons[10] and the regulation of gene expression by various transcription factors within the cell.[11] There was also a great interest in establishing the molecular properties of various cellular structures such as receptors and ion channels, along with neurotransmitter synthesis, re-uptake and release. Neurons were no longer merely seen as cells that generated electrical signals and released neurotransmitters. Rather they were now recognised as incredibly complex chemical factories with the genes housed in the nucleus directing protein synthesis and directing the activity of a multitude of intracellular processes.[12] This type of knowledge was also going to lead to significant advances into our understanding of neurodegenerative diseases (see below). Although it would be wrong to think of all these developments simply stemming from the discovery of DNA in 1953, there is little doubt that it provided the main impetus to the molecular revolution that followed.

Localising the mutation in Huntington's disease

Inevitably, one of the ways DNA changed our understanding of the world was in the field of genetic disease. One inherited disorder that had long interested researchers, ever since its first description by George Huntington in 1872, was Huntington's disease – a degenerative illness of the brain associated with irregular, spasmodic and involuntary movements of the trunk and limbs, leading to paralysis and death. Huntington belonged to a family of doctors who had practised for several generations on Long Island, New York, and initially came across the disease as a young boy when accompanying his father on medical rounds. It caused an indelible impression, for Huntington would later write, 'Driving with my father . . . we suddenly came upon two women, mother and daughter, both tall, thin almost cadaverous, both bowing, twisting, grimacing. I stared in wonderment, almost in fear. What could it mean?' After qualifying as a doctor in 1871, Huntington considered the illness again when returning home for a few months to help in his father's practice. A year later, Huntington presented a short paper on Sydenham's chorea to a medical conference in Ohio, and in its final seven paragraphs described a condition that he called 'hereditary chorea'. Although Huntington was unaware of the existence of genes,[13] he recognised the disease was always passed on to a child by an afflicted parent. Despite this, Huntington realised the inheritance was not inevitable, for if by chance the illness did not manifest itself, then later generations would be free of the affliction. In fact, it would take geneticists another 40 years before the disease was fully recognised as resulting from a single mutated gene that follows an autosomal mode of inheritance. Because a person inherits approximately half their genes from their mother and half from their father, it follows if one parent carries the gene, there will be a 50 per cent chance they will pass it on to their progeny.

One of the great tragedies of Huntington's disease is that it does not normally appear until middle age, by which time a person carrying the mutated gene is likely to have had children, therefore putting them at risk of the disorder. The fact the disease is caused by a single gene mutation, however, at least gave researchers hope it could be identified, allowing a test for its existence to be invented. This became feasible in the 1970s with the emergence of recombinant DNA technology – a development arising when chemicals called restriction enzymes were discovered, which precisely cut DNA at specific base sequences. Although

FIGURE 14.2 The American doctor George Huntington (1850–1916) who first described an inherited excessive movement disorder that now bears his name in 1872.

Source: Wellcome Library, London.

first identified in bacteria where they acted to disable invading viruses, it soon became clear these chemicals could also be used as tools in a laboratory. Thus, DNA could be extracted from a living organism, isolated, and put in a test tube with restriction enzymes. The result would be lots of DNA fragments (or polymorphisms) all neatly snipped at the same place. Although this technique was used to genetically engineer new organisms by splicing together different bits of DNA (see later) it also had other uses. Indeed, some realised it could be used to hunt down the gene responsible for producing Huntington's disease.

One person to think in this way was Nancy Wexler. Although trained as a psychologist, Wexler was herself at risk of Huntington's since her mother suffered from the illness. This situation had also led her father, the psychiatrist Milton Wexler, to campaign for greater research into genetic disorders, which proved successful in 1976 when the US Congress passed a bill with the aim of setting up a commission for further investigation into Huntington's. One of its objectives was to narrow down the exact chromosome carrying the mutated gene – an advance that would allow a test for the illness. However, the task was formidable for researchers had no idea where the gene was located within the human genome. In fact, the only way it could be realistically achieved was to screen large groups of patients and hope a genetic marker could be identified that was specific to the victims. Getting enough subjects to do this was problematical, however, since the disease was extremely rare. Fortunately the Huntington's commission soon learned of a place where a large population of people lived with the disease: this was on the remote shores of Lake Maracaibo in Venezuela where some

700 out of every 100,000 suffered from the disorder – all descendants of a woman who had bought the disease to the area in the 1860s.

In 1979, a team of doctors and scientists led by Nancy Wexler began travelling annually to Lake Maracaibo to collect blood samples and compile family trees. Two years later they had managed to take 570 samples, which they sent back to the US. One scientist to start extracting the genetic material from the blood was a young Canadian called James Gusella who worked at the Massachusetts General Hospital in Boston. His technique was simple: he cut the DNA into small pieces using restriction enzymes and then used a radioactive probe to latch onto the snippets so they could be visualised. He was hoping to find a polymorphism common to all the blood samples from those with Huntington's disease. With luck this snippet might be part of the gene – or at the very least situated close to it on the chromosome. Predicting that it would take at least 100 probes from all parts of the genome to identify a Huntington's fragment, Gusella actually found one with his twelfth probe, which was located near the end of the short arm of chromosome 4. It was the first time a disease-associated gene (or rather its rough whereabouts) had been mapped to a human chromosome.

Although this enabled a blood test for Huntington's to be developed, researchers were also keen to work out the gene's base sequence – information that would give important new insights into how the disease occurred. In fact, the sequencing of the gene would take another ten years, and was described by some as the longest and most frustrating search in the annals of molecular biology. It first entailed developing a set of DNA probes covering chromosome 4 and then establishing which part was transcribed into mRNA. Only then could investigators begin sequencing the gene and compare it with normal controls. This task was finally accomplished in 1993 when a group of over 50 investigators from ten different research institutions called The Huntington's Disease Collaborative Research Group published their findings. Their work showed the gene was relatively large and contained over 300,000 base pairs. However, most striking was the nature of its mutation. The gene contained a triplet CAG – that is, a sequence of the bases comprising of cytosine, adenine and gunaine – which provided the code for the amino acid glutamine. In normal people, this triplet was repeated between 15 and 34 times. However, in those with Huntington's disease, it occurred between 37 and 66 times. The researchers also found a correlation between the number of repeats in the defective gene and disease onset. That is, the more repeats, the earlier the illness was manifested.

The focus turns to Alzheimer's disease

When Alois Alzheimer first described the brain pathology of a rapidly deteriorating illness in a middle-aged women in 1906 (see Chapter 10) he seems to have regarded it as little more than an unusual case of dementia. This position was also taken by the father of modern psychiatry, Emil Kraepelin, who described it as a rare form of presenile dementia that was different from that found in later life. To back up his claim, Kraepelin pointed to the abundance of neurofibrillary tangles and plaques in the brain of Alzheimer victims, along with the rapid progression of the symptoms – both of which were not regarded as typical of later senile dementia. Today, we know Kraepelin was wrong: Alzheimer's disease is primarily an illness of the elderly. There is even a suspicion that Kraepelin was not sincere in his trumpeting of a 'new' disease since all the pathological signs reported by Alzheimer had been seen in dementia before. Whatever the truth of the matter, Kraepelin's great authority meant Alzheimer's disease was regarded as a rare form of presenile dementia. Remarkably, this view only began to change in the late 1960s when researchers[14] working at the Medical Research Council in Newcastle realised many elderly demented patients exhibited the plaques and neurofibrillary tangles typical of Alzheimer's. It suggested that the prevalence of the disease was much more common than had previously been believed. Although some remained sceptical, by the mid 1970s, Alzheimer's disease was no longer seen as a neurological curiosity: rather it was recognised as the most common senile affliction of the elderly, which affected some 5 per cent of people over the age of 65 years and 20 per cent of those over 80. This makes Alzheimer's the fourth most common cause of death in the United States.

Now regarded as a huge public health concern, the attention turned to understanding how the plaques and neurofibrillary tangles arose in the Alzheimer's brain. It had been known since the nineteenth century that the plaques contained amyloid (which means 'starch-like') that is actually a protein. Efforts to isolate and identify this protein culminated in 1984 when George Glenner and Caine Wong, in San Diego, determined that amyloid consisted of a chain of either 40 or 42 amino acids. This also enabled the cloning of the DNA controlling amyloid production – a task accomplished in 1987 when researchers sequenced the amyloid gene. It revealed amyloid was a smaller fragment of a much larger 695 amino acid protein, now known as the beta amyloid precursor protein (β-APP) that existed in the membrane of many cells including neurons. The protein also had a short chain of amino acids that jutted into the cell, and a longer tail projecting out, suggesting that β-APP might act as a receptor molecule of some type. However, why was the β-APP protein being broken down in to smaller fragments? The answer soon became clear: the β-APP protein has a short life. When the protein has served its biological role, it was removed from the membrane by being 'cut' into smaller bits by a class of enzymes called secretases.

By the early 1990s, researchers had established that two types of amyloid (40 and 42) could be formed in the brain – depending on which way the β-APP molecule was 'cut' by enzymes. The 40 amino acid version arose when a secretase (known as alpha) broke down the β-APP protein. This type of amyloid also appeared to be the normal by-product of β-APP metabolism and removed from the brain without causing a problem. However, in some instances the amyloid was cut with two extra amino acids (amyloid 42) by a secretase known as beta, and this was the form that accumulated into hardened sheets and clumps to form the plaques in the Alzheimer's brain. Researchers also began to link this form of amyloid as the initiating event in the subsequent neural degeneration and brain atrophy (perhaps due to faulty metabolism of the β-APP protein). Now known as the amyloid cascade theory, it basically argues the build up of amyloid 42 clumps in the Alzheimer's brain is neurotoxic to the surrounding neurons and with time will produce a series of events leading to marked cell loss. Today, the amyloid cascade theory still provides the most popular theory of Alzheimer's disease, although it is far from being unanimously accepted, with others stressing the involvement of neurofibrillary tangles in the pathological event instead.

Drug treatment for Alzheimer's disease

Faced with an ever growing older population, one of the biggest challenges facing neuroscience today is the need of an effective drug treatment for Alzheimer's disease – a development that would have major economic benefits and improve the lives of millions. Should it be realised, then it may well be the biggest achievement of pharmacological science to date. Unfortunately, as yet, the progress in this field has been slow with drug treatments providing relatively little benefit for Alzheimer patients. Beginning in the 1970s, the main type of therapy to treat dementia was the cholinesterase inhibitors, which increase the levels of acetylcholine in the brain by inhibiting an enzyme called acetylcholinesterase. Although acetylcholine is one of the brain neurotransmitters known to be severely compromised in Alzheimer's disease, and recognised as important for learning and memory, the cholinesterase inhibitors are only effective in the early stages of the illness, and generally do not improve or stabilise symptoms for more than a year. Thus, for the last 50 years or so, we have had no effective way of slowing down this crippling and tragic illness, despite our growing awareness of how it arises in the brain.

However, this situation may well be about to change. One reason for this optimism lies with the discovery of methylthioninium chloride (MHC). Otherwise known as methylene blue, this potential treatment was first identified in the 1980s when Cambridge researcher Claude Wischik accidentally found it dissolved tau protein. This is the main filament protein that becomes 'tangled up' in the neurofibrillary tangles, which frequently occurs alongside amyloid deposition in Alzheimer's disease. Animal studies confirmed MHC slowed down tau formation in the brain leading Wischik to set up the first clinical trials in humans. The results, presented at an international conference in 2008, were impressive with over 300 people with mild Alzheimer's disease taking the drug three times a day showing an 81 per cent reduction in cognitive decline. This led to a further phase of clinical trials being initiated in October 2012, which should take between 12 and 18 months to complete. At the time of writing, the results have not been published, although it would be hugely disappointing if they did not provide an advance in dementia drug treatment.

Another drug, hailed in the national press as the 'turning point' in the fight against Alzheimer's disease during October 2013, was published by a research team led by Professor Giovanna Mallucci at the University of Leicester. These researchers reported that neural degeneration arising from prion disease in mice could be prevented by the oral administration of a substance which inhibited a protein modifying enzyme called a kinase.[15] Although Alzheimer's is not a prion disease, both types of disorder result from the accumulation of misshapen proteins, which leads to neural degeneration. In the case of the prion infection, this causes the brain to shut off the production of new proteins as a defence mechanism, thereby starving it of new healthy proteins, which it needs to survive. The kinase inhibitor, however, prevented the brain from switching off its protein supply to prevent the resulting neural degeneration – the first time a drug had stopped such a process. Remarkably, the mice also showed no signs of behavioural impairment from their treatment. Although it is early days, there is hope a similar substance can be developed for humans over the coming years in the fight against Alzheimer's disease.[16] Thus, there are at least some grounds for optimism that one day dementia will be treatable, or even preventable, through the use of drugs.

Genetic engineering in neuroscience

Although the structure of DNA was revealed in 1953, another 20 years passed before it became possible to sequence the base sequences making up a gene or genome.[17] It would ultimately lead to the completion of the Human Genome Project in April 2003, which mapped out and sequenced our 20,500 genes, or so, along with the genomes of several other species including bacteria, fruit flies and mice. Identifying the base sequence of a gene, however, tells us little about its biological role. To do this other methods are needed, and one approach is to alter its function in some way. This became possible in the 1970s, as we have seen, when recombinant technology allowed DNA to be extracted from a cell and cut into pieces. From this moment, it was only a matter of time before researchers were able to isolate a DNA segment containing a gene taken from one organism, modify it in a laboratory, and then place it back into an embryo of a pregnant female to be bred into future generations. By doing this it became possible to 'knock out' a particular gene by replacing it with an artificial piece of DNA. In fact, the first 'knockout' mouse was created in 1989 by a group led by Martin Evans in Cambridge who created an animal lacking the gene for an enzyme called HPRT, which causes a rare condition called Lesch–Nyhan syndrome. Others would produce animals where new genes were inserted into the genome ('knockin' mice), or altered in some fundamental way ('transgenic' mice).

The use of genetically modified mice has become an important research tool in neuroscience, providing a powerful means of better understanding the genetic determinants of disease, as well as disorders such as schizophrenia, depression and drug addition. Huntington's disease is especially well-suited to this type of investigation since it is caused by a single mutation – and it has led to some surprises. For example, it has been discovered that 'knockout mice', which lack both copies of the Huntington (htt) gene do not complete embryonic development and die within 7–8 days. However, mice engineered to carry one deleted htt gene are normal. This is a curious finding because most instances of Huntington's disease in humans are also caused by a single mutated gene (i.e. the person has one good and one bad gene). It appears, therefore, that the mutated htt gene in the human instead of producing a simple loss of effect (as might be expected) is actually causing a 'gain of function', which results in a new type of toxic reaction. As yet, researchers do not know how this is occurring, although the answer will surely lead to new important insights into the disease process.

Alzheimer's is another disease that can be examined with genetically engineered mice, especially as some forms of the disorder are known to be inherited. Perhaps the most striking finding to emerge from this work is that mice which overexpress the amyloid precursor protein gene, despite showing marked amyloid deposition in cortical and limbic regions of the brain, along with memory impairments, do not exhibit marked neural degeneration, nor do they show the formation of neurofibrillary tangles. Although this finding raises doubts for the amyloid cascade hypotheses (see above), other types of mouse model have also been developed, including those manipulating the so-called pre-senile genes (otherwise known as presenilins) and none of these fully mimic the pathology found in humans either.

Despite these somewhat confusing results, genetic modification studies are now at the centre of intensive neuroscientific investigation and will certainly increase in sophistication over the coming years, especially as the human genome is now mapped, and hundreds of genetically engineered mice are commercially available. This research is also likely to begin using other types of animal, including primates, along with newer technologies such as tissue-specific targeting where a gene is introduced or deleted into a designated brain structure. Indeed, this type of research is already taking place in sheep and pigs that are having genetic material taken from Huntington's patients inserted via cannulae into the striatum.[18] At the very least, we can expect to see some very dramatic developments in the next decade within this field of neuroscience – especially with regard to the understanding and treating of degenerative disease.

Historical landmarks in stem cell biology

Arguably the most profound change in neuroscience over the coming years will be the emergence of stem cell biology – a technology that promises to revolutionize the treatment of disease by growing new body parts or repairing damaged ones. Equally, it will also have a major impact on research leading to a better understanding of disease and many aspects of behaviour. Stem cells are basically immature cells, which have not yet differentiated into the specialised cells making up the fabric of the body. In other words, they are at a stage where they have the potential to become just about any type of cell the investigator chooses. On top of this, stem cells can make copies of themselves indefinitely. Consequently, once created, they are in effect a conveyer belt for making further cells. The most abundant source of stem cells is the embryo, which during its first days is little more than a sphere of identical cells otherwise known as blastocysts. However, stem cells can also be found in fully formed adult tissues such as brain, bone marrow, and liver.[19] Although they often remain in a non-dividing state for many years, these cells can be made to replicate under the right conditions, especially in response to tissue injury. It was once believed that adult stem cells gave rise to a more limited range of cell than those from the embryo. Today, we know this idea is wrong and adult stem cells may well have the potential to differentiate into any type of cell, similar to their embryonic counterparts.

The discovery of stem cells goes back to 1961 when two young Toronto researchers, Ernest McCulloch and James Till, began trying to understand why mice given fresh transplants of bone marrow cells often survived the lethal effects of ionising radiation. Their work soon led them to observe the growth of new nodules in the spleens of the surviving mice, which created reservoirs of new blood cells. Closer examination of these nodes revealed a type of cell which was also producing all the cellular components of blood. It was the first ever demonstration of a stem cell. Despite this, it would not be until the early 1980s when embryonic stem cells were isolated from mouse blastocysts and grown in culture. Later, in 1995, James Thompson at the University of Wisconsin replicated the feat by taking embryonic stem cells from the rhesus monkey. However, this was just a prelude to one of the most significant moments in modern biology when Thompson isolated stem cells from the human embryo just three years later. Despite the huge controversies surrounding the discovery (the cells had been donated from women undergoing in vitro fertilisation) it was a development with the promise to transform the nature of medicine. This was not least because doctors now had an unlimited supply of stem cells with the potential to generate new tissue and organs for human transplantation, or as experimental tools to develop new drugs and therapies.

Stem cells in neuroscience

Part of the great excitement surrounding stem cells is their potential to develop treatments for a wide variety of degenerative conditions. To this end, researchers are exploring two main approaches. One is to take embryonic stem cells from mouse or human tissue to grow specific types of neurons in the lab, which are then transplanted into the brain or nervous system. This type of approach, for example, has been pioneered by Anders Björklund in Sweden, who induced neural stem cells grown in culture to differentiate into primitive dopamine neurons as a possible form of neural replacement in Parkinson's disease. To see if these new neurons would integrate themselves into a 'new' brain, Björklund implanted them into the

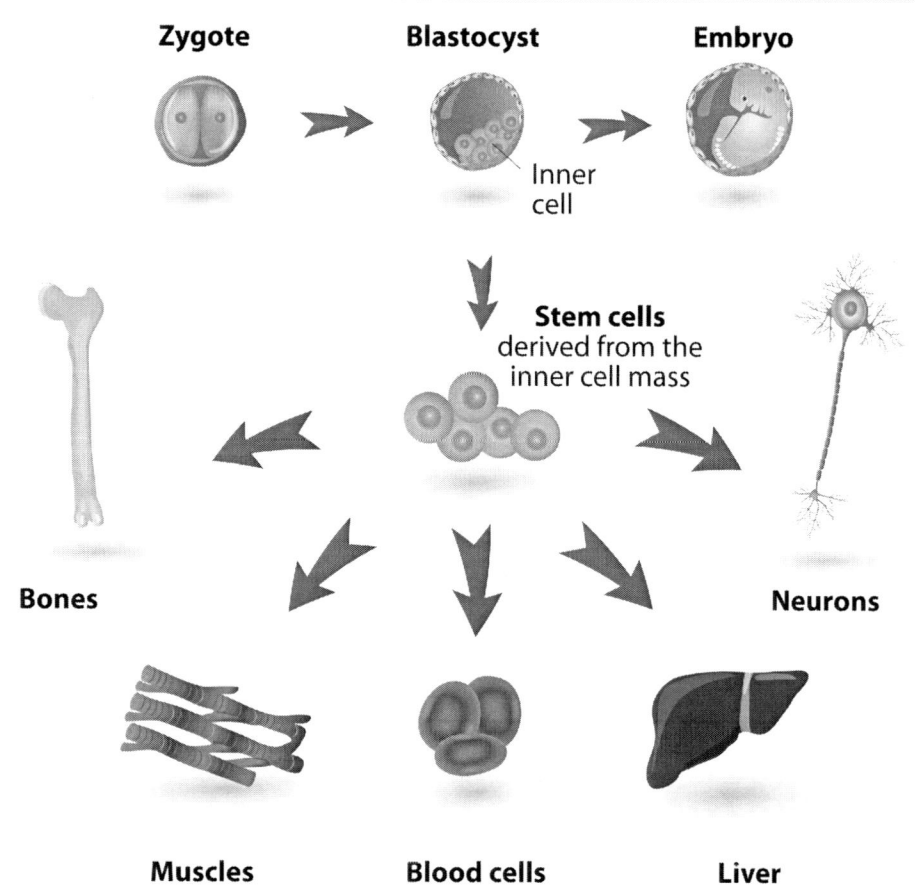

FIGURE 14.3 Diagram showing how stem cells are derived and the tissues that can be produced from their manipulation.

Source: Shutterstock Images.

striatum of adult rats who had received lesions of the substantia nigra (an animal model of Parkinson's disease). Although most of the animals over the course of 90 days showed some new growth of dopamine neurons and behavioural improvement, the results were largely disappointing since a high percentage of the cells died or differentiated into other types. Even more worrying was the discovery that a fifth of the rats also developed lethal tumours. Despite enough positive results to suggest future promise, the results of this early experiment (published in 2002) shows that many problems still have to be overcome before the first clinical trials for embryonic stem cells in the treatment of degenerative disease can be undertaken. Nonetheless, the first studies using these methods with primates are much more encouraging.[20]

Another procedure is to use adult stem cells. Remarkably, researchers can now take fully formed cells from the skin, muscle and other body parts, and turn them back into stem cells resembling those found in the embryo. These are called 'induced pluripotent stem cells', which are then able to be chemically re-engineered into neurons or other specialised cells. Some of the most impressive results have come from Fred Roisen at the University of Louisville who created the biopharmaceutical company RhinoCyte in 2006. This company has pioneered the use of stem cells taken from the lining of the nose, which are then cultured into other types. In one study, Roisen and his colleagues transformed olfactory progenitor cells into dopamine neurons, which were transplanted into the brains of rats with experimentally induced Parkinson's disease. The rats not only showed behavioural improvement as the neurons integrated themselves into the striatum, but they produced dopamine as well – although less than that found in a non-lesioned animal. Encouragingly, none of the transplanted animals developed tumours. This technology also appears to be safe to try on humans for the patient would be both donor and recipient, thereby eliminating the need for immunosuppressive drugs. Indeed, RhinoCyte believes this type of technology will soon lead to effective treatments for spinal cord injuries, motor neurone disease, and diabetes. It is also possible the first clinical trials for Parkinson's disease may start in 2014 or 2015.

Although stem cells have yet to be used to treat degenerative brain disease in humans, they have begun to be used in other neurological conditions. This new era began on 23 January 2009, when a Californian group led by Hans Keirstead, who having first come to prominence in 2010 for creating a retina derived from human embryonic stem cells, initiated clinical trials for transplantation of oligodendrocytes (a glia cell found in the brain and spinal cord) into people with spinal cord injuries. This makes it the first ever stem cell-based clinical trial in the world to be used on humans. While the experimenters point out this trial is aimed largely at testing the safety of their procedures, and can only be performed on paraplegics where no scar tissue has yet formed, some re-growth of myelin and increases in mobility are still likely to occur. If the trial is successful it will undoubtedly lead to further studies involving people with more severe disabilities including those with degenerative disease.

Stem cells for research

I hope I have briefly managed to show some of the ways stem cell technology may lead to advances in treatment for neurological disabilities over the coming years. However, this is not the only way stem cells can be utilised. Just as important, stem cells can be used for research purposes to provide new ways of understanding disease. This is also occurring now. For example, in 1998 researchers led by Ole Isacson at Harvard Medical School took skin cells from Parkinson patients with two rare genetic forms of the illness and cultured them into neural cells.[21] Close examination of these neurons showed them to exhibit certain signs of weakness, including mitochondria abnormalities (i.e. the organelles the cell use to turn oxygen and glucose into energy). Indeed, Parkinson victims with mutations in a gene known as LRRK2 produce mitochondria with lower oxygen consumption than normal. In contrast, those with PINK1 mutations express mitochondria with increased oxygen activity. Both findings are significant for mitochondria are known to produce highly toxic chemicals known as free radicals, which can cause cellular damage. Encouragingly, it has been found that both these types of cultured cells can be protected with the administration of coenzyme Q10. In addition, the immunosuppressant rapamycin was found to protect the LRRK2 containing neurons, although not the cells with the PINK1 gene. These results are not only providing new insights into the causes of the disease, but may lead to personalised forms of treatment for certain individuals, which could delay the onset of Parkinson's disease.

Others have begun to investigate neural stem cells taken from the brain itself. This work has been stimulated in part by the discovery during the 1990s that certain regions of the adult human brain, most notably the hippocampus, is capable of creating new neurons (a phenomenon known as neurogenesis). Following this, potential neural stem cells were found in many other regions of the brain, encouraging researchers to see if they can be coaxed into becoming neurons by using various trophic factors or chemicals. One person to attempt this feat is Jonas Frisén in Sweden. After discovering neural stem cells in the linings of the ventricles in 1999, Frisén founded a biotech firm called NeuroNova to develop drugs which can stimulate neurogenesis by acting on selective brainstem cells. The first disease targeted for treatment was Parkinson's disease and the company reported positive results in 2005 when one of their drugs – clandestinely called sNN0031 – was found to restore paralysis in substantia nigra lesioned rats after only five weeks. These findings indicated that a re-growth of dopamine neurons had occurred, resulting in the first human clinical trials being initiated in 2009. If successful, this approach could revolutionise treatment for stem cells would not need to be grown in a laboratory, but in the brain giving it the potential to heal itself given the right chemical stimulation.

Computers and the development of computerised tomography

Although the single most important scientific discovery of the twentieth century was arguably the elucidation of DNA, the greatest technical accomplishment of this time is surely the digital computer. The invention of the computer can not be creditable to any one person for it began with the invention of the abacus over 2000 years ago in ancient Babylonia. Since then, many inventors have contributed to its development, not least because a modern computer is a complex piece of machinery composed of many parts, each of which can be considered a separate invention. Despite this, most would accept the modern era of computing began when a number of advances took place around the time of the Second World War. Perhaps most important of these was the realisation by Alan Turing[22] that any mathematical calculation is theoretically possible using a machine that can compute algorithms (i.e. a step-by-step logical procedure) provided it has sufficient time and memory. This led him to come up with the idea of a Turing machine in 1937, which utilised binary calculations to solve any problem – a development that greatly helped define the logic behind modern computer science. It also resulted in the Manchester Small-Scale Experimental Machine which was essentially the first working machine to contain all of the elements essential to a modern electronic computer in 1948. The result was a commercially available general-purpose computer appearing in 1951.

This technology would eventually impact on neuroscience, doing so in a most dramatic way during the 1970s with the development of computerised scanning techniques that allowed investigators to see inside the living brain. In fact, such a method was not entirely new for it can be said to have originated with Wilhelm Röntgen's discovery of X-rays in 1895. However, X-rays were poor at visualising the soft tissues of the brain whose density is almost constant throughout. Despite this, there were attempts to use the new technology in a neurological context. The American neurosurgeon Walter Dandy, for example, in 1918, developed a procedure where he removed cerebrospinal fluid from the lateral ventricles and replaced it with air. The lower density of air made it possible for X-rays to highlight the ventricles making it useful for locating tumours or blood clots. Later in 1927, the Portuguese neurologist Egas Moniz (see also previous chapter) injected a radioactive dye into the carotid artery and observed its perfusion by taking X-rays of the brain. This produced pictures, called angiograms, which visualised the brain's blood vessels enabling aneurysms and tumours to be detected.

However, both these methods became redundant with the introduction of computerised axial tomography (CAT) in the early 1970s. Although others had conceived of brain imaging,[23] it was an Englishman called Godfrey Newbold Hounsfield working for the London based company EMI, who is normally most credited with pioneering its development. A weak student who left school to work for a local builder in Nottinghamshire, Hounsfield joined the Royal Air Force at the outbreak of the Second World War where he excelled in radar research. After the hostilities, Hounsfield joined EMI to help develop the UK's first all-transistor computer in the 1950s. This seems to have encouraged him to consider the possibility of medical scanning in 1967, when during a weekend ramble in the countryside, he wondered whether it was possible to take a picture of an object hidden inside a box. He realised it could be done providing the object was scanned with hundreds of gamma rays taken from a large number of different angles. By adding together all the tiny amounts of absorption in a form of a matrix, it was then possible, in theory at least, to reconstruct a three dimensional picture of the object. Recognising this technology would have important practical uses, Hounsfield sought financial support from the Department of Health and Social Security, matched by EMI who were keen to patent the machine.[24]

After experimenting with tissue obtained from the local abattoir, and changing to X-rays, Hounsfield presented the first CAT images to a meeting at Imperial College, London on 20 April 1972. Using a device housed at Atkinson Morley's Hospital, the delegates were amazed to see for the first time internal pictures of the human brain. Although the pictures were basic, one was sufficient to show a circular cyst embedded within the frontal lobes of a female patient, which allowed neurosurgeon James Ambrose to operate and successfully remove it. To obtain the picture, Hounsfield had rotated, degree by degree, an X-ray gun and detector around the patient's head, taking some 28,000 readings. It had also required the patient to lay motionless for 15 hours! After the scan, the data was also taken back to EMI where it was reconstructed by a computer, which took several more hours. Yet the procedure had been a great triumph and clearly visualised the tumour. The delegates seeing this image could have been in little doubt that CAT was going to change their world with the potential to identify all types of brain damage without harming the patient. Within three years EMI was marketing a body scanner, the first of which was installed at Northwick Park Hospital, London.

The invention of CAT was a remarkable achievement – not least, because Hounsfield had established the mathematical rules for transforming hundreds of X-ray beams into patterns of radiodensity absorption from his own somewhat idiosyncratic algebraic techniques. Although sufficient to create simple pictures, or tomograms, unknown to Hounsfield, a South African nuclear physicist called Allan Cormack had also worked out a set of theoretical mathematical algorithms in the early 1960s, which enabled a more accurate image reconstruction. These were soon incorporated into the design of CAT scanning machines. In fact, both men were awarded the Nobel Prize in 1979, with Hounsfield receiving a knighthood in 1981. Today, the technology of CAT scanning has improved beyond all measure from its origins and is an integral feature of any modern hospital with countless millions of scans taking place every year.[25] For investigation brain, the procedure now generally takes less than ten minutes, and a high resolution image (e.g. 1024 × 1024 pixels) reconstructed in a matter of seconds.

Visualising the brain in real-time

The development of CAT soon led to other scanning techniques, including positron emission tomography (PET), which involved injecting rapidly decaying radioactive substances into the bloodstream. The historical roots of PET are very different from CAT. Positrons are positively charged electrons that are created when neutrons break down in the nucleus of the atom. Although they can exist naturally, positrons are produced for experimental purposes by using a cyclotron – essentially, a chamber where certain atoms such as oxygen, carbon or nitrogen are bombarded by protons at high speed. These atoms react by ejecting a neutron in return – a process making their nucleus unstable, causing it to emit positrons. Because of their rapid decay, positrons only stay in the body for a short period of time, which makes them valuable in biological research. Unfortunately, the only way to make positrons is in a cyclotron, which is very expensive to build. Despite this, the first simple PET scanner able to detect brain tumours was made in the early 1950s by physicist Gordon Bromwell and neurosurgeon William Sweet in Massachusetts. Around the same time (1955) the first cyclotron for medical research was installed in Hammersmith Hospital, London.

At first, PET was capable of detecting little more than the approximate source of simple radioactive emissions. However, when the new algorithms for reconstructing CAT images became established in the early 1970s – it quickly led Michael Phelps and Edward Hoffman working in St Louis to construct a more sophisticated type of PET scanner (circa 1975). This was a significant advance for it was now possible to inject substances into the body destined for the brain, which were 'labelled' with a short-lasting radioactive isotope. A scanner could then map the chemical's distribution. One important use of this new technique was to measure blood flow. To do this, a person was given a venous injection of radioactive labelled water, which diffused into the bloodstream of the brain where it emitted positrons. Assuming the amount of blood flow to a given region is a measure of the brain's neural activity,[26] then psychologists had a powerful means of observing a person's brain while engaged in some sort of mental task. Thus, PET was the first computerised scanning technique to track the brain 'in real time' (i.e. as it was thinking). However, PET also proved adaptable in many other ways. For example, in a clinical setting it has assessed the malignancy of brain tumours, identified the sites producing seizures, and monitored the treatment of neurological disorders. In pharmacology PET has visualised the sites where neurotransmitters act, measured receptor occupancy, and timed how long it takes a drug to leave the brain.

Magnetic resonance imaging

Another scanning technique to appear in the 1970s was magnetic resonance imaging (MRI), which used magnetic forces, rather than potentially harmful ionising radiation, to picture the body's anatomy. The basic idea behind MRI first arose in 1946 when Felix Bloch at Stanford and Edward Purcell at Harvard (both of whom would win Nobel Prizes for their work) discovered that if an atom's nucleus was placed in a strong magnetic field, its protons would start to align themselves in an orderly manner along the lines of force. During this state, the protons were also found to be capable of absorbing radio waves, which caused them to resonate (spin). When removed from the magnetic force, however, the protons returned to their previous energy level and in the process emitted radio waves themselves. This was an important discovery because by varying the intensity of the radio stimulation, and measuring the energy released by the resonating nuclei, scientists could determine the structure of molecules that make up a given substance.

It would be another 25 years before MRI was applied to medical diagnosis – a development coming about in 1971 when Raymond Damadian at the University of New York discovered that animal tissue emitted radio signals after being exposed to magnetic forces. Importantly for medical diagnosis, tumours were found to emit different types of radio wave than healthy tissues. At this point, however, magnetic stimulation could only give information about the body's composition and not visualise the body in graphic detail. A solution to this problem came from Paul C. Lauterbur at the University of Illinois who realised imaging of different tissues was possible if a magnetic field was varied in a systematic way through the sample. In this way, each part of the sample received its own magnetic field, causing the nuclei to emit their own unique resonance frequencies. This idea was further improved by Sir Peter Mansfield at the University of Nottingham who showed how gradients of energy produced by magnetic resonance could be mathematically analyzed in order to reconstruct a more accurate image of a body part. This would lead to the first MRI scan taken of the human thorax, revealing the lungs and heart, by Damadian and his colleagues in 1977.[27]

During the 1980s, MRI produced internal images of the body that were far superior to those provided by CAT including the living human brain – a development enabling doctors and surgeons to diagnose its pathology with far greater confidence. Around the same time, researchers began to realise MRI could also be used to examine functional brain activity. One way of doing this was pioneered by Siege Ogawa working at AT&T Bell laboratories in New Jersey who noticed oxygenated and deoxygenate blood gave off different types of radio wave. By comparing this difference, or what was called 'blood oxygen level dependent contrast' (BOLD), Ogawa realised he had found a new way of measuring the brain's energy utilisation. Thus, functional MRI (fMRI) was born allowing investigators to observe the 'working' human brain though its utilisation of oxygen from blood flow. It did not take long for the first fMRI images from the human visual cortex to be published in 1991, by researchers at Massachusetts General Hospital. With this new technique, investigators were able to monitor the blood flow of the brain in high resolution, and see how various regions 'lit up' as they became engaged in some particular task.

MRI had a number of advantages over other scanning techniques. It was not only safer since the use of magnetic waves minimised the small risk of cancers, which can follow X-ray exposure, but had a superior resolution and greater ability to distinguish between similar types of soft tissue. MRI could also be modified to enhance different features in an image – a characteristic beneficial, say, for a doctor interested in multiple sclerosis who wants to highlight the myelin surrounding the axons at the expense of other areas. However, perhaps the biggest advantage of MRI, and one especially relevant for the psychologist interested in examining the cognitive aspects of the brain, was that multiple scans could be run from a single individual. Thus, it was possible to collect a vast amount of data from several experimental tasks in a relatively short space of time. Although the equipment is expensive (machines can cost over a million pounds) the importance of MRI in medical diagnosis has made it a clinical necessity in most areas of medicine. In turn, this has led to the increasing availability of MRI machines capable of functional brain imaging in many hospitals and research laboratories.

Where are brain scanning techniques taking us?

The development of functional scanning techniques has opened up a new chapter into the investigation of the human brain, resulting in a branch of psychology known as cognitive neuroscience[28] dedicated to elucidating the neural substrates of mental activity and thought – or what Michael Gazziniga has defined as understanding how the brain enables the mind. The rise of cognitive neuroscience from its origins in the early 1990s has been a remarkable success story with hundreds of laboratories now having access to scanning machinery and this will increase over the coming years. The popularity of cognitive neuroscience is also reflected in the number of research articles, rising from 146 papers on fMRI in 1990 to almost 7000 papers in 2009.[29] The reasons for this development are not hard to fathom since almost any cognitive function of interest to the psychologist can be examined with the technology. Moreover, it also has the potential to ask some very deep questions about the human brain – including, for example, the perennial issue of whether functions are localised in one place, or whether the brain has alternative ways to perform its mental operations.[30] Nor are cognitive psychologists the only ones interested in fMRI. Others are using it to help people become happy, overcome addiction, battle racism and improve memory (see Clay, R.A., American Psychological Association, 2007). There is also an increasing use of fMRI in a clinical setting with it assisting evaluation, diagnosis and rehabilitation of various disorders. Clearly, brain imaging techniques are here to stay, and psychologists have barely scratched the surface of this rapidly expanding branch of neuroscience.

On first sight, it appears that fMRI provides us with a powerful means of understanding cognitive processes, and perhaps even exposing the 'true mind' of an individual, raising the spectre of all our innermost thoughts and secrets being observed one day by scientists. Although this is a development with many sinister implications, brain scanning technology has already shown it is possible to predict a person's behaviour in several ways. For example, researchers at the University of Los Angeles, after examining activity in the prefrontal cortex, were able to guess with high accuracy whether a person will adopt wearing sunscreen or meaningfully attempt stopping smoking following a health educational film.[31] And, another group predicted how successful an individual was going to be in learning a complex video game called Space Fortress after assessing blood flow in the striatum.[32] Perhaps most interesting of all is the question of whether fMRI can establish if a person is lying or not. In fact, studies examining the anterior cingulate cortex and prefrontal areas have reported that fMRI has an accuracy of between 75 per cent and 90 per cent in lie detection[33] – figures which suggest, despite protestations to the contrary,[34] it will not be long before such evidence is used in a court of law. While lie detection may have its positive uses, other types of information are undoubtedly more questionable and may one day include mental health vulnerabilities, sexual preferences, racial attitudes and even a predilection for violent crime or terrorism. Not surprisingly, this raises a number of concerns – not least because employers, businesses, and the government all have a vested interest in knowing the abilities, personality, truthfulness and other traits of certain people.

It is impossible to know what the future holds for brain imaging, but there is enough evidence to show that the mind will hold onto its innermost secrets for a long time yet. One reason is that PET and MRI have a visual resolution for measuring brain activity, which is very gross. For example, most MRI scanners at the time of writing give pictures of the brain containing volume pixels (or voxels) measuring some $5 \times 5 \times 5$ mm^3 which can be imaged in less than 50 milliseconds, thereby allowing whole brain imaging in less than two–three seconds. While this may sound impressive, a closer examination of the maths shows that a single voxel must contain in the region of some 5.5 million neurons, up to 5.5×10^{10} synapses, and around 220 km of axons. Reducing all this activity to a colour pixel, therefore presents at best a highly simplified measure of brain function, which has been likened to trying to read a book from its cover. If we really want to understand brain images better, then we really need detailed information about the localisation of neurotransmitters and the intricacies of neural circuitry. Of course, simple measures of blood flow do not provide this. Thus, PET and fMRI can tell you where activity is taking place within a large population of cells, but do not provide any meaningful account of how it is occurring. It is perhaps not surprising, therefore, that the use of brain imaging techniques have been criticised as nothing more than a modern version of phrenology.

There are also many other issues that confound the interpretation of brain imaging experiments. For example, while it seems reasonable to assume increased blood flow correlates with increased brain activity, in reality the change in the BOLD signal in fMRI lags behind the neuronal activity by some one to two seconds, and from this point it may take another five–seven seconds to reach its peak. To make matters more complex, unwanted noise produced by the scanner, or random brain activity, can be as big as the signal itself, and require the investigator to repeat the procedure several times to get a more reliable picture. This not only makes statistical analysis difficult, but there are many instances where consistent neural activity does not correlate with performance. And, to confound issues further, it is clear that no two human brains are anatomically or functionally identical. Indeed, this helps explain why fMRI often gives rise to highly variable and non-reproducible data. Thus, as yet, the interpretation of functional brain imaging studies are far from straightforward and it is not clear if this situation will ever be fully resolved.

The promise of artificial intelligence

Artificial intelligence, or the idea that human thought processes can be mechanised, has a long history that goes back to the dawn of civilisation. This is evidenced by the fact that mechanical men appear in Greek mythology and talking statues imbrued with wisdom were revered by the Ancient Egyptians. Since then there have been numerous attempts to invent mechanical devices that act intelligently, bolstered by figures in literature with powers of human thought such as Frankenstein. However, it is only with the emergence of the digital computer in the twentieth century, capable of performing rapid and complex mathematical and logical operations, that a machine seemingly powerful enough to model the human mind has existed. This led to the founding of artificial intelligence (AI) as an academic specialisation in the summer of 1956, when John McCarthy (who coined the term) organised a month long conference at Dartmouth College in New Hampshire to bring together the leading computer experts of the time. They had high hopes, for it was assumed every aspect of human intelligence could be simulated by a computer. Some even predicted a computer with human-like intelligence would exist within a generation, provided enough money was spent on the project to make it come true. Although this has proven to be no more than fanciful speculation, the meeting did lay the groundwork for future AI research and lead to research centres being set up in many universities.

FIGURE 14.4 Alan Turing (1912–1954) who arguably initiated the field of artificial intelligence in 1950 when addressing the question of whether a computer can ever think like a human being.
Source: Royal Society.

However, for some, the real beginning of artificial intelligence began in 1950 when Alan Turing, then working in Manchester, published a paper in the philosophical journal *Mind* that set out a simple but deceptively difficult question: can a machine think like a human being? Realising this question was almost impossible to answer, partly because human thought could not be precisely defined, Turing instead proposed a test. Put simply, he asked whether a computer could hold a meaningful conversation with another human being. In order to test this possibility, Turing imagined a situation where a human was asked to interrogate a hidden entity located in another room through the use of a device such as a keyboard. The person then had to decide, based on the answers they received, whether it was another person or a machine. Turing claimed if the interrogator could not distinguish between the two, then the computer could be credited with the power of intelligent thought.

Over 60 years later, the Turing Test continues to be one of the most important objectives of AI – if not its holy grail. Interestingly, Turing believed the test would be passed around the year 2000 when digital computers existed with a storage capacity of about 10^9. Although accurate with his computer prediction, he was wrong regarding the success of the test, for no computer has come close to mimicking human thought. This is despite the attempts of an American businessman named Hugh Loebner who founded an annual competition in

1991, which pledged a reward (now $100,000) for anyone who could build a computer passing Turing's criteria. Although an annual prize of $2000 for the best attempt remains, it seems most experts have given up the challenge. They are seemingly unable to build a machine that does little better than re-hash words with which they have been programmed. Sadly, Turing would never see any of this for he committed suicide by cyanide poisoning in 1954 at the age of 41 years – an act precipitated by his criminal prosecution in 1952 for homosexuality, which led to him undergoing treatment with female hormones as an alternative to prison.[35]

Not unreasonably, we can ask why has the Turing Test proven so intractable? Have we not yet found the right computer algorithm? Or is there something special about the human mind that can not be modelled with a computer? One person believing the latter is American philosopher John Searle who published an article in 1980 that tried to discredit the idea of 'strong artificial intelligence'. In this paper Searle asked his readers to imagine a situation where a person without any understanding of reading Chinese is sitting in a room. However, the room does contain a book with instructions on how to recognise and manipulate Chinese symbols. Searle now asks what would happen if someone posted through the letterbox a question written in Chinese. Although the person would have no idea of its meaning, Searle argues he would still be able to decipher the message, and return an intelligent enough reply to persuade the recipient they are dealing with a Chinese individual. The central point Searle is making here is that the person inside the room is simply doing what a computer does – that is, manipulating symbols without understanding their meaning. If Searle's scepticism is correct, then it follows the Turing Test will never be passed. In other words, a machine that processes information, no matter how complex will never have that 'extra' ingredient enabling it to have a mind that can comprehend or be conscious.

Whatever the merits of Searle's argument, it does not belittle artificial intelligence since the field has progressed rapidly to become an integral part of the modern world. This is shown by the fact that a huge range of applications exist today that use intelligent computing, including systems for detecting credit card fraud, telephone systems capable of understanding human speech, and search engines such as Google. It has even resulted in programs that have invented new logic theorems,[36] or beaten the world champion at chess. However, whether such programs will be able to think like a human remains open to question. Nonetheless, there are those who believe this will happen one day – not least because computers are becoming increasingly more powerful and sophisticated. One way of understanding how this may occur is to consider Moore's Law. In 1965, the co-founder of Intel, Gordon E. Moore, predicted the number of transistors on integrated circuits (i.e. the components governing speed and memory) would double every two years with cost falling off at a similar rate. This prediction has proven accurate for computers have shrunk from machines the size of large rooms, to chips unable to be seen with the human eye.[37] The calculating power of this new technology is astonishing, and if it continues, a time may arise when artificial intelligence becomes superior to human reasoning. This point has been dubbed 'singularity' (a term first used by mathematician John von Neumann in 1958) and should this ever occur, the future of human history may well become unpredictable or even unfathomable. Maybe only then computers will be able to fathom out a way of making a conscious machine that passes the Turing Test. Some believe singularity is not far away. At the annual 2012 Singularity Summit held in Melbourne, a survey of various experts in the field, taken by Oxford academic Stuart Armstrong, gave predicted dates with a median of 2040.

Brain–computer interfaces

One of the most exciting computing-related advances in neuroscience in recent years has been the emergence of brain–computer interfacing (BCI), which tries to provide a link between the commands of the brain and a device capable of carrying out a desired action. It is a development with the very real promise of being able to restore useful motor function to the severely disabled, or bringing back sight and hearing to the blind or deaf. Although the beginning of this technology can be said to have begun in the 1920s when Hans Berger invented the electroencephalogram (EEG), which revealed the presence of different types of electrical wave within the brain (see Chapter 11), it was not until the late 1960s when ways were found of using this activity to instigate behavioural action. The first demonstrations of this initially came from animal studies when it was found that monkeys could learn to control the deflection of a needle pointer with signals from individual cortical neurons, or generate specific patterns of neural activity in the primary motor cortex. These findings suggested EEG activity could also be used as signals in human-computer communication, and this was realised in 1973 when Jacques Vidal[38] at the University of Los Angeles trained subjects to move a cursor around on a computer monitor, using evoked potentials recorded from the visual cortex obtained by tracking lighted squares on an 8 × 8 checkerboard.

This type of EEG feedback training was soon extended to other brain areas including the sensorimotor and motor cortices where neural activity created by the imagination of planned movements was translated into computer information. Another form of brain activity to be harnessed in a similar way is an EEG event, normally generated by the parietal lobes, known as the P300 wave – so called because it takes place some 300 milliseconds after a person makes a decision. Although it has a small amplitude, the P300 wave is highly reliable, and for this reason provides an excellent signal to control a computer interface. By the end of the 1980s, researchers were using the P300 event-related potential to allow volunteers to spell words on a computer screen. The same procedure would later be taught to paraplegics who performed as successfully as normal subjects.[39]

In recent years BCI has progressed to become a major area of scientific specialisation. Although non-invasive EEG approaches are the most popular, some have implanted devices in the brain itself. The most common way of doing this is to either use a microelectrode array placed on the surface of the brain, or insert microelectrodes into the tissue, both of which are normally used to localise seizure foci in epileptic patients. This has led to some remarkable successes, including the first device to control a prosthetic hand in 2005, pioneered by a team led by the founder of the American company Cyberkinetics, John P. Donoghue. These researchers implanted a 96-electrode microarray into the primary motor cortex of a patient called Matt Nagle who was paralysed from the neck down after being stabbed in the spine. Now by simply imagining the action, Nagle is able to move a robotic arm and hand, which has enabled him to operate a television and open e-mail. Another disability to be overcome by BCI is blindness – pioneered by American artificial vision expert and entrepreneur William Dobelle. In 2000, Dobelle implanted a microarray of 68 electrodes into the visual cortex of a patient simply known as 'Jerry' who had been blind for over 30 years. By getting his patient to wear specially designed spectacles attached to a miniature camera, and feeding the input into a waistband computer, Dobelle was able to stimulate the visual cortex, enabling Jerry to detect changes in light and various shades of grey. Although only twenty of the electrodes were found to work, Jerry was able to read two-inch letters at five feet and navigate around a New York subway.

BCI devices are also currently being used to stimulate the brain in order to control disorders such as Parkinson's disease and epilepsy, and it is easy to imagine that one day they will help to overcome just about any type of disability. Despite their great promise, however, BCI is still very much in its infancy and remains largely experimental. This situation is also not likely to change markedly in the near future as the development of BCI technology for clinical use, especially for paralysis and sensory loss, will be expensive and difficult to implement. It will also inevitably require funding from biomedical companies who will seek to make profit from their investment. Thus, the everyday use of BCIs to treat most types of disability for the average person is still a long way off. Nonetheless, BCIs provide yet another example of how neuroscience has the potential to change the medical landscape with prospects of overcoming the most intractable and distressing disabilities. And, one day it is likely to happen.

A new era begins as big investment targets the brain

As the importance of neuroscience grows in the twenty-first century, so does the amount of expenditure spent on its research. Increasingly, this is involving investments worth vast sums of money from both governmental and private sources – some of which are now being targeted at the so-called 'big questions' of brain science. One of these is the Human Connectome Project, launched in 2009, by the National Institute of Health (NIH) in the US.[40] Although it has long been known the human brain is composed of billions of nerve cells, separated by perhaps one hundred trillion synapses, arranged in large and staggeringly complex networks, there was no way researchers were able to construct a precise map of all this connectivity. At best they only had a generalised map of the human brain – a bit like a large-scale road atlas that does not highlight individual houses or their driveways. However, in 2007, the complete mapping of the brain, neuron by neuron, and synapse by synapse (i.e. a 'connectome') became possible when a new neuroimaging technique involving the creation of fluorescence proteins was developed by Jeff Lichtman and Joshua Sanes at Harvard. This method, known as 'brainbow', utilises a combination of recombinant genetic technologies and cell staining to colour specific types of protein with different ratios of red, green and blue. In turn, this makes it possible to flag each neuron with a distinctive colour. Prior to its discovery the mapping of individual neurons had been a lengthy process. However, brainbow allows over a 100 different neurons to be simultaneously identified (or illuminated) in a single 'staining' procedure. The result is visually striking images enabling investigators to easily differentiate between dendrites and axons – an important requirement if the complete neural connectivity of the brain is to be determined.

With this new technique at their disposal, the NIH awarded grants totalling nearly $40 million to a consortium of universities to build a 'network map' that will illustrate the anatomical and functional connectivity of the healthy human brain, as well as producing data facilitating research into brain disorders such as dyslexia, autism, Alzheimer's disease and schizophrenia.[41] One group receiving funding include the universities of Washington, Minnesota and Oxford who will map the long range fibre connections of the brain in over a thousand healthy adults using a combination of neuroimaging technologies. The goal of this project is not only to build a structural model of the brain in millimetre resolution, but to correlate this information with behavioural testing from fMRI. This project is also examining hundreds of pairs of twins and their siblings to help researchers better understand the effects of genetics on brain function. The second group, led by teams at Harvard and Los Angeles, will use a technique called diffusion MRI, which maps brain fibres by tracking the flow of water. To do this a new MRI scanner was built at Massachusetts General Hospital, which is up to eight times more powerful than most other machines, along with sophisticated computer algorithms enabling the connections of the human brain to be observed in high resolution. As this data becomes available it will provide a reference database and be published on the human connectome website, making it accessible to other scientists.

The Human Connectome Project is likely to have many benefits. For one thing, because the activity of neural networks determines our behaviour, personality, thoughts and memories, then understanding their anatomical structure and function will help more fully explain how the brain works. It should also help elucidate how genetics and environment can act to alter these connections, which affects everything we do from our ability to solve puzzles, play music or become addicted to drugs. Neural networks are also known to change over time and examination of this process will give us greater insights into neural plasticity, ageing and disease. And, perhaps most importantly, the connectome should allow a better understanding of instances where the brain is 'miswired' leading to various mental disorders such as autism, depression and schizophrenia – knowledge that will help treat these brain disorders. It has even been suggested that one day humans will be able to 'upload' their minds into computers, achieving a kind of immortality. Whether the Human Connectome project will achieve these objectives is open to question, although there is little doubt it will provide a treasure trove of information over the coming years.

The Human Connectome Project has also influenced the creation of other large-scale brain research projects. One is the Brain Research Through Advancing Innovative Neurotechnologies (BRAIN) Initiative, or what has become known as the Brain Activity Map Project. This was first outlined by President Barack Obama at the Whitehouse in April 2013 and earmarked to start receiving $100 million in funding in 2014 with extra support provided by various private sector partners and institutions.[42] However, this sum was increased to around $200 million by just a year later. According to John Donoghue, the project attempts to fill a gap in explaining brain function, which lies between measurements at the level of single cells on the one hand, and activity of different brain regions by using such methods such as fMRI on the other. Involving the participation of at least fifteen NIH institutes and centres, the BRAIN initiative has been compared to both the Apollo Space program and the Human Genome project since it will require the development of many new technologies dependent upon the collaboration of scientists and engineers from a multitude of disciplines. Ultimately, its goal is 'to map the circuits of the brain, measure the fluctuating patterns of electrical and chemical activity flowing within those circuits, and understand how their interplay creates our unique cognitive and behavioural capabilities.' Or put another way, it will provide a revolutionary new dynamic picture of the brain that, for the first time, shows how individual cells and complex neural circuits interact in both time and space. This should not only help us understand how the brain is able to think, feel, perceive, learn, decide and act – but also enable researchers to discover new ways to treat and prevent brain disorders. Although the BRAIN initiative is still in an early phase of development, its ambition is immense and will complement much of the research being undertaken on the human connectome.

Another large-scale project with impressive ambitions is the Human Brain Project, which was selected by the European Union as one of its two FET Flagship projects in January 2013.[43] This scheme has its origins in the Blue Brain Project[44] which was a joint effort initiated by IBM Research and the Swiss Federal Institute of Technology in 2005 whose ambition was to build a supercomputer that could simulate a pinhead size column of about 10,000 neurons from the rats cortex. The Human Brain Project[45] aims to go beyond this by simulating the whole brain in a ten-year program which will receive over one billion Euros and involve researchers from at least 15 European member states and nearly 200 research institutes. Instead of mapping the neural connectivity of the human brain piece by piece, as the Americans are doing, the Human Brain Project will attempt to feed data from anatomical investigations and behavioural scanning research into a high-powered computer in an attempt to mimic brain function. By doing this, they hope to run thousands of statistical simulations and use this information to predict how all the neurons of the brain combine together. This should then provide a theoretical model, which can be tested from real data obtained from further research. In theory, therefore, it will allow an understanding of the brain through mathematical analysis. Although the scientists behind the project accept that computer technology is at present insufficient to handle the huge amount of data the project will generate, they are confident it will be available within ten years. If successful, the hope is that such a machine will be able to simulate the activity of a full human brain in biological real time while thinking or engaged in some task. It could also lead to the invention of new supercomputers that think like humans in the future. The dream is certainly ambitious for it has been reported by Professor Steve Furber at the University of Manchester this will involve computing capable of 1,000,000,000,000,000,000 calculations per second and possibly a lot more.

It should be noted the BRAIN initiative and Human Brain Project are not the only large scale research endeavours that are presently making great strides into advancing neuroscience. Other schemes worthy of mention include SyNAPSE[46] which aims to create computer chips that include ten billion electronic neurons joined by 100 trillion synapses; the Big Brain Project[47] which has produced a 3-D model of the human brain at a resolution of 20 microns; and EyeWire[48] which lets members of the public do some of the work in mapping the neural pathways of the brain. There is also major work taking place on the connectivity of the nervous system in different types of animal, and the mapping of gene expression through brain development at the Allen Institute, based in Seattle.[49] The diagnosis of psychiatric disorders is also likely to change in the near future with the Research Domain Criteria Project funded by the National Institute of Mental Health. This has identified a list of 23 core brain illnesses and their associated neural circuitry, neurotransmitters and genes, which one day may replace the current *Diagnostic and Statistical Manual of Mental Disorders* and lead to a revolution in treatment.[50] These are exciting times.

The final frontier

In 1896, the famous naturalist and champion of evolution Thomas Huxley likened consciousness to the sudden appearance of Djin when Aladdin rubbed his lamp. Over a century later, our understanding of consciousness does not seem to have changed significantly from this analogy since many continue to believe the human brain is imbrued with some type of special spiritual essence separate from its mechanical workings. Traditionally known as the mind–body problem, this is an ancient topic in philosophy and touches upon many fundamental questions that neuroscientists have often attempted to avoid. These include, for example: what is the nature of the mind and how is it related to the biology of the brain? The heart of the dilemma is easy to understand. Our bodies are physiological things made of flesh and bone that operate according to fixed biological laws. In contrast, the mind does not appear to have mass or shape, and can think, act and feel, which are not attributes of any other known material object such as a rock or tree. Even more astonishing, the human mind unlike any machine, neural or otherwise, seems capable of free will and self-determination. Because of this, some hold that a soul-like force with thinking properties must exist within our brain – a position known as dualism. This is not only the view held by all the world's great religions, and presumably therefore the majority of the world's population, but also three of the most eminent neuroscientists of the twentieth century – namely Charles Sherrington, John Eccles and Wilder Penfield. However, if dualism is true then it would be simply astonishing for it would demonstrate a new substance in our world which is self-determining and immune to the principles of known science. We would not only have to radically alter our conception of the brain, but the way we understand the whole universe.

It has to be a matter of some concern, if not embarrassment, that modern neuroscience still has no answers to these questions. In fact, Daniel Dennett, one of the most influential philosophers of our times, has gone so far to say that consciousness is just about the last surviving mystery we have to solve. Of course, there have been many other great problems to baffle humans including the origin of the universe, the nature of life and the laws of relativity. But according to Dennett, while we may not have solved these problems completely, we at least have some idea of how to tackle them. The same cannot be said for consciousness that resists definition and explanation. However, in 1990, the co-discoverer of DNA, Francis Crick, along with German Christof Koch, published a rallying call arguing the time was right for neuroscience to attack the problem. In doing this, they argued the way forward was to look for the neural correlates of conscious experience – that is, to identify the brain processes that always accompany self-awareness. There is no doubt this idea has provided a major impetus for researchers, and Crick led the way further by suggesting that consciousness depends crucially on the thalamic connections with the cortex. He also argued the problem could be best examined if investigators turned their attention to the conscious aspects of vision since we know more about this sense than any other. However, a quarter of a century later, there has been no breakthrough in the neuroscience of consciousness. Furthermore, while there are many problems, one seems particularly intractable – that is, the place in our brain where vision and all our senses 'come together' to provide us with our unitary experience of the sensory world. This is known as the binding problem and it is tempting to imagine there is some sort of inner theatre on whose screen all these things converge. Unfortunately, we also know for certain the brain does not work this way.

For some, establishing the physiological basis of conscious experience is actually going to be relatively 'easy' even if we have not accomplished it yet. One person who holds this position is the Australian philosopher David Chalmers who believes that many neural correlates of consciousness will be discovered by scientific progress. However, this will still not solve the real problem of consciousness, for the real difficulty will be trying to understand how the physical processes of the brain give rise to subjective experience. For example, imagine the visual sensation of red, the fresh smell of warm bread, or a sound of a dentist drill. These events are known as qualia and provide the quintessence core of consciousness. Yet, the physical universe does not contain colours, smells or sounds – only certain types of molecules, frequencies and types of force. How the mind constructs the rich variety of subjective experience from the material world is for Chalmers, and many other philosophers, the 'hard' problem, sometimes referred to as the 'explanatory gap'. Sadly, neuroscience has no way of answering it. In fact, Chalmers goes further by arguing it is unlikely we will ever be able to solve the problem or understand the subjective experience of another individual. This point has also been neatly made by philosopher Patrick Nagle who asks what it is like to be a bat? Of course, we can never know, and similarly we will never experience what it will be like to be another person, for they are the only ones with direct access to their own way of being. If this is true, then it shows there are clear limits to what neuroscientists can ever find out about the human mind.

Despite these limitations these are exciting times for neuroscience and there is little doubt we stand on the brink of a new era with advances in genetics, stem cell technology and artificial intelligence (to name a few) all helping to transform our understanding of the brain in the coming years. Hopefully, this progress will also begin to see neuroscience attempting to answer some questions that have traditionally been the prerogative of the philosopher. Perhaps one way these two disparate disciplines can become more fully integrated is to finally establish the true nature of the human mind. Although dualism still has its advocates, most neuroscientists today believe in materialism, holding that the mind is composed of physical matter. Thus, if we take atoms such as carbon, hydrogen and oxygen and arrange them one way we get a rock or a tree; if we alter them differently we create a brain with the capacity for mental functioning and consciousness. Finding a sure way of rejecting dualism (if indeed it is false), however, is not going to be easy. And, even if this happens, further mysteries are certain to exist. For example, what type of materialism best explains the nature of the mind? The simplest version, known as reductive materialism, argues that mental events are nothing more than states of the brain. Thus, if you stimulate the motor cortex you may get movement of the hand, whereas stimulation of the amygdala will produce the sensation of fear and panic. Although many neuroscientists may adhere to this basic idea, there are others who hold that most mental events can not be reduced in this way. In fact, some who advocate non-reductive materialism argue that consciousness is something that 'emerges' from the activity of all their neurons and synapses – in much the same way as water is produced from the gases hydrogen and oxygen. In other words, consciousness is greater than the sum of its parts and operates at a higher level than its basic constituents. Finally, we should not rule out eliminative materialism, which rejects the existence of mental phenomena. Rather, it proposes there is only the physical world and consequently the mind must be some kind of massive delusion. The eliminative materialist even denies we have opinions, desires, or feelings – believing instead that these mentalistic terms will one day be eliminated from brain science. If this conception of materialism turns out to be true then neuroscientists returning to planet Earth in a hundred years time may find their understanding of the brain very different from the one that exists now.

Notes

1 In other words, human knowledge is expanding at a rate that can be represented by 1, 2, 4, 8, 16, etc.
2 If anything, the speed of progress is faster than this. For example, in 1958, some 650 papers were published in brain science, which rose to 6500 in 1978. By 1998, the number was over 17,000 papers, and 2008 saw some 26,500 research publications in neuroscience. A similarly impressive rate of advance is seen in membership of neuroscience societies. The Society for Neuroscience was formed in 1969 with 500 members. Today, it has 42,000 in more than 90 countries. It is also responsible for putting on annual conferences that regularly attract more than 20,000 delegates (See Rose and Abi-Rached 2013).

3 As of 2011, a total of 199 men and 10 women have been awarded the Nobel Prize in Medicine or Physiology since its inception in 1901. Of these, approximately 50 can be regarded as making important contributions to neuroscience. The majority have come in the last 50 years.
4 The substantia nigra was first identified by the French physician Félix Vicq d'Azyr in 1786 who referred to it as the 'locus niger' (i.e. black substance), although it is sometimes credited to Thomas Soemmerring (1755–1830).
5 An open letter to the Nobel Prize award committee complaining of this omission was signed by more than 250 neuroscientists.
6 We now know that the 23 pairs of chromosomes making up the human genome, contain over 3000 million base pairs, which is about 6.5 million base pairs for each chromosome.
7 In fact, there are 64 possible triplets within our DNA, which specify 20 amino acids.
8 CAG is the codon that codes for the amino acid glutamine.
9 RNA is similar to DNA except its sugar is ribose (instead of deoxyribose) and it has a base called uracil that replaces thymine.
10 Signal transduction refers to the processes by which neurotransmitters, hormones and trophic factors are conveyed into biochemical signals within the cell.
11 A transcription factor is a protein that binds to specific DNA sequences, thereby controlling the flow of genetic information from DNA to messenger RNA.
12 To give some idea of the complexity of the neuron, we now know that 1461 proteins are involved in the synapse alone (see Bayes *et al.* 2011).
13 Gregor Mendel's laws of genetic inheritance would not be rediscovered until 1900.
14 Most notably Martin Roth, Gary Blessed and Bernard Tomlinson.
15 More specifically, the substance was a specific inhibitor of the kinase PERK (protein kinase RNA–like endoplasmic reticulum kinase).
16 Perhaps even more importantly, in July 2014 (as this book was being copy-edited) a group of researchers at King's College, London, reported the first ever blood test for Alzheimer's disease – a discovery that will aid prognosis and also help future clinical trials.
17 For the record, the gene to be sequenced was from the bacteriophage MS2 and first published in 1972 by Walter Fiers and his colleagues at the University of Ghent, Belgium. It was followed by the complete genome of the bacteria in 1976.
18 Baxa *et al.* (2013).
19 Actually, the term 'adult stem cell' is a misnomer since such cells are also found in fetuses, placentas, umbilical cord blood and infants.
20 Takagi *et al.* 2005.
21 Although most cases of Parkinson's appear to result from a complex interaction of environmental and inherited factors, about 15 per cent of cases have a family history of this disorder caused by mutations in certain genes.
22 Turing is also now famous for breaking the codes used by the German Enigma machine in the Second World War – arguably the single most important contribution to the allies winning the European conflict in 1945.
23 Special mention should be made to the American William Oldendorf who built a scanning machine from junk parts in the early 1960s, but could not get commercial backing to develop his idea further.
24 Although it is widely reported that the profits from The Beatles, who recorded for EMI, helped fund the CAT project, there is no evidence to support this claim. Apparently EMI Music did not fund EMI Medical.
25 An estimated 72 million scans were undertaken in the United States in 2007.
26 This is a reasonable assumption since all cells obtain energy from oxygen and glucose provided by the blood.
27 Despite his important contributions to the field, it would be Lauterbur and Mansfield who would win the Nobel Prize for their work on MRI in 2003, and not Damadian. The reasons for this are controversial although it probably did not help Damadian's cause that he had a reputation for acting exuberantly at conferences, and was a Young Earth Creationist, who holds that all life on Earth was placed there by God some 5700 to 10,000 years ago.
28 A term invented in the back of a New York taxi cab during the late 1970s by split brain researcher Michael Gazziniga and psychologist George Miller.
29 See Nikolas Rose and Joelle Abi-Rached (2013).

30 This is an important issue. Although it is tempting to view the brain as a social network with a rigid division of labour, it is also possible that the type of connection a neuron makes with other nerve cells, based on past experience, is actually more important than its position. If this idea is correct then we can expect to see overlapping activity among many regions of the brain.
31 Falk *et al.* (2010).
32 Loan *et al.* (2011).
33 Langleben (2008).
34 Rusconi and Mitchenser-Nissen (2013).
35 On 10 September 2009, following a national campaign, Gordon Brown on behalf of the British government, made an official public apology for 'the appalling way he was treated.' He was also grated a statutory pardon by the House of Lords in July 2013.
36 In 1956 Allen Newell, J. C. Shaw and Herbert Simon introduced the first AI program, the Logic Theorist, to find the basic equations of logic as defined in Principia Mathematica by Bertrand Russell and Alfred North Whitehead. For one of the equations, Theorem 2.85, the Logic Theorist surpassed its inventors' expectations by finding a new and better proof.
37 To give an example, in April 2012, Intel launched a chip built with components that are 22 millionths of a millimetre in size and operate at a scale smaller than bacteria.
38 Vidal also coined the term 'brain–computer interface'.
39 Donchin, Spencer and Wijesinghe (2000).
40 http://humanconnectome.org
41 Financial support has also come from various organisations such as the Gatsby Foundation, the Howard Hughes Medical Institute and Microsoft, along with anonymous gifts from various individuals.
42 http:// nih.gov/science/brain/
43 The other winner of the award will investigate the unique properties of a synthetic carbon-based material called graphene, which promises to become the wonder material of the 21st century.
44 http://bluebrain.epfl.ch
45 http://humanbrainproject.eu
46 http://research.ibm.com/cognitive-computing/neurosynaptic-chips.shtml
47 http://bigbrain.loris.ca
48 http://eyewire.org
49 http://alleninstitute.org
50 See Wilson (2014).

Bibliography

Albright, T.D., Jessell, T.M., Kandell, E.R. and Posner, M.I. (2001) Progress in the neural sciences in the century after Cajal (and the mysteries that remain). *Annals of the New York Academy of Sciences*, 929, 11–40.

Ali, F., Stott, S.R.W. and Barker, R.A. (2012) Stem cells and the treatment of Parkinson's disease. *Experimental Neurology*. http://dx.doi.org/10.1016/

Alwasti, H.H., Aris, I. and Janton, A. (2010) Brain computer interface design and applications: challenges and applications. *World Applied Sciences Journal*, 11, 819–825.

Bandettini, P.A. (2009) Functional MRI limitations and aspirations. In Kraft, E., Gulyas, B. and Poppel, E. (eds) *Neural Correlates of Thinking*. Springer-Verlag: Berlin.

Bargmann, C.I. and Marder, E. (2013) From the connectome to brain function. *Nature Methods*, 10, 483–490.

Baxa, M., Hruska-Plochan, M., Juhas, S., Vodicka, P., Pavlok, A., Juhasova, J., Miyanohara, A., Nejime, T., Klima, J., Macakova, M., Marsala, S., Weiss, A., Kubickova, S.P., Vrtel, R., Sontag, E.M., Thompson, L.M., Schier, J., Hansikova, H., Howland, D.S., Cattaneo, E., DiFiglia, M., Marsala, M. and Motlik, J. (2013) A transgenic minipig model of Huntington's disease. *Journal of Huntington's Disease*, 2, 47–68.

Bayés, A.I., van de Lagemaat, L.N., Collins, M.O., Croning, M.D., Whittle, I.R., Choudhary, J.S. and Grant, S.G. (2011) Characterization of the proteome, disease, and evolution of the human postsynaptic density. *Nature Neuroscience*, 14, 19–21.

Bennett, M.R. and Hacker, P.M.S. (2003) *Philosophical Foundations of Neuroscience*. Blackwell: Oxford.

Benraiss, A. and Goldman, S.A. (2011) Cellular therapy and induced neuronal replacement for Huntington's disease, *Neurotherapeutics*, 8, 577–590.

Bloom, F.E. (ed.) (2007) *Best of the Brain from Scientific American*. Dana Press: New York.

Breunig, J.J., Hayder, T.F. and Rakic, P. (2011) Neural stem cells: historical perspectives and future prospects. *Neuron*, 70, 614–625.

Burton, R.A. (2013) *A Skeptics Guide to the Mind*. St Martin's Press: New York.

Cai, D., Cohen, K.B., Luo, T., Lichtman, J.W. and Sanes, J.H. (2013) Improved tools for the brainbow toolbox. *Nature Methods*, 10, 540–547.

Carey, N. (2012) *The Epigenetics Revolution*. Icon: London.

Churchland, P.S. (1989) *Neurophilosophy: Toward a Unified Science of the Mind/Brain*. MIT Press: Boston, MA.

Clay, R.A. (2007) *Functional Magnetic Resonance Imaging: A New Research Tool*. American Psychological Association: Washington, DC.

Connecting Brains and Society (2004). The present and future of brain research: what is possible, what is desirable? *International workshop, 22 and 23 April 2004, The Netherlands Proceedings and Synthesis Report*.

Costandi, M. (2013) *50 Ideas You Really Need to Know the Human Brain*. Quercus: London.

Damadian, R., Goldsmith, M. and Minkoff, L. (1977) NMR in cancer. XVI FONAR image of the live human body. *Physiological Chemistry and Physics*, 9, 97–100.

Dielenberg, R.A. (2013) The speculative neuroscience of the future human brain. *Humanities*, 2, 209–252.

Donchin, E., Spencer, K.M., Wijesinghe, R. (2000) The mental prosthesis: assessing the speed of a P300-based brain computer interface. *IEEE Transactions on Rehabilitation Engineering*, 8, 174–179.

Doty, R.W. (1998) The five mysteries of the mind, and their consequences. *Neuropsychologia*, 36, 1069–1076.

Duvosin, R. (1987) History of Parkinsonism. *Pharmacology and Therapeutics*, 32, 1–17.

Falk, E.B., Berkman, E.T., Mann, T., Harrison, B. and Lieberman, M.D. (2010) Predicting persuasion-induced behavior change from the brain. *The Journal of Neuroscience*, 30, 8421–8424.

Geschwind, D.H. and Konopka, G. (2009) Neuroscience in the era of functional genomics and systems biology. *Nature*, 461, 908–915.

Goetz, C.G. (2012) The history of Parkinson's disease: early clinical descriptions and neurological therapies. *Cold Spring Harbor Perspectives in Medicine*, 1: a008862.

Gonzalez, R. and Berman, M.C. (2010) The value of brain imaging in psychological research. *Acta Psychologica Sinica*, 42, 111–119.

Hall, B., Limaye, A. and Kulkarni, A.B. 2009. Overview: generation of gene knockout mice. *Current Protocols in Cell Biology*, 44: 19.12.1–19.12.17.

Hardy, J. and Allsop, D. (1991) Amyloid deposition as the central event in the aetiology of Alzheimer's disease. *Trends in Pharmacological Sciences*, 12, 383–388.

Henderson, M. (2009) *50 Genetics Ideas You Really Need to Know*. Quercus: London.

Henson, R. (2005) What can functional neuroimaging tell the experimental psychologist? *The Quarterly Journal of Experimental Psychology*, 58, 193–233.

Horstman, J. (2010) *The Scientific American Brave New Brain*. Jossey-Bass.: San Francisco, CA.

Insel, T.R., Landis, S.C. and Collins, E.S. (2013) Research priorities: the NIH initiative. *Science*, 340, 687–688.

Jaworski, T., Dewachter, I., Seymour, C.M., Borghgraef, P., Devijver, H., Kugler, S. and Van Leuven, F. (2010) Alzheimer's disease: old problem, new views from transgenic and viral models. *Biochimica et Biophysica Acta*, 1802, 808–818.

Kevles, B.H. (1997) *Naked to the Bone: Medical Imaging in the Twentieth Century*. Rutgers University Press: New Brunswick, NJ.

Kopin, I.J. (1993) Parkinson's disease: past, present and future. *Neuropsychopharmacology*, 9, 1–12.

Kurzweil, R. (2006) *The Singularity is Near: When Humans Transcend Biology*. Duckworth: London.

Kurzweil, R. (2013) *How to Create a Mind*. Duckworth: London.

Kwint, M. and Wingate, R. (2012) *Brains: The Mind as Matter*. Wellcome Trust: London.

Langleben, D.D. (2008) Detection of deception with fMRI: Are we there yet? *Legal and Criminal Psychology*, 13, 1–9.

Leavitt, D. (2006) *The Man Who Knew Too Much: Alan Turing and the Invention of the Computer*. Phoenix Books: London.

Lebedev, M.A. and Nicolelis, A.L. (2006) Brain-machine interfaces: past, present and future. *Trends in Neurosciences*, 29, 536–546.

Lebedev, M.A., Tate, A.J., Hanson, T.L., Li, Z., O'Doherty, J.E., Winans, J.A., Ifft, P.J., Zhuang, K.Z., Fitzsimmons, N.A., Schwarz, D.A., Fuller, A.M., An, J.H., Miguel, A.L. and Nicollis, M.A.L. (2011) Future developments in brain-machine interface research. *Clinics*, 66, 25–32.

Leergaard T. B., Hilgetag C. C. and Sporns O. (2012) *Mapping the connectome: multi-level analysis of brain connectivity*. Frontiers in Neuroinfomatics, 6, 1–6.

Legrenzi, P. and Umilta, C. (2011) *Neuromania: On the Limits of Brain Science*. Oxford University Press: Oxford.

Lein, E. and Hawrylycz, M. (2014) The genetic geography of the brain. *Scientific American*, April, 57–63.

McCormack, P. (1979) *Machines Who Think*. W.H, Freeman: San Francisco, CA.

Moreno, J.A., Halliday, M., Molloy, C., Radford, H., Verity, N., Axten, J.M., Ortori, C.A., Willis, A.E., Fischer, P.M., Barrett, D.A. and Mallucci, G.R. (2013) Oral treatment targeting the unfolded protein response prevents neurodegeneration and clinical disease in prion-infected mice. *Science Translational Medicine*, 5, Issue 206, p. 206ra138.

Muehllehner, G. and Karp, J.S. (2006) Positron emission tomography. *Physics in Medicine and Biology*, 51, 117–137.

National Institutes of Health (2013) *Interim Report: Brain Research Through Advancing Innovative Neurotechnologies (Brain) Working Group*. 1–58, September 16, 2013.

National Research Council (2002) *Stem Cells and the Future of Regenerative Medicine*. National Academy Press: Washington, DC.

Nicolas-Alonso, L.F. and Gomez-Gil, J. (2012) Brain computer interfaces: a review. *Sensors*, 12, 1211–1279.

Nicolelis, M. (2011) *Beyond Boundaries*. Times Books: New York.

Nutt, R. (2002) The history of positron emission tomography. *Molecular Imaging and Biology*, 4, 11–26.

Ogawa, S., Tank, D.W., Menon, R., Ellermann, J.M., Kim, S.G., Merkle, H. and Ugurbil, K. (1992) Intrinsic signal changes accompanying sensory stimulation: functional brain mapping with magnetic resonance imaging. *Proceedings of the National Academy of Sciences*, 89, 5951–5955.

Otte, A. and Halsband, U. (2006) Brain imaging tools in neuroscience. *Journal of Physiology*, 99, 281–292.

Paigen, K. (2003) One hundred years of mouse genetics: an intellectual history. II. The molecular revolution (1981–2002). *Genetics*, 163, 1227–1235.

Palmer, G.M., Fontanella, A.N., Shan, S. and Dewhurst, M.W. (2012) High-resolution in vivo imaging of fluorescent proteins using window chamber models. *Methods in Molecular Biology*, 872, 31–50.

Phelps, M.E., Hoffman, E.J., Huang, S-C. and Kuhl, D.E. (1978) ECAT: a new computerized tomographic imaging system for positron-emitting radiopharmaceuticals. *Journal of Nuclear Medicine*, 19, 635–647.

Poewi, W. and Seppi, K. (2002) L-dopa therapy in Parkinson's disease: an update after 30 years of clinical use. *Annals of the Academy of Medicine*, 14, 3–6.

Poldrack, R.A. (2011) The future of fMRI in cognitive neuroscience. *Neuroimage*, 62, 1216–1220.

Progress Report on Brain Research (2006) The Dana Alliance for Brain Initiatives: Dana Press.

Raichle, M.E. (1998) Imaging the Mind. *Seminars in Nuclear Medicine*, 4, 278–289.

Raichle, M.E. (1998) Behind the scenes of functional brain imaging: A historical and physiological perspective. *Proceedings of the National Academy of Sciences*, 95, 765–772.

Rose, H. and Rose, S. (2012) *Genes, Cells and Brains*. Verso: New York.

Rose, N. and Abi-Rached, J.M. (2013) *Neuro: The New Brain Science and the Management of the Mind*. Princeton University Press: Princeton, NJ.

Rose, S. (2005) *The Future of the Brain*. Oxford University Press: Oxford.

Rose, S. (2006) *The 21st Century Brain*. Vintage: London.

Rusconi, E. and Mitchener-Nissen, T. (2013) Prospects of functional magnetic resonance imaging as lie detector. *Frontiers in Neuroscience*, 7, doi: 10.3389/finhum.2013.00594.

Savoy, R.L. (2001) History and future directions of human brain mapping and functional neuroimaging. *Acta Psychologica*, 107, 9–42.

Scott, C.T. (2006) *Stem Cell Now: A Brief Introduction to the Coming Medical Revolution*. Plume Books: London.

Selkoe, D. J. (1993) Physiological production of the ß-amyloid protein and the mechanism of Alzheimer's disease. *Trends in Neurosciences*, 16, 403–409.

Seung, S. (2012) *Connectome*. Allen Lane: London.

Simeral, J.D., Kim, S-P., Black, M.J., Donoghue, J.P. and Hochberg, L.R. (2011) Neural control of cursor trajectory and click by a human with a tetraplegia 1000 days after implant of an intracortical microelectrode array. *Journal of Neural Engineering*, 8, 025027.

Slack, J. (2012) *Stem Cells: A Very Short Introduction*. Oxford University Press: Oxford.

Southwell, G. (2013) *50 Philosophy of Science Ideas You Really Need to Know*. Quercus: London.

Sweeney, M.S. (2009) *Brain: The Complete Mind*. National Geographic: New York.

Takagi, Y., Takahashi, J., Saiki. H., Morizane, A., Hayashi, T., Kishi, Y., Fukudal, H., Okamoto, Y., Koyanagil, M., Ideguchi, M., Hayashi, H., Imazato, T., Kawasaki, H., Suemori, H., Omachi, S., Iida, H., Itoh, N., Nakatsuji, N., Sasai, Y. and Hashimoto. N. (2005) Dopaminergic neurons generated from monkey embryonic stem cells function in a Parkinson primate model. *Journal of Clinical Investigation*, 115, 102–109.

Taylor, K. (2012) *The Brain Supremacy: Notes from the Frontiers of Neuroscience*. Oxford University Press: Oxford.

Ter-Pogossian, M.M. (1992) The origins of positron emission tomography. *Seminars in Nuclear Medicine*, 3, 140–149.

Vanderwolf, C.H. (1998) Brain, behavior, and the mind: what do we know and what can we know? *Neuroscience and Biobehavioral Reviews*, 22, 125–142.

Vo, L.T.K., Walther, D.B., Kramer, A.F., Erickson, K.I., Boot, W.R., Voss, M.W., Prakash, R.S. and Lee, H. (2011) Predicting individual's learning success from patterns of pre-learning MRI activity. PLoS ONE 6(1): 16093. Doi: 10: 1371/journal.pone.0016093.

Wagner, H.N. (1998) A brief history of positron emission tomography. *Seminars in Nuclear Medicine*, 3, 213–220.

Wickens, A.P. (1998) *The Causes of Aging*. Harwood: Amsterdam.

Wickens, A.P. (2009) *Introduction to Biopsychology*. Prentice Hall: Harlow.

Wijeyekoon, R. and Barker, R.A. (2011) The current status of neural grafting in the treatment of Huntington's disease. *Frontiers in Integrative Neuroscience*, 5, 114.

Wilson, C. (2014) Psychiatry: the reboot starts here. *New Scientist*, 10 May, 10–11.

AUTHOR INDEX

Adrian, Edgar Douglas 204, 270–275
Aetius 13
Aëtius 50
Akelatis, Andrew 336
Alcmaeon 12–13, 16, 21, 27, 233
Aldini, Giovanni 120–122, 226
Alexander the Great 9, 28
Althaus, Julius 252
Alzheimer, Alois 261–263, 352–353
Anaximander 13
Anaximenes 13
Andén, Nils-Erik 347
Andral, Gabriel 218
Aquinas, Thomas 58
Archimedes 29
Aristotle 2, 16, 18–23, 27, 32, 39, 52, 56, 58, 65, 69, 85, 94, 97
Armstrong, Stuart 366
Aubertin, Ernest 218–220
Augustine, Saint 18, 51–52, 65
Avicenna 52, 53, 60
Axelrod, Jules 313

Babinski, Joseph 256, 257
Babkin, Boris 211
Bacon, Francis 92
Bacon, Roger 58
Barger, George 300, 346
Bartholow, Robert 231–232
Beard, George 254
Beck, Hall P. 209
Bell, Charles 189, 193–196, 292
Bennett, Alexander Hughes 321, 323
Berger, Hans 275, 367
Bernard, Claude 198, 290–291
Bernstein, Julius 267–270, 276–279, 281, 285

Bichat, Francois-Xavier 158
Bielschowsky, Max 262
Binet, Alfred 255
Björklund, Anders 356–357
Bleuler, Eugen 175, 261
Bliss, Timothy 213
Bloch, Felix 361
Blocq, Paul 346
Bogen, Joseph 337
Bol'dyrev, Vasilli 206
Boldrey, Edwin 326–327
Bonet, Théophile 244
Bonnet, Charles 140
Boring, Edwin 150
Bouillaud, Jean-Baptiste 216, 217–218, 220
Bowditch, Henry Pickering 270
Boyle, Robert 92, 94, 95
Bremer, Sydney 348
Brissaud, Edouard 346
Broca, Paul 4, 6, 7, 217, 218–224, 317
Bromwell, Gordon 361
Bronte, Charlotte 147
Brown, Robert 164
Brown-Séquard, Charles 293
Bucknill, John 238
Burckhardt, Gottlieb 332

Caesar, Julius 29
Cajal, Santiago, Ramón y 158, 176–185, 290, 336
Calcar, Jan Stephen van 75
Callcot, John 322
Carlsson, Arvid 309, 312, 346–347
Carpi, Jacopo Berengario da 70–72
Carswell, Robert 251
Castillo, del, Jose 284

Celsus 29
Chalmers, David 371
Charcot, Jean Martin 242–243, 248–258, 346
Charpentier, Paul 311
Cheyne, George 245–246
Clarke, Robert H. 323–324
Cole, Kenneth 277–278, 280
Combe, George 145–147
Constantinus Africanus 52
Copernicus, Nicolaus 72, 91
Corkin, Suzanne 340
Cormack, Allan 360
Costa ben Luca 53
Cotzias, George 347
Crichton-Browne, James 229, 238
Crick, Francis 165, 345, 348–350, 371
Critchley, Macdonald 150
Cruveilhier, Jean 251
Cullen, William 246–247
Curie, Marie 258
Curtis, Howard 277–278
Cushing, Harvey 325
Cuvier, George 151

Dahlström, Annica 310
Dale, Henry Hallett 290, 299–302, 303, 304, 305–306, 308, 346
Damadian, Raymond 362
Dampierre, Marquise de 254
Dandy, Walter 335, 359
Darwin, Charles 118, 150, 165
Dawkins, R. 12
Dax, Gustave 221
Dax, Marc 221
Deiters, Otto 170–171
Dejerine, Joseph 225
Delay, Jean 311
Democritus, 13
Deniker, Pierre 311
Dennett, Daniel 371
Descartes, Rene 83, 84–93, 102, 104, 111, 150, 153, 189, 190–191, 331
Deter, Auguste 262
Dickinson, William Lee 295
Dobelle, William 367
Donoghue, John P. 367
Du Bois Reymond Emil 111, 126–128, 131, 292, 267
Dudley, Harold 305

Eccles, John Carew 282, 291, 306–309, 371
Ehrlich, Paul 299
Elliott, Thomas Renton 298–299, 303
Empedocles 13
Enquist, Olov 258
Eränkö, Olavi 310

Erasistratus of Chios 29, 32–33, 39, 42–43, 44, 48, 56, 91
Eratosthenes 29
Erb, Wilhelm 201
Erlanger, Joseph 272, 273
Euclid 29
Euler, von Ulf 300
Evans, Martin 355
Ewens, James 346
Ewins, Arthur 302

Falck, Bengt 310, 347
Falloppio, Gabrielle 78, 243
Fatt, Paul 283–284
Feldberg, Wilhelm 305–306
Fernel, Jean 204
Ferrier, David 201, 217, 228, 229–236, 317, 321, 328, 332
Fischer, Oskar 262
Flechsig, Paul 258
Flourens, Jean-Marie-Pierre 135, 140, 150–154, 198, 216, 223, 226, 233
Foerster, Otfrid 325
Fontana, Felice 164
Forbes, Alexander 272
Forel, August 175, 181
Forster, Thomas 144
Fossati, Dr 149
Foster, George 122
Foster, Michael 183, 200, 294
Franklin, Benjamin 113, 114
Franz, Shepherd Ivory 209
Freeman, Walter 334–335
Freud, Sigmund 150, 255, 257, 261
Frisén, Jonas 359
Fritsch, Gustave 217, 223, 226–229, 230, 235, 326
Fulton, John 332
Furber, Steve 3700
Fuxe, Kjell 310

Gage, Phineas 235–238, 332
Galen of Pergamon 23, 28, 29, 30, 31, 32, 33–45, 48, 50, 52, 56, 70, 72, 73, 75–76, 78, 91, 97, 151, 190, 193, 293, 335
Galileo 72, 85, 91
Gall, Franz Joseph 134–145, 148–150, 151,152, 154, 217–218
Galvani, Luigi 111, 114–117, 118–120, 131, 268, 282
Gaskell, Walter Holbrook 201, 293–294, 295, 297
Gasser, Herbert 272, 273
Gazziniga, Michael 337–339, 363
Gehrig, Lou 253
Gerald of Cremona 52
Gerlach, Joseph von 170–172, 174

Gilbert, William 112
Gilles de la Tourette, Georges 253–254
Glenner, George 353
Glisson, Francis 102, 103
Godlee, Rickman 317, 321, 323
Goethe, Johann 159
Golgi, Camillo 158, 172–176, 178–179, 184–185
Goltz, Friedrich 201, 234–235, 236, 321, 332
Gordon, John 145
Gotch, Francis 270, 273
Gowers, William 248
Gratiolet, Pierre 220
Gray, Stephen 112
Greengard, Paul 312
Grew, Nehemiah 164
Griesinger, Wilhelm 198
Gross, Charles 70
Gudden, Bernard von 258
Gusella, James 352
Guttenberg, J 64

Hales, Stephen 113
Hall, Chester Moore 158
Hall, Marshall 189, 196–198, 199
Haller, Albrecht von 84, 103–107, 111, 113, 114, 152, 191, 193, 226
Hammurabi, King 8
Harlow, John 236, 237
Harvey, William 91, 97, 158
Hawking, Stephen 253
Hebb, Donald 190, 211–213
Helmholtz, Hermann von 128–131, 198, 267–269
Herder, Johann Gottfried 140
Hermann, Johann 136
Hermann, Ludimar 128
Herophilus of Chalcedon 29–32, 33, 48, 56
Hilarp, Nils-Åke 310, 347
Hill, Archibald 282–283
Hippocrates 2, 13–16, 21, 41, 44, 52, 94, 190, 255
His, Wilhelm 175, 181, 183
Hitzig, Eduard 217, 223, 226–228, 230, 235, 326
Hodgkin, Alan 268, 277–282, 285, 286, 309
Hoffman, Edward 361
Homer 2, 11–12
Hooke, Robert 92, 100, 101, 157, 158, 163–164
Hornykiewicz, Oleh 347
Horsley, Victor 8, 317, 321–324
Hounsfield, Godfrey Newbold 360–361
Humboldt, Alexander von 118, 166
Hunain ibn Ishaq 44
Hunt, Reid 301
Huntington, George 350–351

Huxley, Andrew 268, 277–282, 285, 286, 309
Huxley, Thomas 370

Isacson, Ole 358

Jackson Lister, Joseph 158
Jackson, John Hughlings 220, 221–223, 226, 229, 230, 238, 321–328
Jacobsen, Carlyle 332
James, William 234, 255
Janssen, Hans 100
Janssen, Paul 311–312
Janssen, Zaccharias 100
Jasper, Herbert 326
Jaynes, J. 12

Kandel, Eric 312
Katz, Bernard 279, 282–284
Keats, John 171
Keirstead, Hans 358
Kleist, Ewald Georg von 112
Kline, Nathan 313
Knapp, Thomas C.
Koch, Christof 371
Koch, Robert 201
Koelle, George 310
Kölliker, Rudolf 293
Kölliker, Albert von 163, 167–169, 171–172, 175, 181, 182
Kraepelin, Emil 243, 258–261, 262–263, 352
Kraepelin, Karl 258
Kuhn, Roland 313
Kühne, Wilhelm 183, 202, 292

Laciani, Luigi 232
Lancisi, Giovanni Maria 335
Langley, John Newport 201, 235, 290, 294–299, 303
Laplace, Pierre 150
Lashley, Karl 190, 208, 209–211, 332, 336, 337
Lauterbur, Paul C. 362
Laycock, Thomas 198
Leeuwenhoek, Anton van 100–102, 157, 160, 164
Legallois, Jean-Cesar 153
Leibniz, Gottfried 85
Lesky, E. 150
Leuret, Francois 220
Lewis, Peter 310
Lichtman, Jeff 368
Lima, Almeider 333
Lindqvist, Margit 312
Lister, Joseph 158, 317, 319
Loeb, Jaques 236
Loebner, Hugh 365–366
Loewi, Otto 291, 303–306
Lomo, Terje 213

Lower, Richard 94, 95, 97
Lucas, Keith 270–271
Lugaro, Ernesto 185
Luzzi, Mondino de 61–64, 70

McCarthy, John 364
McCulloch, Ernest 356
Macewen, William 317, 319–321
Magendie, Francois 103, 150, 189, 195–196, 217, 292
Magnus, Albertus 58
Mallucci, Giovanna 354
Malpighi, Marcelo 100, 102, 158, 160
Mansfield, Peter 362
Marcus Aurelius, Emperor 34
Marie de la Condamine, Charles 290
Marinesco, Georges 346
Marinus 38
Matteucci, Carlo 111, 124–125, 126–127, 268
Maxwell, James Clark 129
Melzi, Francesco 65
Mendel, Gregor 165
Merritte, Douglas 209
Meynert, Theodor 224, 232
Mill, John Stuart 85
Milner, Brenda 331, 339
Mistichelli, Domenico 139
Mitchell, J.F. 309
Molaison, Henry Gustav (HM) 318, 339–342
Moniz, Egas 333–334, 360
Montagu, Kathleen 346
Moore, Gordon, E. 366
Morgagni, Giovanni Battista 139, 242, 234–245, 250, 335
Müller, Johannes von 125–126, 165, 198, 270, 275
Munk, Hermann 233–234
Musschenbroek, Pieter van 112
Myers, Ronald 337

Nagle, Matt 367
Nagle, Patrick 372
Nansen, Fridtjof 175–176
Neher, Erwin 285–286
Nemesius 50, 52
Neumann, John von 366
Newton, Isaac 85, 92
Nissl, Franz 262
Niven, David 253
Nobel, Alfred 184
Nobili, Leopoldo 124
Nollet, Jean Antoine 113
Nutton, Vivian 75

Oersted, Hans Christian 123, 268
Ogawa, Siege 362
Oliver, George 297, 301

Ordenstein, Leopold 252
Osler, William 323

Panizza, Bartholomeo 233
Parkinson, James 247–248, 252–253, 346–347
Pavlov, Ivan 190, 205–209, 211
Penfield, Wilder Graves 211, 318, 324–331, 371
Perusini, Gaetano 262–263
Phelps, Michael 361
Piccolhommi, Arcangelo 138
Pinel, Philippe 249
Plato 16–18, 19, 65, 86, 88
Pliny the Elder 56
Plutarch 32
Pordage, Samuel 93, 95, 246
Posidonius 50, 52
Pratensis, Jason 94
Praxagoras 30, 32
Prunières, Barthrélemy P. 4
Purcell, Edward 361
Purkinje, Jan Evangelista 159–163, 165
Pythagoras 13

Quesnay, Francois 318–319

Ranvier, Louis-Antoine 165
Rasmussen, Theodor 331
Rayner, Rosalie 208–209
Reil, Johann Christian 139, 249
Reisch, G. 58, 60
Remak, Robert 165–167
Rembrandt 247
Retzius, Gustaf 172
Richmann, George Wilhelm 113
Ringer, Sydney 282
Roisen, Fred 358
Rolando, Luigi 151, 226
Röntgen, Wilhelm 359
Rose, Caleb 200
Rosenthal, J.F. 162
Rufus of Ephesus 29, 31

Sakmann, Bert 285–286
Sanes, Joshua 368
Santorini, Giovanni 233
Sauvages, Francois Boissier de 129
Scäfer, Edward 297, 301
Schleiden, Matthias 164
Schultz, Max 171
Schwann, Theodor 164
Schweigger, Johann 124
Scoville, William 340
Searle, John 366
Sechenov, Ivan Mikhailovich 190, 198–200, 208, 209
Seleucus, King 32
Shakespeare, William 247

Shaw, Alexander 195
Sheldon, Gilbert Archbishop 84, 94
Shelly, Mary 123
Shelly, Percy 123
Sherrington, Charles 158, 183, 189, 200–205, 235, 238, 275, 293, 307, 325, 331, 371
Shostakovich, Dmitri 253
Shute, Charles 310
Simarro, Don Luis 178
Singer, Charles 65
Skinner, Burrhus Frederick 209, 212
Smith, Edwin 10
Snell, B. 12
Socrates 16, 17
Spencer, Herbert 236
Sperry, Roger Wolcott 318, 336–339
Speusippus 18
Spurzheim, Johan 136, 137, 138–139, 143–145, 147, 148, 154
Squier, Ephraim George 4
Starr, Moss 332
Steno, Nicolas 91, 103
Stol, Martin 8
Stoll, Maximilian 136
Strecker, Andreas 301
Sudhoff, K. 55, 57
Swammerdam, Jan 102–103, 114
Swedenborg, Emanuel 140
Sweet, William 361
Sydenham, Thomas 245
Sylvius (Jaques Dubois) 72, 78

Takamine, Jokichi 297
Tertullian 30
Thales 13, 112
Theophrastus 13
Thomson, James 356
Till, James 356
Titan 75
Torre, Marc Antonio del 65
Trétiakoff, Constantin 346
Turing, Alan 359, 365–366

Valentin, Gabriel 160–161, 165, 167
Valli, Eusebio 119
Valsalva, Antonio 243
Velpeau, Alfred 218
Verrall, Woollgar 183

Verrocchio, Andrea del 64
Vesalius, Andreas 72–80, 91, 95, 96, 97, 243, 335
Vicq d'Azyr, Felix 335
Vidal, Jaques 367
Vigevano, Guido da 63, 64
Vinci, Leonardo da 64–70
Virchow, Rudolf 166, 201
Vogel, Philip 337
Vogt, Marthe 309
Volta, Alessandro 111, 117–120
Vul'fson, Sigizmund 206
Vulpian, Alfred 249, 250, 292

Wada, Juan 331–332
Wagenen, William van 335–336
Waldeyer, Wilhelm 182
Watson, Hewett 147
Watson, James 165, 345, 348–350
Watson, John Broadus 208–209
Watts, James 334–335
Weber, Eduard 293
Weber, Ernst 293
Weinhold, Karl August 123
Weismann, August 202
Weiss, Paul 336
Wellcome, Henry 300
Wepfer, Johann 97
Wernicke, Carl 217, 223–226, 232
Westphal, Carl 201
Wexler, Milton 351
Wexler, Nancy 351–352
Whytt, Robert 84, 106–108, 189, 191–193, 197
Willis, Thomas 84, 93–99, 105–106, 139, 191, 242, 246, 293, 335
Winslow, Jacobus 293
Wischik, Claude 354
Wittman, Blanche Marie 256, 257–258
Wong, Caine 353
Wren, Christopher 92, 94, 95
Wundt, William 258

Yeo, Gerald 235
Yerkes, Robert 208
Young, John Zachary 268, 276–277

Zellar, Albert 313
Zotterman, Yngve 273

SUBJECT INDEX

acetylcholine 284, 290, 301–302, 304–306, 310
action potential 127, 268–270, 277–282; resting potential 268–270, 278–279
adrenaline 289, 297–298, 300
Adveraria Anatomica (Morgagni) 243
Advice for a Young Investigator (Cajal) 185
agnosia 234
Alexander the Great 9, 19, 28, 33
Alexandria 32, 56–57; library 14, 28–29
all-or-nothing law 270–271, 272, 274
Alzheimer's disease 261–263; amyloid 353, 355; amyloid precursor protein 353, 355; drug treatment for 353–354; neurofibrillary tangles 353, 355; plaques 353
American Journal of Medical Sciences, The 231
American Phrenological Journal, The 147
amnesia 339–340
amyotrophic lateral sclerosis 253
An Essay on Health and Long Life (Cheyne) 246
Analytical essay on the faculties of the soul (Bonnet) 140
Anathomia corporis humani (Mondino) 61
Anathomia designata per figures (Guido da Vigevano) 63
Anatomia hepatis (Glisson) 102
Anatomia Nicolai Physici (Anon) 50
anatomical demonstrations 61–62
Anatomie et physiologie du système nerveux (Gall) 138, 140, 143, 145, 150
anatomo-clinical method 245, 249–251
Ancient Egypt: brain, descriptions of 10–11; heart 9; hieroglyphs 8, 9; mummification 9; soul 9
Ancient Greece 11–23
angular gyrus 225, 233–234

animal electricity 114–122
animism 4
anterior commissure 95
antidepressants 312–313
aphasia: Broca's 219–220; first descriptions of 217–218; Hughlings Jackson on 222; Wernicke's 224–225
apoplexy 97, 245
Aristotelian tradition 84, 85, 97, 106, 128, 190
artificial intelligence 364–366
atheism 136, 140, 146
Athens 16, 28
atropine 294, 299, 301, 304
auditory cortex 232
Australopithecus aferensis 2
autonomic or involuntary nervous system 39, 292–297, 299, 302, 303; cranial outflow 294, 296; parasympathetic system 296–297, 299, 303; postganglionic fibres 297, 306; preganglionic fibres 297, 306; sacral outflow 294, 296; sympathetic system 296–297; thoracic/lumbar outflow 295, 295, 296
autopsies 244, 251
axon terminals 183, 283–284, 310
axons 161, 162–163, 169, 171, 172, 173–174, 179, 181–182, 277–282

Babylon 8
Behavior (Watson) 209
behaviorism 208–209
Bell-Magendie law 193–196
Bible, 8, 9
blindness 232–233
blood 10, 15, 22, 33, 34, 39, 70, 76, 102
blood circulation 97, 100, 158

blood oxygen level dependent contrast (BOLD) 362
blood-letting 248
blood pressure 297, 300, 301
blood vessels: arteries 15, 33, 34, 39, 43, 76, 95–96, 98, 158; capillaries 100; veins 33, 43, 73, 76, 158
brain: Alcmaeon on 13; Ancient Eyptians on 10; Aristotle on 20–21; Berengario on 70; Descartes on 89–91; Erasistratus on 33; Flourens on 150–153; Galen on 35–37, 42–44; Gall on 135, 138–140; Galvani on 117; Haller on 104–106; Herophilus on 31; Hippocrtaes on 15; Lashley on 211; Leonardo on 64–69; Plato on 18; Sherrington on 204; Vesalius on 76–78; Willis on 94–97
Brain (journal) 238
brain computer interfaces 367
brain dissection 21, 29–33, 61–71, 76–78, 94–97, 138–140
Brain Mechanisms and Intelligence (Lashley) 211
Brain Research through Advancing Innovative Neurotechnologies (BRAIN) initiative 369–370
brain surgery: beginnings of 318–324; commissurotomy 335–339; epilepsy 323, 324–326, 339; lesioning (removal of tissue) 320, 321, 322–324, 326, 339–340; lobotomy 333–335; psychosurgery 332–335
brainbow 368
Breaths (Hippocrates) 15
Broca's aphasia *see* aphasia
Broca's area 219–220, 328
burials (human) 3–6
Byzantium 49, 64

calcarine fissure 234
Canon of Medicine (Avicenna) 53
carbolic acid 319
carmine 170
caves and cave art 3, 53
cell doctrine 32, 48–58,
cell theory 163–165
cells: binary fission 165–166; blood 165; invention of word 163–164; nucleus 164, 167; ovum 165; protoplasm 163
cerebellum 21, 30, 70, 76, 77, 96–98, 104–105, 141, 151, 152–153, 160, 161–162, 172, 178–179, 181, 183, 191, 193, 346
Cerebral Anatome (Willis) 246
cerebral cortex 30, 33, 96–100, 104–105, 136, 139, 140–141, 152, 153, 160, 173, 180, 198, 203–204, 210–211, 226–228, 229–230, 234–235, 262–263
cerebral dominance 220–223
cerebral hemispheres 15, 21, 37, 76, 97–98, 122, 139, 151, 152–153, 219–223, 333–339

Cerebri anatome (Willis) 93–99
cerebrospinal fluid 21
chemical neurotransmission (first proposals) 292, 298
Chinese room argument 366
chlorpromazine (Thorazine) 311–312, 313
choline 301
choroid plexus 37, 70
Circle of Willis 96
colliculi 37, 76
Commentaria super anatomia Mundani (Berengario) 70
Compendium der Psychiatrie (Kraepelin) 258
computer 345, 359
computerised axial tomography (CAT) 360–361
Conditioned Reflexes (Pavlov) 211
conditioning: operant 209; Pavlovian 206–208, 210
consciousness 371–372
Constitution of Man, The (Combe) 146
corpus callosum 37, 98, 122, 333–339
Corpus Hippocraticum (Hippocrates) 14–15
corpus striatum *see* striatum
cortical localization 216–235, 321
cranial nerves 31, 38, 69, 76–77, 96, 97, 139
Cro-Magnon man 3–4, 53
curare 290–292, 299

De anatomicis adminstrationibus (Galen) 35–36, 41
De anima (Aristotle) 20
De cerebri morbis (Pratensis) 80
De cerebro (Malpighi) 100
De generatione animalium (Aristotle) 56
De humani corporis fabrica (Vesalius) 72–80
De natura hominis (Nemesius) 50
De partibus animalium (Aristotle) 21
De revolutionibus orbium coelestium (Copernicus) 72
De sedibus et causis morborum per anatomen indagatis (Morgagni) 243–245
De sepolturis (Papal proclamation) 61
De usu partium (Galen) 34, 37
De viribus electricitatis in motu musculari commentarius (Galvani) 116
deafness 224, 232
decerebrate rigidity 203
decortication 234
dementia 261–263, 352–354
dementia praecox 261–262
dendrites 161, 175, 179, 181–182; protoplasmic prolongations 161, 171,173–174
deoxyribonucleic acid (DNA) 345, 348–349
depression 312–313
Discourse on Method (Descartes) 85
dissection: human 21, 29–33, 61–71, 73, 76–78, 94–97, 138–140
dopamine 309, 310, 312, 346–347, 356–357, 359

dualism 17, 150, 204, 371–372
dynamic polarization 181–182
dyslexia 225
dyspepsia 245

Edinburgh Phrenological Society, The 145
Edinburgh Review, The 145
Edwin Smith papyrus 10–11
electrical fish 113
electrical stimulation of the brain: animals 104–105, 226–228, 229–230; humans 121–122, 231, 326–330
electricity: animal 114–122; atmospheric 115; bimetallic 115–116, 117, 118; static 112; theories of 113
electro convulsive therapy (ECT) 311
electrode 277
electroencephalograph (EEG) 275, 367
electrophysiology 115, 128
Elementa physiologiae corporis humani (Haller) 103
encephalos 21
end plate potential 283–284
Endeavour of Jean Fernel, The (Sherrington) 204
English Malady, The (Cheyne) 246
epilepsy: seizures 8–9, 15, 221–222, 323, 324–326, 339
epithymetikon 18
Epitome (Vesalius) 78
ergot 299–300
Essay on Shaking Palsy (Parkinson) 248
Essays on the Anatomy of Expression in Painting (Bell) 193
Essays on Phrenology (Combe) 145
Examen de la Phrénologie (Flourens) 153–154
excitatory postsynaptic potential (EPSP) 308

Fasciculo di medicina 62
Fleshes (Hippocrates) 16
foramina 37
fornix 37
Foundations of Psychiatry (Wernicke) 225
Frankenstein 123
frog (brainless) 107, 191
frog leg preparation 103, 114, 118–119, 124–125, 130, 291
frontal cortex 136, 141, 193, 310, 326, 332–335
frontal lobes 217–218, 226–229, 235–238, 332–335
functional magnetic resonance imaging (fMRI) 362–364, 369
Functions of the Brain, The (Ferrier) 238

Galenism 70, 72–73
galvanism 113, 120, 122
galvanometer 123–125, 126–127, 268
genes 348–350, 350–352

genetic engineering 354
giant squid (*Loligo pealili*) 277
God Delusion, The (Dawkins) 12
Golgi stain 172–174, 175, 178
Goulstonian lectures 236
grey matter of brain 76, 97, 98, 100, 104, 139, 161

haloperidol 311–312
Handbuch der Gewebelehre des Menschen (Kölliker) 167, 172
Handbuch der Physiologie (Müller) 125
Handbuch der Physiologischen Optik (Helmholtz) 129
handedness 220–221
heart 9, 18, 21–23, 40, 69, 76, 86, 102, 103, 107, 191, 193, 293
Hebb's law 213
hippocampus 160, 173, 180, 213, 339–342, 359
histamine 300, 305
Histologie du système nerveux de l' homme et des vertébrés (Cajal) 184–185
Historia animalium (Aristotle) 21
Homer: anatomical terms 12; causes of behvaiour 12; *nous* 11; *psyche* 11; *thymos* 11
Homo erectus 2
Homo habilis 2
Homo neanderthalensis see Neanderthals
Homo sapiens 2
Homo sapiens sapiens 2–3
Human Brain Project 369–370
Human Connectome Project 368–369
Human Genome Project 354
humoral theory 14, 32, 34, 245–246
Huntington's disease 350–352, 355
hyoscine 253
hypochondria 245
hysteria 245, 254–257

Illiad (Homer) 11
illustrations: drawings 54–58, 59, 64–68; early depictions 53; five figure series (*funfbilderserie*) 54, 55, 57–58; prints 70–71; woodcuts 58, 60, 70, 74, 75, 79
imagination 50, 69
imipramine (Tofranil) 313
In Search of the Engram (Lashley) 211
Index librorum prohibitorum 72, 86, 148
inferior olive 95, 139, 160, 170
infundibulum 37
inhibitory postsynaptic potential (IPSP) 308
Integrative Action of the Nervous System, The (Sherrington) 200, 203–204, 211
intellect 15, 18, 19, 33, 40, 50, 51, 53, 98, 136
intellectual psyche 19

intracellular recording 307
ion channels 281, 285–286
ions 269, 279, 280, 282–283
iproniazid (Marsilid) 313
irritability 102–107, 117, 191
Isagoge breves (Berengario) 70

jaborandi 294, 299, 301

kalamos 30
kinases 354
knockout mice 355

L-dopa 347
L'Homme (Descartes) 86, 89, 190
language 217–224, 331; *see also* aphasia
laryngeal nerve 40, 41
lateral geniculate nucleus 233
lateral vestibular nucleus 170
Le Monde (Descartes) 85
learning 143, 144
Lectures on the Diseases of the Nervous System (Charcot) 257
Lectures on the Localization of Cerebral and Spinal Diseases (Charcot) 257
lesioning (of brain): Ferrier 232–233, 235; Flourens 150–153; Fritsch and Hitzig 228; Galen 43–44; Goltz 234–235; Lashley 209–211; Munk 233–234; Rolando 151
lesioning technique 151–153, 347
Leyden jar 112–113, 114, 117
ligature 38–39
limbic lobe 219
Literal Meaning of Genesis, The (St Augustine) 51
Little Albert 208–209
liver 9, 30, 73, 76, 102
Localization of Brain Disease, The (Ferrier) 238
localisation of brain functions 98, 105, 136, 141–142, 144, 150–154, 229–230, 326–330, 331, 335–339
locus coeruleus 160, 310
logistikon 18
London Medical Record, The 231
long term potentiation 213
lysergic acid diethylamide (LSD) 309

Macedonia 18, 28
magnetic resonance imaging (MRI) 361–364
Mammalian Physiology: A Course of Practical Exercises (Sherrington) 204
Man and Nature (Sherrington) 204
mania 261
Margarita philosophica (Reisch) 58
materialism 20, 128, 129, 130–131, 135, 140, 149, 151
mechanical universe 85
Meditations of First Philosophy (Descartes) 88

medulla 76, 98, 105, 139, 153, 170, 198, 199, 248
membrane potential 278
memory 50, 53, 69, 90, 98, 210–213, 328–329
meninges 6, 10, 21; dura matter 37, 231; pia matter 37
meningitis 93
Mesopotamia 8
methylthioninium chloride 354
Micrographia (Hooke) 100–101
Micrographia (Hooke) 164
microscope 99, 100, 158–159
microtome 160
Mikroskopische anatomie (Kölliker) 167
minature end plate potentials 283–284
mind–body problem 370
mitrochondria 358
monism 150
monoamine oxidase inhibitors 313
Moore's law 366
motor neuron disease 253
movement (motor) areas of the brain 51, 98, 104, 226–229, 229–230, 321, 326–328
multiple sclerosis 251–252
muscle, skeletal 33, 34, 41, 64, 73, 99, 102, 103, 104, 107, 114, 115, 117, 124, 125, 161, 165, 299
muscle, types of: cardiac 293; hamstrings 201, 308; quadriceps 201, 308; smooth 293
muscle current 124, 125, 127
muscle fibres 117, 271
muscle spindles 202, 273
muscular dystrophy 253
myelin 165, 167, 169, 173, 252
Mystery of the Mind, The (Penfield) 330

Nature of Bones (Hippocrates) 15
Neanderthals 3
negative variation 127
nerve cells: Cajal on 178–184; Deiters on 170–172; discovery of 160, 167; early descriptions 158, 167–169; Forel on 175; Golgi on 172–174, 184; His on 175; independence of 158, 167–169, 175–176; Kölliker on 167–170; Leeuwenhoek on 102; Malpighi on 100; Nansen on 175–176; Purkinje 161; Remak on 165–166; Valentin 160–161
nerve current 124, 127
nerve fibres 16, 90, 99, 139, 161, 165, 167, 175–176, 272, 273
nerve impulse: ion changes in 277–287; patterning 270–271, 273–274; speed of 129–131, 272–273; theories of 44, 89, 113, 117, 125–126, 127–128, 130–131, 268–270, 277–283

nerves: discovery of 13, 15–16, 23
nervous system 107; early descriptions of 1–33, 37–39, 42–43, 63, 73, 88–89, 97; *see also* autonomic, cranial nerves and somatic nervous system
neura 23, 31
neurofibrillary tangles 262
neurology 93, 257
neuromuscular junction 182–183, 283, 291–292
neuron doctrine 176, 181–182; *see also* independence of nerve cells
neurosis 247
neurotransmission 292, 298–299, 303–305
neurotransmitters criteria for 309
New Stone Age (Neolithic) 4–8
nicotine 295, 299
No Man Alone (Penfield) 331
Nobel prize: Adrian (1932) 204, 275; Bloch (1952) 361; Brenner (2002) 384; Cajal (1906) 184; Carlsson (2000) 347; Cormack (1970) 360; Crick (1962) 349; Dale (1936) 306; Eccles (1963) 309; Erlanger (1944) 273; Gasser (1944) 273; Golgi (1906) 184; Hodgkin (1963) 282, 309; Hounsfield (1979) 360; Huxley (1963) 282, 309; Katz (1970) 284; Loewi (1936) 306; Moniz (1949) 333; Nansen (1922) 176; Neher (1991) 286; Pavlov (1904) 206; Purcell (1952) 361; Sakmann (1991) 286; Sherrington (1932) 204, 275; Watson (1962) 349
nodes of Ranvier 165
noradrenaline 300, 305, 309, 310, 313
Novum organum scientiarun (Bacon) 92
nucleus accumbens 310
nutritive psyche 19

Observations on Seven Beheaded Criminals (Weinhold) 123
occipital lobes 233–234
Odyssey (Homer) 11
Oeuvres Complètes (Charcot) 257
olfactory bulb 173
On anatomical procedures (Galen) 73
On Eye Diseases (Herophilus) 30
On the Functions of the Brain (Gall) 138
On the Irritable and Sensible Parts of the Body (Haller) 104
On the Origin of the Species (Darwin) 141
On the Sacred Disease (Hippocrates) 15
optic chiasma 337
optic nerves 13, 21, 76, 233
optic thalamus 191
Organization of Behavior, The (Hebb) 212

paralysis 41, 98, 245, 248, 252, 262, 290
parasympathetic system *see* autonomic nervous system
parietal lobes 231, 233
Parkinson's disease 247–248, 251–252, 309, 346–347, 356–357, 368
Passions of the Soul (Descartes) 91
patch clamp 285–286
periventricular gray substance 170
Phaedro (Plato) 17
Phaedrus (Plato) 18
Philosophiae naturalis principia mathematica (Newton) 92
Philosophical Transactions 229
phrenology: basic assumptions of 135–136, 143–150; criticisms 145, 149–150, 153–154; faculties 141–142, 144; first use of word 144, 145; Fowler brothers 147; organology 135–136, 140, 150; overview 134–135; popularization 143–145; *schädelehre* 135; social implications 143, 145–147
Phrenology Examined (Flourens) 153–154
physiognomical system of Drs. Gall and Spurzheim, The (Spurzheim) 145
physiognomical system of the objections made in Britain against the doctrines of Gall and Spurzheim, The (Spurzheim) 145
physiognomy 136
Physiological Essays (Whytt) 191
physostigmine 304
pineal gland 37, 70, 76, 77, 90–91
pituitary gland 37, 97
pneuma 4, 13, 23, 28, 30, 32, 33, 37, 41–43, 44, 70, 76, 89, 91, 131; *see also* psyche and spirit
pontine area 170
positron emission tomography (PET) 361, 364
Primae lineae physiologiae (Haller) 103
Principles of Psychology (James) 234
printing: invention of 64
proprioception 202
protein synthesis 348–349
psyche 4, 11, 17, 19–20, 21, 22, 43, 89, 112, 190; *see also pneuma* and spirit
psychein 11
psychiatric classification 260
psychiatry 249
Psychology as the Behaviorist Views It (Watson)
psychosis 261
psychosurgery *see* brain surgery
Purkinje cells 161–162, 179
Purkinje shift 159
pyramidal decussation 139

quanta 284

raphe nuclei 310
receptor blockade (antagonism) 312
receptors 299, 302, 303
Recherches expérimentales (Flourens) 152

Recollections of my Life (Cajal) 185
reflex: Descartes on 88–91, 190; Hall on 196–198; Hebb on 211–213; Lashley on 209–211; Pavlov on 205–208; Sechenov on 198–200; Sherrington 200–205; Skinner on 209; Watson on 208–209; Whytt on 107, 191–193; Willis on 98, 191
reflexes: conditioned 200, 206–213; excitatory-motor 196–198; inhibitory 199–200; knee jerk 201, 204; psychic 200; pupil 191; scratch 203; sneezing 199; spinal 192–193, 197–198; stepping 203
Reflexes of the Brain (Sechenov) 200, 205
refractory period 270
Remarks on the Wounds of the Brain (Quesnay) 278
Researches on Animal Electricity (Du Bois Reymond) 128
resting current 127
restriction enzymes 351, 355
resurrection of the dead 120–123
rete mirabile 30, 43, 68, 70, 75–76, 78, 95–96
reticular formation 139, 170
reticulum theory 170–172, 174–176, 184
reverberatory activity 212
Revista trimestral de histologia normal y patologica (Cajal) 178
ribonucleic acid (RNA) 348–349
Rome 34, 44, 49
Royal Society 117, 125, 164, 183, 229, 230
Rules (Descartes) 84

Salpêtrière 249–257
schizophrenia 12, 260–261, 310–312
scholasticism 52
sensation: brain areas for 38, 41, 50, 51, 53, 69, 91, 98, 103–107
sensibility 103–107
sensitive psyche 19
sensus commune 21–22, 65, 67, 69, 98, 105
sentient principle 106–107, 191, 192
Sepulchretum sive anatomica practica (Bonet) 244
serotonin 309, 310, 313
singularity 366
skulls 4–8, 135, 136, 141–142
somatic nervous system 292, 306
somatosensory cortex 327
soul 4, 9, 11, 17–18, 20, 21, 32, 43, 50, 51, 65–69, 76, 88, 99, 104, 106–107, 204, 309
soup versus sparks debate 306–309
Spanish Inquisition 78
specific nerve energies 125–126
speech 217–221, 222–225
spinal cord 39, 41, 63, 107, 165, 167, 172, 173, 193–195, 251, 253; *rami communicantes* 39; spinal nerves 39, 41, 193; spinal roots 194–196, 292

spirit 32–33, 41–43, 53, 70, 78, 87–88, 91, 97, 99, 102–103, 104, 117, 120; *see also* psyche and *pneuma*
split brain experiments 336–339
Statistics on Phrenology (Watson) 147
stem cells 356–359
stereotaxic surgery 323–324
stream of consciousness 328–329
stria terminalis 76, 95
striatum 76, 95, 98, 105, 310, 347
stroke 244
substantia nigra 160, 346–347, 357
Sulla fina anatomia degli organi centrali del sistema nervosa (Golgi)
summation 204
Sur les fonctions du cerveau (Gall) 138
sympathetic ganglion 293
sympathetic system *see* autonomic nervous system
sympathetic trunks 39
sympathy 190, 191, 193
synapse 183–184, 203–204, 292
Synopsis nosolgia methodicae (Cullen) 246

Tabulae anatomicae (Vesalius) 73
temporal lobes 224–225, 232–233, 328–329
tetanus stimulation 125, 127
Textbook of Physiology (Foster) 183
Textbook of Psychiatry (Kraepelin) 258, 260, 262
thalamus 76, 104, 160, 199,
Timaeus (Plato) 18
Torcular Herophili 30
Tourette's syndrome 253–254
tremor 248, 252, 253–254
trepanation 4–8
Turing Test 365–366

uptake blockade 313

vagus nerve 38, 39, 97, 206, 293
ventricles: Avicenna on 53, 60; Berengario on 70; Descartes on 88–91 Erasistratus on 32–33; first description of 21; Galen on 37, 43–44, 50; Herophilius on 30; Leonardo on 65–70; medieval concepts 49–53; Nemesius on 50; Posidonius on 50; Reisch on 60; St. Augustine on 51; Vesalius on 76; Willis on 97
vermis 37, 53
vis insita 104
vis nervosa 104
visual cortex 233, 367
visual pathways 13, 16, 21, 233–234, 337
vitalism 107, 120, 128, 129, 131
voltaic pile 120

Wada technique 331
Walden Two (Skinner) 209
Wellcome Institute 299–300
Wernicke's aphasia *see* aphasia
Wernicke's area 224–225, 328
Wernicke's encephalopathy 225
white matter 76, 97, 98, 100, 104, 140, 161

womb 30
World War I 201
World War II 279, 359
writing, invention of 8

X-rays 359–360

eBooks
from Taylor & Francis
Helping you to choose the right eBooks for your Library

Add to your library's digital collection today with Taylor & Francis eBooks. We have over 50,000 eBooks in the Humanities, Social Sciences, Behavioural Sciences, Built Environment and Law, from leading imprints, including Routledge, Focal Press and Psychology Press.

Choose from a range of subject packages or create your own!

Benefits for you
- Free MARC records
- COUNTER-compliant usage statistics
- Flexible purchase and pricing options
- 70% approx of our eBooks are now DRM-free.

Benefits for your user
- Off-site, anytime access via Athens or referring URL
- Print or copy pages or chapters
- Full content search
- Bookmark, highlight and annotate text
- Access to thousands of pages of quality research at the click of a button.

Free Trials Available

We offer free trials to qualifying academic, corporate and government customers.

eCollections
Choose from 20 different subject eCollections, including:
- Asian Studies
- Economics
- Health Studies
- Law
- Middle East Studies

eFocus
We have 16 cutting-edge interdisciplinary collections, including:
- Development Studies
- The Environment
- Islam
- Korea
- Urban Studies

For more information, pricing enquiries or to order a free trial, please contact your local sales team:

UK/Rest of World: **online.sales@tandf.co.uk**
USA/Canada/Latin America: **e-reference@taylorandfrancis.com**
East/Southeast Asia: **martin.jack@tandf.com.sg**
India: **journalsales@tandfindia.com**

www.tandfebooks.com